Light Element Atom, Molecule and Radical Behaviour in the Divertor and Edge Plasma Regions

International Atomic Energy Agency (IAEA)
Coordinated Research Project from 2009-2013

Journal of Physics: Conference Series Volume 576

Vienna, Austria
Held on various dates between 2009-2013

ISBN: 978-1-63439-909-8
ISSN: 1742-6588

Printed from e-media with permission by:

Curran Associates, Inc.
57 Morehouse Lane
Red Hook, NY 12571

Printed by Curran Associates, Inc. (2015)

Institute of Physics
Dirac House, Temple Back
Bristol BS1 6BE UK

Phone: 44 1 17 929 7481
Fax: 44 1 17 920 0979

techtracking@iop.org

Additional copies of this publication are available from:

Curran Associates, Inc.
57 Morehouse Lane
Red Hook, NY 12571 USA
Phone: 845-758-0400
Fax: 845-758-2634
Email: curran@proceedings.com
Web: www.proceedings.com

TABLE OF CONTENTS

Journal of Physics: Conference Series **576** (2015) 011001 doi:10.1088/1742-6596/576/1/011001

Preface

This volume of Journal of Physics: Conference Series contains contributions by participants in an International Atomic Energy Agency (IAEA) Coordinated Research Project (CRP) on "Light element atom, molecule and radical behaviour in the divertor and edge plasma regions" (in magnetic fusion devices). Light elements are the dominant impurity species in fusion experiments and in the near-wall plasma they occur as atoms or ions and also as hydrides and other molecules and molecular ions. Hydrogen (H or D, and T in a reactor) is the dominant species in fusion experiments, but all light elements He - O and Ne are of interest for various reasons. Helium is a product of the D+T fusion reaction and is introduced in experiments for transport studies. Lithium is used for wall coating and also as a beam diagnostic material. Beryllium is foreseen as a wall material for the ITER experiment and is used on the Joint European Torus (JET) experiment. Boron may be used as a coating material for the vessel walls. Carbon (graphite or carbon-fiber composite) is often used as the target material for wall regions subject to high heat load. Nitrogen may be used as a buffer gas for edge plasma cooling. Oxygen is a common impurity in experiments due to residual water vapor. Finally, neon is another choice as a buffer gas. Data for collisional and radiative processes involving these species are important for plasma modelling and for diagnostics.

The participants in the CRP met 3 times over the years 2009-2013 for a research coordination meeting. Reports and presentation materials for these meetings are available through the web page on coordinated research projects of the (IAEA) Atomic and Molecular Data Unit [1]. Some of the numerical data generated in the course of the CRP is available through the ALADDIN database [2].

The IAEA takes the opportunity to thank the participants in the CRP for their dedicated efforts in the course of the CRP and for their contributions to this volume. The IAEA scientific officers for this project were Mr Bastiaan J. Braams and Ms Hyun-Kyung Chung.

[1] See: https://www-amdis.iaea.org/CRP/.

[2] See: https://www-amdis.iaea.org/ALADDIN/.

Journal of Physics: Conference Series **576** (2015) 011002 doi:10.1088/1742-6596/576/1/011002

Peer review statement

All papers published in this volume of *Journal of Physics: Conference Series* have been peer reviewed through processes administered by the proceedings Editors. Reviews were conducted by expert referees to the professional and scientific standards expected of a proceedings journal published by IOP Publishing.

Electron collisions with beryllium and its ions

I. Bray and D. V. Fursa

ARC Centre for Antimatter-Matter Studies, Curtin University, GPO Box U1987, Perth, WA 6845, Australia

E-mail: I.Bray@curtin.edu.au

Abstract. An overview is provided of the convergent close-coupling calculations for electron-impact on neutral beryllium, as well as its various ionic stages. The calculations were performed from near threshold to high energies with initial states having principal quantum number $n \leq 4$. Excitation to states with $n \leq 5$ and total ionisation cross sections have been calculated with an estimated accuracy of below 10% at all energies considered.

1. Introduction

Electron scattering on atoms and ions is of both fundamental and practical interest. From the fundamental perspective, few-body physics on the atomic scale has only the Coulomb potential governing the interactions, and so has the capacity for accurate solution. Complexities arise from the number of particles involved and the long-ranged nature of the Coulomb potential. From a practical perspective, such interactions are important in the laser, lighting, atmospheric, plasma processing, and fusion energy applications.

During the course of the last two decades there has been immense progress in the field of electron scattering on atoms and ions. Whereas in the early 1990s there were discrepancies between theory and experiment for the fundamental electron-hydrogen scattering problem [1], today much more complex problems are solved routinely, which are accurate for excitation and ionisation processes on the full energy range of interest [2, 3, 4]. The key factor governing the capacity of theory to provide accurate results is that the electron-atom interaction should be dominated by one- or two-electron processes. This is the case for electronic configurations that have one [5] or two [6] valence electrons.

Given that the first wall of ITER will have a substantial amount of beryllium, it is important to have a quantitative knowledge of electron interactions with its neutral and all ionic stages. We write such collision systems as e-Be^{q+}, for $q = 0, 1, 2, 3$. Here we provide an overview of the calculations performed for such systems.

2. Theory

We use the convergent close-couping (CCC) theory for the e-Be^{q+} collisions. The H-like Be^{3+} and the Li-like Be^{+} targets are able to be treated as one valence electron systems [5]. The He-like Be^{2+} and neutral Be require the more complicated two-valence electron target implementation [6]. The details of the CCC method are given in Refs. [5, 6]. Briefly, the underlying building block of the method is a complete Laguerre basis. Diagonalisation of the target Hamiltonian in this basis ensures that all of the target states are square-integrable, yet

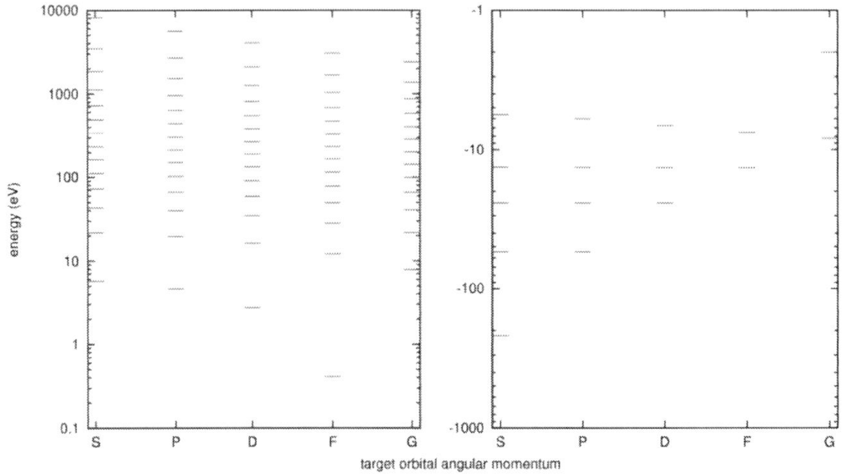

Figure 1. Energy levels in the CCC(85) calculations of e-Be^{3+} scattering.

completeness is approach as the size of the basis is increased. The negative-energy states converge to the true discrete eigenstates, whereas the positive-energy states yield an increasingly dense discretisation of the target continuum. The square-integrable nature of the basis ensures that implementation within the close-coupling formalism is possible with all integrals existing.

The key aspect of the CCC method is that the convergence of the results of interest to a specified accuracy should be able to be established by simply increasing the size of the Laguerre bases N_l, for $l \le l_{\max}$. The Laguerre basis exponential fall-off parameter λ_l may also be varied, but for simplicity of convergence studies we generally fix it at some realistic value, and only vary N_l and $l_{\max}$. Furthermore, we often simplify things further by setting $N_l = N_0 - l$.

2.1. e-Be^{3+} scattering

The Be^{3+} ion is H-like, and as such represents least computational difficulty. The CCC calculations were performed with $\lambda_l = 8$, $l_{\max} = 4$ and $N_l = 20 - l$. These resulted in 85-state calculations, with the energy levels given in figure 1. In Laguerre-based diagonalisation typically only the highest negative-energy state is a pseudostate, with the lower-energy states being good eigenstates. From the figure we see that the calculations have good $n \le 4$ eigenstates, and a rather dense discretisation of the continuum. Accordingly the calculation yields scattering amplitudes for excitation to $n \le 4$ states, and also ionisation from the $n \le 4$ states.

In figure 2 the cross sections for e-Be^{3+} scattering are given for three selected transitions. Comparison is given with the distorted wave approximation (DWBA) and first-order many body theory (FOMBT) [1]. We see generally good agreement between the calculations for both the 2P excitation and total ionisation cross section (TICS). However, only close-coupling based methods are able to resolve any resonance structure at energies below the ionisation threshold.

[1] http://aphysics2.lanl.gov/tempweb/

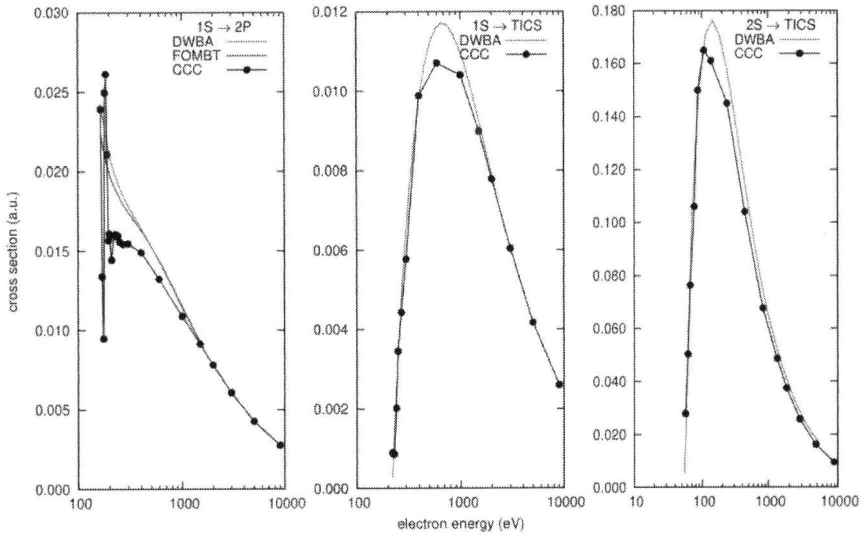

Figure 2. Selected cross sections for e-Be^{3+} scattering calculated using the CCC method, and distorted-wave and first-order many-body theories.

2.2. e-Be^{2+} scattering

The Be^{2+} ion is He-like, and requires a good deal more developmental and computational effort. The CCC calculations were performed with $\lambda_l = 4$, $l_{\max} = 4$ and $N_l = 25 - l$. Here the target states were constructed utilising the orbitals obtained from first diagonalising Be^{3+} in the Laguerre basis, and then diagonalising Be^{2+} in explicitly symmetrised two-electron configurations. The configurations consisted of the frozen-core $\{1s, nl\}$ type, augmented by two s-, p-, and d-orbitals for both electrons to improve the quality of the ground state. The resulting 284 energy levels are given in figure 3. From the figure we see that the calculations have good $n \leq 5$ eigenstates, and again a dense discretisation of the continuum. The extra states with energies a little above 100 eV are due to the doubly excited states arising from the non-frozen-core configurations.

In figure 4 the cross sections for e-Be^{2+} scattering are given for four selected transitions. Once more comparison is given with the DWBA and FOMBT approaches. We see substantial discrepancy between CCC and the first-order methods for the important $1^1S \rightarrow 2^1P$ transition. Surprisingly, agreement considerably improves for the two presented TICS cases. Given the unitarity of the CCC formalism, it is expected to be equally accurate for all transitions.

2.3. e-Be^{+} scattering

The Be^{+} ion is Li-like, and so is only marginally more complicated to work with than the H-like Be^{3+}. The primary difference is that the $1s^2$ core electrons are treated via the self-consistent Hartree-Fock (SCHF) method, with the valence electron moving in the resulting frozen-core Hartree-Fock (FCHF) potential. The CCC calculations were performed with $\lambda_l = 2$, $l_{\max} = 4$ and $N_l = 20 - l$. These resulted in 84-state calculations, with the energy levels given in figure 5, where we see that the calculations have good $n \leq 5$ eigenstates.

In figure 6 the cross sections for e-Be^{+} scattering are given for two selected transitions. As

Journal of Physics: Conference Series **576** (2015) 012001 doi:10.1088/1742-6596/576/1/012001

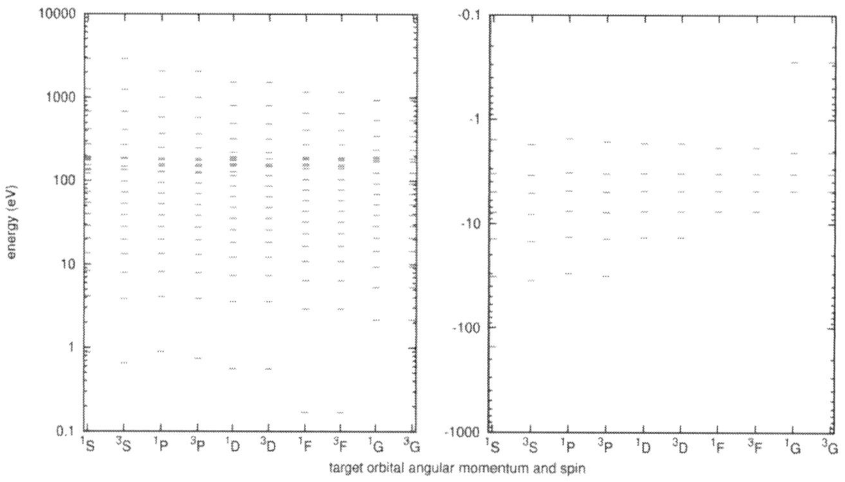

Figure 3. Energy levels in the CCC(284) calculations of e-Be^{2+} scattering.

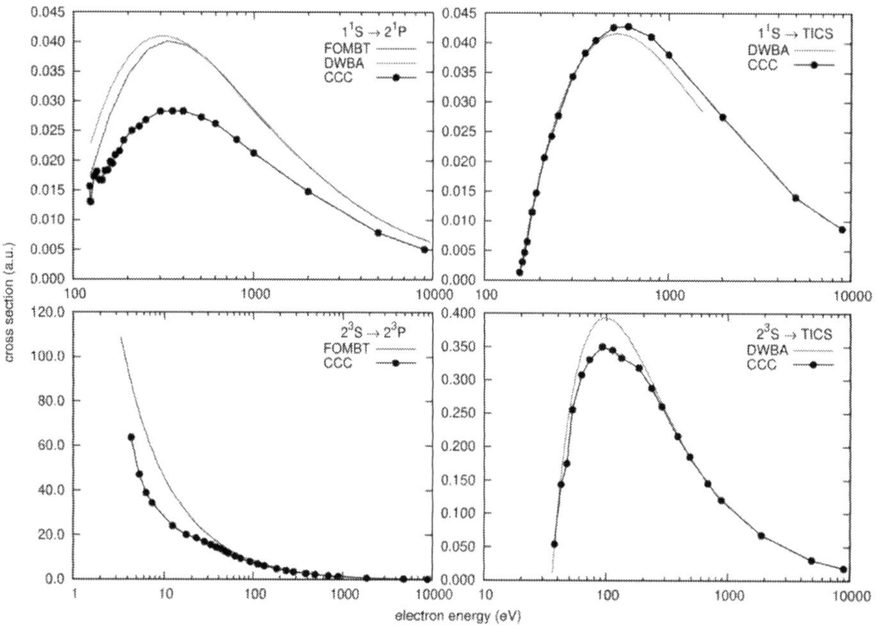

Figure 4. Selected cross sections for e-Be^{2+} scattering calculated using the CCC method, and distorted-wave and first-order many-body theories.

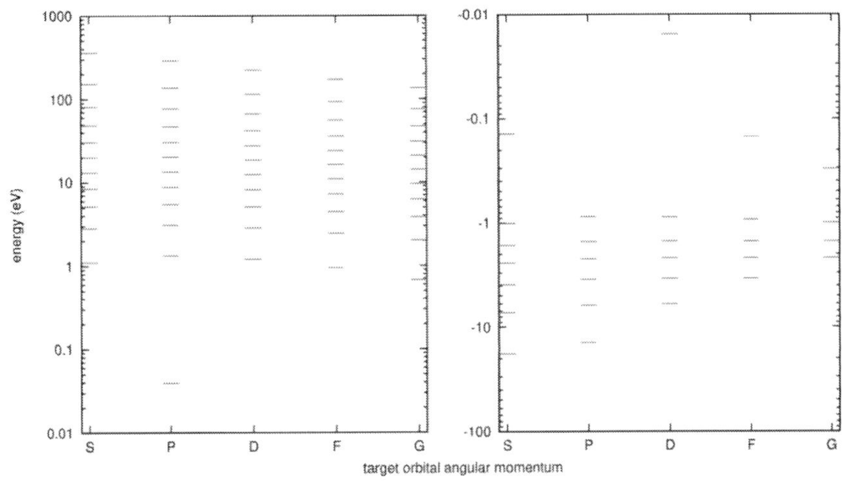

Figure 5. Energy levels in the CCC(84) calculations of e-Be^+ scattering.

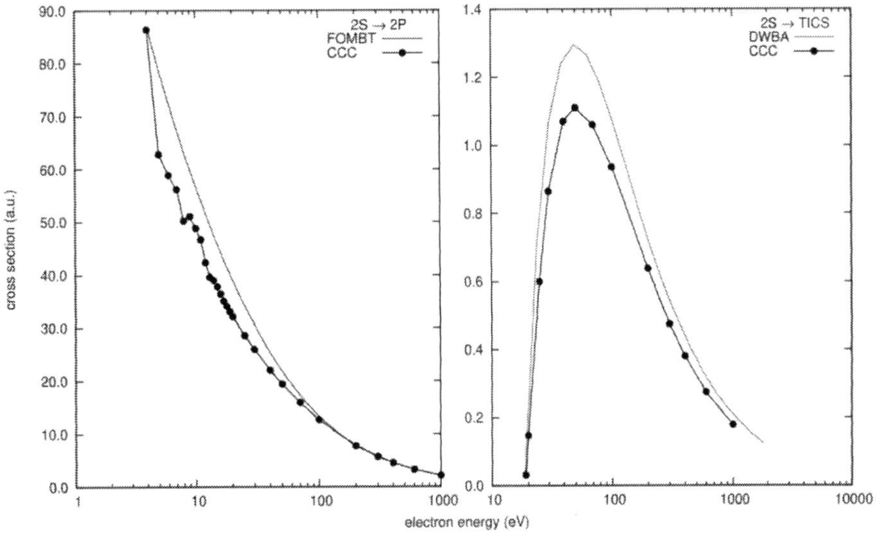

Figure 6. Selected cross sections for e-Be^+ scattering calculated using the CCC method, and distorted-wave and first-order many-body theories.

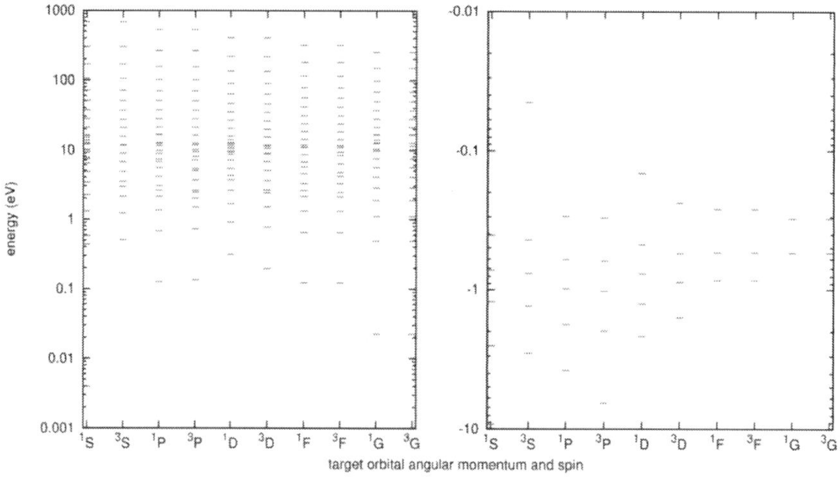

Figure 7. Energy levels in the CCC(284) calculations of e-Be^{2+} scattering.

previously, comparison is given with the DWBA and FOMBT theories. We see generally good agreement between the calculations.

2.4. e-Be scattering

Lastly, neutral Be is the most complex of the targets considered. The core 1s electrons are obtained via the SCHF treatment of Be^+. The orbitals for generating the two-electron configurations were generated by diagonalising the Be^+ Hamiltonian in a Laguerre basis with $\lambda_l = 2$, $l_{\rm max} = 2$ and $N_l = 25 - l$. The frozen core configurations $\{2s, nl\}$ were supplemented by the symmetric ones using two lowest s-, p-, and d-orbitals. The Be Hamiltonian is then diagonalised in these configurations to yield the target states to be used in the close-coupling formalism. The resulting 292 energy levels are given in figure 7. From the figure we see that the calculations have good $n \leq 5$ eigenstates. The doubly excited states are interspersed with the $\{2s, nl\}$ continuum ones around 10 eV.

In figure 8 some fundamental cross sections for e-Be scattering are given for four selected transitions, and compared with the DWBA and FOMBT approaches. We see some substantial discrepancy between CCC and the first-order methods. Given that this is a neutral target perhaps this is not so surprising, with agreement being approached at the higher energies. Interestingly, the total ionisation cross section from the 2^3P state shows unexpectedly good agreement.

3. Conclusions

We have given a brief overview of the recently performed e-Be^{q+}, $q = 0, 1, 2, 3$, scattering calculations using the CCC method. These were a part of the IAEA Coordinated Research Project, motivated by the fact that the first wall of ITER will contain beryllium. The Laguerre basis sizes were sufficiently large to ensure that the accuracy of the results would be within 10%

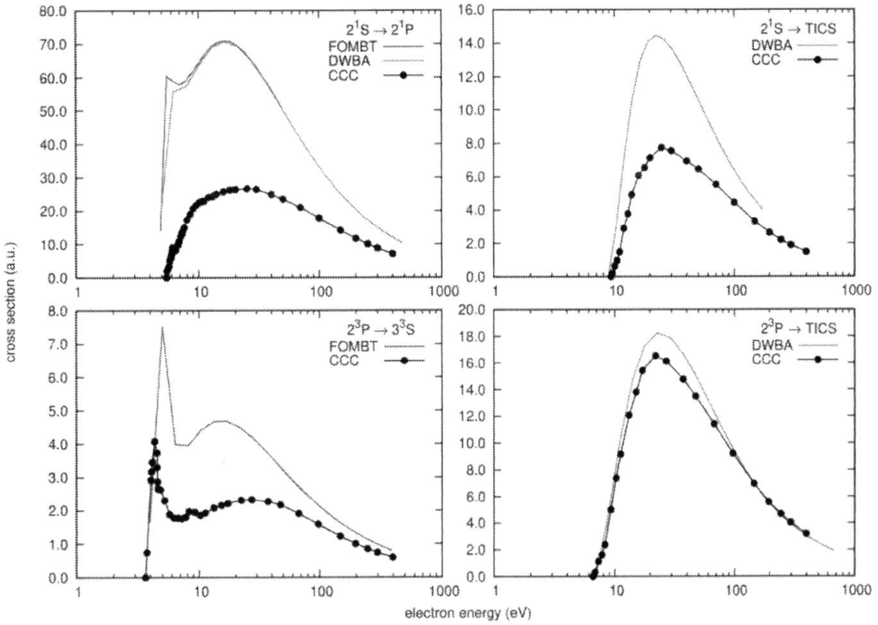

Figure 8. Selected cross sections for e-Be scattering calculated using the CCC method, and distorted-wave and first-order many-body theories.

of the true values. Presently, the resulting cross section data is being used to model radiative transport within the ITER plasma.

Acknowledgments
We would like to thank the IAEA staff Bas Braams and Hyun Chung for initiating the Coordinated Research Project. The work was supported by the Australian Research Council. We are grateful for access to the Australian Partnership for Advanced Computing and its Western Australian node iVEC.

References
[1] Bray I and Stelbovics A T 1992 *Phys. Rev. A* **46** 6995–7011
[2] Bray I, Fursa D V, Kheifets A S and Stelbovics A T 2002 *J. Phys. B* **35** R117–R146
[3] Pindzola M S, Robicheaux F, Loch S D, Berengut J C, Topcu T, Colgan J, Foster M, Griffin D C, Ballance C P, Schultz D R, Minami T, Badnell N R, Witthoeft M C, Plante D R, Mitnik D M, Ludlow J A and Kleiman U 2007 *J. Phys. B* **40** R39–R60
[4] Bray I, Fursa D V, Kadyrov A S, Stelbovics A T, Kheifets A S and Mukhamedzhanov A M 2012 *Physics Reports* **520** 135–174
[5] Bray I 1994 *Phys. Rev. A* **49** 1066–1082
[6] Fursa D V and Bray I 1997 *J. Phys. B* **30** 5895–5913

Calculation of total cross sections for electron capture in collisions of Carbon ions with H(D,T)(1s)

L. F. Errea, Clara Illescas, Alba Jorge, L. Méndez, I. Rabadán and J. Suárez

Laboratorio Asociado al CIEMAT de Física Atómica y Molecular en Plasmas de Fusión. Departamento de Química, Universidad Autónoma de Madrid, Madrid-28049, Spain.

E-mail: `l.mendez@uam.es`

Abstract. The calculations of total cross sections of electron capture in collisions of C^{q+} with H(1s) are reviewed. At low collision energies, new calculations have been performed, using molecular expansions, to analyze isotope effects. The Classical Trajectory Monte Carlo method have been also applied to discuss the accuracy of previous calculations and to extend the energy range of the available cross sections.

1. Introduction

Carbon ions are one of the main impurities in present tokamak plasmas, where carbon composites are used in first wall tiles, specially in the divertor. It is known that these materials are not appropriate for D-T plasmas because of the tritium deposition problem, and it is planned that ITER first wall will be completely made of Be and W tiles. However, small amounts of carbon impurities will be present in ITER, and it is possible that future devices will include materials which will release C^{q+} ions. The importance of carbon impurities has stimulated many theoretical and experimental works on collisions of these ions with Hydrogen. In particular, the electron capture reactions (EC) were reviewed in 2006 by Suno and Kato [1]; they reviewed the bibliography and proposed a set of recommended data for both total and state-selective electron capture cross sections.

The aim of the present work is to discuss the existing calculations on the EC reactions:

$$C^{q+} + H(1s) \rightarrow C^{(q-1)+} + H^+. \qquad (1)$$

The collisions involving the fully stripped projectile C^{6+} have been studied in many works because their relevance in charge exchange diagnostics [2] and also because, being a one-electron system, it is relatively easier to describe theoretically than the collisions with other Carbon ions, which require the use of sophisticated quantum chemistry techniques to obtain accurate results. The collision C^{4+} + H, can also be treated as a one-electron system, where the interaction of the active electron with the C^{4+} core is described by means of an effective potential.

In order to discuss the accuracy of previous calculations and to fill up some gaps of the existing database, we have carried out new calculations employing classical trajectory Monte Carlo (CTMC) methods for energies above 10 keV/u. We have applied the CTMC method to H collisions with partially stripped projectiles by employing a model potential to represent the interaction of the active electron with the ionic core, as previously employed in the classical calculations of Stancil *et al* [3].

At low energies, we have employed the molecular-functions-close-coupling (MFCC) method[1]. our calculations aim to extend the energy range of the computed EC total cross sections and to discuss the isotopic effect. These two aspects have been less studied because they are not relevant in core plasma diagnostics, but can be important in passive diagnostics and in plasma modeling. For collisions of low charged (q=1,2) Carbon ions, the EC reaction is endothermic, the EC channels $C^{(q-1)+}$ are closed at low energies, and the total cross section decreases rapidly as the collision energy, E, decreases. However, for q=3-6, the reaction (1) is exothermic for some EC channels and, consequently, there is not a threshold in the EC total cross sections, which increase approximately as $E^{-1/2}$ as E decreases, as predicted by the Langevin model [4, 5], leading to large cross sections at energies below $E \approx 10$ eV/u. Another relevant aspect of low-energy EC is the presence of shape resonances, found in several theoretical works [6, 7, 8, 9], although the precision of the experimental techniques has precluded to observe them hitherto.

2. Computational methods

2.1. The MFCC method

The MFCC calculations have been carried out using the method described ion detail in previous publications [8, 10]. In this method, the scattering wave function Ψ^J is solution of the stationary Schrödinger equation:

$$\widehat{H}\Psi^J = E\Psi^J \tag{2}$$

where E is the collision energy and $\widehat{H}$ is the total Hamiltonian of the system given by

$$\widehat{H} = -\frac{1}{2\mu}\nabla_R^2 + \widehat{H}_{\rm el} \tag{3}$$

with μ the reduced mass of the nuclei and $\widehat{H}_{\rm el}$ the Born-Oppenheimer electronic Hamiltonian. Ψ^J is expanded in a molecular basis set, $\{\phi_j\}$:

$$\Psi^J(\boldsymbol{r},\boldsymbol{\xi}) = \sum_j \chi_j^J(\boldsymbol{\xi})\phi_j(\boldsymbol{r};\xi) \tag{4}$$

In this expression, J is the total angular momentum and $\boldsymbol{r}$ are the electronic coordinates. $\boldsymbol{\xi}$ is a common reaction coordinate [11], a linear combination of the internuclear vector $\boldsymbol{R}$ and the electronic coordinates that ensures that a truncated expansion fulfills the boundary conditions. We have employed a common reaction coordinate based on the switching function of reference [12] for the many-electron systems, while for C^{4+} and C^{6+} + H collisions the common reaction coordinate is defined by means of the switching function of references [13, 14]. The molecular functions ϕ_j are (approximate) eigenfunctions of $\widehat{H}_{\rm el}$:

$$\widehat{H}_{\rm el}\phi_j(\boldsymbol{r};R) = \epsilon_j(R)\phi_j(\boldsymbol{r};R) \tag{5}$$

For a N-electron system, $\widehat{H}_{\rm el}$ has the form:

$$\widehat{H}_{\rm el} = -\frac{1}{2}\sum_{k=1}^{N}\nabla_k^2 + V(\boldsymbol{r},R) \tag{6}$$

[1] Since in this method the collision wave function is expanded in terms of many-electron wave functions, we name these basis functions "molecular functions" instead of the commonly-used "molecular orbitals" name.

Journal of Physics: Conference Series **576** (2015) 012002 doi:10.1088/1742-6596/576/1/012002

Table 1. Values of the parameter α of equation (8) for $C^{(q-1)+}$.

	C	C^{2+}	C^{3+}	C^{4+}
α	1.18	1.85	3.72	5.67

where $V(\boldsymbol{r}, R)$ is the potential that includes electron-nuclei and electron-electron interactions.

Substitution of the expansion (4) into the Schrödinger equation leads to system of differential equations for the nuclear functions χ_j^J, which is solved numerically. The asymptotic behavior of these functions allows us to evaluate the **S**-matrix, as explained in detail in [15], and the total cross section for the transition between two states $\phi_i \to \phi_j$, σ_{ij}, is computed from the **S**-matrix elements using the equation:

$$\sigma_{ij} = \frac{\pi}{k_i^2} \sum_J (2J+1) \left|\delta_{ij} - S_{ij}^J\right|^2 \tag{7}$$

The molecular orbitals of the one-active electron systems CH^{4+} and CH^{6+} were obtained using the method of Power [16]. For the many-electron systems we have computed the molecular wave functions, the potential energy curves and non-adiabatic couplings by applying the multi-reference configuration interaction (MRCI) methods. In the present work we have employed the package MOLPRO [17] for calculating the molecular functions of the four-active electron system CH^{+}, and the package MELD [18] for the three- and two-active electron quasimolecules CH^{2+} and CH^{3+}.

2.2. The CTMC method

At $E > 10$ keV/u, we have employed the CTMC method. In our treatment the nuclei follow rectilinear trajectories ($\boldsymbol{R} = \boldsymbol{b} + \boldsymbol{v}t$) and the electronic motion is described by means of a classical distribution function $\rho(\boldsymbol{r}, \boldsymbol{p}, t)$ for an ensemble of $\mathcal{N}$ ($\mathcal{N} \approx 10^5$ in our calculations) independent electron trajectories, $(\boldsymbol{r}(t), \boldsymbol{p}(t))$, which are obtained by integration of the Hamilton equations determined by the electronic Hamiltonian. The standard CTMC method [19] is restricted to one-electron systems, but it can be applied to one-effective electron systems by employing an effective potential. In our calculations we have used the model potential:

$$V(r_C) = -\frac{6 - N_c}{r_C} - \frac{N_c}{r_C}(1 + \alpha\, r_C) \exp(-2\alpha\, r_C) \tag{8}$$

to represent the interaction of the active electron with the C^{q+} ion. In equation (8), r_C is the electron distance to the C nucleus, N_c the number or core electrons and the parameter α (see table 1) has been obtained by fitting the ionization energy of the $C^{(q-1)+}$ ion.

The total EC cross section is given in this method by:

$$\sigma^{EC}(E) = 2\pi \int_0^\infty b P^{EC}(b) \mathrm{d}b \tag{9}$$

where the EC transition probability is calculated as:

$$P^{EC} = \frac{\mathcal{N}^{EC}}{\mathcal{N}}, \tag{10}$$

being $\mathcal{N}^{EC}$ the number of trajectories where the electron is bound to the projectile at the end of the collision ($vt_{fin} = 500$ a.u. in our calculation); i.e., those with negative energy with respect to the projectile and positive energy with respect to the target.

Table 2. Total cross sections in 10^{-16} cm^2 for $C^{q+} + H(1s) \to C^{(q-1)+} + H^+$ as functions of the impact energy in keV/u.

E	$q=1$	$q=3$	$q=4$	$q=5$	$q=6$
10	5.23	21.4	31.1	40.3	49.2
25	2.73	17.0	25.8	34.3	42.6
50	0.946	7.71	13.2	19.4	26.3
100	0.266	1.59	3.09	5.02	7.37
200	0.0565	0.173	0.341	0.599	0.926
300	0.0186	0.0396	0.0714	0.126	0.209

At low collision energies, the accuracy of the CTMC method is determined by the quality of the initial distribution, $\rho^i(\boldsymbol{r},\boldsymbol{p})$, that, in our case, is the classical distribution for the ground state of the Hydrogen atom. The standard CTMC treatment employs a microcanonical distribution, where all electron trajectories have the energy −0.5 a.u. of the quantal level. In order to compare with the quantal distributions, one can introduce the radial and momentum distributions:

$$\rho^i_r(\boldsymbol{r}) = \int \mathrm{d}\boldsymbol{p}\rho^i(\boldsymbol{r},\boldsymbol{p});\; \rho^i_p(\boldsymbol{p}) = \int \mathrm{d}\boldsymbol{r}\rho^i(\boldsymbol{r},\boldsymbol{p}) \tag{11}$$

It is well known (e.g. [20]) that, although the microcanonical distribution leads to a momentum distribution identical to the quantal one, it is unable to describe the tail of the radial quantal distribution, which includes electron-nucleus distances in the classically forbidden region. At low collision energies, this limitation is specially relevant because the EC probability is high for the initially loosely bound electron trajectories. Some alternatives have been suggested in order to solve this limitation of the microcanonical distribution. In particular, we have employed the so-called hydrogenic distribution [21, 22], which is a linear combination of several (ten in our calculations) microcanonical distributions with different energies; the coefficients of the combination are obtained by fitting the quantal radial distribution with the restriction that the mean energy of this distribution is approximately equal to −0.5 a.u.. In practice, it is found that the hydrogenic spatial and momentum distributions agree with their quantal counterparts.

Previous calculations on collisions Hydrogen atoms and fully stripped ions with $q \geq 6$ have shown that the CTMC EC cross sections at high impact energy ($E \gtrsim 500$ keV/u) are larger than those obtained with perturbative treatments, which are expected to be accurate in this energy range. Therefore, we present in table 2 our CTMC cross sections for $E < 300$ keV/u.

3. C^+ + H collisions

Janev *et al* [23] deduced a recommended total cross section for the reaction

$$C^+(1s^2 2s^2 2p\,^2P) + H(1s) \to C + H^+ \tag{12}$$

based on the experimental data of references [24, 25] and [26]. Stancil *et al* [3] have performed a combined experimental and theoretical study of this process. In order to cover a wide energy range, they reported merged-beam experiments and calculations with several methods: MFCC, CTMC, multielectron hidden crossing and the decay model. Although the new results agree with the high-energy measurements of references [24] and [26], they disagree with those of [25] at collision energies between 0.3 and 1 keV/u. However, the MFCC results of Stancil *et al* [3] agree with the merged-beam experimental cross sections reported in the same paper, which lead to the authors to suggest a new recommended cross section for reaction (12) that clearly differs from that of reference [23] for energies below 1 kev/u. As explained in reference [3], the potential energy curves of the CH^+ system do not show any avoided crossing between the potential energy curves of the entrance channels and those of the molecular states leading to EC. The main mechanism of reaction (12) involves transitions from the state $a^3\Pi$ to $2^3\Pi$, with a energy difference of about 0.2 a.u., that yield a cross section that decays rapidly as E decreases.

We have performed a MRCI calculation with an aug-cc-pVQZ basis set [27]. The molecular orbitals were obtained by a 5-state CASSCF procedure, and the ensuing MRCI calculation included up to 5×10^5 configuration state functions. We have employed these molecular wave functions to calculate the EC total cross section, which is practically identical to that of reference [3]. Moreover, the merged-beams experiment of Stancil et al. [3] was carried out with Deuterium, while that of Nutt et al. [25] was performed with Hydrogen. We have checked that the isotopic dependence of the calculated cross section for the reaction (12) is negligible above $E = 10$ eV/u and does not explain the difference between both measurements.

On the other hand, the CTMC results of Stancil *et al* [3] underestimate the experimental data for energies between 1 and 25 keV/u. These can be due to the inadequacy of the microcanonical classical distribution for H(1s) [19, 20], which leads to a decrease of the EC total cross section at low energies (see e.g. reference [28]). The use of a model potential, whose functional form is not shown in the paper of reference [3], might also limit the validity of that calculation. For $E < 1$ keV the CTMC calculation overestimates the experiment, which is explained by the authors as due to the trajectories leading to capture with energies below those in the phase space bin associated to the 2p level and that would correspond to the electron capture into quantum levels that are already occupied. The above-mentioned difficulties are probably not important at $25 < E < 200$ keV/u, where the CTMC calculation agrees with the available experiments [24, 26]. The recommended data are not supported by either calculations or experiments at $E > 200$ keV/u. To further study the workings of the CTMC method we have calculated the total cross section with the eikonal CTMC method and with the model potential of equation (8). A hydrogenic distribution [22] has been used to represent the initial distribution of the H(1s). The comparison of the present calculation with the recommended data of Stancil *et al* [3] and the CTMC data from the same reference (figure 1) indicates that the underestimation of the total cross section for energies below 25 keV/u is reduced by using the hydrogenic distribution; our results with the microcanonical distribution, not shown in figure 1, are indistinguishable from those from reference [3]. In the energy range of figure 1 we have not found a sizable effect of trajectories leading to nonphysical EC.

4. C^{2+} + H collisions

The system has been studied theoretically by Gu *et al* [29] and Errea *et al* [30], both calculations used a MFCC method within a semiclassical treatment. This system was studied experimentally by Nutt *et al* [31], Phaneuf *et al* [24], Goffe *et al* [26], Gardner *et al* [32] and Voulot *et al* [33]. The calculation of reference [30] tabulated total and state-selective cross section for the EC reaction:

$$C^{2+}(1s^2 2s^2\,{}^1S) + H(1s) \rightarrow C^+ + H^+ \tag{13}$$

From the experimental point of view, the main difficulty to study this system is the presence of unknown quantities of metastable ions $[C^{2+}(1s^2 2s2p\,{}^3P)]$ in the beams. However, references [33] and [30] pointed out that the experimental total cross sections are not modified by a contamination of the beam as high as 20% for energies above 0.1 keV/u. Nonetheless, the calculated cross sections overestimate the experimental ones at $E < 1$ keV/u (see figure 2). Errea *et al* [30] suggested that the discrepancies could be due to a normalization problem of the experiment, but neither new experiments nor calculations have been performed. At $E > 1$ keV/u, target ionization is not negligible, and the molecular close-coupling treatment, which does not include continuum wave functions, is not applicable. To our knowledge, no other theoretical methods have been applied to this particular reaction.

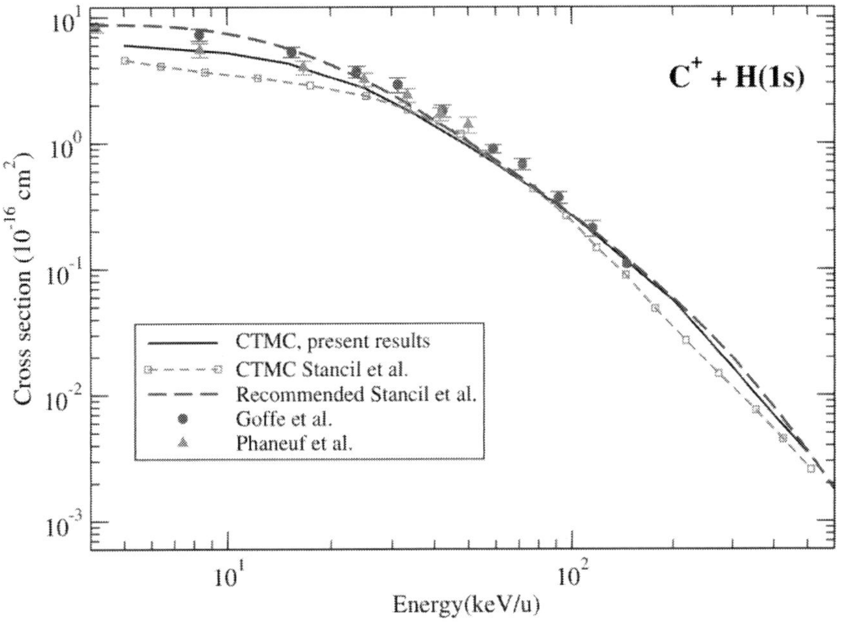

Figure 1. Total cross section for reaction (12) as a function of the impact energy. CTMC results are compared with previous calculations of Stancil *et al* [3], recommended data of reference Stancil *et al* [3] and experimental data [24, 26].

At low energies ($E < 100$ eV/u), the eikonal approximation employed in references [29] and [30] is not accurate, and we have carried out a fully quantal calculation. Since energy differences are critical in low-energy calculations, we have recalculated the potential energy curves and dynamical couplings of the doublet molecular states of the CH^{2+} quasimolecule. In our calculation, the asymptotic energy difference between the states dissociating into $C^{2+}(1s^2 2s^2\,{}^1S)$ + H(1s) (the entrance channel) and $C^+(1s^2 2s 2p^2\,{}^2D)$ (the main exit channels) is 0.0577 a.u., which has to be compared to the value 0.0384 a.u. reported in [30] and the experimental value 0.0547 a.u. [34]. The cross sections for reaction (13) are shown in figure 2. It is clear that the new MFCC calculation with rectilinear trajectories reproduces the results of Errea *et al* [30], and the difference with the experiment of Nutt *et al* [31] is essentially due to the population of the state $C^+(2s^2p^2\,{}^2S)$. In fact, the 3-state calculation, including the entrance channel and the most populated ${}^2\Sigma^+$ states [those dissociating into $C^+(2s^2 2p\,{}^2P)$ + H^+ and $C^+(2s2p^2\,{}^2D)$ + H^+], yields a cross section practically identical to the experimental one and to that reported by Gu *et al* [29], who only included the populations of ${}^2P$ and ${}^2D$ exit channels.

At energies below 0.1 keV/u, the trajectory effects are sizable, but a relatively simple 2-state calculation can be carried out, as can be deduced from the comparison of 2- and 3-state cross sections. The two-state quantal calculation leads to the cross sections for different isotopes shown in figure 2. As for C^+ + H collisions, the EC cross section of figure 2 shows a threshold at low E, as expected for a endothermic reaction. The isotope effect is small but noticeable, and with a behavior ($\sigma_T > \sigma_D > \sigma_H$) already observed in other systems (see [35, 36]). This isotopic dependence is typical of systems where the electron capture reaction takes place at short internuclear distances where the ion-atom interaction potential that defines the trajectory is positive. We do not present CTMC results for reaction (13) because this collision is not accurately described by means of a single electron treatment given that two-

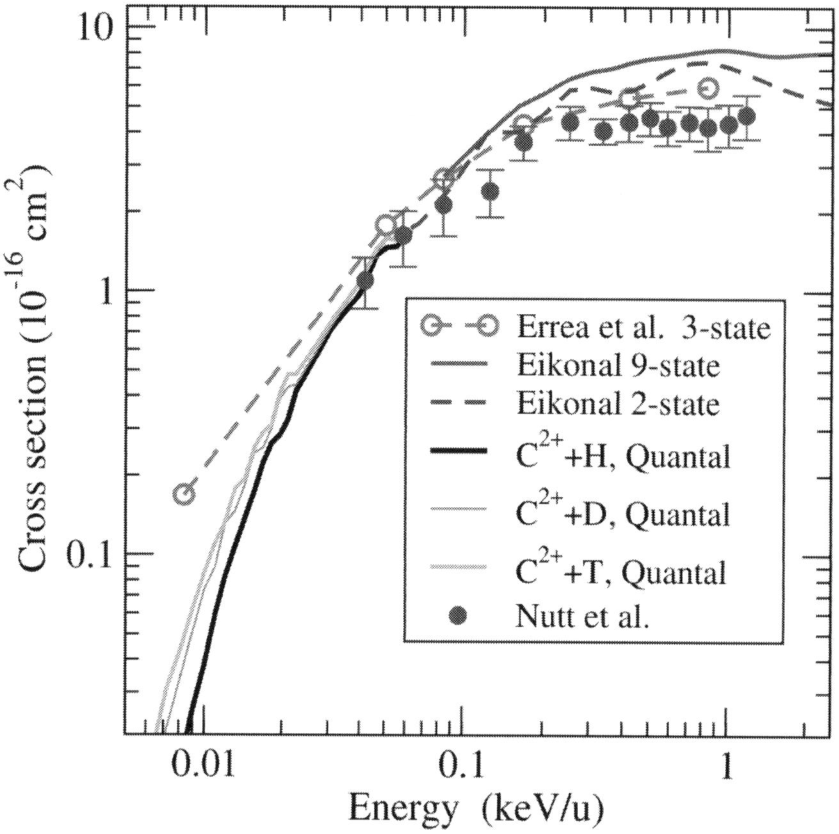

Figure 2. Total cross section for electron capture in C^{2+} + H(D,T)(1s) collisions, reaction (13), as function of the impact energy. Semiclassical results with 9- and 2-state basis are compared to 2-state quantal calculations for collisions with H, D and T. Also included are the eikonal result of Errea *et al* [30] and the experimental results of Nutt *et al* [31].

electron processes are known to play a significant role. In particular, the formation of the main exit channel $C^+(1s^2 2s 2p^2\,{}^2D)$ of the EC reaction involves a two-electron process from the initial state, $C^{2+}(1s^2 2s^2\,{}^1S)$ + H(1s), in which one electron is captured and another one excited.

5. C^{3+} + H collisions

Total cross sections for electron capture in C^{3+} + H collisions were measured by Phaneuf *et al* [24, 37], Goffe *et al* [26], Crandall *et al* [38], Gardner *et al* [32], Ćirić *et al* [39] and Havener *et al* [40]. Previous calculations have been carried out using MFCC treatments within semiclassical [41] and quantal formalisms [42, 43, 44], atomic orbital close-coupling (AOCC) treatments [45] and the so-called Electron-Nuclear-Dynamics method [46]. The results of previous works are plotted in figure 3. In the present work we have performed a MFCC calculation with MRCI wave functions. As a confirmation of the quality of the molecular wave functions, the asymptotic energy differences between the most important channels differ in about 0.01 a.u. from the spectroscopic values, which is a significant improvement with respect to those employed by Errea *et al* [41] and Herrero *et al* [44].

In the present work, we have evaluated the EC cross section by using an improved set of 33

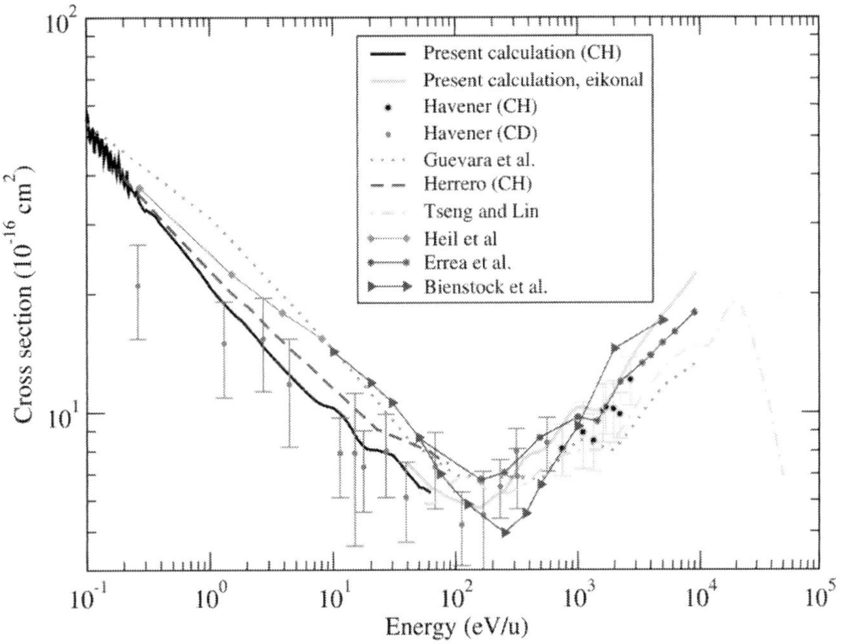

Figure 3. Total cross section for electron capture in C^{3+} + H(1s) collisions as function of the impact energy. Full lines: present quantal (thin line) and eikonal (thick line) results. Previous calculations of references [42, 43, 41, 44, 45, 46] and experiments [40] are also included, as indicated in the figure.

molecular functions (15 singlets and 18 triplets) in the eikonal calculation and 10 functions (5 singlets and 5 triplets) in the quantal one. Our quantal total cross section is lower than that of Herrero *et al* [44] and the eikonal one is somewhat lower than that of Errea *et al* [41]; the differences are due to the changes in the potential energy curves and couplings with respect to those of the valence bond calculation [47] used in previous MFCC calculations. The new results show very good agreement with the merged-beams experiment [40]. At energies above $E \approx 2$ keV the molecular expansion converges slowly, and the AOCC results of Tseng and Lin [45] are probably more accurate. Moreover, the AOCC calculation agrees with the experimental data and the cross section computed by Guevara *et al* [46].

At low velocities, the EC cross section (see figure 4) increases rapidly as E decreases, following approximately the Langevin model [4], where the cross section is proportional to $E^{-1/2}$. As already pointed out by Herrero *et al* [44], the isotopic dependence is small in the triplet subsystem, which leads to a small but noticeable isotopic dependence of the total cross section for $v < 2 \times 10^{-3}$ a.u. ($E < 0.1$ eV/u). One can also note the spikes in the cross sections due to the presence of quasi-stationary vibrational states in the effective potential of the initial molecular state. This effective potential is obtained by adding the centrifugal term and the electronic energy, whose asymptotic behavior:

$$\epsilon \sim -\frac{q^2\alpha}{2R^4}, \tag{14}$$

corresponds to the ion-induced-dipole interaction, with α the atom polarizability. One can also note the oscillations of the EC cross section for $0.1 < E < 0.3$ eV/u, caused by the interferences

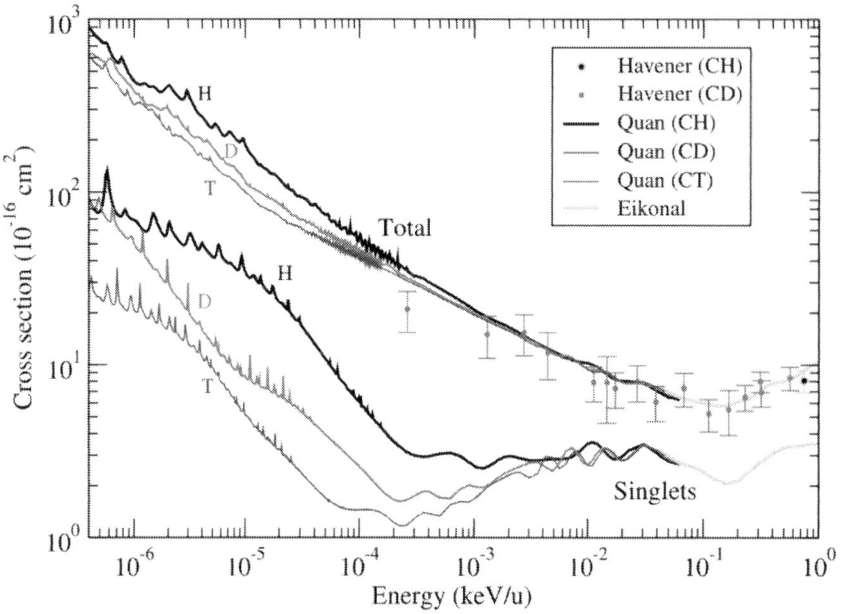

Figure 4. Total cross section for electron capture in C^{3+} + H(D,T)(1s) collision as function of the collision energy. Present results corresponding to different isotopes are plotted with solid lines. The contributions of the singlet subsystem to the total cross sections are also shown. Experimental data by Havener *et al* [40] are included.

between the nuclear wave functions in two avoided crossings between the potential energy curves of the entrance and the main exit channel in the triplet subsystem.

At high energy (see figure 5), there are no theoretical data available, and we have performed a calculation similar to that for C^+ + H (figure 1). The calculation employs the eikonal-CTMC method with the hydrogenic initial distribution and the model potential of equation (8). At $E > 30$ keV/u, our cross section shows good agreement with the experiments and the recommended data. For $E < 30$ keV/u the calculation agrees with the experimental results of Goffe *et al* [26], but these energies are probably too low to employ the CTMC method to asses the accuracy of the experimental values.

6. C^{4+} + H collisions

Several calculations have been carried out of EC total cross sections in C^{4+} + H collisions. In particular, MFCC calculations of [48, 49], expansions in terms of atomic orbitals [50], wave-packet treatments [51] and the hyperspherical close coupling method [52]. The low-energy behavior of the cross section has been studied in reference [9]. The EC cross sections is almost constant for collision energies between 100 eV/u and 10 keV/u. For $E > 10$ keV/u, target ionization is the dominant process and the EC cross section decreases rapidly as E increases. The cross section at high energies was measured by Phaneuf *et al* [24] and Goffe *et al* [26], but there is no theoretical counterpart of these measurements.

At $E < 10$ keV/u, there is a remarkable good agreement between different calculations, but the agreement with the experiments [53, 54] is less satisfactory. In this respect, Liu *et al* [52] concluded that the low-energy measurements are not supported by the theories, and that further experimental work is needed. In our opinion, the calculated cross sections are more accurate

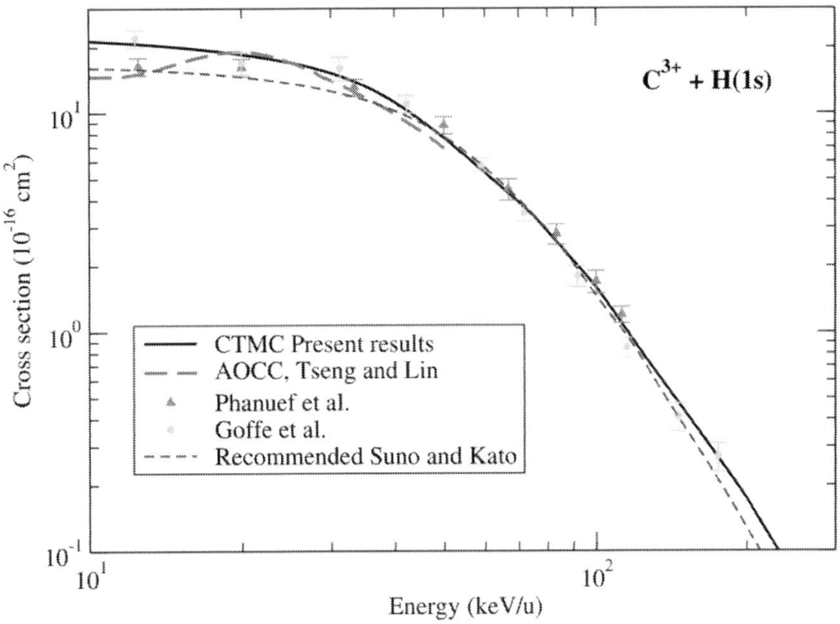

Figure 5. Total cross section for electron capture in C^{3+} + H (1s) collisions as function of the impact energy. Our CTMC results are compared to the experimental results of references [24, 26], the AOCC calculation of Tseng and Lin [45] and the recommended data of Suno and Kato [1].

than the experimental ones for this particular system, in contrast with the recommended values of reference [1] that follow the experimental data at $E < 100$ eV/u. One should also note the excellent agreement between different calculations (see reference [52]) for the partial cross sections for populating $C^{3+}(3l)$ states.

In this work we have recalculated the total EC cross section at very low energies using the molecular basis set of Barragán *et al* [9] to study the isotope effect. Our results are displayed in figure 6. In contrast with the dependence illustrated in figure 2, characteristic of reactions that take place at internuclear distances where the interaction potential is repulsive, the isotopic dependence ($\sigma_T > \sigma_D > \sigma_H$) shown in figure 6 is similar to that found for C^{3+} + H and comes from trajectory effects for the relative motion in the attractive ion-induced dipole interaction potential [36]. Our CTMC results are compared to the experimental results in figure 7, where one can note the excellent agreement between experiments and calculations.

7. C^{5+} + H collisions

C^{5+} + H collisions have been considered in MFCC calculations within both quantal [55, 56] and semiclassical [57, 56] formalisms. CTMC calculations have been reported in reference [57]. Total cross sections have been measured in references [38, 58, 26]. Recently, Draganić *et al* [59] have performed merged-beams experiments. The MFCC calculations lead to a Langevin-type cross section in satisfactory agreement with the experimental data, although the energy grid of the calculation does not allow to observe the spikes due to shape resonances.

We have applied the CTMC method with the corresponding model potential to evaluate the EC total cross section (figure 8), which shows a general good agreement with experiments [26] and the calculation of reference [57], although our calculation extends over a larger energy

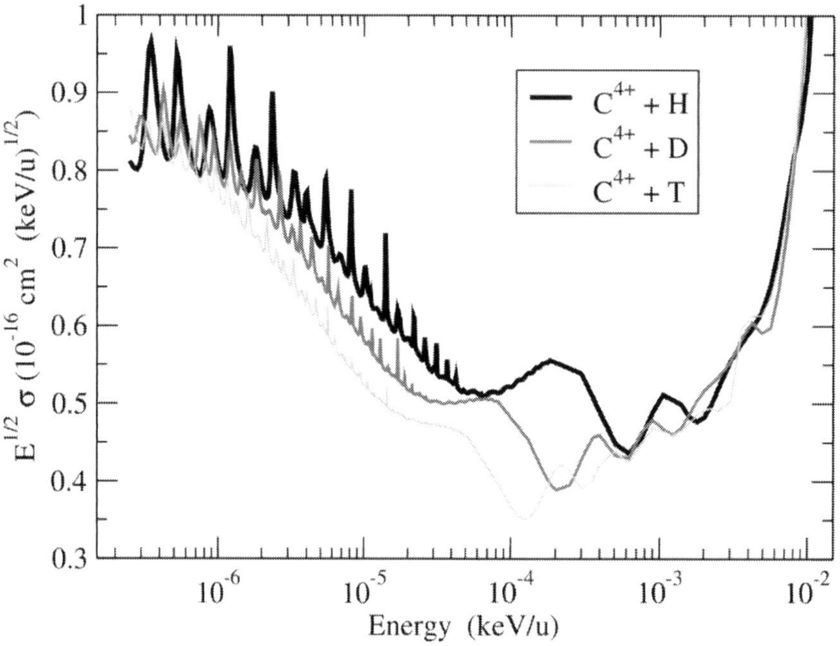

Figure 6. Total cross section for electron capture in C^{4+} + H(D,T)(1s) collisions, multiplied by $E^{1/2}$, as function of E.

range than the previous one, It can be noted that the energy dependence of our cross section is identical to that of the recommended data of Suno and Kato.

8. C^{6+}+ H collisions

As already mentioned, EC in C^{6+} + H(1s) collisions has been studied in several publications. Experimental data have been reported in references [26], [58] and [60]. High-energy calculations include the first Born approximation [61] and the eikonal impulse approximation [62]. The AOCC method has been employed in references [63, 64, 65, 66] and [67]. CTMC calculations have been carried out in references [22, 67, 68] and [69]. MFCC calculations were performed in references [70, 71] and [72]. The hyperspherical-close-coupling method was applied in the reference [73]. The mentioned calculations provide accurate values of the EC total cross section for energies above 10 eV/u together with state-selective cross sections. At E <10 eV/u, the work of Liu *et al* [73] has pointed out the relevance of the EC into $C^{5+}(n = 5)$ and the Langevin-type behavior of the total cross section, but values of this cross section have not been reported. The EC reaction at low energies takes place via transitions between the molecular orbitals 6σ0 and 4σ0 (in the united atom notation), at the avoided crossing at large internuclear distances ($R \approx 21.3$ bohr). In the present work, we have calculated the EC total cross section by using a molecular basis set that includes the entrance channel and the exit channels dissociating into $C^{5+}(n = 5)$ + H^+; the cross section is shown in figure 9. We have also checked that EC into $C^{5+}(n = 4)$ + H^+ is negligible for the collision energies of figure 9. We have also plotted in figure 9 the total cross section for collisions with H, D and T to show that the isotope effect is almost unnoticeable for this collision system, in accordance with the fact that, at the large internuclear distances where the transitions take place, the trajectory effects that give rise to the isotopic dependence are very small.

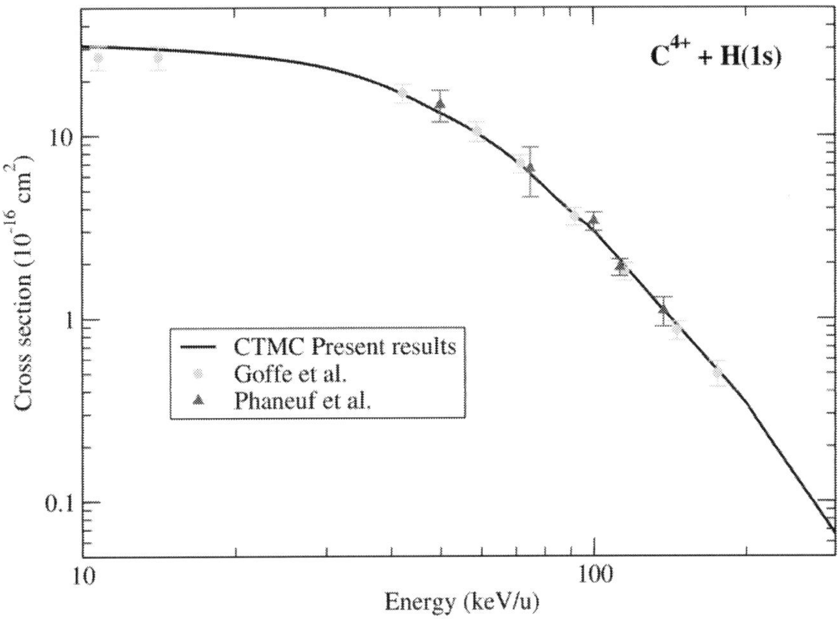

Figure 7. Total cross section for electron capture in C^{4+} + H(1s) collisions as function of the collision energy. Our CTMC results are compared to the experimental results of Goffe *et al* [26] and Phaneuf *et al* [24].

Recently, we have carried out CTMC calculations for this collision [68], and the results are displayed in figure 10, which also contains previous calculations and experimental data. Our results agree satisfactorily with the available experimental data. We also find a good agreement with the AOCC calculation of Igenbergs *et al* [67] for $E < 100$ keV/u. At higher energies, the CTMC calculation agrees with the first Born approximation [61], but it overestimates the results of the reference [62] (eikonal impulse approximation), which may be due to a limitation of the CTMC method to describe the EC into low n shells in the high-velocity limit.

9. Concluding remarks

In this work we have reviewed the calculations of electron capture total cross sections in collisions of Carbon ions with H(D,T)(1s). Although there exist several calculations for C^{6+} + H(1s) with different methods, few calculations have considered collisions of partially stripped ions, and the recommended data are based on experimental data, with the exception of C^{4+}+ H(1s); this system can be accurately described by employing a one-active electron treatment, and the agreement between different calculations is conspicuous. For the other collision systems, molecular calculations have been carried out systematically, employing multireference-configuration-interaction wave functions. The atomic expansion is difficult to apply to many-electron systems and the CTMC method has not been employed routinely until now to evaluate the EC total cross sections.

The new calculations presented in this work are focused on two points. First, the low-energy behavior of the EC cross section. It has been shown that the cross sections for Hydrogen collisions with the ions with $q \geq 3$ show the Langevin-type energy dependence, and the isotopic effect can be explained as due to trajectory effects. In the particular case of C^{6+}+ H(1s), the isotopic effect is negligible. Although the EC cross section at $E < 20$ eV/u for the collision

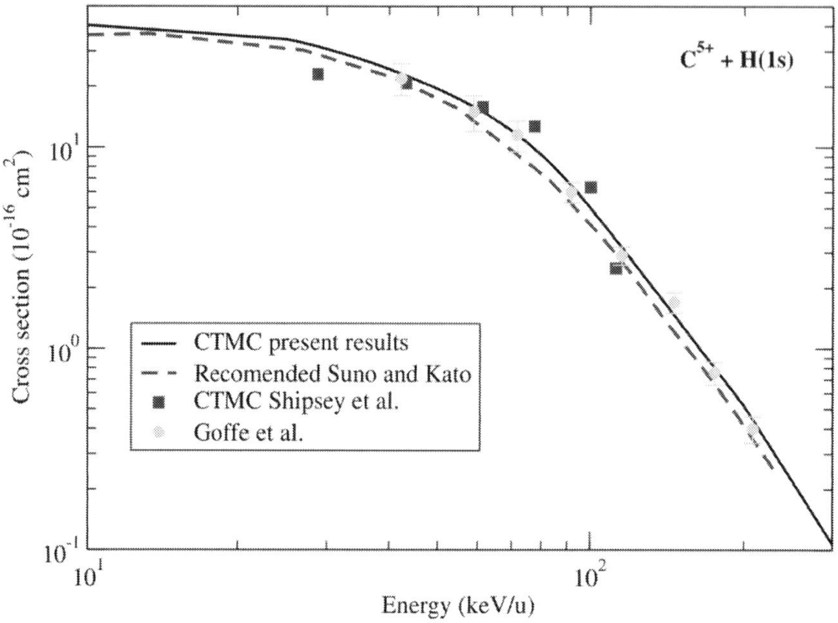

Figure 8. Total cross section for electron capture in C^{5+} + H(1s) collisions as function of the collision energy. Our CTMC results are compared to the experimental results [26], the CTMC calculation of Shipsey *et al* [57] and the recommended data of Suno and Kato [1].

C^{2+}+ H(1s) is small, its isotopic dependence is noticeable.

The second aspect covered by our work is the calculation of EC cross sections at energies $E \gtrsim 25$ keV/u. We present CTMC calculations for q = 1, 3, 4, 5 and 6, with an excellent agreement with experiments. Our results support the usefulness of CTMC method and show that the one-electron approximation, with the model potential of equation (8), provides a remarkably good representation of EC in these collisions at high energies. In particular, the comparison with the experimental data indicates that capture into occupied shells is not significant. The method is not applicable to treat C^{2+}+ H(1s) collisions, and new calculations are required for this system, for instance using a many-electron AOCC expansion, to compute the EC cross section for energies above 1 keV/u.

Acknowledgments
This work was carried out as a part of an IAEA Coordinated Research Project on "Light Element Atom, Molecule and Radical Behaviour in the Divertor and Edge Plasma Regions", and was supported by the project ENE2011-28200 (Secretaría de Estado de I+D+i, Spain). We thank the Centro de Computación Científica of UAM for computational support and Prof. B. Pons for providing us with the potential energy curves and dynamical couplings for the CH^{6+} system.

References
[1] Suno H and Kato T 2006 *Atomic Data and Nuclear Data Tables* **92** 407 – 455
[2] Isler R C 1994 *Plasma Phys. Control. Fusion* **36** 171
[3] Stancil P C, Gu J P, Havener C C, Krstic P S, Schultz D R, Kimura M, Zygelman B, Hirsch G, Buenker R J and Bannister M E 1998 *J. Phys. B: At. Mol. Opt. Phys.* **31** 3647
[4] Langevin P 1905 *Ann. Chim. Phys* **5** 245
[5] Gioumousis G and Stevenson D P 1958 *J. Chem. Phys.* **29** 294

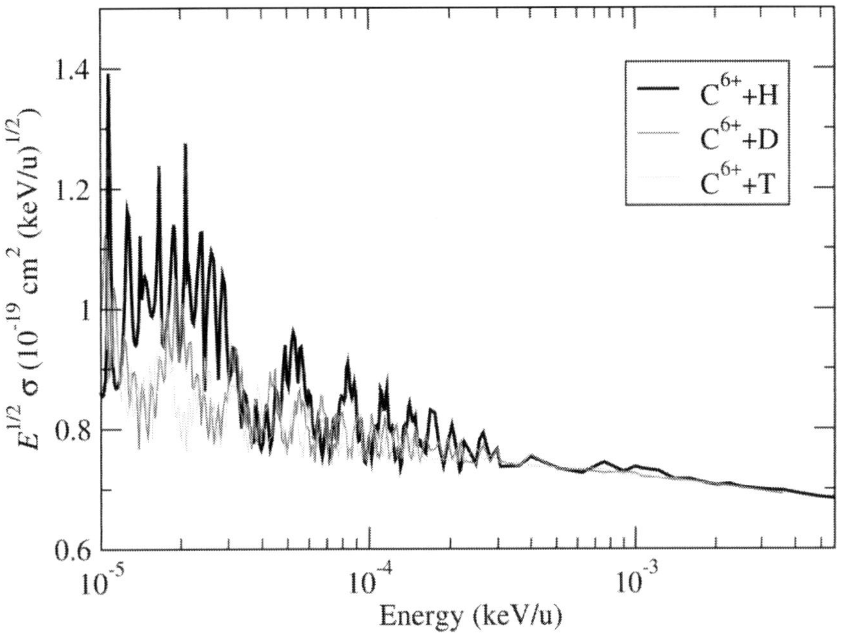

Figure 9. Total cross section for electron capture in C^{6+} + H(D,T)(1s) collisions multiplied by $E^{1/2}$, as function of E.

[6] Rittby M, Elander N, Brändas E and Bárány A 1984 *J. Phys. B: At. Mol. Phys.* **17** L677
[7] Barragán P, Errea L F, Méndez L, Rabadán I and Riera A 2006 *ApJ* **636** 544
[8] Barragán P, Errea L F, Méndez L, Rabadán I and Riera A 2008 *J. Phys. B: At. Mol. Opt. Phys.* **41** 225202
[9] Barragán P, Errea L F, Guzmán F, Méndez L, Rabadán I and Ben-Itzhak I 2010 *Phys. Rev. A* **81** 062712
[10] Barragán P, Errea L F, Méndez L, Macías A, Rabadán I and Riera A 2004 *Phys. Rev. A* **70** 022707
[11] Thorson W R and Delos J B 1978 *Phys. Rev. A* **18** 117
[12] Errea L F, Méndez L and Riera A 1982 *J. Phys. B: At. Mol. Opt. Phys.* **15** 101
[13] Harel C and Jouin H 1988 *J. Phys. B: At. Mol. Opt. Phys.* **21** 859
[14] Harel C and Jouin H 1991 *J. Phys. B: At. Mol. Opt. Phys.* **24** 3219
[15] McCarroll R 2003 *The Physics of Multiply and Highly Charged Ions* vol 2 ed Currell F J (Kluwer Academic Publishers, Dordrecht) pp 275–308
[16] Power J D 1973 *Phil. Soc. Trans. R. Soc* **274** 663
[17] *MOLPRO is a package of ab initio programs written by H. J. Werner and P. J. Knowles, with contributions from J. Almlöf, R. D. Amos, M. J. O Deegan, S. T. Elbert, C. Hampel, W. Meyer, K. Peterson, R. Pitzer, A. J. Stone, P. R. Taylor and R. Lidth.*
[18] Davidson E R 1990 *MOTECC, Modern Techniques in Computational Chemistry* ed Clementi E (ESCOM Publishers B. V., Leiden)
[19] Abrines R and Percival I C 1966 *Proc. Phys. Soc.* **88** 861
[20] Peach G, Willis S L and McDowell M R C 1985 *J. Phys. B: At. Mol. Phys.* **18** 3921
[21] Hardie D J W and Olson R E 1983 *J. Phys. B: At. Mol. Phys.* **16** 1983
[22] Illescas C and Riera A 1999 *Phys. Rev. A* **A60** 4546
[23] Janev R, Phaneuf R and Hunter H 1988 *Atomic Data and Nuclear Data Tables* **40** 249 – 281
[24] Phaneuf R A, Meyer F W and McKnight R H 1978 *Phys. Rev. A* **17** 534–545
[25] Nutt W L, McCullough R W and Gilbody H B 1979 *Journal of Physics B: Atomic and Molecular Physics* **12** L157
[26] Goffe T V, Shah M B and Gilbody H B 1979 *J. Phys. B: At. Mol. Opt. Phys.* **12** 3763
[27] Dunning, Jr T 1989 *J. Chem. Phys.* **90** 1007
[28] Errea L F, Illescas C, Méndez L, Pons B, Riera A and Suárez J 2004 *J. Phys. B: At. Mol. Opt. Phys.* **37** 4323–4338

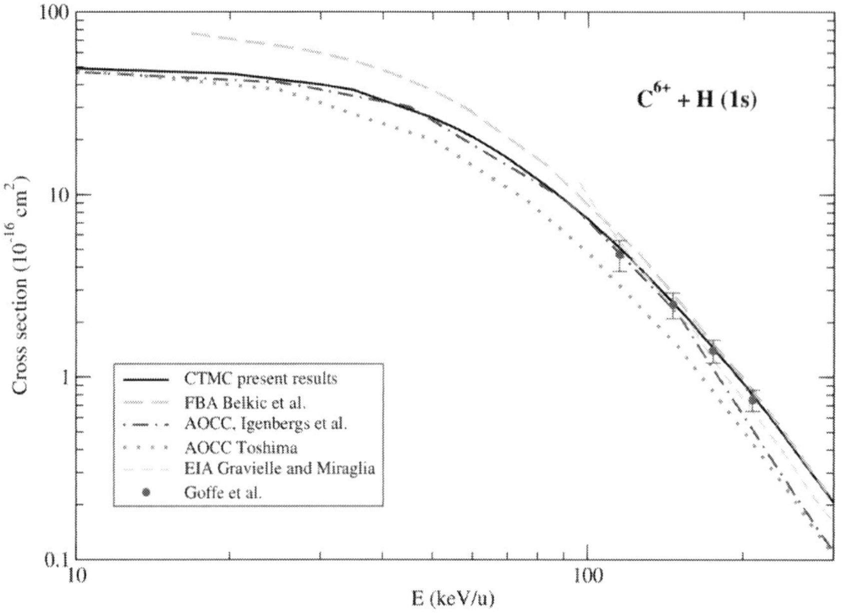

Figure 10. Total cross section for electron capture in C^{6+} + H(1s) collisions as function of the collision energy. Our CTMC results are compared to the experimental results [26], the AOCC calculations of Toshima [65] and Igenbergs *et al* [67], the eikonal impulse approximation (EIA) [62] and first Born approximation (FBA) [61].

[29] Gu J P, Hirsch G, Buenker R J, Kimura M, Dutta C M and Norlander P 1998 *Phys. Rev. A* **57** 4483
[30] Errea L F, Macías A, Méndez L and Riera A 2000 *J. Phys. B: At. Mol. Opt. Phys.* **33** 1369
[31] Nutt W L, McCullough R W and Gilbody H B 1978 *J. Phys. B: At. Mol. Phys.* **11** L181
[32] Gardner L D, Bayfield J E, Koch P M, Sellin I A, Pegg D J, Peterson R S and Crandall D H 1980 *Phys. Rev. A* **21** 1397–1402
[33] Voulot D, Gillen D R, Thompson W R, Gilbody H B, McCullough R W, Errea L, Macías A, Méndez L and Riera A 2000 *J. Phys. B: At. Mol. Opt. Phys.* **33** L187–L192
[34] Ralchenko Y, Jou F C, Kelleher D, Kramida A, Musgrove L, Reader J, Wiese W and Olsen K 2007 *NIST Atomic Spectra Database (version 3.1.2), [Online]. Available: http://physics.nist.gov/asd3 [2007, June 26]. National Institute of Standards and Technology, Gaithersburg, MD.*
[35] Errea L F, Guzmán F, Méndez L, Pons B and Riera A 2008 *Phys. Rev. A* **77** 012706
[36] Barragán P, Errea L F, Méndez L and Rabadán I 2010 *Phys. Rev. A* **82** 030701
[37] Phaneuf A, Alvarez I, Meyer F W and Crandall D H 1982 *Phys. Rev. A* **26** 1892
[38] Crandall D H, Phaneuf R A and Meyer F W 1979 *Phys. Rev. A* **19** 504–514
[39] Ciric D, Brazuk A, Dijkkamp D, de Heers F J and Winter H 1985 *J. Phys. B: At. Mol. Phys.* **18** 3629
[40] Havener C C, Müller A, Zeijlmans van Emmichoven P A and Phaneuf R A 1995 *Phys. Rev. A* **51** 2982–2988
[41] Errea L F, Herrero B, Méndez L and Riera A 1991 *J. Phys. B: At. Mol. Opt. Phys.* **24** 4061
[42] Heil T G, Butler S E and Dalgarno A 1981 *Phys. Rev. A* **23** 1100–1109
[43] Bienstock S, Heil T G, Bottcher C and Dalgarno A 1982 *Phys. Rev. A* **25**(5) 2850–2852
[44] Herrero B, Cooper I L, Dickinson A S and Flower D R 1995 *J. Phys. B: At. Mol. Opt. Phys.* **28** 711
[45] Tseng H C and Lin C D 1999 *J. Phys. B: At. Mol. Opt. Phys.* **32** 5271
[46] Guevara N L, Teixeira E, Hall B, Öhrn Y, Deumens E and Sabin J R 2011 *Phys. Rev. A* **83** 052709
[47] Errea L F, Herrero B, Méndez L, Mó O and Riera A 1991 *J. Phys. B: At. Mol. Opt. Phys.* **24** 4049
[48] Gargaud M, McCarroll R and Valiron P 1987 *J. Phys. B: At. Mol. Phys.* **20** 1555
[49] Errea L F, Gorfinkiel J D, Harel C, Jouin H, Macías A, Méndez L, Pons B and Riera A 1999 *J. Phys. B: At. Mol. Opt. Phys.* **32** L673
[50] Tseng H C and Lin C D 1998 *Phys. Rev. A* **58** 1966
[51] Vaeck N, Desouter-Lecomte M and Lievin J 1999 *J. Phys. B: At. Mol. Opt. Phys.* **32** 409–428

[52] Lee T G, Le A and Lin C D 2003 *J. Phys. B: At. Mol. Opt. Phys.* **36** 4081
[53] Dijkkamp D, Ciric D, Vlieg E, de Boer A and de Heer J 1985 *J. Phys. B: At. Mol. Opt. Phys.* **18** 4763
[54] Bliek F W, Hoekstra R, Bannister M E and Havener C C 1997 *Phys. Rev. A* **56** 426
[55] Bottcher C and Heil T G 1982 *Chem. Phys. Lett.* **86** 506
[56] Shimakura N, Koizumi S, Suzuki S and Kimura M 1992 *Phys. Rev. A* **45**(11) 7876–7882
[57] Shipsey E J, Browne J C and Olson R E 1981 *Journal of Physics B: Atomic and Molecular Physics* **14** 869
[58] Panov M N, Basalev A A and Lozhkin K 1983 *Phys. Scr.* **T3** 124
[59] Draganić I N, Seely D G and Havener C C 2011 *Phys. Rev. A* **83**(5) 054701
[60] Meyer F W, Howald A M, Havener C C and Phaneuf R A 1985 *Phys. Rev. A* **32** 3310
[61] Belkić D, Saini S and Taylor H S 1987 *Phys. Rev. A* **36**(4) 1601–1617
[62] Gravielle M S and Miraglia J E 1995 *Phys. Rev. A* **51** 2131
[63] Fritsch W and Lin C D 1984 *J. Phys. B: At. Mol. Phys.* **17** 3271
[64] Kimura M and Lin C D 1985 *Phys. Rev. A* **32** 1357–1362
[65] Toshima N 1994 *Phys. Rev. A* **50** 3940
[66] Caillat J, Dubois A and Hansen J P 2000 *J. Phys. B: At. Mol. Opt. Phys.* **33** L715
[67] Igenbergs K, Schweinzer J, Veiter A, Perneczky L, Frühwirth E, Wallerberger M, Olson R E and Aumayr F 2012 *J. Phys. B: At. Mol. Opt. Phys.* **45** 065203
[68] Jorge A, Errea L F, Illescas C, Méndez L and Suárez J 2013 *Physica Scripta* **2013** 014032
[69] Jorge A, Errea L F, Illescas C and Méndez L 2013 *J. Phys. B: At. Mol. Opt. Phys.* submitted
[70] Green T A, Shipsey E J and Browne J C 1982 *Phys. Rev. A* **25** 1364–1373
[71] Green T A, Riley M E, Shipsey E J and Browne J C 1982 *Phys. Rev. A* **26**(6) 3668–3671
[72] Harel C, Jouin H and Pons B 1998 *At. Data. Nucl. Data Tables* **68** 279
[73] Liu C N, Cheng S C, Le A T and Lin C D 2005 *Phys. Rev. A* **72**(1) 012717

Journal of Physics: Conference Series **576** (2015) 012003 doi:10.1088/1742-6596/576/1/012003

Reactive collisions involving the BeH molecular system

Åsa Larson, Johanna B. Roos, Hanae Hedberg
Dept. of Physics, Stockholm University, SE-106 91 Stockholm, Sweden

Ann E. Orel
Dept. of Chemical Engineering and Materials Science, University of California, Davis, USA

Abstract. In the divertor and fusion edge plasma regions reactive collisions involving the BeH molecular system are taking place. Theoretical *ab initio* quantum studies of electron collisions with BeH^+ resulting in either dissociative recombination, vibrational excitation or dissociative excitation are performed for ions in different vibrational states as well for different isotopologues. Furthermore, the mutual neutralization reaction in collisions of H^- with Be^+ is studied semiclassically.

1. Introduction

At the ITER fusion reactor in France, there are plans for having beryllium on the primary walls [1]. The Be will enter the plasma and Be-containing atomic and molecular ions will be formed in the cooler divertor region of the fusion reactor. Within a coordinated research project initiated by IAEA (International Atomic Energy Agency) with the title *"Light Element Atom, Molecule and Radical Behavior in the Divertor and Edge Plasma Regions"* [2] we have performed theoretical studies of collisional processes involving the BeH molecular system. In electron collisions with the BeH^+ molecular ion a neutral resonant state is formed. If the system dissociates into neutral fragments the process is dissociative recombination (DR) [3]. The system can also autoionize (re-emit the electron). If a bound vibrational state of the ion is formed the reaction results in vibrational excitation (or de-excitation) (VE), while if the system autoionizes into a state above the dissociation limit of the ion, the process is called resonant dissociative excitation (DE). These processes have here been studied *ab initio* and using a quantum mechanical description. Furthermore, the possible presence of the H^- anion in the fusion plasma [4, 5] motivated studies of the mutual neutralization process taking place when H^- collides with Be^+. This study has been carried out using a semi-classical model. The collisional processes in the divertor plasma occur at relative low energies (1 meV to 50 eV). Furthermore, the molecular ions present in the plasma may be vibrationally excited. Different istopologues of the ions are present. We study the effect of vibration as well as isotopic substitution on the above mentioned reactive collisional processes. State specific cross sections as well as final state distributions for these reactions are presented.

Journal of Physics: Conference Series **576** (2015) 012003 doi:10.1088/1742-6596/576/1/012003

2. Theory

2.1. Calculation of relevant electronic states

In processes such as dissociative recombination, vibrational or dissociative excitation, the electronic molecular states involved are often electronic resonant states. The potential energy curves of the resonant states are situated above the ground state potential of the ion. These states interact with the ionization continuum and the interaction give rise to the autoionization width and the finite lifetime of the state. To compute electronic resonant states, electron scattering calculations are employed. We use the complex-Kohn variational method [6] to obtain the energy positions and autoionization widths of the resonant states. The electron scattering calculations are combined with consistent structure calculations of bound electronic states. For BeH, we used [3] the multi-reference configuration interaction method with a reference space consisting all excitations of the electrons among the seven lowest molecular orbitals and additional single external excitations.

The electronic states are "quasi-diabatized" [3]. The diabatization is based on the assumption that the configurations of the resonant electronic states can be followed as the internuclear distance changes. The Rydberg states converging to the ground state of BeH^+ as well as the ionization continuum, all have dominant configurations of the form $(1\sigma)^2(2\sigma)^2(n\lambda)^1$, where $n\lambda$ is a more or less diffuse orbital. The configurations of the resonant states are different from this and by following the configurations of the resonant states as the internuclear distance changes, an initial approximation for the quasidiabatic states can be obtained. To compute the electronic couplings, the adiabatic states are transformed using an orthogonal transformation matrix that is constructed by successive two-by-two rotations. The transformation matrix has a number of parameters that are optimized by fitting the diagonal terms of the potential energy matrix to the guessed quasidiabatic potential energy curves. In Fig. 1 the quasidiabatic potential energy curves of BeH are displayed together with the ground state potential of the ion.

The potential energy curves are computed for internuclear distances smaller than 10 a_0. The ion-pair state associated with the $Be^+ + H^-$ limit asymptotically has $^2\Sigma^+$ symmetry. In this study we did not compute the electronic states involved in the avoided crossings occurring at larger distances when the ion-pair state interacts with the covalent states associated with Be^*+H limits.

2.2. Nuclear dynamics

In electron recombination processes the nuclear dynamics on coupled electronic states are described using wave packet propagation by a direct integration of the time-dependent Schrödinger equation

$$i\frac{\partial}{\partial t}\Psi = \left(-\frac{1}{2\mu}\frac{\partial^2}{\partial R^2} + \mathbf{U}^d\right)\Psi. \tag{1}$$

When several coupled states are involved $\mathbf{U}^d$ is the diabatic non-diagonal potential energy matrix. Autoionization from resonant states is included using the local approximation [7]. The potential energy curves of the resonant states are given by

$$U^d_{ii}(R) = V_i(R) - i\frac{1}{2}\Gamma_i(R), \tag{2}$$

where $\Gamma_i(R)$ is the autoionization width of the resonant state. The electron recombination with the molecular ion is described using the initial condition for the wave packets

$$\Psi_i(R, t = 0) = \sqrt{\frac{\Gamma_i(R)}{2\pi}}\chi_v(R), \tag{3}$$

where $\chi_v(R)$ is the initial wave function of the ion. Thus, in the present study, only the electronic couplings between the ionization continuum and the resonant states are included and

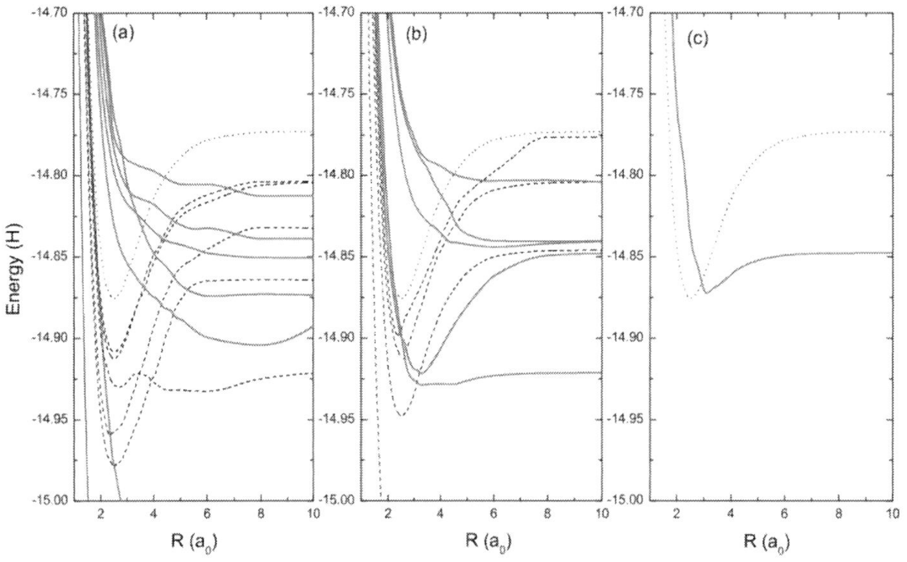

Figure 1. Quasidiabatic potential energy curves of BeH of (a) $^2\Sigma^+$, (b) $^2\Pi$ and (c) $^2\Delta$ symmetries. The potential energy curve of the ground state of the ion is displayed with the thin dotted curve. The bound states are the dashed black curves, while the thick red curves show the potential energy curves of the resonant states.

the direct capture into the Rydberg states by non-adiabatic couplings is not considered. This indirect pathway to dissociative recombination and vibrational excitation has recently [8] been incorporated for the e^--BeH^+ scattering processes using multi-channel quantum defect theory.

The cross section of dissociative recombination is computed [7, 9] by projecting the asymptotic wave packet onto energy-normalized eigenstates of the fragments

$$\sigma_i(R) = \frac{2\pi^3}{E} g \left| \langle \Phi_E^i(R) | \Psi_i(t_\infty, R) \rangle \right|^2. \tag{4}$$

Here g is the ratio of multiplicity between the final and initial states and the total DR cross section is obtained by summing the contributions from all states.

The cross section for vibrational excitation is obtained by computing the transition matrix element from each resonant states [7, 10]

$$T_{v'v}^i(E) = -i \int_0^\infty e^{iEt} \left\langle \chi_{v'}(R) | \sqrt{\frac{\Gamma_i(R)}{2\pi}} | \Psi_i(t, R) \right\rangle dt. \tag{5}$$

The total transition matrix is obtained from a coherent sum of the T-matrix elements for all resonant states of the same symmetry

$$\sigma_{v'v}(E) = \frac{2\pi^3}{E} \left| \sum_i T_{v'v}^i(E) \right|^2. \tag{6}$$

Table 1. Parameters used in the Landau-Zener study of mutual neutralization in Be^+ + H^- collisions.

Asymptotic limit	Binding energy (eV)	R_x (a_0)	Coupling strength (eV)
H+Be^+	0		
H^- + Be^+	0.7542		
H + Be($2s4s\ ^1S$)	1.233	56.83	$7.465 \cdot 10^{-5}$
H + Be($2s4s\ ^3S$), H +Be($2s3d\ ^1D$)	1.330	47.33	$8.025 \cdot 10^{-4}$
H + Be($2s3d\ ^3D$)	1.629	31.22	$9.252 \cdot 10^{-3}$
H + Be($2s3p\ ^1P$)	1.860	24.77	$2.867 \cdot 10^{-2}$
H + Be($2s3p\ ^3P$)	2.019	21.74	$8.155 \cdot 10^{-2}$
H + Be($2p^2\ ^1D$)	2.270	18.27	$8.189 \cdot 10^{-2}$
H + Be($2s3s\ ^1S$)	2.544	15.64	$1.127 \cdot 10^{-1}$
H + Be($2s3s\ ^3S$)	2.866	13.46	$2.782 \cdot 10^{-1}$
H + Be($2s2p\ ^1P$)	4.045	9.358	$2.820 \cdot 10^{-1}$

When the transition matrix for dissociative excitation is computed, the wave packet is instead projected onto the energy normalized continuum functions of the ground ionic state. The total DE cross section is obtained by integrating over all energetically open continuum functions

$$\sigma_{DE}(E) = \frac{2\pi^3}{E} \int |T_{\epsilon v}|^2 \, d\epsilon. \tag{7}$$

2.3. *Semi-classical Landau-Zener study*

In order to compute the cross section for mutual neutralization when Be^+ collides with H^- a semi-classical Landau-Zener [11, 12] study has been performed. Here the avoided crossings occurring between the ionic and covalent states at large internuclear distances ($R > 9$ a_0) are considered. Since the adiabatic potential energy curves of the involved states are not computed *ab initio*, semi-classical expressions of quasidiabatic potential energy curves and couplings are applied. At large internuclear distances, the ion-pair state is assumed to have a potential of the form

$$U_{ip}(R) = U_{th} - \frac{1}{R} - \frac{\alpha}{2R^4}, \tag{8}$$

where U_{th} is the asymptotic threshold energy and α is the polarizability of H^- [13]. The diabatic covalent states are assumed to have potential energy curves with constant energies (independent on R). The electronic couplings between the ionic and covalent states are computed using the approach developed by R K Janev based on an one-electron approximation [14, 15, 16]. The parameters of the potentials and couplings are summarized in Table 1. When two curve crossings occur close to each other, an effective state is used where the energy and crossing point are obtained by the average of those involved and the coupling strength is given by the roots of the sums of the squares of the corresponding coupling elements [16].

By applying the semi-classical Landau-Zener model described by Bates and Lewis [17], the cross section for mutual neutralization is given by

$$\sigma(E) = \frac{\pi}{k_i^2} \sum_{\ell} (2\ell + 1) P_\ell(E). \tag{9}$$

Here, k_i is the wave number of the reactants and $P_\ell(E)$ is the multi-state Landau-Zener probability that the reacting ion-pair forms neutral fragments after passing through the crossings between ionic and covalent states both on the way in and out. This probability depends on the angular quantum number ℓ of the system since the centrifugal barrier $\frac{\ell(\ell+1)}{2\mu R^2}$ will be added

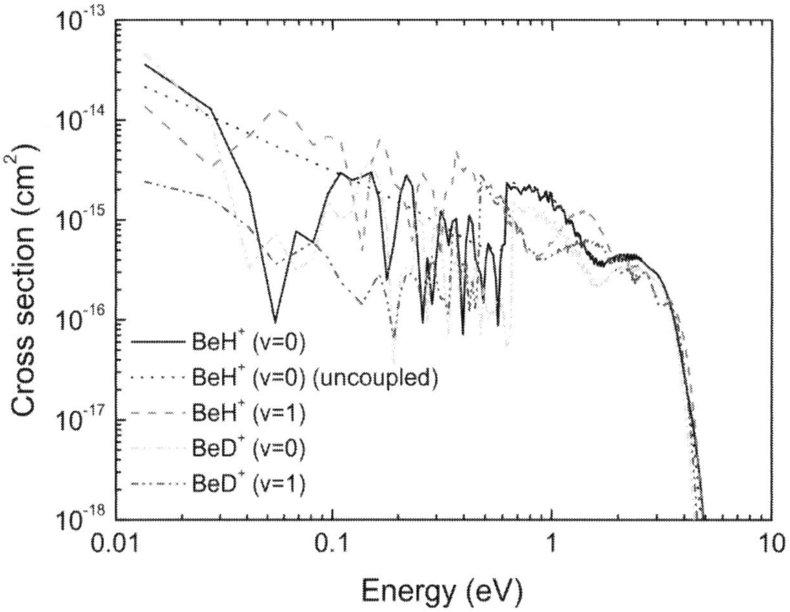

Figure 2. Total cross section for the direct mechanism of dissociative recombination of BeH^+ and BeD^+ in the $v = 0$ and $v = 1$ vibrational states. Also the cross section computed using uncoupled quasidiabatic potential energy curves for BeH^+ ($v = 0$) is displayed.

to the diabatic potentials. For large enough ℓ, the crossing points R_x at smaller distances might not be reached [17]. When calculating the cross section for mutual neutralization, the contributions from the different pathways must be added together. Here the contributions are added incoherently.

3. Results and Discussion

3.1. Dissociative recombination

The direct dissociative recombination cross section of BeH^+ is computed using the wave packet approach described above. The results [3] are summarized in the Fig. 2. The electronic couplings between the neutral states do not significantly change the magnitude of the cross section, but create sharp oscillations that can be interpreted as Feshbach resonances formed when the states open for dissociation are coupled to nuclear bound states [3]. As mentioned above, the indirect mechanism of dissociative recombination, i.e., the direct capture to bound Rydberg states via non-adiabatic couplings and subsequent predissociation is here not included. In a separate study [8] using multi-channel quantum defect theory, both the direct and indirect pathways are considered.

By changing the initial state of the target ion to $v = 1$, the cross section for vibrationally excited ions is computed. Due to the change in the vibrational energy of the target ion, the high energy peak in the cross section is shifted toward smaller energies. Furthermore, the larger capture probability for the vibrationally excited target ion, causes a slightly larger cross section at low energies.

Finally by changing the reduced mass of the system to the one of BeD, the cross section for

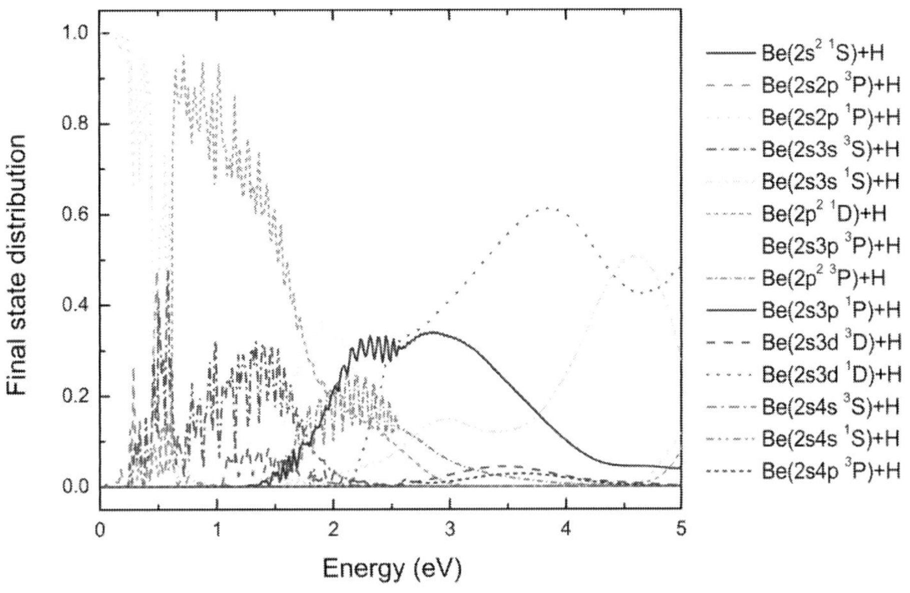

Figure 3. Final state distributions in DR of BeH^+.

DR of BeD^+ is computed. The heavier isotopologue has a DR cross section similar to the one of BeH^+ both with respect to the magnitude and the overall shape. The high-energy peak is slightly reduced. At the same collision energy, the velocity of the heavier isotopologue is slower, which increases the time in the regions where autoionization can occur. This will increase the probability for autoionization and lower the cross section.

For modeling the fusion plasma of the divertor not only the total cross section is important, but also the final state distribution. It is essential to know what fragments are formed in the reaction. In the present calculation of DR, twelve states of $^2\Sigma$ symmetry, nine states of $^2\Pi$ symmetry and one state of $^2\Delta$ symmetry are included. Asymptotically these are correlated with Be*+H. As mentioned above, the adiabatic states are quasidiabatized and electronic couplings for internuclear distances $< 10\ a_0$ are included. By analyzing the dissociating fluxes on respective electronic states, the final state distributions can be computed. We thus neglect any interactions between the electronic states outside the edge of our grid. We do not consider ionic-covalent couplings occurring at large internuclear distances. (This is reasonable since the $^2\Sigma^+$ electronic states only have minor contributions to the total DR cross section at low energies [3]).

The resulting final state distributions as a function of the collision energy are displayed in Fig. 3. It should be noted that at low energies (< 0.5 eV), dissociation into Be($2s2p\ ^1P$)+H dominates. The lowest resonant state of $^2\Pi$ symmetry dissociates into this limit and this state dominates the low energy DR. At higher energies, the Be($2p^2\ ^1D$)+H channel becomes important, which can be understood from the opening of the resonant state of $^2\Delta$ symmetry. At even higher energies, the uncertainties in the computed branching ratios are larger since it becomes harder to identify the asymptotic limits of the excited states.

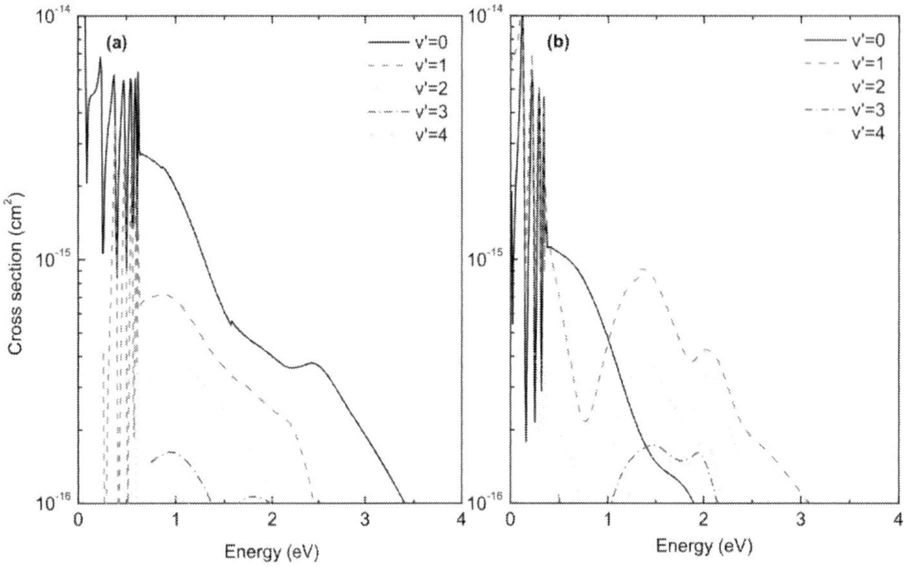

Figure 4. Cross section for vibrational excitation in electron collisions with BeH^+ in (a) $v = 0$ and (b) $v = 1$.

3.2. Vibrational excitation

The cross section for vibrational excitation (or de-excitation) [$BeH^+(v)+e^- \rightarrow BeH^* \rightarrow BeH^+(v')+e^-$] is computed using a model where the couplings between the neutral states are neglected. This is justified from the assumption that autoionization from the resonant state occurs at small distances where the couplings can be neglected. The cross section is computed for initial target ions in $v = 0$ (a) and $v = 1$ (b) and for final states of $v' = 0, 1, ..., 4$.

The computed cross sections displayed in Fig. 4 show thresholds at the energies of the opening of respective vibrational channel. The sharp structures in the cross sections are due to resonances due to bound vibrational states. The structures disappear above the dissociation limit of the ion.

3.3. Dissociative excitation

The cross section for resonant dissociative excitation [$BeH^+(v)+e^- \rightarrow BeH^* \rightarrow Be^++H+e^-$] is obtained from the model where electronic couplings between the neutral states are neglected. The cross section is computed for vibrational states of the target ion $v = 0, 1, ..., 5$. As can be seen in Fig. 5 the cross sections are peaked at higher energies (around 3 eV) and increases with increasing vibrational state of the ion. The cross section for DE is dominated by capture into resonant states of $^2\Sigma^+$ symmetry.

3.4. Mutual neutralization

When Be^+ collides with H^-, the non-adiabatic couplings at large internuclear distances causes electron transfer and formation of neutral fragments, Be^* + H. In the present semi-classical

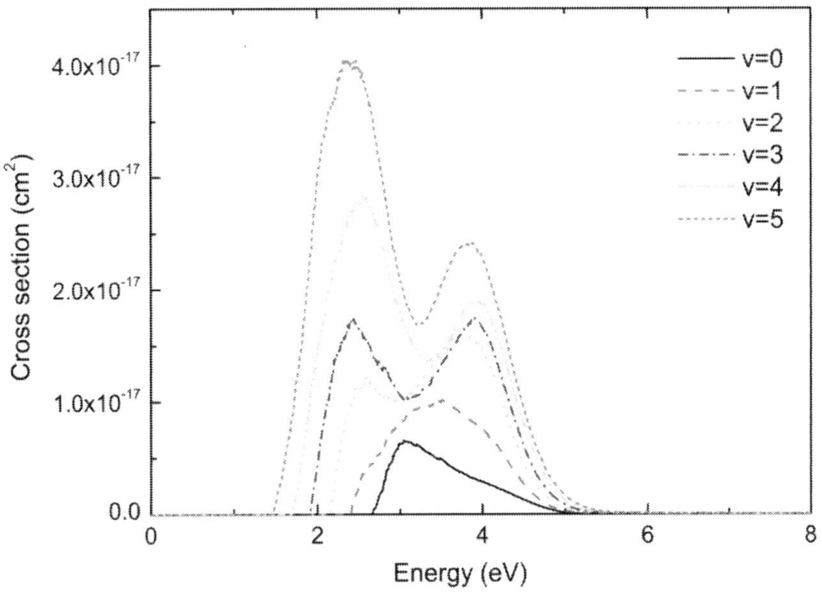

Figure 5. Cross section for resonant dissociative excitation of $BeH^+(v)$.

Landau-Zener study all electronic couplings at small distances ($R < 9$ a_0) are neglected. Furthermore, autoionization from the neutral molecular system is not included. The calculations are performed for reduced masses corresponding to the BeH, BeD and BeT molecular systems. As can be seen in Fig 6, the computed cross sections show an E^{-1} energy dependence with a very small isotope effect. By analyzing the cross sections for each final state, it is found that the low energy mutual neutralization is dominated of formation by Be($2s3p$ ^1P)+H [60%] and Be($2s3d$ ^3D)+H [40%].

4. Summary

Using a quantum *ab initio* approach dissociative recombination, resonant vibrational excitation and dissociative excitation of BeH^+ have been studied. We have computed total cross sections and final state distributions for these processes. The influences of initial vibrational state of the ion and isotopic substitution have been investigated. Furthermore, using a semi-classical Landau-Zener study we report the cross section for mutual neutralization in collisions of Be^+ with different isotopes of H^-. These are molecular processes that could play an important role for the modeling of the fusion plasma of the divertor at ITER.

Acknowledgments

Å Larson acknowledges support from the Swedish Research Council. Some of this material was based on work supported by the National Science Foundation Grant No. PHY-1252953, while A E Orel was working at the Foundation. Any opinion, finding and conclusions or recommendations expressed in this material are those of the author and do not necessarily reflect the views of the National Science Foundation. This work was carried out as a part of an IAEA Coordinated

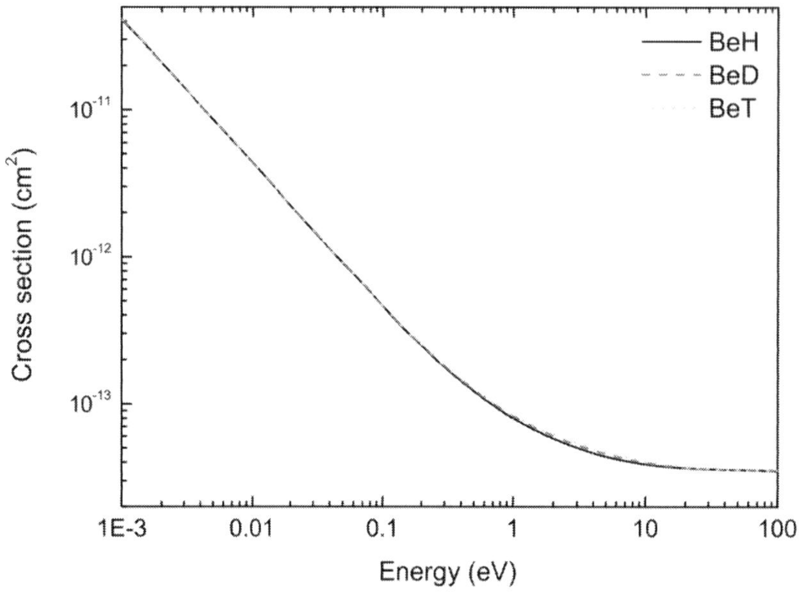

Figure 6. Total cross section for mutual neutralization in collision of Be^+ with H^-, D^- and T^-.

Research Project on "Light Element Atom, Molecule and Radical Behaviour in the Divertor and Edge Plasma Regions"

References

[1] The ITER project, see: www.iter.org
[2] Braams B J (ed.) *"Light Element Atom, Molecule and Radical Behaviour in the Divertor and Edge Plasma Regions"'* Research Coordination Meeting, INDC International Nuclear Data Committee, INDC(NDC)-0564, IAEA, Vienna. http://www-amdis.iaea.org/CRP/LightElement/public.html
[3] Roos J B, Larsson M, Larson Å, and Orel A E 2009 *Phys. Rev. A* **80** 012501
[4] Private communication with Prof. Janev R K.
[5] Fantz U, Wünderlich D 2006 *New. J. Phys* **8** 301
[6] Rescigno T N, McCurdy C W, Orel A E, and Lengsfield III B H 1995 *The Complex Kohn Variational Method* in *Computational methods for electron-molecule scattering* eds. Huo W H and Gianturco F A (Plenum New York)
[7] McCurdy C W, and Turner J L 1983 *J. Chem. Phys.* **78** 6773
[8] Niyonzima S, Lique F, Chakrabarti K, Larson Å, Orel A E, and Schneider I F 2013 *Phys. Rev.* **87** 022713
[9] Kulander K C and Heller E J 1978 *J. Chem. Phys.* **69** 2439
[10] Royal J and Orel A E 2007 *Phys. Rev. A* **75** 052706
[11] Landau L, 1932 *Phys. Z Sowjetunion* **2** 46
[12] Zener C 1932 *R. Soc. London Ser. A* **137** 696
[13] Holka F, Neogrády P, Kellö V, Urban M and Diercksen G H F 2007 *Mol. Phys.* **103** 2747
[14] Janev R K 1976 *Adv. At. Mol. Phys.* **12** 1
[15] Wang J G *et al.* 2006 *Chin. Phys. Soc.* **15** 2006
[16] Janev R K, Liu C L, Wang J G and Yan J 2006 *Europhys Lett.* **74** 616
[17] Bates D R and Lewis J T 1955 *Proc. Phys. Soc. A* **68** 173

Journal of Physics: Conference Series **576** (2015) 012004 doi:10.1088/1742-6596/576/1/012004

Generalized Collisional Radiative Model for Light Elements: C: Data for the B Isonuclear Sequence

S. D. Loch, C. P. Ballance, and M.S. Pindzola
Department of Physics, Auburn University, Auburn, AL 36849

D. C. Griffin
Department of Physics, Rollins College, Winter Park, FL 32789

J. P. Colgan
Los Alamos National Laboratory, Los Alamos, NM 87545

N. R. Badnell and M. G. O'Mullane
Department of Physics, University of Strathclyde, Glasgow, UK G40NG

Abstract. A first stage collision database is assembled which contains electron-impact excitation, ionization, and recombination rate coefficients for B, B^+, B^{2+}, B^{3+}, and B^{4+}. The first stage database is constructed using the R-matrix with pseudo-states, time-dependent close-coupling, and perturbative distorted-wave methods. A second stage collision database is then assembled which contains generalized collisional-radiative ionization, recombination, and power loss rate coefficients as a function of both temperature and density. The second stage database is constructed by solution of the collisional-radiative equations in the quasi-static equilibrium approximation using the first stage database. Both collision database stages reside in electronic form at the IAEA Labeled Atomic Data Interface (ALADDIN) database and the Atomic Data Analysis Structure (ADAS) open database.

1. Introduction

Accurate atomic and molecular databases underpin current research efforts in a variety of scientific and engineering areas: including controlled fusion energy, astrophysics, radiation biophysics, fluorescent lamps, and atmospheric pollutant removal [1]. For laboratory plasmas boronization has been used for many years, in particular for tokamak facilities, to reduce impurity contamination and increase plasma confinement times [2]. In this paper we report on the construction of an electron-ion collision database for the modelling of boron emission in plasmas, for both fusion and astrophysical applications. We present all the fundamental data, along with a brief description of each dataset, and references to the original calculations from which the data was compiled. The fundamental data has been processed through a collisional-radiative modeling suite of codes [3] to produce coefficients that can be used directly in impurity transport and plasma modeling codes.

The fundamental atomic data has been processed into Generalized Collisional-Radiative (GCR) coefficients. Such GCR coefficients are widely used in impurity transport codes such as DEGAS-2 [4], SOLPS [5], and TRANSP [6]. It has been known for some time that impurity transport codes need to track both the ground and metastable populations in their modeling, and that the role of the excited states should be taken into account. These effects are included in GCR coefficients, which are functions both of electron temperature and density. The importance of such coefficients has been demonstrated on various experiments, including Li transport modeling at DIII-D [7].

In this paper we first describe the electron-impact excitation, ionization, and recombination data that is used to produce the new GCR coefficients for B, a number of questions are then explored using the new data.

A range of theoretical methods are used to calculate the atomic data for this work, with the data being taken from a number of previous papers. Effective collision strengths are obtained from recent *R*-matrix with pseudo-states (RMPS) calculations of excitation cross sections for B [8], B^+ [9], and B^{4+} [10]. Previously unpublished RMPS calculations of excitation cross sections for B^{2+} and B^{3+} are also used to obtain effective collision strengths.

Ionization rate coefficients are obtained from recent *R*-matrix with pseudo-states (RMPS) and time-dependent close-coupling (TDCC) calculations [11] of ionization cross sections for ground and excited states of B, B^+, and B^{2+}. The recent RMPS and TDCC calculations [11] compare well with previous RMPS and TDCC calculations for the ground and metastable states of B [12], B^+ [13], and B^{2+} [14]. Ionization rate coefficients are obtained from recent distorted-wave (DW) calculations of ionization cross sections for ground and excited states of B^{3+} and B^{4+}. The recent DW calculations compare well with previous convergent close-coupling (CCC) calculations for the ground and metastable states of B^{3+} [15] and previous RMPS calculations for the ground and metastable states of B^{4+} [16].

Recombination rate coefficients are obtained from previous distorted-wave (DW) calculations for first row ionization states of all atoms [17]; in particular, recombination for Be-like [18], Li-like [19], He-like [20], and H-like [22] atomic ions.

Making use of the complete electron-ion collision database for boron, the collisional-radiative codes found in the Atomic Data and Analysis Structure (ADAS) package [23, 3] are employed to derive Generalized Collisional-Radiative (GCR) and radiated power loss coefficients for the ground and metastable terms of each ionization stage of boron. These generalized collisional-radiative coefficients can be used by subsequent plasma transport codes to track the evolution of the ground and metastable term populations of the B ions. Of particular interest for this study is the role of excited state ionization on the effective ionization rate coefficients. It was previously found for Li that the excited states can increase the effective ionization by up to a factor of 10 over the pure ground state ionization rate coefficient [24]. The current data for B provides a useful opportunity to study the trends of the excited state ionization contribution along an iso-nuclear sequence. For example, at what ion stage does the excited state ionization contribution become negligible? Calculations of ionization cross sections using non-perturbative methods can be very large, so knowing which ion stages and n-shells are the most important for excited state ionization contributions would help to guide future calculations.

The basic collisional rate coefficients and all the ADAS derived generalized collisional-radiative and radiated power loss coefficients are stored in electronic form at the IAEA Labeled Atomic Data Interface (ALADDIN) database [25] and the Atomic Data Analysis Structure (ADAS) open database [26]. The present paper is the third in a series providing generalized collisional-radiative data for the modelling of light species. The first data paper was on lithium [24], the second data paper on beryllium [27], and with work underway on papers for carbon and other light elements.

The remainder of this paper is organized as follows. In Section II we give a brief review of the electron-impact excitation, ionization, and recombination cross sections and rates forming the first stage collision database including references to the original works. In Section III we describe the generalized collisional-radiative data which is generated from the first stage data. In Section IV we investigate the role of excited state ionization on the effective ionization rate coefficients. In Section V we summarize our findings and discuss future directions.

2. First Stage Collision Database

2.1. Electron-Ion Excitation

Effective collision strengths (dimensionless) versus temperatures (in K) are stored in the first stage collision database. The effective collision strengths, obtained by Maxwellian convolution of the cross sections, are given by:

$$\Upsilon_{i\to f}(T_e) = \int_0^\infty w_i \frac{\epsilon_i}{I_H} \frac{\sigma_{i\to f}}{\pi a_0^2} e^{-(\frac{\epsilon_f}{kT_e})} d(\frac{\epsilon_f}{kT_e}) , \quad (1)$$

where w_i is the statistical weight of the initial LS term, ϵ_i and ϵ_f are initial and final electron energies (in eV), $I_H = 13.6$ eV, $\sigma_{i\to f}$ is the excitation cross section (in cm^2), kT_e is the temperature (in eV), and $a_0 = 5.29 \times 10^{-9}$ cm. One can easily transform the stored effective collision strengths into excitation rate coefficients (in cm^3/sec) given by:

$$q_{i\to f}(T_e) = 2\sqrt{\pi}\alpha c a_0^2 \frac{1}{w_f} \frac{I_H}{kT_e} \Upsilon_{i\to f}(T_e) , \quad (2)$$

where $\alpha = 1/137$ and $c = 3.0 \times 10^{10}$ cm/sec. The effective collision strengths, rather than the rate coefficients, are archived as they are slowly varying and easily interpolated. Considering each of the B ions in turn, we briefly describe the calculations that were used to produce the excitation cross sections.

For neutral B, excitation cross sections for the 55 transitions among 11 of the 13 LS terms of the $1s^22s^22p$, $1s^22s2p^2$, $1s^22s^23l$ (l=0-2) $1s^22s^24l$ (l=0-3), and $1s^22s^25s$ configurations were calculated using the RMPS method. The 640 LS term RMPS calculation [8] involved the $1s^22s^2nl$ (n=2-11, l=0-4), $1s^22s2pnl$ (n=2-11, l=0-4), and $1s^22p^2nl$ (n=2-11, l=0-3) configurations. As shown in Figure 1 for the $1s^22p^22p$ (^2P) $\to$ $1s^22s^23s$ (^2S) transition and in Figure 2 for the $1s^22s2p^2$ (^4P) $\to$ $1s^22s^23d$ (^2D) transition, the 640 LS term RMPS cross sections are slightly lower than previous smaller RMPS calculations [28] and do not show the pseudoresonance problems above (see Figure 1) and straddling (see Figure 2) the ionization threshold. Effective collision strengths at 8 temperatures from 4,000 K to 50,000 K were obtained using Eq.(1) for all 55 transitions and are found in the first stage electronic collision databases.

For B$^+$, excitation cross sections for the 190 transitions among the 20 LS terms of the $1s^22s^2$, $1s^22s2p$, $1s^22p^2$, $1s^22s3l$ (l=0-2), and $1s^22s4l$ (l=0-3) configurations were calculated using the RMPS method. The 114 LS term RMPS calculation [9] involved the $1s^22snl$ (n=2-10, l=0-4) and $1s^22pnl$ (n=2-5, l=0-4) configurations. The study [9] compared this 114 LS term calculation with a 134 LS term calculation [9], showing little difference in the cross sections, demonstrating convergence in the pseudostate expansion. The 134 LS term RMPS calculation [9] included additional $1s^22snl$ (n=11-12, l=0-4) configurations. Effective collision strengths at 10 temperatures from 800 K to 800,000 K were obtained using Eq.(1) for all 190 transitions and are found in the first stage electronic collision database.

For B^{2+}, excitation cross sections for the 36 transitions among the 9 LS terms of the $1s^22s$, $1s^22p$, $1s^23l$ (l=0-2), and $1s^24l$ (l=0-3) configurations were calculated using the RMPS method. The 41 *LS* term RMPS calculation involved the $1s^2nl$ (n=2-12, l=0-3) configurations. The same RMPS calculations were reported for other members of the Li isonuclear sequence [29]. Effective collision strengths at 12 temperatures from 1,800 K to 9,000,000 K were obtained using Eq.(1) for all 36 transitions and are found in the first stage electronic collision database.

For B^{3+}, excitation cross sections for the 171 transitions among the 19 *LS* terms of the $1s^2$, $1s2s$, $1s2p$, $1s3l$ (l=0-2), and $1s4l$ (l=0-3) configurations were calculated using the RMPS method. The 89 *LS* term RMPS calculation involved the $1s^2$ and $1snl$ (n=2-11, l=0-4) configurations. Effective collision strengths at 10 temperatures from 16,000 K to 16,000,000 K were obtained by using Eq.(1) for all 171 transitions and are found in the first stage electronic collision database. These calculations were performed for the ADAS database, but are previously unpublished.

For B^{4+}, excitation cross sections for the 105 transitions among the 15 *LS* terms of the $1s$, $2s$, $2p$, $3l$ (l=0-2), $4l$ (l=0-3), and $5l$ (l=0-4) configurations were calculated using the RMPS method. The 39 *LS* term RMPS calculation [10] involved the $1s$ and nl (n=2-9, l=0-5) configurations. Effective collision strengths at 12 temperatures from 50,000 K to 12,500,000 K were obtained by using Eq.(1) for all 105 transitions and are found in the first stage electronic collision database.

2.2. Electron-Ion Ionization

Scaled ionization rate coefficients (in cm^3/sec) versus temperatures (in K) are stored in the first stage collision database. The scaled ionization rate coefficients are given by:

$$\bar{S}_{i\to f}(T_e) = \mathscr{S}_{i\to f}(T_e)e^{-(\frac{I_p}{kT_e})} , \tag{3}$$

where I_p is the ionization potential (in eV) of the level being ionized (i.e. not simply the ionization potential of the ground state). Ionization rate coefficients are given by:

$$\mathscr{S}_{i\to f}(T_e) = \frac{2\alpha c}{\sqrt{\pi}}(\frac{1}{I_H})^{\frac{1}{2}}(\frac{1}{kT_e})^{\frac{3}{2}} \int_{I_p}^{\infty} \epsilon_i \sigma_{i\to f} e^{-(\frac{\epsilon_i}{kT_e})} d\epsilon_i , \tag{4}$$

where $\sigma_{i\to f}$ is the ionization cross section (in cm^2). The scaled ionization rate coefficients are stored since they are more easily interpolated than the rate coefficients, particularly at low temperatures. We briefly describe here the sources and methods used for the ionization data.

For neutral B, ionization cross sections for the 8 *LS* terms of the $1s^2 2s^2 2p$, $1s^2 2s^2 3l$ (l=0-2), and $1s^2 2s^2 4l$ (l=0-3) configurations were calculated using the RMPS method. The 1036 *LS* term RMPS calculation [11] involved the $1s^2 2s^2 2p$, $1s^2 2s^2 nl$ (n=3-13, l=0-5), $1s^2 2s 2p^2$, $1s^2 2s 2pnl$ (n=3-13, l=0-5), $1s^2 2p^3$, and $1s^2 2p^2 nl$ (n=3-13, l=0-5) configurations. The $1s^2 2s^2 2p$ (^2P) cross sections are in good agreement with previous 476 *LS* term RMPS and TDCC calculations [12]. As shown in Figure 3 for the ionization of the 7 *LS* terms of the $1s^2 2s^2 3l$ and $1s^2 2s^2 4l$ configurations, the RMPS calculations for the $1s^2 2s^2 3s$ (^2S) term are in good agreement with TDCC calculations [11]. Scaled ionization rate coefficients at 8 temperatures from 4,000 K to 50,000 K were obtained using Eqs.(3)-(4) for all 8 *LS* terms and are found in the first stage electronic collision database.

For B^+, ionization cross sections for the 17 *LS* terms of the $1s^2 2s^2$, $1s^2 2s 2p$, $1s^2 2s 3l$ (l=0-2), and $1s^2 2s 4l$ (l=0-3) configurations were calculated using the RMPS method. The 486 *LS* term RMPS calculation [11] involved the $1s^2 2s^2$, $1s^2 2s 2p$, $1s^2 2snl$ (n=3-14, l=0-5), $1s^2 2p^2$, and $1s^2 2pnl$ (n=3-14, l=0-5) configurations. The $1s^2 2s^2$ (^1S) cross sections are in good agreement with previous 414 *LS* term RMPS and TDCC calculations [13]. Scaled ionization rate coefficients at 10 temperatures from 800 K to 800,000 K were obtained using Eqs.(3)-(4) for all 17 *LS* terms and are found in the first stage electronic collision database.

For B^{2+}, ionization cross sections for the 14 *LS* terms of the $1s^2 2s$, $1s^2 2p$, $1s^2 3l$ (l=0-2), $1s^2 4l$ (l=0-3), and $1s^2 5l$ (l=0-4) configurations were calculated using the RMPS method. The 68 *LS* term RMPS calculation [11] involved the $1s^2 2s$, $1s^2 2p$, and $1s^2 nl$ (n=3-14, l=0-5) configurations. The $1s^2 2s$ ^2S cross sections are in good agreement with previous 68 *LS* term RMPS and TDCC calculations [14]. Scaled ionization rate coefficients at 12 temperatures from 1,800 K to 9,000,000 K were obtained using Eqs.(3)-(4) for all 14 *LS* terms and are found in the first stage electronic collision database.

For B^{3+}, ionization cross sections for the 19 *LS* terms of the $1s^2$, $1s2s$, $1s2p$, $1s3l$ (l=0-2), and $1s4l$ (l=0-3) configurations were calculated using the DW method. The $1s^2$

and $1s2s$ ^3S cross sections are in good agreement with previous CCC calculations [15]. Scaled ionization rate coefficients at 10 temperatures from 16,000 K to 16,000,000 K were obtained using Eqs.(3)-(4) for all 19 LS terms and are found in the first stage electronic collision database.

For B^{4+}, ionization cross sections for the 15 LS terms of the $1s$, $2s$, $2p$, $3l$ (l=0-2), $4l$ (l=0-3), and $5l$ (l=0-4) configurations were calculated using the DW method. The $1s$ and $2l$ (l=0-1) cross sections are in good agreement with previous RMPS calculations [16]. Scaled ionization rate coefficients at 12 temperatures from 50,000 K to 12,500,000 K were obtained using Eqs.(3)-(4) for all 15 LS terms and are found in the first stage electronic collision database.

Note that DW cross sections are used for ion stages B^{3+} and B^{4+}. It is well known that DW calculations become increasingly appropriate with charge state, while at the same time they may become inaccurate with increasing n-shell. It was not clear which of these two effects dominates along a sequence. However, recently calculations on the excited states of C^{3+} [21] show that DW calculations are valid for ionization from the 5s subshell. Thus, we expect that DW calculations should be appropriate for both the ground and excited states of B^{3+} and B^{4+}.

As part of the set of nonperturbative calculations for B, B^+, and B^{2+}, the possibility of n-scaling the low n-shell data to higher n-values was investigated. For all of the ions it was discovered that the bundled-n data converged onto an n^4-scaling, see Figure 4. The aim in the extrapolation is to provide recommended rates for the higher n-shells present in the projection matrix part of the collisional-radiative modeling, see Section 2.4 for a description of the projection matrix. The projection matrix data uses n-shell resolved data, separated by the spin of the parent, the bundled-nS approach. Note that applying the n^4 scaling is a little more subtle for ion stages that have more than one spin system, in which case one should extrapolate data based upon each spin system separately, see [11] for more details. For example, in B^+ one would n-scale the triplet states separately from the singlet states. This is consistent with the bundled-nS data used in the projection matrix.

2.3. Electron-Ion Recombination

For each of the B ions, the dielectronic recombination (DR) Maxwellian rate coefficients have aleady been calculated as part of the DR project [17]. Our final recommended recombination data consists of this DR data combined with radiative recombination rate coefficients.

Dielectronic recombination rate coefficients (in cm^3/sec) versus temperatures (in K) are stored in the first stage collision database. The Maxwellian recombination rate coefficients are given by:

$$\mathscr{R}_{i\to f}(T_e) = \frac{2\alpha c}{\sqrt{\pi}}(\frac{1}{I_H})^{\frac{1}{2}}(\frac{1}{kT_e})^{\frac{3}{2}} \int_0^\infty \epsilon_i \sigma_{i\to f} e^{-(\frac{\epsilon_i}{kT_e})} d\epsilon_i \ , \tag{5}$$

where $\sigma_{i\to f}$ is the recombination cross section (in cm^2). The recombination rate coefficients generally vary smoothly with temperature.

Dielectronic recombination cross sections for 11 of the 13 LS terms of the $1s^22s^22p$, $1s^22s2p^2$, $1s^22s^23l$ (l=0-2), $1s^22s^24l$ (l=0-3), and $1s^22s^25s$ configurations of B coming from the $1s^22s^2\ ^1S$ and $1s^22s2p\ ^3P$ terms of B^+ were calculated [18] using the DW method. Recombination rate coefficients at 8 temperatures from 4,000 K to 50,000 K were obtained using Eq.(5) for all 22 transitions and are found in the first stage electronic collision databases.

Dielectronic recombination cross sections for the 20 LS terms of the $1s^22s^2$, $1s^22s2p$, $1s^22p^2$, $1s^22s3l$ (l=0-2), and $1s^22s4l$ (l=0-3) configurations of B^+ coming from the $1s^22s\ ^2S$ term of B^{2+} were calculated [19] using the DW method. Recombination rate coefficients at 10 temperatures from 800 K to 800,000 K were obtained using Eq.(5) for all 20 transitions and are found in the first stage electronic collision databases.

Dielectronic recombination cross sections for the 9 LS terms of the $1s^22s$, $1s^22p$, $1s^23l$ (l=0-2), and $1s^24l$ (l=0-3) configurations of B^{2+} coming from the $1s^2\ ^1S$ and $1s2s\ ^3S$ terms of B^{3+} were calculated [20] using the DW method. Recombination rate coefficients at 12 temperatures from 1,800 K to 9,000,000 K were obtained using Eq.(5) for all 18 transitions and are found in the first stage electronic collision databases.

Dielectronic recombination cross sections for the 19 LS terms of the $1s^2$, $1s2s$, $1s2p$, $1s3l$ (l=0-2), and $1s4l$ (l=0-3) configurations of B^{3+} coming from the $1s\ ^2S$ term of B^{4+} were calculated [22] using the DW method. Recombination rate coefficients at 10 temperatures from 16,000 K to 16,000,000 K were obtained using Eq.(5) for all 19 transitions and are found in the first stage electronic collision databases.

For all ions, radiative recombination was calculated using a Gaunt factor approach [30]. RR rate coefficients were evaluated for the same LS terms as those calculated for the DR calculations described above, and were on the same temperature grid.

For B^{5+} recombining to B^{4+} only radiative recombination is possible. Thus, the Gaunt factor calculations included recombination into the 15 LS terms of the $1s$, $2s$, $2p$, $3l$ (l=0-2), $4l$ (l=0-3), and $5l$ (l=0-4) configurations of B^{4+}. Recombination rate coefficients at 12 temperatures from 50,000 K to 12,500,000 K were obtained using Eq.(5)

for all 15 transitions and are found in the first stage electronic collision databases.

Note that three-body recombination can be calculated from the electron-impact ionization rate coefficients using the detailed balance relations, and so is included in our collisional-radiative modeling.

2.4. Hign n-Shell Data

As with our previously published Li [24] and Be [27] datasets, we included approximate atomic data for high n-shells. It can be important to include data for these higher n-shells for a number of reasons. Under low temperature or higher density conditions, the total ionization losses from an ion stage can be dominated by stepwise ionization and inclusion of these higher n-shells in the population modelling becomes important. Also, at low densities dielectronic recombination into the higher n-shells is a strong populating mechanism, and must be accounted for.

Our high n-shell data is included through the use of a projection matrix. Our data consists of semi-empirical expressions, based on the Gaunt factor and ECIP expressions of Burgess and Summers [30]. For B, B^+, and B^{2+} we use the scaled R-matrix ionization rate coefficients as described in Section 2.2. We note that the dielectronic recombination data in our projection matrix data is based on the ICDW calculations reported in Section IIC, and as such are the only part of the projection matrix data that is not semi-empirical or n-scaled.

3. Second Stage Collision Database

The first stage atomic collision data presented in Section II can be used directly to calculate term populations via the collisional-radiative equations:

$$\begin{aligned}\frac{dN_j^z}{dt} = & -\sum_{j'<j} A_{j\to j'} N_j^z - N_e \sum_{j'\neq j} q_{j\to j'} N_j^z \\ & + \sum_{j'>j} A_{j'\to j} N_{j'}^z + N_e \sum_{j'\neq j} q_{j'\to j} N_{j'}^z \\ & - N_e \sum_k \mathscr{S}_{j\to k} N_j^z + N_e \sum_k \mathscr{R}_{k\to j} N_k^{z+1} \ , \end{aligned} \tag{6}$$

where $A_{j\to j'}$ are radiative decay rates, $\mathscr{S}_{j\to k}$ are electron-impact ionization rates, $\mathscr{R}_{k\to j}$ are electron-impact recombination rates, and N_e is the electron density. The electron-impact excitation rates, $q_{i\to j}$, are proportional [14] to the effective collision strengths.

For many plasma conditions, the ground and metastable term populations evolve on a much slower timescale than the excited term populations, allowing the time derivative of the excited term populations to be set to zero. In the quasi-static equilibrium approximation, the time-dependence of the excited states are set to zero and one can extract generalized coefficients that connect the ground and metastable states for each ion stage. These generalized coefficients can be used to evaluate a time-dependent ionization balance, as shown in Eq. (7), or can be used directly in plasma transport codes. As an illustrative example, consider the connections between three adjacent ion stages, labelled α, β, and γ. Each ion stage has metastable states, given by α', β', and γ'. The center ion stage (with β and β') contains excited states given by j and j'. The time-dependence of the ground state β can be calculated from:

$$\begin{aligned}\frac{dN_\beta^z}{dt} = & - N_e \sum_{\beta'\neq\beta} (X_{\beta\to\beta'} + Q_{\beta\to\beta'}) N_\beta^z \\ & + N_e \sum_{\beta'\neq\beta} (X_{\beta'\to\beta} + Q_{\beta'\to\beta}) N_{\beta'}^z \\ & - N_e \sum_\gamma S_{\beta\to\gamma} N_\beta^z + N_e \sum_\alpha S_{\alpha\to\beta} N_\alpha^{z-1} \\ & + N_e \sum_\gamma R_{\gamma\to\beta} N_\gamma^{z+1} - N_e \sum_\alpha R_{\beta\to\alpha} N_\beta^z \ . \end{aligned} \tag{7}$$

The cross-coupling rate coefficients are given by:

$$X_{\beta\to\beta'} = (\mathcal{C}_{\beta'\beta}^z - \sum_j \mathcal{C}_{\beta' j}^z \sum_{j'} (\mathcal{C}_{jj'}^z)^{-1} \mathcal{C}_{j'\beta}^z)/N_e, \tag{8}$$

and the parent cross-coupling rate coefficients are given by:

$$Q_{\beta\to\beta'} = N_e \sum_i \mathscr{S}_{i\to\beta'} \sum_{i'} (\mathcal{C}_{ii'}^{z-1})^{-1} \mathscr{R}_{\beta\to i'} \ , \tag{9}$$

where the collisional-radiative matrix elements for ion stage z are given by:

$$\mathcal{C}_{jj}^z = \sum_{j'<j} A_{j\to j'} + N_e \sum_{j'\neq j} q_{j\to j'}, \tag{10}$$

$$\mathcal{C}^z_{jj'} = -A_{j'\to j} - N_e q_{j'\to j} \qquad for \quad j' > j, \tag{11}$$

$$\mathcal{C}^z_{jj'} = -N_e q_{j'\to j} \qquad for \quad j' < j. \tag{12}$$

The generalized ionization rate coefficients are given by:

$$S_{\beta\to\gamma} = \mathscr{S}_{\beta\to\gamma} - \sum_j \mathscr{S}_{j\to\gamma} \sum_{j'} (\mathcal{C}^z_{jj'})^{-1} \mathcal{C}^z_{j'\beta}, \tag{13}$$

$$S_{\alpha\to\beta} = \mathscr{S}_{\alpha\to\beta} - \sum_i \mathscr{S}_{i\to\beta} \sum_{i'} (\mathcal{C}^{z-1}_{ii'})^{-1} \mathcal{C}^{z-1}_{i'\alpha}. \tag{14}$$

The generalized recombination rate coefficients are given by:

$$R_{\gamma\to\beta} = \mathscr{R}_{\gamma\to\beta} - \sum_j \mathcal{C}^z_{\beta j} \sum_{j'} (\mathcal{C}^z_{jj'})^{-1} \mathscr{R}_{\gamma\to j'}, \tag{15}$$

$$R_{\beta\to\alpha} = \mathscr{R}_{\beta\to\alpha} - \sum_i \mathcal{C}^{z-1}_{\alpha i} \sum_{i'} (\mathcal{C}^{z-1}_{ii'})^{-1} \mathscr{R}_{\beta\to i'}. \tag{16}$$

These generalized collisional-radiative coefficients are a function of both temperature and density. Consider for example Eqn (13) and the effective ionization from β to γ. The GCR ionization rate coefficient includes both the direct ionization from β to γ, as well as excited state ionzization (i.e. collisional-excitation fromβ to j', collisional redistribution to j and then ionization from j).

Our second stage database contains the generalized collisional-radiative coefficients for all of the B ion stages. We note that it is these generalized collisional-radiative coefficients that are most useful in plasma transport codes, since they contain the effects of the excited states on the ground and metastables, without the need for a large number of coefficients to be archived.

To illustrate the role of excited state ionization on the B ions, we show a comparison of the ground state ionization rate coefficient with the effective ionization rate coefficient at an electron density of N_e=1× 10^{14}cm^{-3}. Figure 5 shows the absolute ionization rate coefficient for neutral B. Figure 6 shows the ratio of the ground state ionization rate coefficient to the GCR ionization rate coefficient calculated at N_e=1× 10^{14}cm^{-3} for the first three charge states of B. We show the results here using just the low-n data (i.e. no projection matrix) to highlight the sensitivity of the GCR coefficients to the non-perturbative rate coefficients calculated as part of this work. We have chosen a typical density for a tokamak divertor plasma (N_e = 1×10^{14} cm^{-3}) and an electron temperature at the lower range of expected divertor conditions (T_e = 1 eV), as representative of the types of conditions that B ions sputtered into the divertor region might experience. For neutral B, the excited terms contribute up to a factor of 10 more than the ground state, similar to previous findings for neutral Li [24]. For B^+ the excited states contribute about 20% to the total ionization rate and for B^{2+} the effect is less than 5%. As one would expect, due to the increasing radiative rates, the role of the excited states is becoming less pronounced as one progresses to the higher charge states. For charge states greater than B^{2+}, the excited states do not contribute significantly to the effective ionization rate coefficient. We note that these effects become larger for higher densities or for lower temperatures, as one might expect.

As a final consideration, we discuss briefly the role of uncertainties on our atomic data. Assigning uncertainties to the final GCR coefficients is a complex problem, as the uncertainties on the fundamental cross section data carry through the collisional-radiative modeling to the final temperature and density dependent coefficients. However, the uncertainties in the fundamental cross sections could be estimated from the spread of non-perturbative cross section data for each of the atomic processes considered here. Inspection of the B ionization cross section papers shows that the TDCC, RMPS and CCC data are usually within 15% of each other for ionization cross sections. Uncertainties on the excitation and recombination cross sections are more difficult to estimate, due to the important role of resonant processes. Clearly the uncertainties will be largest at the lowest temperatures and reduce as the temperature increases. A systematic method to generate uncertainties on excitation and recombination data is currently under development. We note that once uncertainties are assigned to the excitation, ionization and recombination rate coefficients, a Monte-Carlo method can be used to propagate those uncertainties through to uncertainties on the final GCR coefficients.

Journal of Physics: Conference Series **576** (2015) 012004 doi:10.1088/1742-6596/576/1/012004

4. Radiated Power Loss

The radiated power loss function from a single excited term, due to all possible bound-bound spectral line emissions, is given by:

$$RPL_j^z = \sum_{j'<j} \frac{A_{j\to j'}^z \Delta E_{jj'}^z N_j^z}{N_e N_{tot}} , \tag{17}$$

where $\Delta E_{jj'}^z$ is the transition energy. On the other hand, the total radiated power loss function from an ionization stage, due to spectral line emission, is given by:

$$RPL^z = \sum_{\beta} \frac{P_{exc,\beta}^z N_\beta^z}{N_{tot}} + \sum_{\gamma} \frac{P_{rec,\gamma}^z N_\gamma^{z+1}}{N_{tot}} . \tag{18}$$

Note that to use this last equation, one needs to know the fractional abundances of each of the ion stages, either from an ionization balance calculation, or from an impurity transport calculation. The generalized excitation power loss coefficients are given by:

$$P_{exc,\beta}^z = \sum_j \sum_{j'<j} \frac{A_{j\to j'}^z \Delta E_{jj'}^z}{N_e} \left[-\sum_{j''} \mathcal{C}_{jj''}^{-1} \mathcal{C}_{j''\beta} \right] , \tag{19}$$

and the generalized recombination power loss coefficients are given by:

$$P_{rec,\gamma}^z = \sum_j \sum_{j'<j} \frac{A_{j\to j'}^z \Delta E_{jj'}^z}{N_e} \left[N_e \sum_{j''} \mathcal{C}_{jj''}^{-1} \mathscr{R}_{j''\gamma} \right] . \tag{20}$$

The power loss coefficients for all charge states of B are also archived as part of this work.

Journal of Physics: Conference Series **576** (2015) 012004 doi:10.1088/1742-6596/576/1/012004

5. Summary

An atomic collision dataset for boron is presented consisting of the highest quality data available for electron impact excitation, ionization and recombination. We used RMPS data for excitation, RMPS, TDCC, and DW data for ionization, and DW data for recombination. For processes involving higher n-shells, we used various approximate expressions for the excitation, ionization and recombination processes, including an extrapolation of the lower n-shell ionization data for B, B+, and B^{2+}. As well as presenting the fundamental data, we have presented the derived atomic coefficients necessary to calculate ionization balance and radiative power loss. These coefficients are referred to as generalized collisional-radiative and radiated power loss coefficients. The first and second stage collision databases will be archived in electronic form at the IAEA/ALADDIN database [25] and the ADAS open database [26].

References

[1] *Atomic and Molecular Data and Their Applications*, editors: D. R. Schultz, P. S. Krstic, and F. Ownby, AIP Conference Proceedings **636**, 2002.
[2] J. WInter, Plasma Physics and Controlled Fusion, **38** 1503 (1996)
[3] *The ADAS User Manual*, H. P. Summers (2001), *http://adas.phys.strath.ac.uk*
[4] D.P. Stotler, CFF. Karney, Contrib. Plasma Phys. **34** 392 (1994)
[5] http://www.ipp.mpg.de/~dcp/solps.html
[6] http://w3.pppl.gov/transp/
[7] J.P. Allain, D.G. Whyte, and J.N. Brooks, Nucl. Fusion **44**, 655 (2004).
[8] C. P. Ballance, D. C. Griffin, K. A. Berrington, and N. R. Badnell, J. Phys. B **40**, 1131 (2007).
[9] N. R. Badnell, D. C. Griffin, and D. M. Mitnik, J. Phys. B **36**, 1337 (2003).
[10] C. P. Ballance, N. R. Badnell, and E. S. Smyth, J. Phys. B **36**, 3707 (2003).
[11] T. G. Lee, S. D. Loch, C. P. Ballance, J. A. Ludlow, and M. S. Pindzola, Phys. Rev. A **82**, 042721 (2010).
[12] J. C. Berengut, S. D. Loch, M. S. Pindzola, C. P. Ballance, and D. C. Griffin, Phys. Rev. A **76**, 042704 (2007).
[13] J. C. Berengut, S. D. Loch, M. S. Pindzola, C. P. Ballance, D. C. Griffin, M. Fogle, and M. E. Bannister, Phys. Rev. A **78**, 012704 (2008).
[14] O. Woitke, N. Djuric, G. H. Dunn, M. E. Bannister, A. C. H. Smith, B. Wallbank, N. R. Badnell, and M. S. Pindzola, Phys. Rev. A **58**, 4512 (1998).
[15] A. C. Renwick, I. Bray, D. V. Fursa, J. Jacobi, S. Schippers, and A. Muller, J. Phys. B **42**, 175203 (2009).
[16] D. C. Griffin, C. P. Ballance, M. S. Pindzola, F. Robicheaux, S. D. Loch, J. A. Ludlow, M. C. Witthoeft, J. Colgan, C. J. Fontes, and D. R. Schultz, J. Phys. B **38**, L199 (2005).
[17] N. R. Badnell, M. G. O'Mullane, H. P. Summers, Z. Altun, M. A. Bautista, J. Colgan, T. W. Gorczyca, D. M. Mitnik, M. S. Pindzola, and O. Zatsarinny, Astronomy and Astrophysics **406**, 1151 (2003).
[18] J. Colgan, M. S. Pindzola, A. D. Whiteford, and N. R. Badnell, Astronomy and Astrophysics **412**, 597 (2003).
[19] J. Colgan, M. S. Pindzola, and N. R. Badnell, Astronomy and Astrophysics **417**, 1183 (2004).
[20] M. A. Bautista and N. R. Badnell, Astronomy and Astrophysics **466**, 755 (2007).
[21] M. S. Pindzola, C. P. Ballance, and S. D. Loch, Phys. Rev. A **83**, 062705 (2011)
[22] N. R. Badnell, Astronomy and Astrophysics **447**, 389 (2006).

[23] H.P. Summers, W.J. Dickson, M.G. O'Mullane, N.R.Badnell, A.D. Whiteford, D.H. Brooks, J. Lang, S.D. Loch, and D.C. Griffin, Plasma Phys. Control. Fusion **48**, 263 (2006).

[24] S. D. Loch, J. Colgan, M. C. Witthoeft, M. S. Pindzola, C. P. Ballance, D. M. Mitnik, D. C. Griffin, M. G. O'Mullane, N. R. Badnell, and H. P. Summers, Atomic Data and Nuclear Data Tables **92**, 813 (2006).

[25] IAEA/ALADDIN, *www-amdis.iaea.org/aladdin*

[26] EU/ADAS, *www.adas.ac.uk/openadas*

[27] S. D. Loch, M. S. Pindzola, C. P. Ballance, D. C. Griffin, J. Colgan, N. R. Badnell, M. G. O'Mullane, and H. P. Summers, Atomic Data and Nuclear Data Tables **94**, 257 (2008).

[28] P. J. Marchalant and K. Bartschat, J. Phys. B **30**, 4373 (1997).

[29] D. C. Griffin, N. R. Badnell, and M. S. Pindzola, J. Phys. B **33**, 1013 (2000).

[30] A. Burgess and H.P. Summers, Mon. Not. R. Astr. Soc. **174**, 345 (1976).

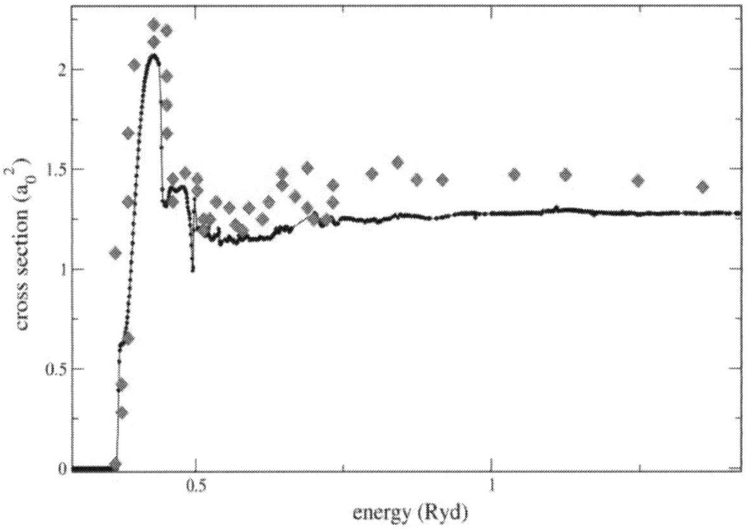

Figure 1. Excitation cross sections for the $1s^22s^22p$ (^2P) → $1s^22s^23s$ (^2S) transition in B. Solid line: RMPS method [8], diamonds: RMPS method [28] (1.0 a_0^2 = 2.80 × 10^{-17} cm^2 and 1.0 Ryd = 13.6 eV). Figure taken from [8]

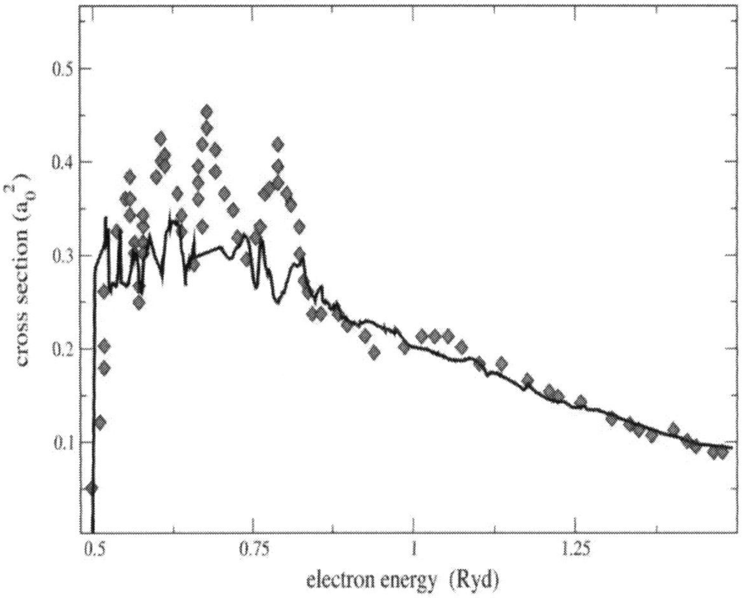

Figure 2. Excitation cross sections for the $1s^22s2p^2$ (^4P) → $1s^22s^23d$ (^2D) transition in B. Solid line: RMPS method [8], diamonds: RMPS method [28] (1.0 a_0^2 = 2.80 × 10^{-17} cm^2 and 1.0 Ryd = 13.6 eV). Figure taken from [8]

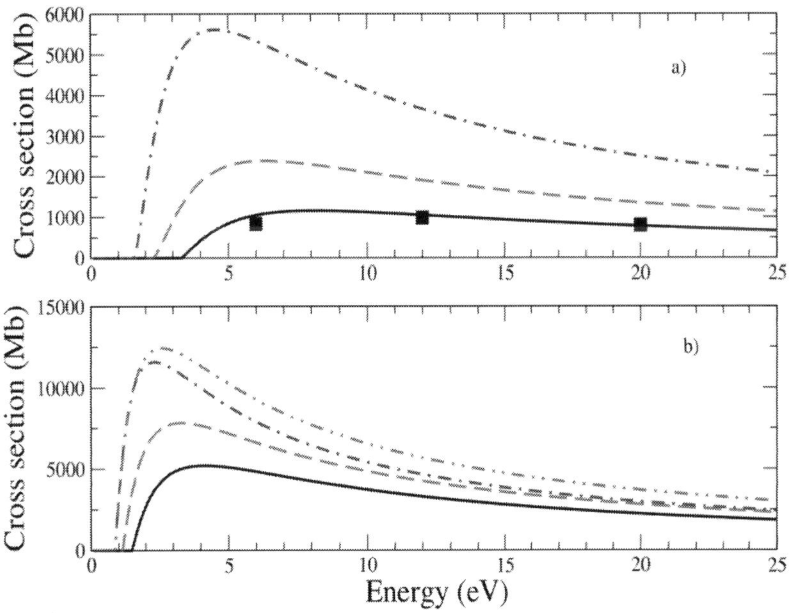

Figure 3. Ionization cross sections for the nl excited states of B. All RMPS and TDCC data are taken from [11]. (a) Solid line: RMPS method for $3s$, dashed line: RMPS method for $3p$, dot-dashed line: RMPS method for $3d$, solid squares: TDCC method for $3s$. (b) Solid line: RMPS method for $4s$, dashed line: RMPS method for $4p$, dot-dashed line: RMPS method for $4d$, double-dot dashed line: RMPS method for $4f$ (1.0 Mb = 1.0×10^{-18} cm^2). Figure taken from [11]

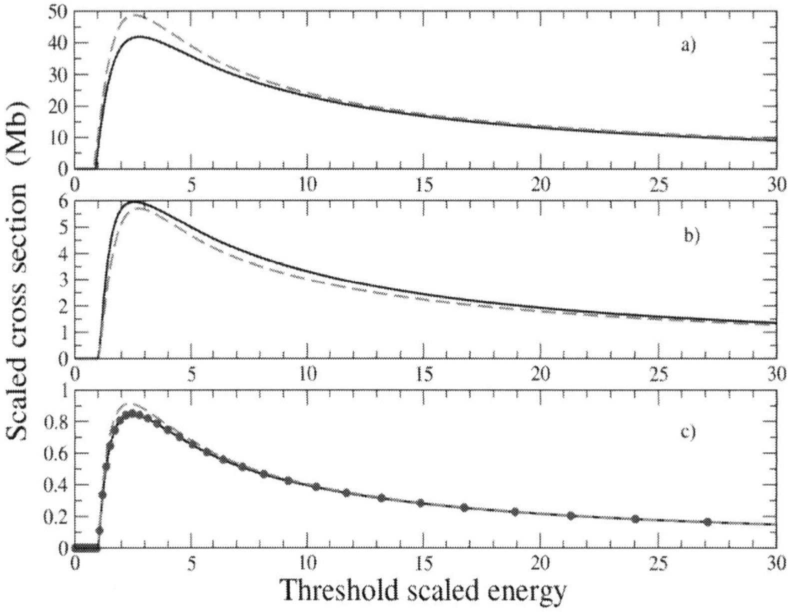

Figure 4. Electron-impact-ionization cross sections divided by n^4 vs threshold scaled energy for the n-bundled excited states of (a) B, (b) B^+,and (c)B^{2+}. In all plots the solid line shows the n = 3 RMPS data, the dashed line shows the n = 4 RMPS data and in panel (c) the solid circles show the n = 5 RMPS data (1 Mb = 10^{18} cm^2). Figure taken from [11]

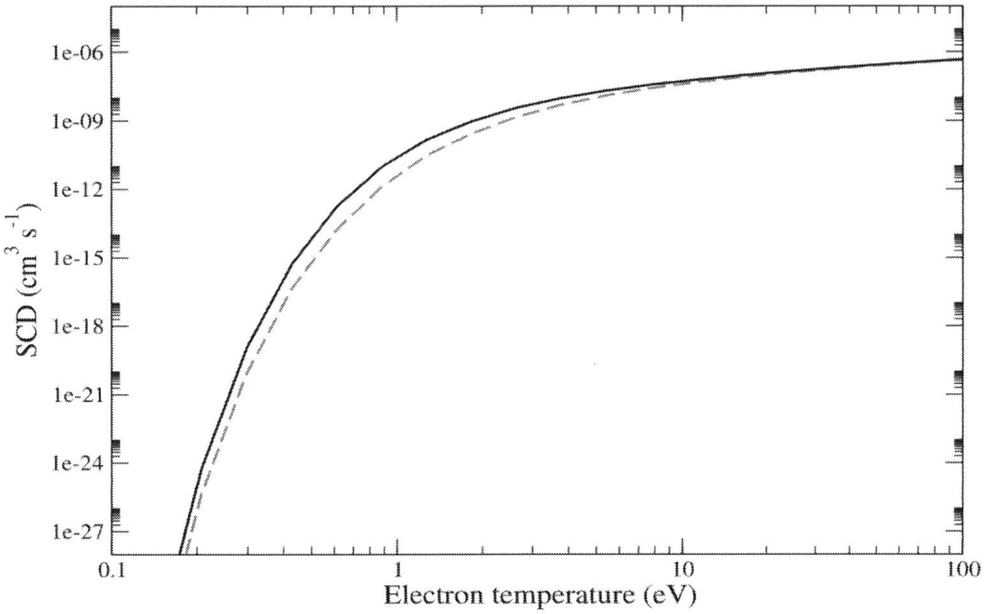

Figure 5. Effective ionization rate coefficient for B at a density of N_e=1 × 10^6 cm^{-3} (dashed line) and N_e=1 × 10^{14} cm^{-3} (solid line).

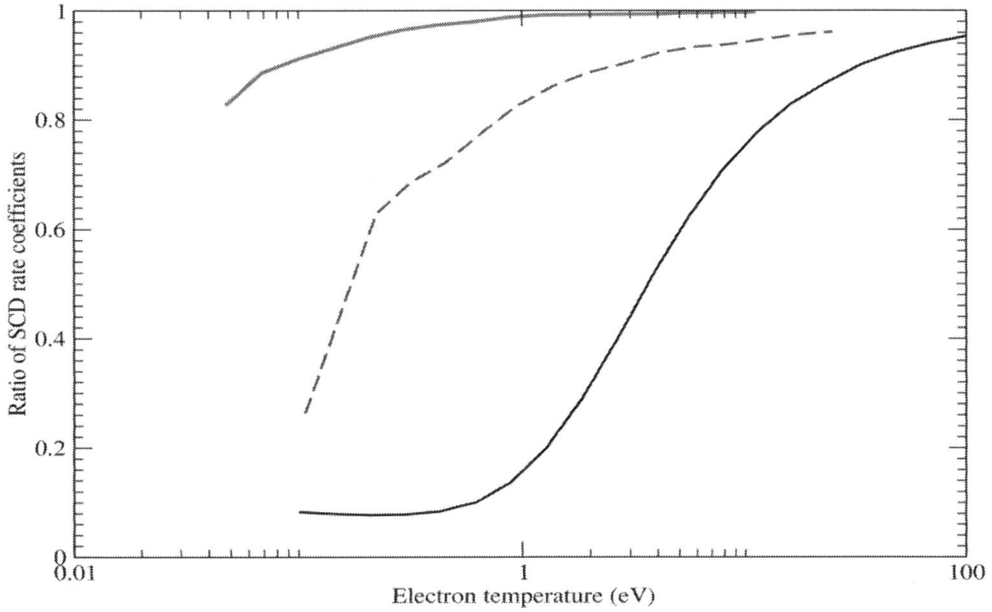

Figure 6. Ratio of the effective ionization rate coefficient at N_e=1 × 10^6 cm^{-3} to the value at N_e=1 × 10^{14} cm^{-3}. The lowest line shows the results for B, the dashed line shows the results for B^+ and the upper solid line shows the results for B^{2+}.

Electronic and photonic reactive collisions in edge fusion plasma and interstellar space: Application to H_2 and BeH systems

J. Zs. Mezei[1,2,3], S. Niyonzima[1,4], D. Backodissa[1], N. Pop[5], F. O. Waffeu Tamo[1], K. Chakrabarti[6], O. Motapon[7], A. Wolf[8], J. Robert[2], O. Dulieu[2], Å. Larson[9], A. E. Orel[10], A. Bultel[11] and I. F. Schneider[1,*]

[1]LOMC, CNRS-UMR-6294, Normandie Université, Université du Havre, F-76058 Le Havre, France
[2]LAC, CNRS-UPR-3321, Univ. Paris-Sud and Ecole Normale Supérieure de Cachan, F-91405 Orsay, France
[3]Institute of Nuclear Research of the Hungarian Academy of Sciences, P.O. Box 51, Debrecen H-4001, Hungary
[4]Département de Physique, Faculté des Sciences, Université du Burundi, B.P. 2700 Bujumbura, Burundi
[5]Department of Physical Foundations of Engineering, University Politehnica of Timisoara, Bv. Vasile Parvan No. 2, 300223, Timisoara, Romania
[6]Dept. of Mathematics, Scottish Church College, 1 & 3 Urquhart Sq., Kolkata 700 006, India
[7]Physics Department, University of Douala, P. O. Box 24157, Douala, Cameroon
[8]Max-Planck-Institute für Kernphysik, Saupfrecheckweg 1, Heidelberg D-69117, Germany
[9]Department of Physics, Stockholm University, AlbaNova University Center, S-106 91 Stockholm, Sweden
[10]Department of Chemical Engineering and Materials Science, University of California, Davis, California 95616, USA
[11]CORIA-UMR 6614 - Normandie Université CNRS, Université and INSA de Rouen, Campus Universitaire du Madrillet, 76800 Saint Etienne du Rouvray, France
[*]Part of this work has been performed while IFS was visitor of Laboratoire Aimé–Cotton, Univ. Paris-Sud

E-mail: `ioan.schneider@univ-lehavre.fr`

Abstract. Reactive collisional and radiative elementary processes rate coefficients have been either computed using multichannel-quantum-defect theory methods, or measured in merged-beam (storage ring) and crossed-beam experiments. The reaction mechanisms are explained and output data are displayed in ready-to-be-used form, appropriate for the modeling of the kinetics of the edge fusion plasma and of the interstellar molecular clouds.

1. Introduction

The collision of an electron of energy ε with a molecular cation initially in its ground electronic state on its ro-vibrational level (N_i^+, v_i^+) can cause the formation of various superexcited states, decaying eventually by autoionization and predissociation to different products:

$$AB^+(N_i^+, v_i^+) + e^-(\varepsilon) \longrightarrow AB^*, AB^{*(c)}, AB^{**}, AB^{**(c)} \longrightarrow \begin{cases} [A+B]_{(\epsilon)} & \text{(1a)} \\ AB^+(N_f^+, v_f^+) + e^-(\varepsilon_f). & \text{(1b)} \\ [A+B^+]_{(\epsilon)} + e^-(\varepsilon_f) & \text{(1c)} \end{cases}$$

According to the type of these products, different processes take place: Dissociative Recombination (DR (1a)), Elastic, Inelastic or Superelastic Collisions (EC, IC, SEC (1b)), corresponding to the final energy of the electron ε_f equal, smaller or larger than ε respectively, and, if ε exceeds the dissociation energy of the target cation, electron impact ion dissociation, i.e. Dissociative Excitation (DE (1c))[1, 2, 3, 4, 5]. Here AB^* stands for bound states from both electronic and vibrational point of view, $AB^{*(c)}$ for states from the electronic continuum (superscript (c)), AB^{**} for bound states from the electronic point of view but dissociative (i.e. from the vibrational continuum), $AB^{**(c)}$ for states from the electronic continuum *and* dissociative, and ϵ for the relative kinetic energy release of the heavy products. Eqs. (1) are appropriate for diatomic systems, but can be generalized for polyatomic ones.

The same superexcited states as those appearing in the common part of eqs. (1) and, consequently, the same products, occur through the photoabsorption by a neutral molecule:

$$AB(N_i, v_i) + h\nu \longrightarrow AB^*, AB^{*(c)}, AB^{**}, AB^{**(c)} \longrightarrow \begin{cases} [A+B]_{(\epsilon)} & \text{(2a)} \\ AB^+(N_f^+, v_f^+) + e^-(\varepsilon_f), & \text{(2b)} \\ [A+B^+]_{(\epsilon)} + e^-(\varepsilon_f) & \text{(2c)} \end{cases}$$

resulting in Photodissociation (PhD) (2a), Photoionization (PhI) (2b) [6, 7, 8, 9] and Dissociative Photoionization (DPhI) (2c) [10].

And finally, the dissociative states AB^{**} are intermediate stages of excitation in the high-energy electron-impact dissociation of molecules [11]:

$$AB(N_i, v_i) + e^-(\varepsilon) \longrightarrow AB^{**} + e^-(\varepsilon_f) \longrightarrow [A+B]_{(\epsilon)} + e^-(\varepsilon_f)\,. \quad (3)$$

All the above processes drive the particle densities in fusion plasmas close to the walls and in the divertor region [12, 13, 14, 15], in various media of astrophysical interest [16, 17, 18, 19], in the hypersonic entry plasmas [20], and in many other environments of fundamental or technological relevance.

The present paper collects the most relevant data - cross sections and rate coefficients - concerning these reactions and available today as output of our studies performed in the frame of the CRP "Light Element Atom, Molecule and Radical Behaviour in the Divertor and Edge Plasma Regions" with IAEA. It is organized as follows: Sections 2, 3 and 4, devoted to the processes - (1), (2), (3) - respectively, have all the same structure, namely after an outline of the relevant theoretical approach and/or the experimental techniques, the cross sections and the rate coefficients are displayed, directly in graphs and, when available, in tables containing the parameters for quick generation by interpolation. Section 5 contains the conclusions and the perspectives.

2. Electron/molecular cation reactive collisions

2.1. The stepwise MQDT-type approach

The dynamics of the infinite series of AB^* states, often called 'Rydberg resonances', appearing in (1), is elegantly and efficiently modeled by a method [21, 22, 23, 24, 25] based on the Multichannel Quantum Defect Theory (MQDT) [26, 27, 28, 29]. The processes (1) result from the coupling between ionization and dissociation *channels*, i.e. groups of states characterized by a common set of quantum numbers and by the same fragmentation threshold (either for ionization, or for dissociation), having the energy below or above this threshold. More specifically, within a *quasi-diabatic* representation [22, 23, 30], an *ionization* channel is built starting from the ground electronic state of the ion on one of its ro-vibrational levels N^+v^+, and is completed by gathering all the single-electron states of a given orbital quantum number l, describing an 'optical' electron. These single-electron states describe, with respect to the N^+v^+ threshold, either a 'free' electron - in which case the total state $AB^{*(c)}$ corresponds to (auto)ionization - or a bound one - in which case the total state AB^* corresponds to a temporary capture into a Rydberg state. Meanwhile, a *dissociation* channel relies on an electronically bound state AB^{**} whose potential energy in the asymptotic limit is situated below the total energy of the system.

Accordingly, the ionization channels gather together AB^* and $AB^{*(c)}$ states, and the dissociation channels correspond to AB^{**} states. As for those of $AB^{**(c)}$ type, they can be classified either into dissociation, or into ionization channels, but the latter option has been preferred so far. Given the total energy of the molecular system, a channel is *open* if this energy is higher than the energy of its threshold, and *closed* in the opposite case. In the modeling of the reactions (1)-(3), relying exclusively on the open channels means accounting for the *direct* mechanism only. The inclusion of the closed channels - including AB^* states - allows for the *indirect* mechanism, which interferes with the direct one resulting in the *total* process.

In order to illustrate the concept of open and closed channels, and to be more specific, we address the case of the HD^+/HD system, for which a relevant energy diagram - restricted to the ${}^1\Sigma_g^+$ symmetry of the neutral - is given in Figure 1. Obviously, the only dissociation channel present here, represented by the HD^{**} ${}^1\Sigma_g^+$ $(2p\sigma_u^2)$ state, corresponding to the red curve in this figure, is open, since the asymptotic limit of this curve is situated below the ground vibrational level of the cation. For simplicity, we will assume that a single partial wave of the electron is dominant in the ionization channels. The thick, violet, horizontal lines stand for possible values of the *total* energy of the system. E_1 corresponds to the case where the only *open* ionization channel is that associated to the ion ground vibrational state ($v^+ = 0$), all the other channels, which are of ionization type, associated to $v^+ = 1, 2, \ldots$ are *closed*, and correspond to families of states of the type HD^*. A Born-Oppenheimer - and therefore, *approximate* - picture of some of these states, i.e. those labelled by ${}^1\Sigma_g^+$ $(1s\sigma_g 5s\sigma_g)$, $v = 0, 1, 2, \ldots$ is shown in green in Figure 1. E_2 represents a total energy for which ionization channels associated to $v^+ = 0 - 7$ are open, and those associated to $v^+ \geq 8$ are closed.

Our MQDT approach starts with the building of the interaction matrix $\boldsymbol{\mathcal{V}}$, performed in the 'A-region' [31], where the Born-Oppenheimer context is appropriate for the description of the collision system. The good quantum numbers in this region are N, M, and Λ, associated respectively to the total angular momentum and its projections on the z-axis of the laboratory-fixed and of the molecule-fixed frames. In the A-region, the states belonging to an ionization channel may be modeled reasonably well with respect to hydrogenic states in terms of the quantum defect μ_l^Λ, which is dependent on the internuclear distance R, but assumed to be *independent of energy*. An ionization channel is coupled to a dissociation one, labeled d_j, on *electronic* level first, through an R-dependent *scaled* 'Rydberg-valence' interaction term, $V_{d_j,l}^{(e)\Lambda}$, which is assumed to be *independent of the energy* of the electronic states pertaining to the ionization channel. Integrating this coupling over the internuclear distance gives elements of the interaction matrix $\boldsymbol{\mathcal{V}}$:

$$\mathcal{V}_{d_j,lN^+v^+}^{NM\Lambda}(E,E) = \langle \chi_{Nd_j}^\Lambda | V_{d_j,l}^{(e)\Lambda} | \chi_{N^+,v^+}^\Lambda \rangle, \tag{4}$$

where E is the total energy and $\chi_{d_j}^\Lambda$ and χ_{N^+,v^+}^Λ are the nuclear wave-functions corresponding to a dissociative state and to an ionization channel, respectively. This procedure applies in each Λ-subspace, and results in a block-diagonal global interaction matrix. Starting from the

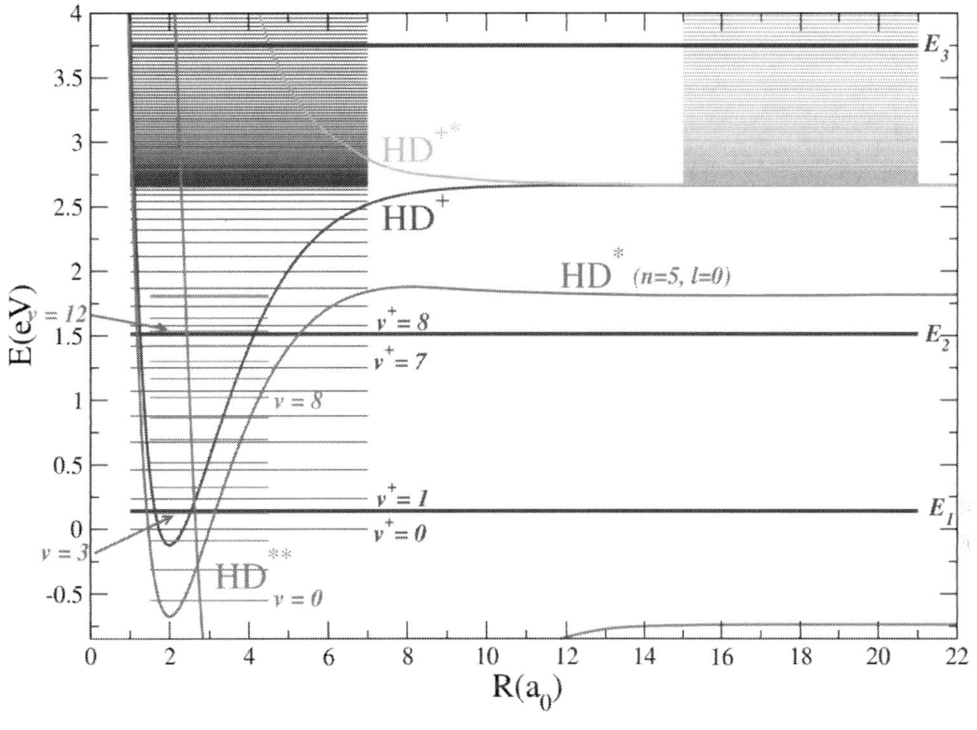

Figure 1. HD^+/HD states (potential energy curves and vibrational levels) relevant for the electron/HD^+ reactive collisions: HD^+ $^2\Sigma_g^+(1s\sigma_g)$ (blue), HD^+ $^2\Sigma_u^+(2p\sigma_u)$ (orange), HD^{**} $^1\Sigma_g^+(2p\sigma_u^2)$ dissociative state (red), and HD^* $^1\Sigma_g^+(1s\sigma_g, 5s\sigma_g)$ Rydberg state (green). The violet lines represent illustrative values of the *total* energy of the system, see the text.

interaction matrix $\boldsymbol{\mathcal{V}}$ and from the zero-order Hamiltonian $\boldsymbol{H_0}$, we build the reaction K-matrix, which satisfies the Lippmann-Schwinger equation [32]:

$$\boldsymbol{\mathcal{K}} = \boldsymbol{\mathcal{V}} + \boldsymbol{\mathcal{V}} \frac{1}{E - \boldsymbol{H_0}} \boldsymbol{\mathcal{K}}. \tag{5}$$

In order to express the result of the short-range interaction in terms of phase-shifts, we perform a unitary transformation of our initial basis into a new one, corresponding to *eigenchannels*, via the diagonalization of the reaction matrix $\boldsymbol{\mathcal{K}}$:

$$\boldsymbol{\mathcal{K}}\mathbf{U} = -\frac{1}{\pi} \tan(\boldsymbol{\eta})\mathbf{U}. \tag{6}$$

In the external 'B-region' [31] the Born-Oppenheimer model is no longer valid for the neutral molecule, and a frame transformation [33, 34, 35, 36] is performed, via the projection coefficients (having neglected the partial wave mixing [36, 37]):

$$\begin{aligned} \mathcal{C}_{lN^+v^+,\Lambda\alpha} &= \left(\tfrac{2N^++1}{2N+1}\right)^{1/2} \langle l\,(\Lambda - \Lambda^+)\,N^+\Lambda^+|lN^+N\Lambda\rangle \\ &\times \tfrac{1+\tau^+\tau(-1)^{N-l-N^+}}{\left[2(2-\delta_{\Lambda^+,0})(1+\delta_{\Lambda^+,0}\delta_{\Lambda,0})\right]^{1/2}} \times \textstyle\sum_v U^{\Lambda}_{lv,\alpha} \langle \chi^{\Lambda^+}_{N^+v^+}|\cos(\pi\mu^{\Lambda}_l(R) + \eta^{\Lambda}_{\alpha})|\chi^{\Lambda}_{Nv}\rangle, \end{aligned} \tag{7}$$

Journal of Physics: Conference Series **576** (2015) 012005 doi:10.1088/1742-6596/576/1/012005

$$\mathcal{C}_{d_j,\Lambda\alpha} = U^{\Lambda}_{d_j\alpha}\cos\eta^{\Lambda}_{\alpha}, \tag{8}$$

which can be organized in a matrix $\boldsymbol{\mathcal{C}}$. The other projection coefficients organized in a matrix $\boldsymbol{\mathcal{S}}$, are obtained as $\mathcal{S}_{lN^+v^+,\Lambda\alpha}$ and $\mathcal{S}_{d_j,\Lambda\alpha}$ by replacing cosine with sine in Eqs. (7) and (8). In these equations, the quantities τ^+ and τ are related to the reflection symmetry of the ion and neutral wave function respectively, and take the values $+1/-1$ for symmetric/antisymmetric states respectively.

Matrices $\boldsymbol{\mathcal{C}}$ and $\boldsymbol{\mathcal{S}}$ are the building blocks of the generalized scattering matrix $\boldsymbol{X}$, involving all the channels, open ('o') and closed ('c'), and organized in 4 sub-matrices:

$$\boldsymbol{X} = \frac{\boldsymbol{\mathcal{C}} + i\boldsymbol{\mathcal{S}}}{\boldsymbol{\mathcal{C}} - i\boldsymbol{\mathcal{S}}}, \qquad \boldsymbol{X} = \begin{pmatrix} \boldsymbol{X}_{oo} & \boldsymbol{X}_{oc} \\ \boldsymbol{X}_{co} & \boldsymbol{X}_{cc} \end{pmatrix}. \tag{9}$$

Imposing boundary conditions leads to the physical scattering matrix [26]:

$$\boldsymbol{S} = \boldsymbol{X}_{oo} - \boldsymbol{X}_{oc}\frac{1}{\boldsymbol{X}_{cc} - \exp(-\mathrm{i}2\pi\boldsymbol{\nu})}\boldsymbol{X}_{co}, \tag{10}$$

where the diagonal matrix $\exp(-\mathrm{i}2\pi\boldsymbol{\nu})$ is formed with the effective quantum numbers $\nu_{N^+v^+} = [2(E_{N^+v^+} - E)]^{-1/2}$ (in atomic units) associated with each vibrational threshold $E_{N^+v^+}$ of the ion situated *above* the current energy E (and consequently labeling a *closed* channel).

For a molecular ion initially in the level $N_i^+v_i^+$ and recombining with an electron of kinetic (collision) energy ε, the cross section of capture into *all* the dissociative states d_j of the same symmetry is given by

$$\sigma^{N,\,\mathrm{sym}}_{\mathrm{diss}\leftarrow N_i^+v_i^+} = \frac{\pi}{4\varepsilon}\frac{2N+1}{2N_i^+ + 1}\rho^{\mathrm{sym}}\sum_{l,\Lambda,j}|S^{N\Lambda}_{d_j,lN_i^+v_i^+}|^2. \tag{11}$$

On the other hand, the cross section for a ro-vibrational transition to the final level $N_f^+v_f^+$, giving elastic scattering or (de-)excitation, writes:

$$\sigma^{N,\,\mathrm{sym}}_{N_f^+v_f^+\leftarrow N_i^+v_i^+} = \frac{\pi}{4\varepsilon}\frac{2N+1}{2N_i^+ + 1}\rho^{\mathrm{sym}} \times \sum_{l,l',\Lambda,j}\left|S^{N\Lambda}_{N_f^+v_f^+l',N_i^+v_i^+l} - \delta_{N_f^+N_i^+}\delta_{v_f^+v_i^+}\delta_{l'l}\right|^2. \tag{12}$$

Here ρ^{sym} is the ratio between the multiplicities of the neutral and the target ion. After performing the MQDT calculation for all accessible total rotational quantum numbers N and for all the relevant symmetries, one has to add up the corresponding cross sections in order to obtain the global cross section for dissociative recombination or ro-vibrational (de-)excitation as a function of the electron collision energy ε.

When rotational excitation and rotational couplings are neglected - the so-called *"non-rotational case"* - the formalism becomes much simpler. Moreover, one has to perform separate calculation within each symmetry block Λ and eventually sum over this quantum number the resulting cross sections. For a given Λ, the previous formulae become:

$$\mathcal{V}^{\Lambda}_{d_j,lv^+}(E,E) = \langle\chi^{\Lambda}_{d_j}|V^{(e)\Lambda}_{d_j,l}|\chi_{v^+}\rangle, \tag{13}$$

$$\mathcal{C}_{lv^+,\Lambda\alpha} = \sum_{v} U^{\Lambda}_{lv,\alpha}\langle\chi_{v^+}(R)|\cos(\pi\mu^{\Lambda}_{l}(R) + \eta^{\Lambda}_{\alpha})|\chi_{v}(R)\rangle, \qquad \mathcal{C}_{d,\Lambda\alpha} = U^{\Lambda}_{d\alpha}\cos\eta^{\Lambda}_{\alpha}, \tag{14}$$

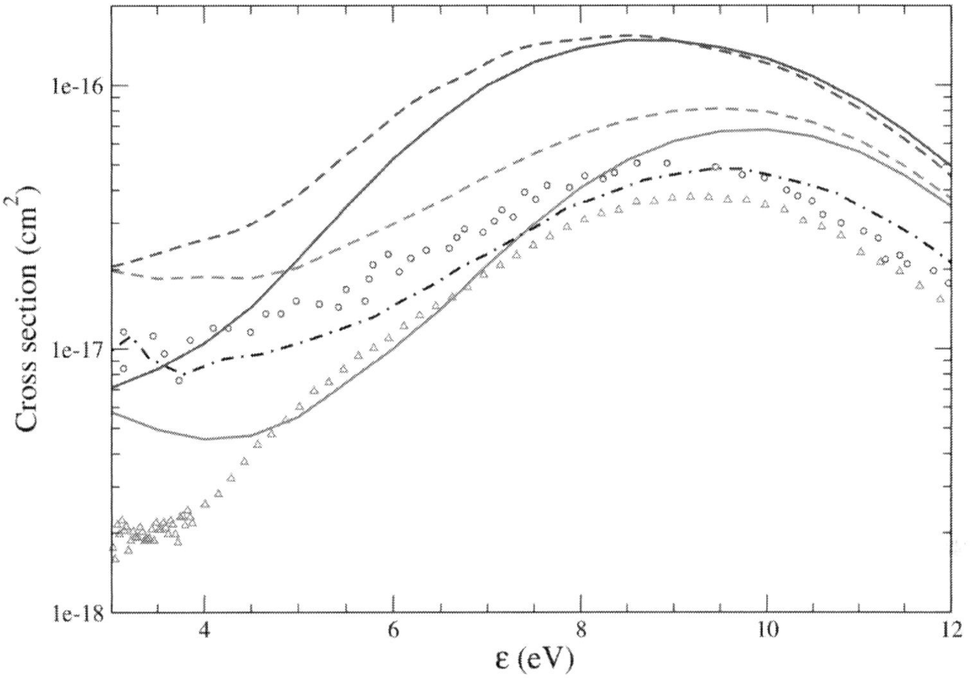

Figure 2. Dissociative recombination of $^2\Sigma_g^+(1s\sigma_g)$ v_i^+=0 HD^+ ion. Blue (upper)/red (lower) curves: dissociative excitation neglected/accounted, dashed/solid corresponding to first-order/second-order calculations respectively. Black dot-dashed curve: computation of Takagi [36]. Circles: Experiment by Tanabe *et al* [37]. Triangles: Experiment by Al-Khalili *et al* [38] (See Ref [39] for more details).

$$\sigma^{sym,\Lambda}_{diss\leftarrow v_i^+} = \frac{\pi}{4\varepsilon}\rho^{sym,\Lambda}\sum_{l,j}|S^{\Lambda}_{d_j,lv_i^+}|^2, \qquad \sigma^{sym}_{diss\leftarrow v_i^+} = \sum_{\Lambda}\sigma^{sym,\Lambda}_{diss\leftarrow v_i^+}. \tag{15}$$

The cross section for vibrational transition - elastic scattering or vibrational (de-)excitation - is expressed as:

$$\sigma^{sym,\Lambda}_{v_f^+\leftarrow v_i^+} = \frac{\pi}{4\varepsilon}\rho^{sym,\Lambda}\sum_{l,l'}|S_{l'v_f^+,lv_i^+} - \delta_{l'l}\delta_{v_f^+v_i^+}|^2, \tag{16}$$

$$\sigma_{v_f^+\leftarrow v_i^+} = \sum_{\Lambda,sym}\sigma^{sym,\Lambda}_{v_f^+\leftarrow v_i^+}. \tag{17}$$

At energies higher than the dissociation threshold of the ion - e.g. E_3 in Figure 1 - we have to allow for the autoionization with respect to ion states from the continuum part of the vibrational spectrum, i.e. *dissociative excitation* (DE) [39]. For the relevant electronic states of the ionic core, we discretize their vibrational continua by adding a high potential energy wall at a large internuclear distance. With respect to the formalism outlined above, this increases the number of the ionization channels pertaining to the ground electronic state of the ion - see the quasi-continuum spectrum of vibrational levels in blue in Figure 1, responsible in our model for the "Dissociative Excitation of the first kind", DE1. Moreover, if excited electronic states of the ion are available, they will imply the inclusion of new series of ionization channels - see the quasi-continuum spectrum of vibrational levels in orange in the same figure, supporting the

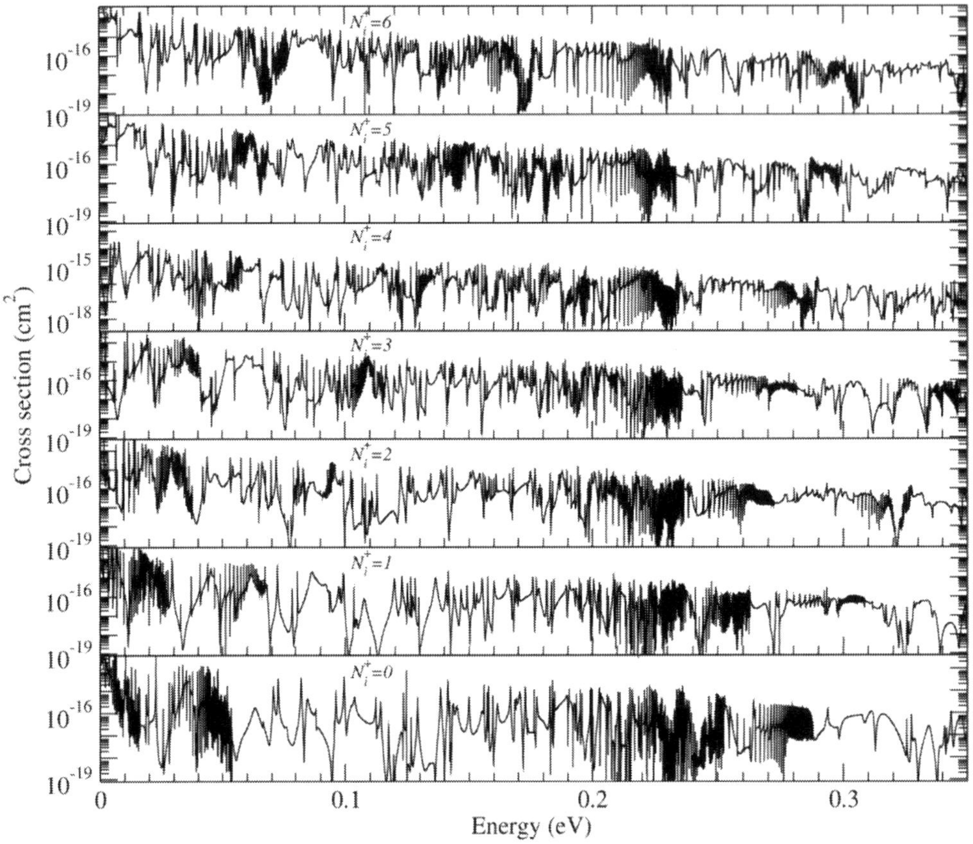

Figure 3. DR cross sections of HD^+ initially in one of its lowest rotational level N_i^+ (vibrational ground state).

"Dissociative Excitation of the second kind", DE2. Consequently, significantly more channels *and* their couplings with respect to the already existing ones at low energy have to be added in the V-, K-, X- and S-matrices displayed in eqs. (4)-(10).

2.2. Electron/HD$^+$ reactive collisions

2.2.1. High energy (above ground-state-ion dissociation threshold)

We have performed extensive calculations of DE-assisted DR cross section for capture into all the dissociative states within all the relevant symmetries, ${}^1\Sigma_g^+$, ${}^{1,3}\Sigma_u^+$, ${}^{1,3}\Pi_{u,g}$, and for a broad range of energy, $0-12$ eV [39]. The computations have been carried out in the first and second order of the K-matrix, and relied on molecular data previously used [25, 40, 41].

Whereas the capture into Rydberg resonant states AB^* (appearing in the general eq. (1)) is excluded in this energy range, the contribution of the $AB^{*(c)}$ and $AB^{**(c)}$ states, related to DE1 and DE2 respectively, is decisive. Their inclusion in the model is performed, as shown in the previous section, by the discretization of the vibrational continua associated to the lowest electronic states of the ionic core, ${}^2\Sigma_g^+(1s\sigma_g)$ and ${}^2\Sigma_u^+(2p\sigma_u)$, by adding a 15 eV high potential energy wall at $R = 25$ a$_0$. Our final best result, the full red curve in Fig. 2, is in reasonable agreement with previous theoretical estimations - black dashed-dotted line on the same figure -

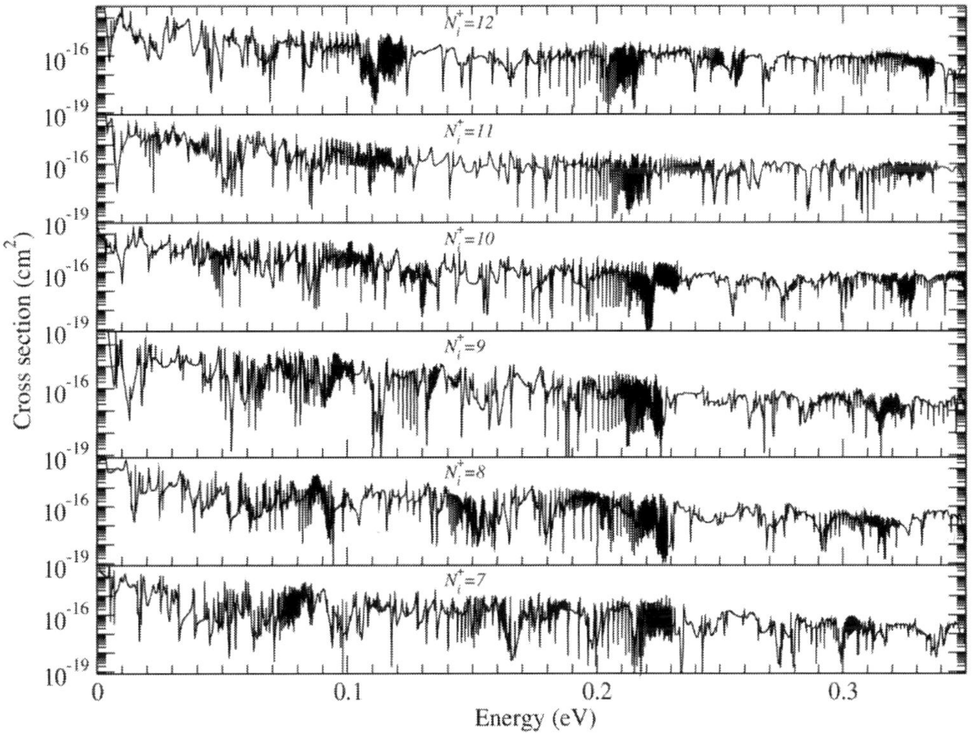

Figure 4. DR cross sections of HD^+ initially in one of its lowest rotational level N_i^+ (vibrational ground state).

[36] and storage-ring measurements: circles [37] and triangles [38].

2.2.2. Low energy (below ground state ion dissociation threshold)

When the total energy of our molecular system is situated in the discrete part of the vibrational spectrum of the cation, closed channels become available and, consequently, temporary capture into bound AB^* states occur. This induces local resonances in the shape of the cross section. For example, for HD, one may notice that the levels $v = 3$ and $v = 12$ of the Rydberg state having as potential the green curve appearing in Figure 1 are situated at the energies E_1 and E_2 respectively, and such resonant features are likely to appear at these total energies.

However, in order to produce accurate low-energy DR rate coefficients, the energy diagram shown in Figure 1 is not sufficiently detailed, since it completely neglects the rotational structure of both HD and HD^+.

Figures 3 and 4 display our results at incident electron energies close to zero, after taking into account the rotational structure and the corresponding interactions, for the case of the cation in its ground vibrational state, but on the lowest 12 rotational states. One may notice the very rich resonant structures, which we have discussed in detail in Ref. [42]. In experiments [42] at the TSR heavy ion storage ring at the Max Planck Institute für Kernphysik in Heidelberg, the DR rate of HD^+ ions was measured using merged electron and ion beams as a function of the electron collision energy and compared to the rotationally averaged theoretical results. In the range $0.05 - 0.2$ eV, theory was found to reproduce the magnitude and resonant structure of the cross section very well (typical collision energy spread $0.005 - 0.01$ eV). Above 0.05 eV collision

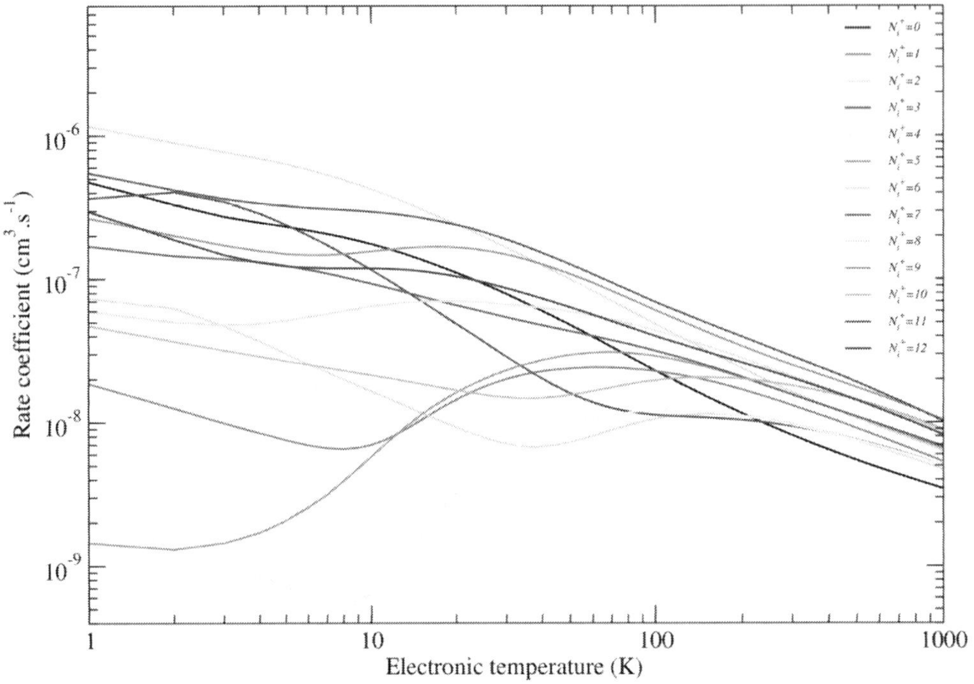

Figure 5. DR rate coefficients of HD^+ initially in one of its lowest rotational level N_i^+ (vibrational ground state).

energy, some experimental peaks were similarly well reproduced as at higher energies, with some deviations in the structure remaining. Yet, the overall magnitude of the DR rate was reproduced very well also in this energy region. The corresponding Maxwell rate-coefficients are given in Figure 5. A large scatter of the rates at low electron temperature occurs between the rotational level, which becomes smaller for higher electron temperatures. Experimental studies of this rotational dependence of the low-energy electron rate coefficient by state specific measurements in a cryogenic ion storage ring [43] are planned in future work.

2.3. *Electron/BeH$^+$ reactive collisions*

After producing the molecular data - potential energy curves and autoionization widths - relevant for the study of the collisions of electrons with diatomic Beryllium Hydride cation, two of the present authors (A. L. and A. E. O.) have performed a time-dependent wave-packet study of the DR [44]. Later, we used this data as input for a series of MQDT calculations of DR, vibrational IC and vibrational SEC cross sections, taking into account the direct and indirect processes [45]. We have considered the target molecular ion in its lowest four vibrational levels ($v_i^+ = 0 - 3$) of its ground electronic state. The cross sections are calculated separately for each symmetry (involving all the relevant dissociative states within the symmetry) and are summed up to give the global cross section.

We have explored the energy range $10^{-3} - 2.7$ eV of the incident electron, which corresponds to a total energy of the system below the dissociation limit of its ground electronic state. The rotational structure and interactions have been neglected, which is completely justified by the weak role of the indirect process [35] for this ion. Whereas extensive information is given in Ref.

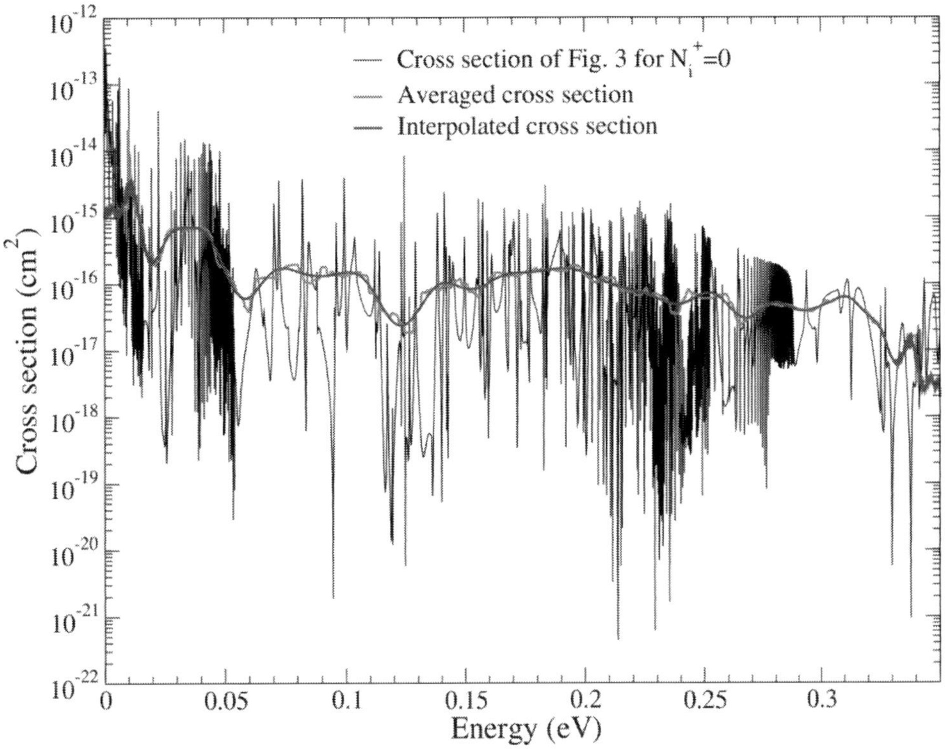

Figure 6. Dissociative recombination of $^2\Sigma_g^+(1s\sigma_g)$ v_i^+=0 HD^+ ion. Black curve: data displayed on Fig. 3 for $N_i^+ = 0$. Red curve: Averaged calculated cross section of the black curve over $\Delta\epsilon = 0.02$ eV between $\epsilon_{min} = 0$ eV and $\epsilon_{max} = 0.35$ eV. Blue curve: Interpolation of averaged cross section by eq. (A.2) (see Tables A1, A2, and A3).

[45] on the cross sections, we focus here on the Maxwell rate coefficients up to 5000 K.

Figure 7 shows the global DR, IC and SEC rate coefficients for the three lowest excited vibrational levels of the ion, in comparison with that corresponding to the ground vibrational state. The rate coefficient for the Vibrational Transitions (VT) - IC and SEC - result from a sum on all the possible final vibrational levels accessible by the reactive scattering. This figure illustrates a relatively fast DR of BeH^+ in its $v_i^+ = 2$ level, as well as a relatively fast vibrational de-excitation for $v_i^+ = 3$, at low temperature.

Figure 8 gives relevant information on the complete set of state-to-state rate coefficients available for the detailed kinetics of BeH^+ on its lowest four vibrational states.

2.4. Interpolation and averaging

In order to allow a simple use of the complicated cross sections and rate coefficients, we have averaged them using standard plasma kinetics procedures. They are presented for both electron/HD^+ and electron/BeH^+ systems in the Appendix.

3. Photoionization of H_2

In the present section we report theoretical photoionization cross sections for excitation from excited vibrational levels of the ground state. Our calculations are based on MQDT [29] which

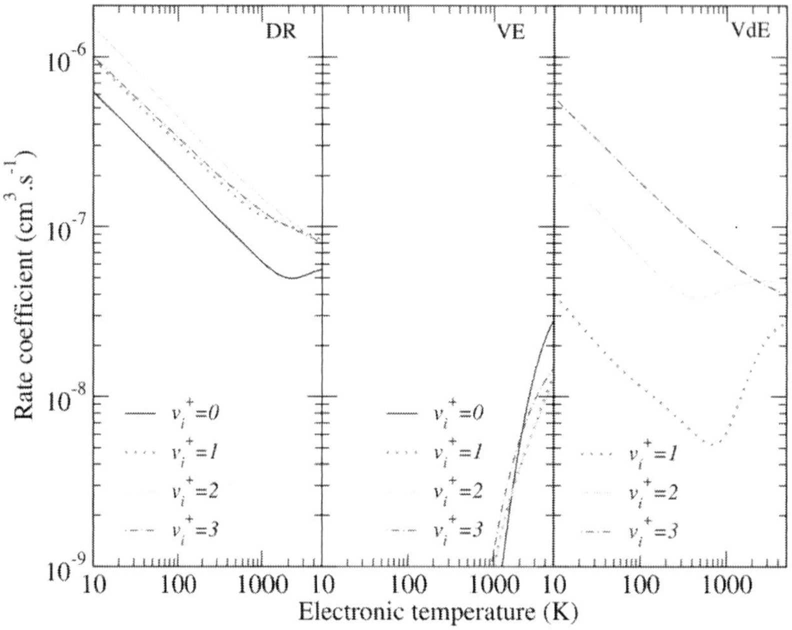

Figure 7. Vibrational dependence of the global DR, VE and VdE rate coefficients of BeH^+ (v_i^+ stands for the vibrational quantum number of the target ion).

allows us to take account of the full manifold of Rydberg states and their interactions with the electronic continuum. We show that the photoionization cross section is actually *dominated by resonance effects*, in the sense that autoionizing resonances produce a major fraction of the averaged cross section. The same kind of patterns are present in the dissociative recombination of the molecular cations with low energy electrons, whenever the DR is determined by resonant capture into Rydberg states.

The present report deals with the Q transitions ($\Delta N = 0$) which account for roughly one third of the total photoabsorption cross section. We focus in particular on the role of the resonances in the photoionization cross sections, and we choose $v'' = 1, N'' = 1$ to exemplify this. Analogous results dealing with the Q, P and R optical dipole transitions will be published soon, including an extensive ensemble of transition energies, Einstein A coefficients as well as photoionization cross sections and Rydberg resonance parameters for various v'', N'' values.

3.1. The global MQDT-type approach

In this section, we closely follow the derivations given in Ref [9]. The manifold of $^1\Pi_u$ excited states of H_2 is assumed to represent a single unperturbed $np\pi$ Rydberg series converging to the $X^2\Sigma_g^+$ ground state of H_2^+. The quantum defects may thus be extracted directly from the clamped-nuclei (Born-Oppenheimer) potential energy curves (PEC) by use of the one-channel Rydberg equation, written here in atomic units as a geometry-dependent function:

$$U_n(R) = U^+(R) - \frac{1}{2\left[n - \mu_n(R)\right]^2}. \tag{18}$$

The index n of the quantum defect μ indicates that this quantity may vary slightly with the principal quantum number n, i.e., it is energy-dependent. This energy dependence is formally

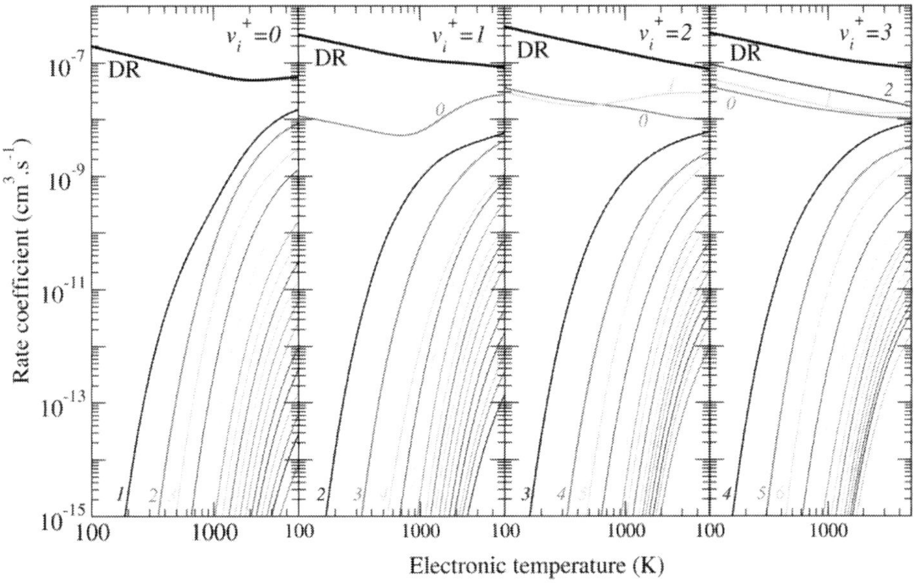

Figure 8. Dissociative recombination and state-to-state vibrational (de-)excitation rate coefficients of BeH^+ initially in one of its four lowest vibrational levels, v_i^+. For each initial vibrational state of BeH^+, the final vibrational quantum numbers are indicated for de-excitation and for the first three (from the left to the right) vibrational excitation curves. For the remaining unlabeled curves, their order corresponds to the successive increase of the vibrational quantum numbers of the target cation.

parametrized as an expansion in terms of the electron binding energy ϵ (in a.u.):

$$\mu(\epsilon, R) = \mu^{(0)}(R) + \epsilon\mu^{(1)}(R) + \frac{1}{2}\epsilon^2\mu^{(2)}(R)] + \frac{m}{M}\mu^{\rm spec}(R). \tag{19}$$

The three coefficients $\mu^{(k)}(R), k = 0, 1, 2$ that appear in Eq. (19) are determined by means of Eq. (18) from the three clamped nuclei PEC, $U_n(R)$, of the $np\pi, n = 2, 3, 4$ Rydberg states computed by Wolniewicz and co-workers [46, 47, 48], combined with the PEC, $U^+(R)$, of the ion ground state $X^2\Sigma_g^+$. The clamped-nuclei electron binding energy is given for each R-value by the difference $\epsilon = \epsilon(R) = U_n(R) - U^+(R)$. $\mu^{\rm spec}(R)$ is related to the mass polarization term arising from the cross-term $H_3' = -(m/4M)\nabla_1\nabla_2$ of the molecular Hamiltonian [49] (m electron mass, M nuclear reduced mass). $H_3'(R)$ has also been evaluated *ab initio* by Wolniewicz and co-workers [48].

The electronic dipole transition moments constitute a second set of input data required for the description of radiative electronic transitions. Energy dependent clamped-nuclei electronic channel transition moments are parametrized in analogy with Eq. (19) as

$$d(\epsilon, R) = d^{(0)}(R) + \epsilon d^{(1)}(R) + \frac{1}{2}\epsilon^2 d^{(2)}(R). \tag{20}$$

These quantities are energy normalized and have the dimension [length] × [energy]$^{-1/2}$. The coefficients $d^{(q)}(R), q = 0, 1, 2$, may be derived directly from the *ab initio* clamped-nuclei dipole transition moments of Wolniewicz and co-workers [47, 48], see Refs. [9, 50].

Journal of Physics: Conference Series **576** (2015) 012005 doi:10.1088/1742-6596/576/1/012005

Upper state energy levels and/or continuum states and the rovibronic dipole transitions leading to them are computed using standard MQDT techniques [9, 27, 28, 29]. The transition of the excited electron from the molecule-fixed reference frame to the space-fixed frame is described by a frame transformation which takes account of the non-adiabatic effects. In the case of the Q(N) spectral lines corresponding to upper states of pure $^1\Pi_u^-$ symmetry there is no rotational non-adiabatic coupling so that, neglecting spin-orbit and hyperfine effects, the frame transformation reduces to evaluating integrals involving the initial and final state vibrational wave functions:

$$\mu^{(q,N,d)}_{v^+N^+,v^{+\prime}N^{+\prime}} = \int \chi_{v^+N^+}(R)\mu^{(q)}(R)\chi_{v^{+\prime}N^{+\prime}}(R)dR, \tag{21}$$

for the vibronic quantum defect, and:

$$d^{(q,N,d)}_{v^+N^+,v''N''} = \int \chi_{v^+N^+}(R)d^{(q)}(R)\chi_{v''N''}(R)dR, \tag{22}$$

for the vibronic transition moments. Here $q = 0, 1, 2$ as in Eqs. (19) and (20). The Q(N) transitions studied here require that the total angular momenta of the molecule N', N'' and of the molecular ion $N^+, N^{+\prime}$ all be equal, i.e., $N = N^+ = N^{+\prime} = N''$. d refers to the Kronig's symmetry label, designating levels that have total parity $-(-1)^N$.

The energy dependences of the vibronic quantum defects and dipole moment matrix elements of Eqs. (21) and (22) have the same functional form as those given in the Eqs. (19) and (20), with the difference that instead of ϵ one must use a vibrationally averaged value, $\bar{\epsilon}$. For the quantum defects the appropriate choice is $\bar{\epsilon} = \frac{1}{2}\left[(E - E^+_{v^+N^+}) + (E - E^+_{v^{+\prime}N^{+\prime}})\right]$ [29], where E denotes the total energy and $E^+_{v^+N^+}$ are the vibration-rotation levels of the ion ground state. For the transition moments one chooses $\bar{\epsilon} = (E - E^+_{v^+N^+})$ [29]. Once the rovibronic quantum defect have been evaluated they must be converted into the so called $\mathcal{S}$ and $\mathcal{C}$ matrices which are related to the reaction matrix $\mathbf{K}$ according to $\mathbf{K} = \mathcal{S}\mathcal{C}^{-1}$. This is achieved by means of procedures described in Refs. [51] and [52]. The "eigenchannel" formulation of MQDT [29] rewrites these matrices in the form $\mathcal{S} = \mathbf{U}\sin\pi\mu\mathbf{U}^{\mathrm{T}}$ and $\mathcal{C} = \mathbf{U}\cos\pi\mu\mathbf{U}^{\mathrm{T}}$, where $\mathbf{U}$ is the eigenvector matrix of the reaction matrix $\mathbf{K}$ while $\tan\pi\mu_\alpha$ are its eigenvalues and their arguments μ_α are the associated eigenquantum defects. (The indices α indicate that these quantities are not to be confused with the quantum defects that appear in Eqs. (19) and (21)).

Applying the asymptotic boundary conditions to the total wave function, one arrives at the generalized eigenvalue system [29]:

$$\mathbf{\Gamma B} = \tan\beta\mathbf{\Lambda B}\,, \tag{23}$$

where

$$\begin{aligned}
\Gamma_{ii'} &= \sin\beta_i\mathcal{C}_{ii'} + \cos\beta_i\mathcal{S}_{ii'} && i \in \text{closed}\\
\Lambda_{ii'} &= 0 && i \in \text{closed}\\
\Gamma_{ii'} &= \mathcal{S}_{ii'} && i \in \text{open.}\\
\Lambda_{ii'} &= \mathcal{C}_{ii'} && i \in \text{open}\\
\beta_i &= -\pi\nu_i && i \in \text{closed}\\
\beta_i &= +\pi\tau_i && i \in \text{open}\,.
\end{aligned} \tag{24}$$

If bound levels and discrete line intensities are to be evaluated (because there are no open channels or because open channels are neglected), Eq. (24) reduces to a homogenous set of linear equations, namely

$$\sum_k \left[\cos(\pi\nu_j)\mathcal{S}_{jk} + \sin(\pi\nu_j)\mathcal{C}_{jk}\right] B_k(E) = 0, \tag{25}$$

where $\nu_j(E) = \sqrt{-\mathcal{R}hc/(E - E_j^+)}$ are the channel effective quantum numbers, B_k are the channel mixing coefficients and $\mathcal{R}$ is the Rydberg constant. The indices j and k run over all vibrational channels v^+N^+ corresponding to the given N^+ value. By searching for zeros of Eq. (25) one finds the energies E_n of the bound levels and the corresponding channel mixing coefficients. The former, in wave number units and defined relative to the ionization thresholds, are

$$\left[\left(E_n - E^+_{v^+N^+}\right)/hc\right] = \frac{\mathcal{R}_{H_2}}{\left[\nu_{v^+N^+}(E)\right]^2}. \tag{26}$$

$\mathcal{R}_{H_2}$ here is the mass-corrected Rydberg constant. The effective transition moment to the bound Rydberg state n is given by the following superposition of channel amplitudes:

$$D_n = \frac{1}{\mathcal{N}} \sum_k d_{kk''}(E_n) B_k(E_n)\,, \tag{27}$$

where k stands for the ionization channels (v^+N^+) and k'' stands for the lower state $(v''N'')$, respectively. $d_{kk''}(E_n)$ is the frame transformed vibronic dipole moment. $\mathcal{N}$ is the overall normalization factor of the bound-state wave functions, see Ref. [29]. It has the dimension $[\text{energy}]^{-1/2}$ which compensates the dependence $[\text{energy}]^{-1/2}$ contained in the channel moments $d_{kk''}$. The transition moment D_n for each spectral line is finally converted into an upper state emission probability according to [29]:

$$A_{n\to v''N''} = \frac{4mc^2\alpha^5}{\hbar} \left(\frac{1}{2N'+1}\right) \left(\frac{E_n - E_{v''N''}}{2\mathcal{R}hc}\right)^3 \left|\frac{D_n}{a_0}\right|^2, \tag{28}$$

where α here denotes the fine structure constant.

In order to calculate photoabsorption/photoionization cross section, σ_a/σ_i, we have to consider the open ionization channels in addition to their closed counterparts. The quantization condition for an open channel j, see Eqs. (23) and (24), reads [29]

$$\sum_k \left[\cos(\pi\tau_\rho)\mathcal{S}_{jk} - \sin(\pi\tau_\rho)\mathcal{C}_{jk}\right] B_k^{(\rho)}(E) = 0\,, \tag{29}$$

where $\pi\tau_\rho$ is an open-channel ionization eigenphase and ρ is a solution index. In all, there are as many eigenphases as there are open channels, N_P, and there is a set of channel mixing coefficients, $B_k^{(\rho)}$, corresponding to each particular eigenphase. The total photoionization intensity is expressed in terms of a set of real dipole amplitudes which replace Eq. (27):

$$D^{(\rho)}(E) = \sum_k d_{kk''}(E) B_k^{(\rho)}(E)\,. \tag{30}$$

The apparent disagreement between Eqs. (27) and (30) reflects the different normalization of the discrete and continuum wave functions: $D^{(\rho)}(E)$ in Eq. (30) gives the dipole amplitude *per unit energy* and has dimension [length] $\times$ $[\text{energy}]^{-1/2}$. The total photoabsorption cross section, σ_a (equal to the photoionization cross section σ_i when no competing decay channels are present), becomes [29]

$$\sigma_a(E) = \frac{4\pi^2\alpha}{2N''+1}\left(E - E^+_{v^+N^+}\right)\left[D(E)\right]^2, \tag{31}$$

where $[D(E)]^2 = \sum_{\rho=1}^{N_P} [D^{(\rho)}(E)]^2$.

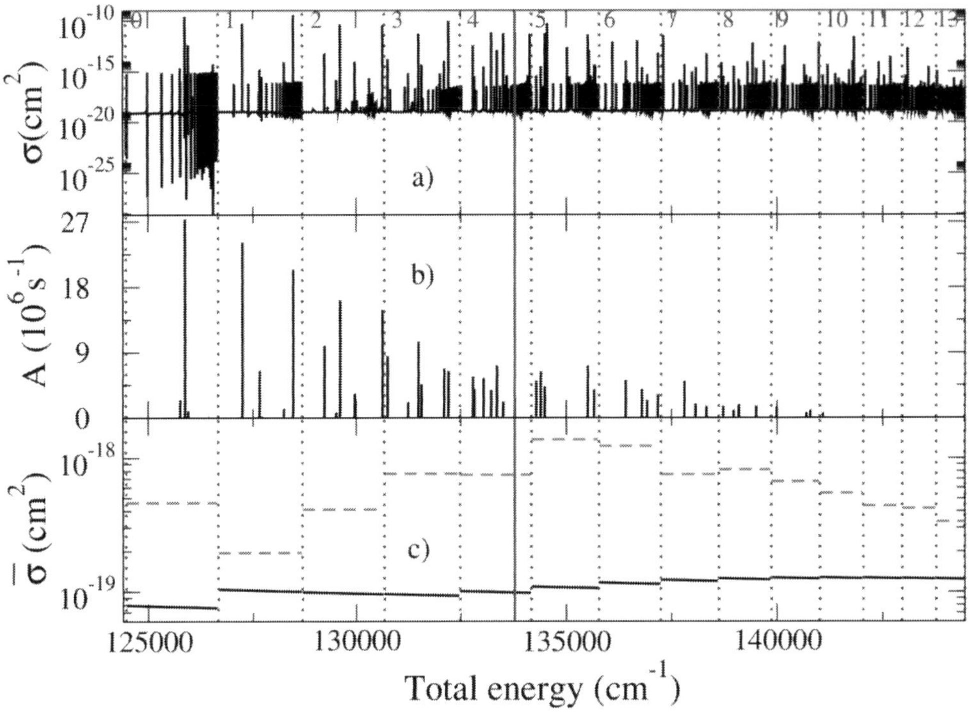

Figure 9. Cross sections and line intensities for the Q(1) transitions from the $X^1\Sigma_g^+, v'' = 1$ ground state of the H_2 molecule, compiled from [9]. The vertical dotted lines represent the vibrational ionization thresholds, starting with $v^+ = 0$ and ending with $v^+ = 14$. The vertical continuous line (blue online) indicates the position of the H(1) + H(n = 3) dissociation limit of H_2. a) Continuum photoionization cross section. b) Spontaneous emission coefficients for the transitions listed in Ref. [9], equivalent to the resonant part of the cross section. c) Photoionization: continuous black lines: background cross section; dashed line (red online): averaged full cross section.

3.2. Results on H_2 photoionization

A more elaborated presentation of the theory can be found in [29, 9], while the methodology of the calculations is presented in [9]. In what follows we will focus on the presentation of theoretical photoionization cross sections for Q transitions ($\Delta N = 0$) from the ground electronic state ($v'' = 1, N'' = 1$). The results obtained rely on the use of the MQDT package developed by Ch. Jungen.

By the inclusion of the open channels in addition to the closed ones, and solving Eq. (23) in its most general form, the continuum phase shifts, channel mixing coefficients, effective transition moments are provided and, as the final result, the photoionization cross sections as a function of the transition energy are calculated. Strictly speaking, the theoretical photoabsorption spectrum is produced, which is broadened by interaction with the open ionization continua, in other words photoabsorption eventually results in ionization whenever the two additional decay channels (dissociation and fluorescence) are neglected. Fig. 9a displays the corresponding continuous distribution. In order to calculate the *true* photoionization cross section, the continuous spectrum has been corrected for dissociation and fluorescence processes. This was performed by using the data from Refs. [7, 8], where absolute cross sections for the competing decay-channels

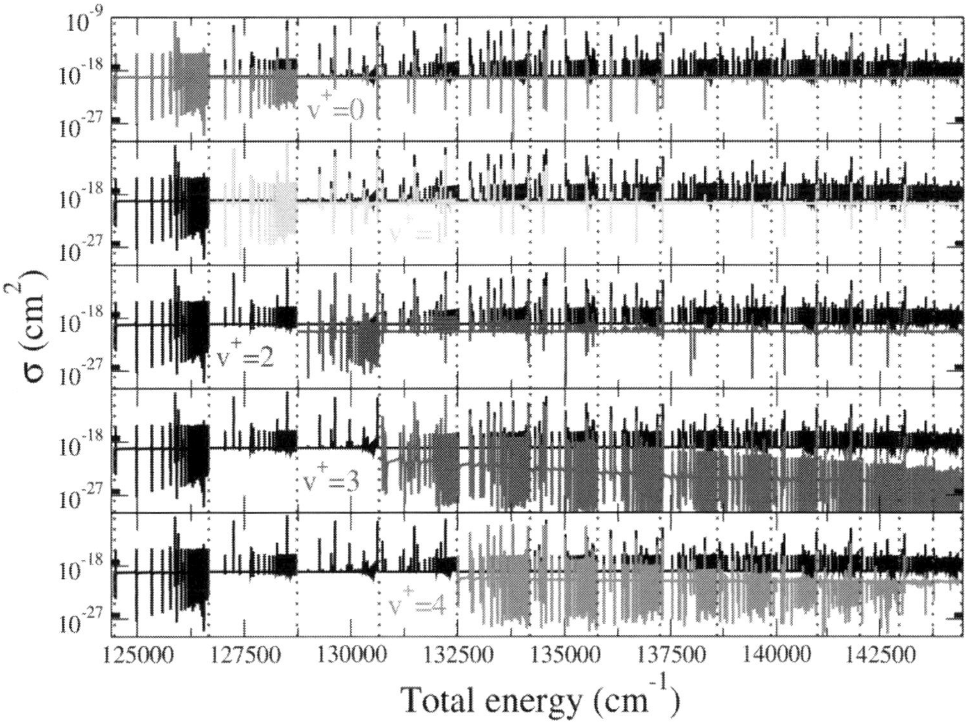

Figure 10. Final state vibrationally resolved photoionization cross sections for the lowest five ionic vibrational levels. Black continuous line is the total cross section, while the colored ones are the partial cross sections, corresponding to ionization towards H_2^+ $v^+ = 0 - 4$, from top to bottom in increasing order. The vertical dotted lines represent the vibrational ionization thresholds, starting with $v^+ = 0$ and ending with $v^+ = 14$.

fluorescence, dissociation, and ionization of photoexcited long-lived superexcited H_2 molecular levels have been reported from the ionization threshold up to the $n = 3$ dissociation limit. Essentially, we have *renormalized* the continuous spectrum by multiplying each resonance profile with its corresponding photoionization yield taken from Ref. [8], and thus obtain the photoionization cross section σ_i.

The plot in Fig. 9a shows that the cross section is dominated by the presence of resonance structures corresponding to excitation of various vibrational levels of bound electronic states which lie above the ionization threshold. A logarithmic intensity scale has been used in order to highlight the density of autoionization structures that arise in the cross section. By representing every discrete upper level by a sharp line with the intensity corresponding to its spontaneous emission probability one obtains the spectrum as a sequence of Feshbach resonances. This can be seen in Figure 9b, where only the lines corresponding to states up to $n = 6$ are given, whereas the total photoionization cross section shown in Fig. 9a contains the states with higher ($n \sim 70$) quantum numbers as well. The overall contribution of the Rydberg resonances of the lowest lying four excited electronic states to the photoionization cross section are summarized in Table II of [9]. It turns out that more than 42% of the total photoionization cross section are produced by the four lowest electronic states.

Fig. 9c is designed to demonstrate the remarkable importance of resonances in the Q(1)

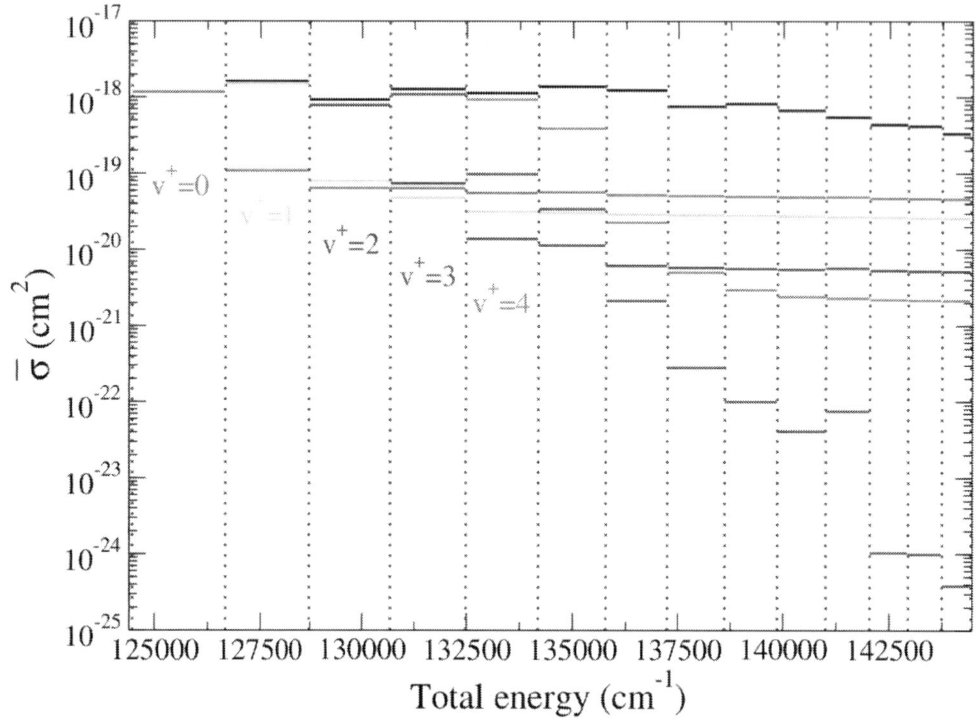

Figure 11. Final state vibrationally resolved averaged photoionization cross sections (cf. eq. (32)) for the lowest five ionic vibrational levels. Black continuous line is the total cross section, while the colored ones are the partial cross sections, corresponding to ionization towards H_2^+ $v^+ = 0 - 4$ (same color code is used as in Fig. 10). The vertical dotted lines represent the vibrational ionization thresholds, starting with $v^+ = 0$ and ending with $v^+ = 14$.

photoionization cross sections. In order to assess the importance of the resonances, we have - for each vibrational interval - calculated the averaged photoionization cross section:

$$\bar{\sigma}_i(v^+) = \frac{\int_{E_{v^+}}^{E_{v^++1}} \sigma_i(E)dE}{E_{v^++1} - E_{v^+}}, \tag{32}$$

indicated in the figure by dashed (red online) line. The black solid line represents the background cross section (direct ionization), obtained when all closed channels are excluded in the calculation and hence all resonances disappear from the spectrum.

We found that at certain photon energies, just above the $n = 3$ dissociation threshold (blue continuous vertical line), the resonance enhancement exceeds the background cross section by more than one order of magnitude, while at lower energy (for instance, in the range between the $v^+ = 1$ and $v^+ = 2$ thresholds) as well as at higher energies this is still more than a factor of two larger than the latter. The total enhancement on the whole energy range exceeds with more then a factor of 6 the background cross section.

The surprising feature that the most intense lines of the spectrum lying at low energies contribute less to the photoionization cross section can be understood from the fact that, as

Journal of Physics: Conference Series **576** (2015) 012005 doi:10.1088/1742-6596/576/1/012005

displayed in Figure 8 of [8] for the lowest lying excited states, the fluorescence dominates for $n = 3$ (D state) and is also present in the $n = 4$ and 5 (D' and D" states). Photodissociation is significant ($\approx$ 15 to 20%) for $n = 3$, but is absent for $n = 4$ and 5. Photoionization is dominant for $n \geq 5$. Thus, for instance, in the energy range between the $v^+ = 1$ and $v^+ = 2$ thresholds, the most intense lines, corresponding to the $3p\pi$D state, result from a strong decay by fluorescence and consequently, do not contribute to the photoionization.

An other key issue in these photo-induced reactions is to produce results which are resolved with respect to the vibrational level (v^+) of the resulting H_2^+ ion. These partial cross sections are in particularly important in the kinetic models of the primordial universe chemistry [16].

Figures 10 and 11 display the final state vibrationally resolved photoionization cross sections for the lowest five vibrational levels of the ground state of the molecular ion. The black curve labels the total photoionization cross section, while the colored ones those with $v^+ = 0 - 4$ from the top to the bottom. Due to favorable Frank-Condon overlaps, the partial cross sections are not restricted to a single energy interval (belonging to the same ionization threshold), but can be spread out in the higher energy regions as well (belonging to higher ionization thresholds).

4. Electron/H_2 reactive collisions

4.1. The experimental method

We present experimental results on electron-impact dissociation of H_2 molecules. We have used a molecular nozzle beam in order to produce molecules with cold internal degrees of freedom with high intensity flux conditions.

The basic layout of the experimental setup, whose detailed description is given in [11] is equivalent to the classical one [54] for time-of-flight (TOF) studies of $H(2^2S_{1/2})$ produced by electron impact on H_2, i.e. a hydrogen jet crossing an electron beam at right angle. The H_2 beam is produced by a Campargue-type supersonic jet source [55], allowing to get a reduced velocity spread and a well defined initial molecular state and target density. In the chamber where the interaction with the electrons takes place, the pressure is maintained at 5×10^{-7} torr , when the supersonic beam is not in operation, and 1.2×10^{-6} torr, when the supersonic beam is in operation, by a 2000 l/s diffusion pump. The measured average velocity of the hydrogen molecules in the supersonic jet is 2.7 km/s with $\Delta v/v \sim 1\%$. The beam has a rotational temperature of about 1 K and a vibrational temperature of about 10 K. Due to the expansion, essentially all hydrogen molecules are in the $v = 0$ vibrational level, with relative rotational populations that depend drastically on the detail of the expansion dynamics but are mainly restricted to the states $J = 0$, parahydrogen and $J = 1$, orthohydrogen [56, 57, 58]. In the region where the molecular beam crosses the electron beam, its diameter is approximately 3 mm, and the flow is estimated to be about 10^{22} molecules per steradian per second, which corresponds to $(2.3 \pm 0.2) \times 10^3$ molecules per microsecond per cubic millimeter in the interaction volume. The pulsed electron gun consists of a 150 μm diameter thoriated-tungsten filament, a grid and a collimator with a circular aperture of 2 mm radius, and produces an electron beam with an angular spread FWHM of 5°. The electron beam is detected by a Faraday cup at a distance of 70 mm and produces a typical average current of 2 mA in continuous operation, for 120 V acceleration. The electron beam can be pulsed by applying a negative step voltage, whose duration can be chosen in the range 0.2 to 2 μsec, to the filament's DC circuit, which is otherwise maintained positive in order to reduce the electron background. The typical pulse rate is 10 kHz. The electron energy is given by the difference between the potential of the filament and that of the interaction region, which has to be grounded in order to avoid the disappearance of $H(2^2S_{1/2})$ atoms through the very efficient coupling with radiative $H(2^2P_{1/2})$ states. The energy spread of the electron beam is estimated to be of about 6 eV, taking into account the voltage drop in the directly heated filament. This low resolution has no significant effect at high electron energies, as we do not intend to perform precise cross section measurements near threshold. One may also notice that

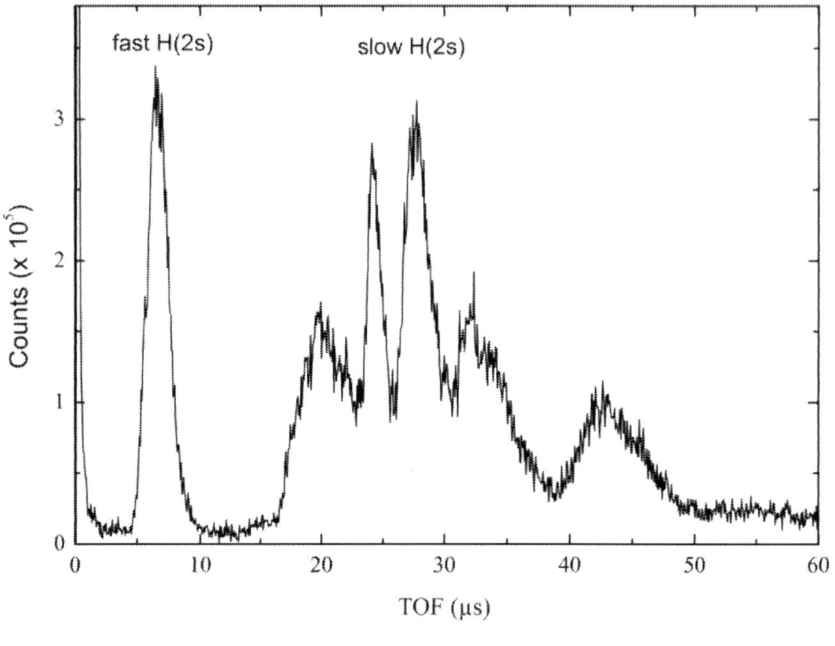

Figure 12. Typical Time of flight spectrum [11] for 126 eV electron energy.

the coherent energy width corresponding to $\Delta E = \hbar/\Delta t$ is of the order of 10 eV for electron energies of 100 eV, assuming a size of the hydrogen molecules of few atomic units.

The detection of the metastable $H(2^2S_{1/2})$ atoms is performed by a specially devised detection system optimized in order to get a well defined detection zone whose distance was chosen to be 230 mm and 254 mm away from the collision zone. We have cross checked the TOF measurements with two detectors working at the same time, which reduces the uncertainties concerning the exact determination of the flight distances. Within this distance range excited $H(2^2P_{1/2})$ atoms do not reach the detection zone, as their lifetime is 1.7 ns, to be compared with typical TOF of few μsec for the fastest atoms. As these experiments were designed in order to search for coincidence events between pairs of twin atoms, care is taken to place the detectors in positions consistent with the expected recoil of the superexcited H_2 molecules after collision with the electrons, before dissociation.

A collimator of 2 mm diameter is placed at the entrance of each detection system, and is followed by a pair of needles normal to the direction of $H(2^2S_{1/2})$ atoms. The potential difference applied to these needles was chosen to be 200 V and produces an electrostatic field of maximum value 4×10^4 V/m that mixes the $2^2S_{1/2}$ and the $2^2P_{1/2}$ states of the atoms, inducing the decay to the $1^2S_{1/2}$ state and consequently producing Lyman-α radiation (1216 Å). In order to lower the direct photon counts originating from the de-excitation of short lived atomic states and of the radiative molecular states, a channel electron multiplier detector of 10 mm aperture is oriented at right angles with respect to the atom trajectories and to the needles. It is located 4 mm away from the needles. The emitted Lyman-α radiation passes through a grid and a MgF_2 plate that shields the channeltron against massive species such as electrons and ions. The efficiency of the detection system is of the order of 0.3%, including intrinsic detector efficiency, effective solid angle seen by the Lyman-α radiation, plate transparency and grid transmission. The quenching

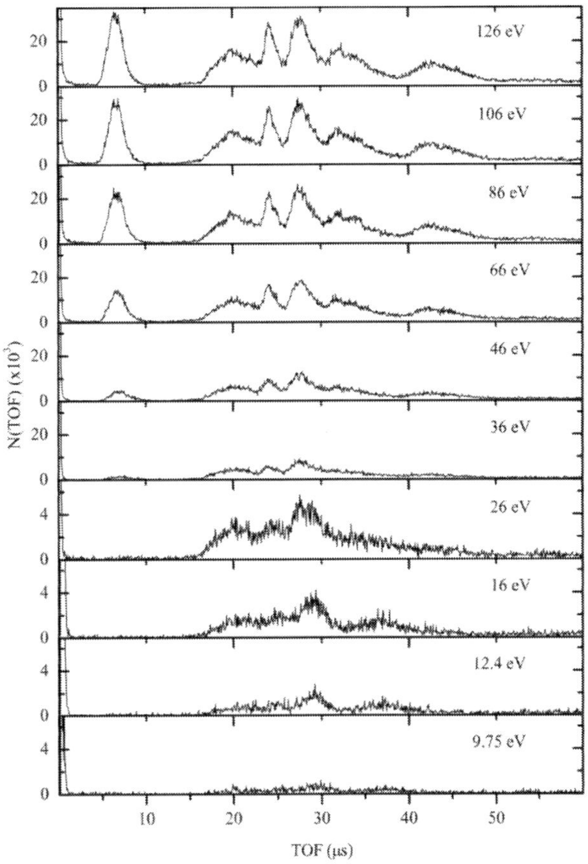

Figure 13. Time-of-flight spectra [11] of $H(2^2S)$ atoms dissociated from H_2.

electric field profile inside the detection system acts in a very localized region of the order of 10 μm and located about 10 μm in front of the needle axis for atoms with velocities in the range of a few dozens of km/s and needle voltages in the 10 – 500 V range. The uncertainty on the total flight length is given by the size of the interaction zone between the electron and the molecular beams, *i.e.* 2 mm, and consequently the main uncertainty in the time of flight is due to the electron pulse duration: our best resolved peak involves an electron pulse of 0.25 μs duration with a flight length distance of 254 mm, which leads to a peak with a maximum around 8.0 μs and consequently to a TOF uncertainty of 0.125 $\mu s/8.0\ \mu s \approx 1.7\%$.

Pulses from the detectors are separately pre-amplified and amplified by standard NIM electronics. After discrimination from electronic noise, the signals are fed as stop pulses of a "FAST ComTec" multichannel analyzer card. The detector pulses are also analyzed in parallel, after pre-amplification, by a fast oscilloscope. Particular care was taken that the average pulse heights had a uniform non saturated value over all the observation range. Indeed, saturation effects such as bleaching can arise because of the huge peak associated with the electron pulse

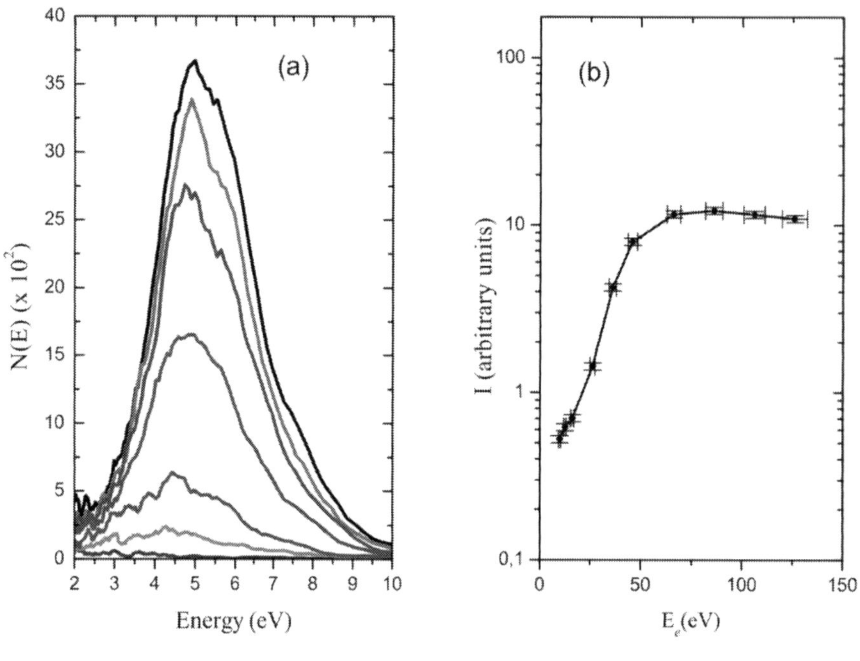

Figure 14. (a) Energy spectra for the fast atoms with the same counting time intervals, for different electron collision energy. From bottom to top: 26, 36, 46, 66, 86, 106 and 126 eV. (b) We display (vertical axis in arbitrary units) $I \times E_e$, where $I = \int_{\text{Fast peak}} N(E)dE/V_{\text{Faraday cup}}$, which is proportional to the H(2^2S) production, from [11].

itself. As a consequence all experimental spectra appear on a very small and uniform background.

4.2. Results on electron-impact /H_2 dissociation

The experimental time-of flight spectrum obtained for electron collision energy of 126 eV that we have used for calibration of our data is presented in Fig.12 for the detector at 254 mm from the collision center. In this spectrum, one can distinguish three main regions corresponding to increasing time-of-flights: (i) a very bright peak related to the electron gun pulse, in width and intensity, (ii) a contribution of fast atoms, and (iii) a contribution of slow atoms at a larger time-of-flight, which is quite as well resolved as the spectra obtained by photoabsorption [59] and whose structures are analyzed in [63].

Experimental TOF spectra obtained for electron collision energies between 10 and 126 eV are presented in Fig. 13. They correspond to the measurements of one detector, normalized both to electron beam current and acquisition time, for the detector at 230 mm from the collision center.

In these spectra the traditional classification of the metastable H(2^2S) atoms into "slow" (velocities of about 10 km/s) and "fast" (velocities of about 35 km/s) is evident, as well as their energetic thresholds: while the slow atoms arising from the singly excited states are visible in our entire energy range, the fast atoms, originating from the doubly excited states, require electron energies greater than 30 eV.

In order to express the data of Fig. 13 as a function of E_H, the kinetic energy of the H(2^2S) atoms in the center-of-mass reference frame, we need to describe the kinematics of the

Table 1. Maximum location of the fast atoms peak [11].

Date	Authors	Maximum position(eV)	Peak width (eV)	Electron energy (eV)
1967	Leventhal, Robiscoe & Lea [54]	4.7±0.7	2.6	60
1972	Misakian & Zorn [60]	4.5±0.5	2.8	70
1977	Carnahan & Zipf [61]	5.1±0.4	3.1	75
1977	Carnahan & Zipf [61]	5.3±0.4	3.4	100
1977	Hazi & Wiemers [62]	4.8±0.4	2.6	100
1980	Spezeski, Kalman & McIntyre Jr. [53]	5.6±0.4	3.7	98
2011	Medina *et al* [11]	5.6±0.2	3.3	126
2011	Medina *et al* [11]	5.5±0.5	3.2	106
2011	Medina *et al* [11]	5.4±0.5	3.3	86
2011	Medina *et al* [11]	5.4±0.5	3.4	66
2011	Medina *et al* [11]	5.0±0.5	3.8	46
2011	Medina *et al* [11]	4.9±0.5	3.1	36

Experimental conditions:
ref. [54]: $\theta = 77^o$ and 90°, $\tau_e = 0.2$ μs and $L = 10$ cm.
ref. [60]: $\theta = 80^o$, $\tau_e = 0.5$ μs and $L = 13.14 \pm 0.24$ cm.
ref. [61]: $\theta = 90^o$, $\tau_e = 0.4$ μs and $L = 28$ cm.
ref. [53]: $\theta = 90^o$, $\tau_e = 0.1$ μs and $L = 18.3 \pm 0.3$ cm.
ref. [11]: for $E_e = 126$ eV; $\theta = 90^o$, $\tau_e = 0.25$ μs and $L = 25.4 \pm 0.1$ cm.
ref. [11]: for $E_e = 106 - 36$ eV; $\theta = 90^o$, $\tau_e = 0.5$ μs and $L = 23.0 \pm 0.1$ cm,
with E_e the electron energy of the electron beam, θ the angle between the direction where the metastable atoms are detected and the electron-beam axis, τ_e the time duration of the electron pulse and L the distance between the collision region and the quenching region or the detector itself (TOF distance). The uncertainties displayed above, in the maximum position (in eV), are a consequence of the uncertainties in these quantities and of the electron pulse width, and their values were extracted from ref. [53]. The pressure used in all works was in the range of $10^{-4} - 10^{-5}$ torr.

dissociation. Frame transformation for the velocity vectors are detailed in [11].

In our experiments, the position of the detectors was chosen so that to select hydrogen atoms produced by dissociation with velocities orthogonal to the plane defined by molecular and the electron beams. The inversion of the spectra requires some care for the slow atom part [63] but is easy for the fast atom part that we will focus on.

In these spectra - Fig. 14 - where the components of the peak are not resolved, one can assume a smooth evolution of the location of the maximum of the peak from low to high energies, which saturates at energies higher than 60 eV (about twice the mean threshold energy). The TOF spectra are not sensitive to the temperature of the H_2 gas as one can see from previous works on these processes summarized with our contributions in Table 1.

In these results, the production of the fast metastable hydrogen atoms has been studied as a function of the variation of the electron energy with two detectors. Using uncorrelated detectors, we have observed that the signal intensities are angle-independent. This will be no longer the case for future coincidence experiments with correlated detectors, because this configuration corresponds to the selection of definite values for the direction of the interatomic vector in the center-of-mass frame. This angular dependence will reflect the angular momentum conservation rules of the electron-molecule excitation process. For instance it will allow us to determine the dependence of the production of the atom pairs with respect to the forward (small electron-molecule impact parameters) or backward (high electron-molecule impact parameters) processes.

5. Conclusion and perspectives

Using the Multichannel Quantum Defect Theory, we have computed cross sections and rate coefficients for dissociative recombination, vibrational excitation and de-excitation for HD^+ and BeH^+, and dissociative excitation for HD^+. The computed cross sections for HD^+/HD system are in good agreement with the measurements. We plan to complete the data set for this system, carrying out calculations with a careful state-to-state analysis for a broad range of energies. This represents a significant computational effort which is currently in progress.

By calculating the photoionization cross section from the vibrationally excited ground electronic state of the hydrogen molecule, we have shown that in contrast to what had been assumed previously, autoionization resonances dominate the cross section. Moreover we have determined the final state vibrationally resolved cross sections in order to estimate the level populations of the produced molecular ions.

The raw data corresponding to the figures related to the electron-molecular cation reactive collisions and to the photoionization (Figs. 2–10) are available on demand.

We have revisited previous work on dissociation by electron impact of the H_2 molecule leading to one $H(2^2S_{1/2})$ atom. Uncorrelated detectors lead to isotropic spectra. Coincidence experiments will select specific interatomic directions related to an unique molecule for each event. Accurate analysis of the data with detectors selectively sensitive to protons or to metastable or Rydberg atoms will provide refined and quantitative information on the dynamics of these superexcited states versus the Coulomb explosion mechanism [65].

6. Acknowledgements

This work was carried out as a part of an IAEA Coordinated Research Project on "Light Element Atom, Molecule and Radical Behaviour in the Divertor and Edge Plasma Regions". We thank Ch. Jungen for his constant scientific support and fruitful discussions, and critical reading of this article. We acknowledge the scientific and financial support from EFDA, the French Research Federation for Fusion Studies, the European Spatial Agency/ESTEC 21790/08/NL/HE, the CNRS/INSU/'Physique et Chimie du Milieu Interstellaire' programme, the Triangle de la Physique/PEPS/'Physique théorique et ses interfaces', the CPER Haute-Normandie/CNRT/'Energie, Electronique, Matériaux', IEFE:Rouen-Le Havre, and Conseil Régional de la Haute Normandie. This work has been supported by the Agence Nationale de la Recherche under Contracts No. 09-BLAN-02090 "SUMOSTAI" and ANR-12-BS05-0011-01"HYDRIDES". This work is supported by CAPES, FAPERJ, CNPq and the Scientific Cooperation Agreement CAPES/COFECUB between France and Brazil, project number Ph 636/09. AEO acknowledges support from the National Science Foundation Grant No's PHY-11-60611. ÅL acknowledges support from the Swedish research council.

Appendix

In order to allow a simple use of the complicated physical observables like cross sections or rate coefficients shown for example in Figures 3 and 4, we have averaged them by using the simple formulas

$$\bar{\sigma}^{DR}_{(AB^+),N_i^+,\Delta\epsilon}(\epsilon < \epsilon_{max}) = \frac{1}{\Delta\epsilon}\int_{\epsilon-\frac{\Delta\epsilon}{2}}^{\epsilon+\frac{\Delta\epsilon}{2}} \sigma^{DR}_{(AB^+),N_i^+}(\epsilon')\,d\epsilon', \tag{A.1}$$

between $\epsilon_{min} = 0$ eV and $\epsilon_{max} = 0.35$ eV, where $\Delta\epsilon = 0.02$ eV. This results in smooth cross sections, as for example that shown for the *Dissociative Recombination of HD*$^+$ in Fig. 6, which can be eventually generated by a polynomial interpolation formula:

$$\overline{\sigma}^{DR}_{(AB^+),N_i^+,\Delta\epsilon}(\epsilon < \epsilon_{max}) = \overline{\overline{\sigma}}^{DR}_{(AB^+),N_i^+,\epsilon_{min}\rightarrow\epsilon_{max}} \exp\left[\sum_{j=0}^{45} C_{N_i^+}(j)\,(\epsilon - \overline{\epsilon}_{\epsilon_{min}\rightarrow\epsilon_{max}})^j\right], \quad (A.2)$$

where the different coefficients for the HD^+ ions are tabulated in the Tables A1, A2, and A3. The term $\overline{\overline{\sigma}}^{DR}_{(AB^+),N_i^+,\epsilon_{min}\rightarrow\epsilon_{max}}$ is the cross section averaged over the whole energy range $[\epsilon_{min}, \epsilon_{max}]$. And finally, if Maxwell distribution can be assumed, the thermal rate coefficients for the same system displayed in Fig. 5 can be produced by the interpolation formula:

$$k^{DR}_{(HD^+),N_i^+}(T_e) = A_{N_i^+}\, T_e^{\alpha_{N_i^+}} \exp\left[-\sum_{j=1}^{7} \frac{B_{N_i^+}(j)}{j\, T_e^j}\right], \quad (A.3)$$

over the electron temperature range 10 K $\leq T_e \leq$ 5000 K. The parameters $A_{N_i^+}$, $\alpha_{N_i^+}$ and $B_{N_i^+}(j)$ are listed in Table A4.

The calculated rate coefficients of the *Dissociative Recombination of BeH*$^+$ in each of its first four vibrational states have been interpolated under the mathematical form, similar to eq. (A.3)

$$k^{DR}_{(BeH^+),v_i^+}(T_e) = A_{v_i^+}\, T_e^{\alpha_{v_i^+}} \exp\left[-\sum_{j=1}^{7} \frac{B_{v_i^+}(j)}{j\, T_e^j}\right], \quad (A.4)$$

over the electron temperature range 100 K $\leq T_e \leq$ 5000 K. The parameters $A_{v_i^+}$, $\alpha_{v_i^+}$ and $B_{v_i^+}(j)$ are listed in Table A5.

Using again the (A.3) general formula, the calculated rate coefficients of the *Vibrational Excitation of BeH*$^+$ have been interpolated under the form

$$k^{VE}_{(BeH^+),v_i^+\rightarrow v_f^+}(T_e) = A_{v_i^+\rightarrow v_f^+}\, T_e^{\alpha_{v_i^+\rightarrow v_f^+}} \exp\left[-\sum_{j=1}^{7} \frac{B_{v_i^+\rightarrow v_f^+}(j)}{j\, T_e^j}\right] \quad (A.5)$$

over the electron temperature range 300 K $\leq T_e \leq$ 5000 K. The parameters $A_{v_i^+\rightarrow v_f^+}$, $\alpha_{v_i^+\rightarrow v_f^+}$ and $B_{v_i^+\rightarrow v_f^+}(j)$ are listed in Tables A6 and A7.

The values interpolated using the equations (A.3)–(A.5) reproduce the reference values within a few percent, for the listed different rate coefficients.

Table A1. List of the parameters used in Eq. (A.2) for the DR cross sections of HD^+ for ($N_i^+ = 0, 1, 2, 3$ and 4), Figures 3, 4 and 6.

N_i^+	$\overline{\overline{\sigma}}^{DR}_{(HD^+),N_i^+,\epsilon_{min}\to\epsilon_{max}}$	$\overline{\epsilon}_{\epsilon_{min}\to\epsilon_{max}}$	$C_{N_i^+}(0)$	$C_{N_i^+}(1)$	$C_{N_i^+}(2)$	$C_{N_i^+}(3)$	$C_{N_i^+}(4)$
	$C_{N_i^+}(5)$	$C_{N_i^+}(6)$	$C_{N_i^+}(7)$	$C_{N_i^+}(8)$	$C_{N_i^+}(9)$	$C_{N_i^+}(10)$	$C_{N_i^+}(11)$
	$C_{N_i^+}(12)$	$C_{N_i^+}(13)$	$C_{N_i^+}(14)$	$C_{N_i^+}(15)$	$C_{N_i^+}(16)$	$C_{N_i^+}(17)$	$C_{N_i^+}(18)$
	$C_{N_i^+}(19)$	$C_{N_i^+}(20)$	$C_{N_i^+}(21)$	$C_{N_i^+}(22)$	$C_{N_i^+}(23)$	$C_{N_i^+}(24)$	$C_{N_i^+}(25)$
	$C_{N_i^+}(26)$	$C_{N_i^+}(27)$	$C_{N_i^+}(28)$	$C_{N_i^+}(29)$	$C_{N_i^+}(30)$	$C_{N_i^+}(31)$	$C_{N_i^+}(32)$
	$C_{N_i^+}(33)$	$C_{N_i^+}(34)$	$C_{N_i^+}(35)$	$C_{N_i^+}(36)$	$C_{N_i^+}(37)$	$C_{N_i^+}(38)$	$C_{N_i^+}(39)$
	$C_{N_i^+}(40)$	$C_{N_i^+}(41)$	$C_{N_i^+}(42)$	$C_{N_i^+}(43)$	$C_{N_i^+}(44)$	$C_{N_i^+}(45)$	
0	$0.134679643\times10^{-14}$	$0.175000000\times10^{+00}$	$-0.221709000\times10^{+01}$	$0.228019286\times10^{+01}$	$0.184529240\times10^{+03}$	$0.120133841\times10^{+06}$	$-0.522441032\times10^{+07}$
	$-0.313225196\times10^{+09}$	$0.138002871\times10^{+11}$	$0.301684129\times10^{+12}$	$-0.170406589\times10^{+14}$	$-0.129836651\times10^{+15}$	$0.121543196\times10^{+17}$	$0.149275681\times10^{+17}$
	$-0.564779157\times10^{+19}$	$0.106424954\times10^{+20}$	$0.184050570\times10^{+22}$	$-0.625019627\times10^{+22}$	$-0.440797417\times10^{+24}$	$0.178582021\times10^{+25}$	$0.800197268\times10^{+26}$
	$-0.338563500\times10^{+27}$	$-0.112433893\times10^{+29}$	$0.466133549\times10^{+29}$	$0.124023263\times10^{+31}$	$-0.485877114\times10^{+31}$	$-0.108393579\times10^{+33}$	$0.392130182\times10^{+33}$
	$0.754366511\times10^{+34}$	$-0.248011593\times10^{+35}$	$-0.418549594\times10^{+36}$	$0.123553344\times10^{+37}$	$0.184613621\times10^{+38}$	$-0.484458504\times10^{+38}$	$-0.642522545\times10^{+39}$
	$0.148541835\times10^{+40}$	$0.174100330\times10^{+41}$	$-0.351454610\times10^{+41}$	$-0.359348214\times10^{+42}$	$0.627478460\times10^{+42}$	$0.545433115\times10^{+43}$	$-0.815098531\times10^{+43}$
	$-0.573607590\times10^{+44}$	$0.724363987\times10^{+44}$	$0.373182615\times10^{+45}$	$-0.391977302\times10^{+45}$	$-0.113100881\times10^{+46}$	$0.967889038\times10^{+45}$	
1	$0.451929974\times10^{-15}$	$0.175000000\times10^{+00}$	$-0.128223279\times10^{+01}$	$-0.345535940\times10^{+02}$	$0.109941706\times10^{+04}$	$0.248747875\times10^{+06}$	$-0.913644908\times10^{+07}$
	$-0.587889214\times10^{+09}$	$0.207795716\times10^{+11}$	$0.723226565\times10^{+12}$	$-0.245078574\times10^{+14}$	$-0.558110456\times10^{+15}$	$0.176810636\times10^{+17}$	$0.293528425\times10^{+18}$
	$-0.850893920\times10^{+19}$	$-0.110117662\times10^{+21}$	$0.288655286\times10^{+22}$	$0.304319390\times10^{+23}$	$-0.717559086\times10^{+24}$	$-0.635524843\times10^{+25}$	$0.134471467\times10^{+27}$
	$0.102340175\times10^{+28}$	$-0.193981533\times10^{+29}$	$-0.129064066\times10^{+30}$	$0.218660065\times10^{+31}$	$0.128914445\times10^{+32}$	$-0.194568347\times10^{+33}$	$-0.102743857\times10^{+34}$
	$0.137476406\times10^{+35}$	$0.655882427\times10^{+35}$	$-0.772750230\times10^{+36}$	$-0.335443084\times10^{+37}$	$0.344738150\times10^{+38}$	$0.136954526\times10^{+39}$	$-0.121196574\times10^{+40}$
	$-0.442775265\times10^{+40}$	$0.331383973\times10^{+41}$	$0.111786897\times10^{+42}$	$-0.689619466\times10^{+42}$	$-0.215539213\times10^{+43}$	$0.105459732\times10^{+44}$	$0.306292226\times10^{+44}$
	$-0.111671536\times10^{+45}$	$-0.302152091\times10^{+45}$	$0.731131343\times10^{+45}$	$0.184702423\times10^{+46}$	$-0.222882825\times10^{+46}$	$-0.526736007\times10^{+46}$	
2	$0.390733561\times10^{-15}$	$0.175000000\times10^{+00}$	$-0.117255827\times10^{+01}$	$0.139244433\times10^{+02}$	$-0.109082997\times10^{+04}$	$-0.125197646\times10^{+06}$	$0.936348646\times10^{+07}$
	$0.202827114\times10^{+09}$	$-0.211746156\times10^{+11}$	$-0.992865056\times10^{+11}$	$0.229937522\times10^{+14}$	$-0.508577285\times10^{+14}$	$-0.148886262\times10^{+17}$	$0.888761434\times10^{+17}$
	$0.636497760\times10^{+19}$	$-0.526795287\times10^{+20}$	$-0.191205421\times10^{+22}$	$0.187310981\times10^{+23}$	$0.421234322\times10^{+24}$	$-0.456883426\times10^{+25}$	$-0.701597913\times10^{+26}$
	$0.813456160\times10^{+27}$	$0.902933640\times10^{+28}$	$-0.109582683\times10^{+30}$	$-0.911721060\times10^{+30}$	$0.114246183\times10^{+32}$	$0.729606926\times10^{+32}$	$-0.935076831\times10^{+33}$
	$-0.465367762\times10^{+34}$	$0.605848168\times10^{+35}$	$0.236952750\times10^{+36}$	$-0.311768851\times10^{+37}$	$-0.960602768\times10^{+37}$	$0.127247186\times10^{+39}$	$0.307777491\times10^{+39}$
	$-0.409256214\times10^{+40}$	$-0.768997947\times10^{+40}$	$0.102412373\times10^{+42}$	$0.146592990\times10^{+42}$	$-0.195186952\times10^{+43}$	$-0.205817605\times10^{+43}$	$0.273621538\times10^{+44}$
	$0.200512737\times10^{+44}$	$-0.265890154\times10^{+45}$	$-0.121017296\times10^{+45}$	$0.159947728\times10^{+46}$	$0.340701689\times10^{+45}$	$-0.448588080\times10^{+46}$	
3	$0.106613017\times10^{-14}$	$0.175000000\times10^{+00}$	$-0.218400607\times10^{+01}$	$-0.107962076\times10^{+02}$	$-0.180773454\times10^{+04}$	$0.310649105\times10^{+05}$	$0.756920375\times10^{+07}$
	$0.313620284\times10^{+08}$	$-0.970717727\times10^{+10}$	$-0.168500835\times10^{+12}$	$0.539567182\times10^{+13}$	$0.210666871\times10^{+15}$	$-0.118181004\times10^{+16}$	$-0.141615798\times10^{+18}$
	$-0.211118626\times10^{+18}$	$0.611552212\times10^{+20}$	$0.228643132\times10^{+21}$	$-0.184292442\times10^{+23}$	$-0.800974032\times10^{+23}$	$0.407141652\times10^{+25}$	$0.176157105\times10^{+26}$
	$-0.681247917\times10^{+27}$	$-0.276437617\times10^{+28}$	$0.882945587\times10^{+29}$	$0.326189586\times10^{+30}$	$-0.900190680\times10^{+31}$	$-0.297342900\times10^{+32}$	$0.729269164\times10^{+33}$
	$0.212479293\times10^{+34}$	$-0.472097840\times10^{+35}$	$-0.119839700\times10^{+36}$	$0.244576770\times10^{+37}$	$0.533813443\times10^{+37}$	$-0.101125245\times10^{+39}$	$-0.186813305\times10^{+39}$
	$0.331219098\times10^{+40}$	$0.507539867\times10^{+40}$	$-0.847841721\times10^{+41}$	$-0.104837903\times10^{+42}$	$0.165921609\times10^{+43}$	$0.159053417\times10^{+43}$	$-0.239605521\times10^{+44}$
	$-0.167060855\times10^{+44}$	$0.240512338\times10^{+45}$	$0.108500162\times10^{+45}$	$-0.149799931\times10^{+46}$	$-0.328173880\times10^{+45}$	$0.435839991\times10^{+46}$	
4	$0.302446453\times10^{-15}$	$0.175000000\times10^{+00}$	$-0.907756878\times10^{+00}$	$0.991178465\times10^{+01}$	$-0.931432336\times10^{+02}$	$-0.116405632\times10^{+06}$	$-0.572763265\times10^{+07}$
	$0.117790279\times10^{+09}$	$0.173143394\times10^{+11}$	$-0.214730421\times10^{+11}$	$-0.214940887\times10^{+14}$	$-0.731348365\times10^{+13}$	$0.152530416\times10^{+17}$	$-0.991815567\times10^{+16}$
	$-0.707634269\times10^{+19}$	$0.139403154\times10^{+20}$	$0.230736440\times10^{+22}$	$-0.722060838\times10^{+22}$	$-0.553017582\times10^{+24}$	$0.223598195\times10^{+25}$	$0.100386845\times10^{+27}$
	$-0.475181523\times10^{+27}$	$-0.140905948\times10^{+29}$	$0.739069466\times10^{+29}$	$0.155137924\times10^{+31}$	$-0.872057596\times10^{+31}$	$-0.135246642\times10^{+33}$	$0.797501942\times10^{+33}$
	$0.938478914\times10^{+34}$	$-0.572306469\times10^{+35}$	$-0.519018352\times10^{+36}$	$0.324164931\times10^{+37}$	$0.228144510\times10^{+38}$	$-0.144954972\times10^{+39}$	$-0.791204616\times10^{+39}$
	$0.508947718\times10^{+40}$	$0.213606716\times10^{+41}$	$-0.138635373\times10^{+42}$	$-0.439256227\times10^{+42}$	$0.286932603\times10^{+43}$	$0.664220468\times10^{+43}$	$-0.435912637\times10^{+44}$
	$-0.695893953\times10^{+44}$	$0.458236220\times10^{+45}$	$0.451028307\times10^{+45}$	$-0.297712140\times10^{+46}$	$-0.136178060\times10^{+46}$	$0.900420725\times10^{+46}$	

Journal of Physics: Conference Series **576** (2015) 012005 doi:10.1088/1742-6596/576/1/012005

Table A2. Same as Table A1 for ($N_i^+ = 5, 6, 7$ and 8), Figures 3, 4 and 6.

N_i^+	$\overline{\overline{\sigma}}^{DR}_{(HD^+),N_i^+,\epsilon_{min}\to\epsilon_{max}}$	$\overline{\epsilon}_{\epsilon_{min}\to\epsilon_{max}}$	$C_{N_i^+}(0)$	$C_{N_i^+}(1)$	$C_{N_i^+}(2)$	$C_{N_i^+}(3)$	$C_{N_i^+}(4)$
	$C_{N_i^+}(5)$	$C_{N_i^+}(6)$	$C_{N_i^+}(7)$	$C_{N_i^+}(8)$	$C_{N_i^+}(9)$	$C_{N_i^+}(10)$	$C_{N_i^+}(11)$
	$C_{N_i^+}(12)$	$C_{N_i^+}(13)$	$C_{N_i^+}(14)$	$C_{N_i^+}(15)$	$C_{N_i^+}(16)$	$C_{N_i^+}(17)$	$C_{N_i^+}(18)$
	$C_{N_i^+}(19)$	$C_{N_i^+}(20)$	$C_{N_i^+}(21)$	$C_{N_i^+}(22)$	$C_{N_i^+}(23)$	$C_{N_i^+}(24)$	$C_{N_i^+}(25)$
	$C_{N_i^+}(26)$	$C_{N_i^+}(27)$	$C_{N_i^+}(28)$	$C_{N_i^+}(29)$	$C_{N_i^+}(30)$	$C_{N_i^+}(31)$	$C_{N_i^+}(32)$
	$C_{N_i^+}(33)$	$C_{N_i^+}(34)$	$C_{N_i^+}(35)$	$C_{N_i^+}(36)$	$C_{N_i^+}(37)$	$C_{N_i^+}(38)$	$C_{N_i^+}(39)$
	$C_{N_i^+}(40)$	$C_{N_i^+}(41)$	$C_{N_i^+}(42)$	$C_{N_i^+}(43)$	$C_{N_i^+}(44)$	$C_{N_i^+}(45)$	
5	$0.512662068\times10^{-15}$	$0.175000000\times10^{+00}$	$-0.199653797\times10^{+01}$	$-0.202471466\times10^{+02}$	$0.378123542\times10^{+04}$	$-0.162489080\times10^{+06}$	$-0.630587520\times10^{+07}$
	$0.543204153\times10^{+09}$	$0.632286305\times10^{+10}$	$-0.702401308\times10^{+12}$	$-0.510358393\times10^{+13}$	$0.518444989\times10^{+15}$	$0.335728772\times10^{+16}$	$-0.252715786\times10^{+18}$
	$-0.167848399\times10^{+19}$	$0.877333745\times10^{+20}$	$0.615386479\times10^{+21}$	$-0.226872417\times10^{+23}$	$-0.165708202\times10^{+24}$	$0.449803794\times10^{+25}$	$0.333081204\times10^{+26}$
	$-0.697239292\times10^{+27}$	$-0.508954001\times10^{+28}$	$0.856504963\times10^{+29}$	$0.600468985\times10^{-30}$	$-0.841467841\times10^{+31}$	$-0.553393912\times10^{+32}$	$0.664853094\times10^{+33}$
	$0.401375394\times10^{+34}$	$-0.423455801\times10^{+35}$	$-0.229859308\times10^{+36}$	$0.217213432\times10^{-37}$	$0.103821666\times10^{+38}$	$-0.893323591\times10^{+38}$	$-0.367614691\times10^{+39}$
	$0.291976475\times10^{+40}$	$0.100796869\times10^{+41}$	$-0.747490916\times10^{+41}$	$-0.209588625\times10^{+42}$	$0.146523710\times10^{+43}$	$0.319291370\times10^{+43}$	$-0.212145477\times10^{+44}$
	$-0.335979682\times10^{+44}$	$0.213624351\times10^{+45}$	$0.218150625\times10^{+45}$	$-0.133512006\times10^{+46}$	$-0.658433644\times10^{+45}$	$0.389817187\times10^{+46}$	
6	$0.349378176\times10^{-14}$	$0.175000000\times10^{+00}$	$-0.370761973\times10^{+01}$	$-0.237057965\times10^{+02}$	$0.629636322\times10^{+04}$	$0.280561506\times10^{+05}$	$-0.159971710\times10^{+08}$
	$0.667549859\times10^{+08}$	$0.170007560\times10^{+11}$	$-0.141940590\times10^{+12}$	$-0.974016822\times10^{+13}$	$0.133983474\times10^{+15}$	$0.329522183\times10^{+16}$	$-0.848463981\times10^{+17}$
	$-0.654746294\times10^{+18}$	$0.387539152\times10^{+20}$	$0.591307990\times10^{+20}$	$-0.129526906\times10^{+23}$	$0.510516945\times10^{+22}$	$0.321304499\times10^{+25}$	$-0.247439096\times10^{+25}$
	$-0.602166062\times10^{+27}$	$0.391717956\times10^{+27}$	$0.867415743\times10^{+29}$	$-0.357759632\times10^{+29}$	$-0.974061611\times10^{+31}$	$0.178926095\times10^{+31}$	$0.861495774\times10^{+33}$
	$-0.151609673\times10^{+31}$	$-0.603886040\times10^{+35}$	$-0.772373780\times10^{+34}$	$0.336259295\times10^{-37}$	$0.707821138\times10^{+36}$	$-0.148441067\times10^{+39}$	$-0.377068657\times10^{+38}$
	$0.515983591\times10^{+40}$	$0.137453616\times10^{+40}$	$-0.139416059\times10^{+42}$	$-0.354913796\times10^{+41}$	$0.286591556\times10^{+43}$	$0.642965142\times10^{+42}$	$-0.432827203\times10^{+44}$
	$-0.780883723\times10^{+43}$	$0.452581561\times10^{+45}$	$0.572573526\times10^{+44}$	$-0.292596471\times10^{+46}$	$-0.191947396\times10^{+45}$	$0.880834578\times10^{+46}$	
7	$0.103287236\times10^{-14}$	$0.175000000\times10^{+00}$	$-0.203397598\times10^{+01}$	$0.410997375\times10^{-02}$	$0.172771158\times10^{+04}$	$-0.320378359\times10^{+06}$	$-0.752435588\times10^{+07}$
	$0.745837307\times10^{+09}$	$0.142330593\times10^{+11}$	$-0.911510637\times10^{+12}$	$-0.145226836\times10^{+14}$	$0.695136998\times10^{+15}$	$0.919248409\times10^{+16}$	$-0.361357011\times10^{+18}$
	$-0.394048124\times10^{+19}$	$0.134597133\times10^{+21}$	$0.120812060\times10^{+22}$	$-0.371340800\times10^{+23}$	$-0.274674304\times10^{+24}$	$0.777769548\times10^{+25}$	$0.475026079\times10^{+26}$
	$-0.126042883\times10^{+28}$	$-0.636496788\times10^{+28}$	$0.160344440\times10^{+30}$	$0.669518037\times10^{-30}$	$-0.161816426\times10^{+32}$	$-0.557754039\times10^{+32}$	$0.130446089\times10^{+34}$
	$0.369810642\times10^{+34}$	$-0.842954530\times10^{+35}$	$-0.195379336\times10^{+36}$	$0.436687134\times10^{-37}$	$0.820193105\times10^{+37}$	$-0.180687672\times10^{+39}$	$-0.271554199\times10^{+39}$
	$0.592300081\times10^{+40}$	$0.699658422\times10^{+40}$	$-0.151688638\times10^{+42}$	$-0.137255345\times10^{+42}$	$0.296816980\times10^{+43}$	$0.197923800\times10^{+43}$	$-0.428251074\times10^{+44}$
	$-0.197668103\times10^{+44}$	$0.429133820\times10^{+45}$	$0.122078259\times10^{+45}$	$-0.266596383\times10^{+46}$	$-0.351088034\times10^{+45}$	$0.773043438\times10^{+46}$	
8	$0.972568122\times10^{-15}$	$0.175000000\times10^{+00}$	$-0.148034836\times10^{+01}$	$-0.175077264\times10^{+02}$	$-0.419379741\times10^{+04}$	$0.380354905\times10^{+06}$	$0.116578039\times10^{+08}$
	$-0.104906916\times10^{+10}$	$-0.133308355\times10^{+11}$	$0.126649950\times10^{+13}$	$0.749946154\times10^{-13}$	$-0.906534986\times10^{+15}$	$-0.199895897\times10^{+16}$	$0.434655371\times10^{+18}$
	$0.461733212\times10^{+16}$	$-0.149064633\times10^{+21}$	$0.186620270\times10^{+21}$	$0.380883398\times10^{-23}$	$-0.728709748\times10^{+23}$	$-0.745254590\times10^{+25}$	$0.163163232\times10^{+26}$
	$0.113819925\times10^{+28}$	$-0.253437852\times10^{+28}$	$-0.137519986\times10^{+30}$	$0.291950725\times10^{-30}$	$0.132654249\times10^{+32}$	$-0.257639588\times10^{+32}$	$-0.102736825\times10^{+34}$
	$0.177200866\times10^{+34}$	$0.640370708\times10^{+35}$	$-0.957768865\times10^{+35}$	$-0.320991801\times10^{+37}$	$0.407454184\times10^{+37}$	$0.128834554\times10^{+39}$	$-0.135805987\times10^{+39}$
	$-0.410491482\times10^{+40}$	$0.350585469\times10^{+40}$	$0.102353518\times10^{+42}$	$-0.686744595\times10^{+41}$	$-0.195275042\times10^{+43}$	$0.986347869\times10^{+42}$	$0.275047301\times10^{+44}$
	$-0.979312118\times10^{+43}$	$-0.269362991\times10^{+45}$	$0.600424663\times10^{+44}$	$0.163711165\times10^{-46}$	$-0.171235604\times10^{+45}$	$-0.464854259\times10^{+46}$	

Table A3. Same as Table A1 for ($N_i^+ = 9, 10, 11$ and 12), Figures 3, 4 and 6.

N_i^+	$\overline{\overline{\sigma}}^{DR}_{(HD^+),N_i^+,\epsilon_{min}\to\epsilon_{max}}$ $C_{N_i^+}(5)$ $C_{N_i^+}(12)$ $C_{N_i^+}(19)$ $C_{N_i^+}(26)$ $C_{N_i^+}(33)$ $C_{N_i^+}(40)$	$\overline{\epsilon}_{\epsilon_{min}\to\epsilon_{max}}$ $C_{N_i^+}(6)$ $C_{N_i^+}(13)$ $C_{N_i^+}(20)$ $C_{N_i^+}(27)$ $C_{N_i^+}(34)$ $C_{N_i^+}(41)$	$C_{N_i^+}(0)$ $C_{N_i^+}(7)$ $C_{N_i^+}(14)$ $C_{N_i^+}(21)$ $C_{N_i^+}(28)$ $C_{N_i^+}(35)$ $C_{N_i^+}(42)$	$C_{N_i^+}(1)$ $C_{N_i^+}(8)$ $C_{N_i^+}(15)$ $C_{N_i^+}(22)$ $C_{N_i^+}(29)$ $C_{N_i^+}(36)$ $C_{N_i^+}(43)$	$C_{N_i^+}(2)$ $C_{N_i^+}(9)$ $C_{N_i^+}(16)$ $C_{N_i^+}(23)$ $C_{N_i^+}(30)$ $C_{N_i^+}(37)$ $C_{N_i^+}(44)$	$C_{N_i^+}(3)$ $C_{N_i^+}(10)$ $C_{N_i^+}(17)$ $C_{N_i^+}(24)$ $C_{N_i^+}(31)$ $C_{N_i^+}(38)$ $C_{N_i^+}(45)$	$C_{N_i^+}(4)$ $C_{N_i^+}(11)$ $C_{N_i^+}(18)$ $C_{N_i^+}(25)$ $C_{N_i^+}(32)$ $C_{N_i^+}(39)$
9	$0.179206017\times10^{-14}$	$0.175000000\times10^{+00}$	$-0.214739614\times10^{+01}$	$-0.449051952\times10^{+02}$	$-0.480640739\times10^{+04}$	$0.243478175\times10^{+06}$	$0.920525991\times10^{+07}$
	$-0.582813972\times10^{+09}$	$-0.625746100\times10^{+10}$	$0.665542515\times10^{+12}$	$0.444644664\times10^{-12}$	$-0.458871942\times10^{+15}$	$0.191054492\times10^{+16}$	$0.213524580\times10^{+18}$
	$-0.143219316\times10^{+19}$	$-0.711649375\times10^{+20}$	$0.576708036\times10^{+21}$	$0.176025064\times10^{-23}$	$-0.156747627\times10^{+24}$	$-0.330822477\times10^{+25}$	$0.310957031\times10^{+26}$
	$0.480314704\times10^{+27}$	$-0.467944591\times10^{+28}$	$-0.545049334\times10^{+29}$	$0.546055598\times10^{-30}$	$0.487093648\times10^{+31}$	$-0.500608314\times10^{+32}$	$-0.344073409\times10^{+33}$
	$0.363142890\times10^{+34}$	$0.192026465\times10^{+35}$	$-0.208943000\times10^{+36}$	$-0.842284150\times10^{+36}$	$0.951636115\times10^{+37}$	$0.287014021\times10^{+38}$	$-0.340733649\times10^{+39}$
	$-0.743867191\times10^{+39}$	$0.946744511\times10^{+40}$	$0.141127395\times10^{+41}$	$-0.199802169\times10^{+42}$	$-0.181455109\times10^{+42}$	$0.309283577\times10^{+43}$	$0.127770987\times10^{+43}$
	$-0.330946116\times10^{+44}$	$0.381890063\times10^{+42}$	$0.218620772\times10^{+45}$	$-0.828279405\times10^{+44}$	$-0.671517989\times10^{+45}$	$0.469401069\times10^{+45}$	
10	$0.630755327\times10^{-15}$	$0.175000000\times10^{+00}$	$-0.187982455\times10^{+01}$	$-0.221048093\times10^{+02}$	$0.176084787\times10^{+03}$	$0.806373023\times10^{+05}$	$-0.350230523\times10^{+07}$
	$-0.146951465\times10^{+09}$	$0.965442010\times10^{+10}$	$0.128373652\times10^{+12}$	$-0.123592368\times10^{+14}$	$-0.581380654\times10^{+14}$	$0.921805753\times10^{+16}$	$0.832746187\times10^{+16}$
	$-0.447618372\times10^{+19}$	$0.539464332\times10^{+19}$	$0.151680399\times10^{+22}$	$-0.388397142\times10^{+22}$	$-0.375624037\times10^{+24}$	$0.132066108\times10^{+25}$	$0.701649186\times10^{+26}$
	$-0.296621386\times10^{+27}$	$-0.101057635\times10^{+29}$	$0.482580311\times10^{+29}$	$0.113935361\times10^{-31}$	$-0.593097117\times10^{+31}$	$-0.101546757\times10^{+33}$	$0.563362485\times10^{+33}$
	$0.719397630\times10^{+34}$	$-0.418863391\times10^{+35}$	$-0.405692963\times10^{+36}$	$0.245206250\times10^{-37}$	$0.181633133\times10^{+38}$	$-0.113049103\times10^{+39}$	$-0.640866119\times10^{+39}$
	$0.408271247\times10^{+40}$	$0.175844471\times10^{+41}$	$-0.114131616\times10^{+42}$	$-0.367135735\times10^{+42}$	$0.241899745\times10^{+43}$	$0.563110640\times10^{+43}$	$-0.375582275\times10^{+44}$
	$-0.597852790\times10^{+44}$	$0.402748328\times10^{+45}$	$0.392321087\times10^{+45}$	$-0.266459029\times10^{+46}$	$-0.119830971\times10^{+46}$	$0.819370807\times10^{+46}$	
11	$0.262100159\times10^{-14}$	$0.175000000\times10^{+00}$	$-0.354114889\times10^{+01}$	$-0.540347374\times10^{+01}$	$0.273609684\times10^{+04}$	$0.111280085\times10^{+05}$	$-0.103854490\times10^{+08}$
	$0.263039015\times10^{+08}$	$0.184973205\times10^{+11}$	$-0.148093959\times10^{+12}$	$-0.188270312\times10^{+14}$	$0.201473100\times10^{+15}$	$0.122333189\times10^{+17}$	$-0.143983367\times10^{+18}$
	$-0.544634499\times10^{+19}$	$0.656399628\times10^{+20}$	$0.174391919\times10^{+22}$	$-0.208995558\times10^{+23}$	$-0.415641920\times10^{+24}$	$0.489335836\times10^{+25}$	$0.755797988\times10^{+26}$
	$-0.870169682\times10^{+27}$	$-0.106733896\times10^{+29}$	$0.120047449\times10^{+30}$	$0.118533982\times10^{-31}$	$-0.130309521\times10^{+32}$	$-0.104376064\times10^{+33}$	$0.112291919\times10^{+34}$
	$0.731994060\times10^{+34}$	$-0.771819625\times10^{+35}$	$-0.409173485\times10^{+36}$	$0.423493954\times10^{-37}$	$0.181744087\times10^{+38}$	$-0.184919883\times10^{+39}$	$-0.636578638\times10^{+39}$
	$0.637636031\times10^{+40}$	$0.173467995\times10^{+41}$	$-0.171279593\times10^{+42}$	$-0.359797756\times10^{+42}$	$0.350615167\times10^{+43}$	$0.548363257\times10^{+43}$	$-0.527959981\times10^{+44}$
	$-0.578617138\times10^{+44}$	$0.550954225\times10^{+45}$	$0.377421152\times10^{+45}$	$-0.355740519\times10^{+46}$	$-0.114603251\times10^{+46}$	$0.107014325\times10^{+47}$	
12	$0.146229000\times10^{-14}$	$0.175000000\times10^{+00}$	$-0.311835327\times10^{+01}$	$0.360867102\times10^{-02}$	$0.109234001\times10^{+04}$	$-0.877127984\times10^{+05}$	$-0.164093501\times10^{+07}$
	$0.816664127\times10^{+08}$	$0.192656746\times10^{+10}$	$-0.552371327\times10^{+11}$	$-0.903681573\times10^{+12}$	$0.328594671\times10^{+14}$	$-0.116516074\times10^{+15}$	$-0.142448980\times10^{+17}$
	$0.329660558\times10^{+18}$	$0.382048479\times10^{+19}$	$-0.178547830\times10^{+21}$	$-0.506169933\times10^{+21}$	$0.562447596\times10^{+23}$	$-0.263017990\times10^{+23}$	$-0.121374965\times10^{+26}$
	$0.267794376\times10^{+26}$	$0.191742186\times10^{+28}$	$-0.625550068\times10^{+28}$	$-0.229730465\times10^{+30}$	$0.908278764\times10^{+30}$	$0.213140110\times10^{+32}$	$-0.943363328\times10^{+32}$
	$-0.154983661\times10^{+34}$	$0.737303116\times10^{+34}$	$0.888275327\times10^{+35}$	$-0.443524961\times10^{+36}$	$-0.401364105\times10^{+37}$	$0.207185942\times10^{+38}$	$0.142212838\times10^{+39}$
	$-0.751303888\times10^{+39}$	$-0.390475264\times10^{+40}$	$0.209634639\times10^{+41}$	$0.813791197\times10^{-41}$	$-0.441755613\times10^{+42}$	$-0.124386625\times10^{+43}$	$0.680206207\times10^{+43}$
	$0.131461084\times10^{+44}$	$-0.722231562\times10^{+44}$	$-0.858202875\times10^{+44}$	$0.472712805\times10^{-45}$	$0.260693193\times10^{+45}$	$-0.143748741\times10^{+46}$	

Journal of Physics: Conference Series **576** (2015) 012005 doi:10.1088/1742-6596/576/1/012005

Table A4. List of the parameters used in Eq.(A.3) for the DR rate coefficients of HD^+, Fig. 5.

N_i^+	$A_{N_i^+}$	$\alpha_{N_i^+}$	$B_{N_i^+}(1)$	$B_{N_i^+}(2)$	$B_{N_i^+}(3)$	$B_{N_i^+}(4)$	$B_{N_i^+}(5)$	$B_{N_i^+}(6)$	$B_{N_i^+}(7)$
0	0.222071374×10^{-5}	-0.932864010×10^{0}	0.739191072×10^{2}	-0.154789113×10^{5}	0.103315553×10^{7}	-0.316311988×10^{8}	0.493047262×10^{9}	$-0.380248534\times10^{10}$	0.115087036×10^{11}
1	0.128851757×10^{-4}	-0.110718597×10^{1}	0.171547468×10^{3}	-0.133254014×10^{5}	0.636973066×10^{6}	-0.162724883×10^{8}	0.225760856×10^{9}	$-0.161473237\times10^{10}$	0.465576516×10^{10}
2	0.114310768×10^{-4}	-0.109891626×10^{1}	0.228874916×10^{3}	-0.626220862×10^{4}	-0.292604786×10^{6}	0.162301810×10^{8}	-0.292959848×10^{9}	0.235215411×10^{10}	$-0.713445323\times10^{10}$
3	0.210694935×10^{-4}	-0.115741172×10^{1}	0.387330067×10^{3}	-0.458061947×10^{5}	0.215862337×10^{7}	-0.524142418×10^{8}	0.692418139×10^{9}	$-0.472234646\times10^{10}$	0.130101406×10^{11}
4	0.104194981×10^{-4}	-0.107653008×10^{1}	0.259272046×10^{3}	-0.225402550×10^{5}	0.113744859×10^{7}	-0.302083722×10^{8}	0.434933765×10^{9}	$-0.321416426\times10^{10}$	0.951009824×10^{10}
5	0.134920614×10^{-4}	-0.109088522×10^{1}	0.147709496×10^{3}	-0.107365043×10^{5}	0.588051475×10^{6}	-0.175373466×10^{8}	0.276671733×10^{9}	$-0.217615510\times10^{10}$	0.671291314×10^{10}
6	0.874218683×10^{-5}	-0.104397770×10^{1}	0.847564491×10^{2}	-0.147034346×10^{5}	0.931407905×10^{6}	-0.280229117×10^{8}	0.434010714×10^{9}	$-0.333951526\times10^{10}$	0.101015040×10^{11}
7	0.130857676×10^{-4}	-0.107990536×10^{1}	0.164277599×10^{3}	-0.169491618×10^{5}	0.867333556×10^{6}	-0.237117821×10^{8}	0.350008362×10^{9}	$-0.262310582\times10^{10}$	0.781247654×10^{10}
8	0.206886625×10^{-4}	-0.111130085×10^{1}	0.161516618×10^{3}	-0.169829856×10^{5}	0.933477114×10^{6}	-0.266848249×10^{8}	0.404489322×10^{9}	$-0.308099132\times10^{10}$	0.927053336×10^{10}
9	0.323544041×10^{-4}	-0.114880459×10^{1}	0.190230296×10^{3}	-0.277588793×10^{5}	0.167583946×10^{7}	-0.488514291×10^{8}	0.739258454×10^{9}	$-0.559128198\times10^{10}$	0.166941689×10^{11}
10	0.474515481×10^{-4}	-0.120225797×10^{1}	0.320379237×10^{3}	-0.242842906×10^{5}	0.891353217×10^{6}	-0.189565820×10^{8}	0.237461583×10^{9}	$-0.161098316\times10^{10}$	0.452761220×10^{10}
11	0.439397909×10^{-4}	-0.119221433×10^{1}	0.177276579×10^{3}	-0.239180085×10^{5}	0.135806621×10^{7}	-0.381412189×10^{8}	0.563259564×10^{9}	$-0.418763908\times10^{10}$	0.123461437×10^{11}
12	0.246800711×10^{-4}	-0.113548321×10^{1}	0.210755163×10^{3}	-0.267666526×10^{5}	0.148126308×10^{7}	-0.412669275×10^{8}	0.608552008×10^{9}	$-0.452848138\times10^{10}$	0.133722848×10^{11}

Journal of Physics: Conference Series **576** (2015) 012005 doi:10.1088/1742-6596/576/1/012005

Table A5. Parameters used in Eq. (A.4) for the DR rate coefficients of BeH^+ displayed in Fig. 8.

v_i^+	$A_{v_i^+}$	$\alpha_{v_i^+}$	$B_{v_i^+}(1)$	$B_{v_i^+}(2)$	$B_{v_i^+}(3)$	$B_{v_i^+}(4)$	$B_{v_i^+}(5)$	$B_{v_i^+}(6)$	$B_{v_i^+}(7)$
0	0.253513120×10^{-9}	0.592469479	-0.190649410×10^{4}	0.123315004×10^{7}	-0.388342503×10^{9}	0.647265585×10^{11}	$-0.566722264\times10^{13}$	0.232068987×10^{15}	$-0.295078806\times10^{16}$
1	0.314462330×10^{-5}	-0.407538250	0.958577714×10^{3}	-0.135957934×10^{7}	0.725436198×10^{9}	$-0.192640935\times10^{12}$	0.271812541×10^{14}	$-0.194765942\times10^{16}$	0.557221814×10^{17}
2	0.114678715×10^{-5}	-0.322376187	-0.265630394×10^{3}	0.219601686×10^{6}	-0.916549130×10^{8}	0.207255861×10^{11}	$-0.259613563\times10^{13}$	0.169396874×10^{15}	$-0.449074035\times10^{16}$
3	0.849641548×10^{-6}	-0.275140774	0.171673351×10^{3}	-0.451956583×10^{6}	0.285096525×10^{9}	$-0.827403486\times10^{11}$	0.123588751×10^{14}	$-0.921680323\times10^{15}$	0.271612346×10^{17}

Journal of Physics: Conference Series **576** (2015) 012005 doi:10.1088/1742-6596/576/1/012005

Table A6. Parameters for the VE and VdE rate coefficients of BeH^+ ($v_i^+ = 0$ and 1), Fig. 8.

v_f^+	$A_{v_i^+\to v_f^+}$	$\alpha_{v_i^+\to v_f^+}$	$B_{v_i^+\to v_f^+}(1)$	$B_{v_i^+\to v_f^+}(2)$	$B_{v_i^+\to v_f^+}(3)$	$B_{v_i^+\to v_f^+}(4)$	$B_{v_i^+\to v_f^+}(5)$	$B_{v_i^+\to v_f^+}(6)$	$B_{v_i^+\to v_f^+}(7)$
1	$0.124244699\times10^{-01}$	$-0.133548647\times10^{+01}$	$0.125068733\times10^{+05}$	$-0.119462867\times10^{+08}$	$0.612265338\times10^{+10}$	$-0.159115695\times10^{+13}$	$0.221261867\times10^{+15}$	$-0.156815975\times10^{+17}$	$0.444766919\times10^{+18}$
2	$0.129961950\times10^{-04}$	$-0.661035442\times10^{+00}$	$0.875512695\times10^{+04}$	$-0.266357941\times10^{+07}$	$0.108710933\times10^{+10}$	$-0.242430358\times10^{+12}$	$0.306483606\times10^{+14}$	$-0.205871435\times10^{+16}$	$0.569537410\times10^{+17}$
3	$0.142379982\times10^{-04}$	$-0.764830977\times10^{+00}$	$0.101101728\times10^{+05}$	$-0.205253723\times10^{+07}$	$0.137208247\times10^{+10}$	$-0.493346150\times10^{+12}$	$0.974091099\times10^{+14}$	$-0.991580495\times10^{+16}$	$0.405560372\times10^{+18}$
4	$0.613107973\times10^{-05}$	$-0.715948096\times10^{+00}$	$0.116459658\times10^{+05}$	$0.180037625\times10^{+07}$	$-0.380655693\times10^{+10}$	$0.289048070\times10^{+13}$	$-0.104565345\times10^{+16}$	$0.181732642\times10^{+18}$	$-0.121943739\times10^{+20}$
5	$0.454423919\times10^{-05}$	$-0.724833034\times10^{+00}$	$0.144167691\times10^{+05}$	$0.187042671\times10^{+07}$	$-0.553846921\times10^{+10}$	$0.525679261\times10^{+13}$	$-0.233886685\times10^{+16}$	$0.498406730\times10^{+18}$	$-0.410074688\times10^{+20}$
6	$0.779183743\times10^{-05}$	$-0.847008699\times10^{+00}$	$0.181829327\times10^{+05}$	$-0.181366766\times10^{+07}$	$-0.132640151\times10^{+10}$	$0.318338532\times10^{+13}$	$-0.206370313\times10^{+16}$	$0.581977475\times10^{+18}$	$-0.613361897\times10^{+20}$
7	$0.180900968\times10^{-05}$	$-0.769516882\times10^{+00}$	$0.174642032\times10^{+05}$	$0.933492644\times10^{+07}$	$-0.211638248\times10^{+11}$	$0.224415719\times10^{+14}$	$-0.123418778\times10^{+17}$	$0.339594431\times10^{+19}$	$-0.369092511\times10^{+21}$
8	$0.885677466\times10^{-06}$	$-0.757779635\times10^{+00}$	$0.179047867\times10^{+05}$	$0.166541738\times10^{+08}$	$-0.378380799\times10^{+11}$	$0.425126855\times10^{+14}$	$-0.252095351\times10^{+17}$	$0.753876184\times10^{+19}$	$-0.894639753\times10^{+21}$
9	$0.292847848\times10^{-05}$	$-0.931863593\times10^{+00}$	$0.202925021\times10^{+05}$	$0.165263943\times10^{+08}$	$-0.417425440\times10^{+11}$	$0.513399746\times10^{+14}$	$-0.332126562\times10^{+17}$	$0.108344862\times10^{+20}$	$-0.140382835\times10^{+22}$
10	$0.375829669\times10^{-17}$	$0.172904099\times10^{+01}$	$-0.949241046\times10^{+04}$	$0.152679710\times10^{+09}$	$-0.333798077\times10^{+12}$	$0.390903195\times10^{+15}$	$-0.250274637\times10^{+18}$	$0.823914165\times10^{+20}$	$-0.108923336\times10^{+23}$
11	$0.261968821\times10^{-13}$	$0.774835837\times10^{+00}$	$0.117881450\times10^{+04}$	$0.118759130\times10^{+09}$	$-0.275480710\times10^{+12}$	$0.343445630\times10^{+15}$	$-0.235068413\times10^{+18}$	$0.830049285\times10^{+20}$	$-0.117994706\times10^{+23}$
12	$0.881455067\times10^{-10}$	$-0.113146035\times10^{+00}$	$0.119234535\times10^{+05}$	$0.792457916\times10^{+08}$	$-0.194518373\times10^{+12}$	$0.257071238\times10^{+15}$	$-0.187454839\times10^{+18}$	$0.708187002\times10^{+20}$	$-0.108036847\times10^{+23}$
13	$0.144309667\times10^{-20}$	$0.225970970\times10^{+01}$	$-0.174823864\times10^{+05}$	$0.228436065\times10^{+09}$	$-0.564780467\times10^{+12}$	$0.765207700\times10^{+15}$	$-0.575653243\times10^{+18}$	$0.224947435\times10^{+21}$	$-0.355386076\times10^{+23}$
14	$0.515928934\times10^{-33}$	$0.500054420\times10^{+01}$	$-0.527361648\times10^{+05}$	$0.412102666\times10^{+09}$	$-0.104124985\times10^{+13}$	$0.144983244\times10^{+16}$	$-0.112378255\times10^{+19}$	$0.453177248\times10^{+21}$	$-0.739667532\times10^{+23}$
15	$0.409733697\times10^{-30}$	$0.424682138\times10^{+01}$	$-0.444964251\times10^{+05}$	$0.384057361\times10^{+09}$	$-0.100210836\times10^{+13}$	$0.144676938\times10^{+16}$	$-0.116586218\times10^{+19}$	$0.489738407\times10^{+21}$	$-0.833887502\times10^{+23}$
16	$0.817690729\times10^{-15}$	$0.703951856\times10^{+00}$	$0.112434814\times10^{+04}$	$0.157806613\times10^{+09}$	$-0.429889131\times10^{+12}$	$0.650940867\times10^{+15}$	$-0.552117504\times10^{+18}$	$0.244767149\times10^{+21}$	$-0.440734326\times10^{+23}$
17	$0.318362667\times10^{-07}$	$-0.112905511\times10^{+01}$	$0.258463367\times10^{+05}$	$0.286089905\times10^{+08}$	$-0.817025631\times10^{+11}$	$0.130297321\times10^{+15}$	$-0.116894583\times10^{+18}$	$0.549998801\times10^{+20}$	$-0.105390504\times10^{+23}$
0	$0.119995521\times10^{-01}$	$-0.133145724\times10^{+01}$	$0.949370743\times10^{+04}$	$-0.119324786\times10^{+08}$	$0.611542737\times10^{+10}$	$-0.158919832\times10^{+13}$	$0.220977965\times10^{+15}$	$-0.156607293\times10^{+17}$	$0.444156376\times10^{+18}$
2	$0.153405809\times10^{-09}$	$0.434544617\times10^{+00}$	$0.126900527\times10^{+03}$	$0.316420456\times10^{+07}$	$-0.162425671\times10^{+10}$	$0.428949781\times10^{+12}$	$-0.607065893\times10^{+14}$	$0.437214022\times10^{+16}$	$-0.125745509\times10^{+18}$
3	$0.438915959\times10^{-09}$	$0.363985856\times10^{+00}$	$0.410517739\times10^{+04}$	$0.114700075\times10^{+07}$	$-0.465492198\times10^{+09}$	$0.111603719\times10^{+12}$	$-0.156948862\times10^{+14}$	$0.118915651\times10^{+16}$	$-0.372436135\times10^{+17}$
4	$0.236206473\times10^{-08}$	$0.152020035\times10^{+00}$	$0.104026260\times10^{+05}$	$-0.487621366\times10^{+07}$	$0.343240714\times10^{+10}$	$-0.110471209\times10^{+13}$	$0.171900079\times10^{+15}$	$-0.114545501\times10^{+17}$	$0.179064658\times10^{+18}$
5	$0.152191494\times10^{-09}$	$0.345225244\times10^{+00}$	$0.513753422\times10^{+04}$	$0.178355178\times10^{+08}$	$-0.225274397\times10^{+11}$	$0.138279944\times10^{+14}$	$-0.439165696\times10^{+16}$	$0.693264288\times10^{+18}$	$-0.429807741\times10^{+20}$
6	$0.101510366\times10^{-10}$	$0.631699375\times10^{+00}$	$0.288258591\times10^{+04}$	$0.345014862\times10^{+08}$	$-0.493043937\times10^{+11}$	$0.351403621\times10^{+14}$	$-0.131109063\times10^{+17}$	$0.244599334\times10^{+19}$	$-0.179832560\times10^{+21}$
7	$0.503759081\times10^{-14}$	$0.136346953\times10^{+01}$	$-0.282054702\times10^{+04}$	$0.655513089\times10^{+08}$	$-0.104669273\times10^{+12}$	$0.846366689\times10^{+14}$	$-0.361353660\times10^{+17}$	$0.775876551\times10^{+19}$	$-0.659112966\times10^{+21}$
8	$0.104460699\times10^{-22}$	$0.329471379\times10^{+01}$	$-0.222066050\times10^{+05}$	$0.146482864\times10^{+09}$	$-0.248918167\times10^{+12}$	$0.219701804\times10^{+15}$	$-0.103979287\times10^{+18}$	$0.249845144\times10^{+20}$	$-0.238977274\times10^{+22}$
9	$0.770482688\times10^{-19}$	$0.233273067\times10^{+01}$	$-0.149429814\times10^{+05}$	$0.136734073\times10^{+09}$	$-0.251527270\times10^{+12}$	$0.242731069\times10^{+15}$	$-0.126489623\times10^{+18}$	$0.336140544\times10^{+20}$	$-0.356623146\times10^{+22}$
10	$0.438603837\times10^{-43}$	$0.788846091\times10^{+01}$	$-0.730962596\times10^{+05}$	$0.371153627\times10^{+09}$	$-0.701373815\times10^{+12}$	$0.704940585\times10^{+15}$	$-0.384794261\times10^{+18}$	$0.107432846\times10^{+21}$	$-0.119977129\times10^{+23}$
11	$0.320761937\times10^{-24}$	$0.347485090\times10^{+01}$	$-0.283733940\times10^{+05}$	$0.215297915\times10^{+09}$	$-0.436678198\times10^{+12}$	$0.473346729\times10^{+15}$	$-0.279759433\times10^{+18}$	$0.848241442\times10^{+20}$	$-0.103105452\times10^{+23}$
12	$0.113204823\times10^{-03}$	$-0.131225290\times10^{+01}$	$0.246991254\times10^{+05}$	$0.431579951\times10^{+07}$	$-0.157723943\times10^{+11}$	$0.223768013\times10^{+14}$	$-0.159834373\times10^{+17}$	$0.568566052\times10^{+19}$	$-0.800164550\times10^{+21}$
13	$0.104988047\times10^{-74}$	$0.148966581\times10^{+02}$	$-0.159227178\times10^{+06}$	$0.808002769\times10^{+09}$	$-0.174489353\times10^{+13}$	$0.203780599\times10^{+16}$	$-0.130694229\times10^{+19}$	$0.432028955\times10^{+21}$	$-0.574347050\times10^{+23}$
14	$0.180099766\times10^{-89}$	$0.181631747\times10^{+02}$	$-0.201181188\times10^{+06}$	$0.102436792\times10^{+10}$	$-0.228735959\times10^{+13}$	$0.277407933\times10^{+16}$	$-0.185325477\times10^{+19}$	$0.639479025\times10^{+21}$	$-0.888701509\times10^{+23}$
15	$0.888540666\times10^{-98}$	$0.199435296\times10^{+02}$	$-0.226741706\times10^{+06}$	$0.117101483\times10^{+10}$	$-0.269183299\times10^{+13}$	$0.337382650\times10^{+16}$	$-0.233573986\times10^{+19}$	$0.836822518\times10^{+21}$	$-0.120912586\times10^{+24}$
16	$0.328338427\times10^{-101}$	$0.206397435\times10^{+02}$	$-0.238922323\times10^{+06}$	$0.125253606\times10^{+10}$	$-0.294342481\times10^{+13}$	$0.378225945\times10^{+16}$	$-0.268992435\times10^{+19}$	$0.991346617\times10^{+21}$	$-0.147482671\times10^{+24}$
17	$0.174113501\times10^{-52}$	$0.940244289\times10^{+01}$	$-0.108030577\times10^{+06}$	$0.673355963\times10^{+09}$	$-0.167173275\times10^{+13}$	$0.227801968\times10^{+16}$	$-0.172196403\times10^{+19}$	$0.675100081\times10^{+21}$	$-0.106830896\times10^{+24}$

Table A7. Parameters for the VE and VdE rate coefficients of BeH^+ ($v_i^+ = 2$ and 3), Fig. 8.

v_f^+	$A_{v_i^+\to v_f^+}$	$\alpha_{v_i^+\to v_f^+}$	$B_{v_i^+\to v_f^+}(1)$	$B_{v_i^+\to v_f^+}(2)$	$B_{v_i^+\to v_f^+}(3)$	$B_{v_i^+\to v_f^+}(4)$	$B_{v_i^+\to v_f^+}(5)$	$B_{v_i^+\to v_f^+}(6)$	$B_{v_i^+\to v_f^+}(7)$
0	$0.126308415\times10^{-09}$	$0.456305622\times10^{+00}$	$-0.282698010\times10^{+04}$	$0.324128858\times10^{+07}$	$-0.166236562\times10^{+10}$	$0.438820087\times10^{+12}$	$-0.620881147\times10^{+14}$	$0.447096361\times10^{+16}$	$-0.128575131\times10^{+18}$
1	$0.108956019\times10^{-04}$	$-0.641702757\times10^{+00}$	$0.276723409\times10^{+04}$	$-0.249602163\times10^{+07}$	$0.977115209\times10^{+09}$	$-0.206865990\times10^{+12}$	$0.246613751\times10^{+14}$	$-0.155686663\times10^{+16}$	$0.404532125\times10^{+17}$
3	$0.296926175\times10^{-08}$	$0.124816242\times10^{+00}$	$0.173836549\times10^{+04}$	$0.908486506\times10^{+06}$	$-0.424294717\times10^{+09}$	$0.108080533\times10^{+12}$	$-0.150733302\times10^{+14}$	$0.107908466\times10^{+16}$	$-0.309653342\times10^{+17}$
4	$0.233152145\times10^{-08}$	$0.122367759\times10^{+00}$	$0.452862967\times10^{+04}$	$0.653277887\times10^{+06}$	$-0.273049775\times10^{+09}$	$0.670734633\times10^{+11}$	$-0.944014877\times10^{+13}$	$0.700586130\times10^{+15}$	$-0.211863471\times10^{+17}$
5	$0.179746490\times10^{-09}$	$0.385415212\times10^{+00}$	$0.455113975\times10^{+04}$	$0.557366075\times10^{+07}$	$-0.437760675\times10^{+10}$	$0.181187931\times10^{+13}$	$-0.404354650\times10^{+15}$	$0.458962441\times10^{+17}$	$-0.207392120\times10^{+19}$
6	$0.662229481\times10^{-13}$	$0.115712028\times10^{+01}$	$0.161972309\times10^{+04}$	$0.182461133\times10^{+08}$	$-0.191412396\times10^{+11}$	$0.106134357\times10^{+14}$	$-0.314224909\times10^{+16}$	$0.468199997\times10^{+18}$	$-0.275398722\times10^{+20}$
7	$0.145068779\times10^{-16}$	$0.194335259\times10^{+01}$	$-0.679220386\times10^{+04}$	$0.602954203\times10^{+08}$	$-0.819656147\times10^{+11}$	$0.556924052\times10^{+14}$	$-0.198433419\times10^{+17}$	$0.354151824\times10^{+19}$	$-0.249461077\times10^{+21}$
8	$0.409680689\times10^{-28}$	$0.462578650\times10^{+01}$	$-0.314098439\times10^{+05}$	$0.145520499\times10^{+09}$	$-0.209844542\times10^{+12}$	$0.156055016\times10^{+15}$	$-0.618820134\times10^{+17}$	$0.124075458\times10^{+20}$	$-0.987496244\times10^{+21}$
9	$0.124226786\times10^{-18}$	$0.235220787\times10^{+01}$	$-0.124786214\times10^{+05}$	$0.101019453\times10^{+09}$	$-0.163557194\times10^{+12}$	$0.137832705\times10^{+15}$	$-0.622423124\times10^{+17}$	$0.142528576\times10^{+20}$	$-0.129773407\times10^{+22}$
10	$0.245518460\times10^{-28}$	$0.454190007\times10^{+01}$	$-0.339017796\times10^{+05}$	$0.187717660\times10^{+09}$	$-0.321080290\times10^{+12}$	$0.288460789\times10^{+15}$	$-0.139570052\times10^{+18}$	$0.343550312\times10^{+20}$	$-0.337037720\times10^{+22}$
11	$0.587221095\times10^{-38}$	$0.668854662\times10^{+01}$	$-0.585692446\times10^{+05}$	$0.300404084\times10^{+09}$	$-0.549700752\times10^{+12}$	$0.532010351\times10^{+15}$	$-0.278620007\times10^{+18}$	$0.744613338\times10^{+20}$	$-0.794724821\times10^{+22}$
12	$0.282820034\times10^{-48}$	$0.900692240\times10^{+01}$	$-0.854015521\times10^{+05}$	$0.421766486\times10^{+09}$	$-0.803276765\times10^{+12}$	$0.814702124\times10^{+15}$	$-0.449137228\times10^{+18}$	$0.126738498\times10^{+21}$	$-0.143144660\times10^{+23}$
13	$0.265515367\times10^{-54}$	$0.103422905\times10^{+02}$	$-0.102161017\times10^{+06}$	$0.507355535\times10^{+09}$	$-0.100269394\times10^{+13}$	$0.106072719\times10^{+16}$	$-0.611987469\times10^{+18}$	$0.181152972\times10^{+21}$	$-0.214995200\times10^{+23}$
14	$0.261195739\times10^{-62}$	$0.121198250\times10^{+02}$	$-0.123754196\times10^{+06}$	$0.614670324\times10^{+09}$	$-0.125449510\times10^{+13}$	$0.137621206\times10^{+16}$	$-0.825669890\times10^{+18}$	$0.254634769\times10^{+21}$	$-0.315273928\times10^{+23}$
15	$0.337950444\times10^{-70}$	$0.138489790\times10^{+02}$	$-0.145442136\times10^{+06}$	$0.725912499\times10^{+09}$	$-0.152624068\times10^{+13}$	$0.173091740\times10^{+16}$	$-0.107614767\times10^{+19}$	$0.344467648\times10^{+21}$	$-0.443138527\times10^{+23}$
16	$0.890704375\times10^{-70}$	$0.136862264\times10^{+02}$	$-0.146142169\times10^{+06}$	$0.748211873\times10^{+09}$	$-0.161520665\times10^{+13}$	$0.188583754\times10^{+16}$	$-0.120927715\times10^{+19}$	$0.399702625\times10^{+21}$	$-0.531334515\times10^{+23}$
17	$0.150098120\times10^{-51}$	$0.940497017\times10^{+01}$	$-0.999890655\times10^{+05}$	$0.566857445\times10^{+09}$	$-0.126885981\times10^{+13}$	$0.154015811\times10^{+16}$	$-0.102880135\times10^{+19}$	$0.354673172\times10^{+21}$	$-0.492093817\times10^{+23}$
0	$0.255293462\times10^{-08}$	$0.141696630\times10^{+00}$	$-0.108678652\times10^{+04}$	$0.970062410\times10^{+06}$	$-0.455079294\times10^{+09}$	$0.116124140\times10^{+12}$	$-0.162068475\times10^{+14}$	$0.116059371\times10^{+16}$	$-0.333086705\times10^{+17}$
1	$0.454371529\times10^{-09}$	$0.361457499\times10^{+00}$	$-0.148626220\times10^{+04}$	$0.100937278\times10^{+07}$	$-0.353342852\times10^{+09}$	$0.694367809\times10^{+11}$	$-0.772450214\times10^{+13}$	$0.453934615\times10^{+15}$	$-0.109451051\times10^{+17}$
2	$0.485973467\times10^{-05}$	$-0.651029662\times10^{+00}$	$0.804957988\times10^{+03}$	$-0.857704662\times10^{+06}$	$0.412563914\times10^{+09}$	$-0.104493488\times10^{+12}$	$0.143877589\times10^{+14}$	$-0.101687756\times10^{+16}$	$0.288548713\times10^{+17}$
4	$0.232467658\times10^{-06}$	$-0.302405394\times10^{+00}$	$0.384019801\times10^{+04}$	$-0.156393045\times10^{+07}$	$0.782465204\times10^{+09}$	$-0.197911019\times10^{+12}$	$0.269171925\times10^{+14}$	$-0.187489378\times10^{+16}$	$0.524590130\times10^{+17}$
5	$0.440932239\times10^{-06}$	$-0.461349565\times10^{+00}$	$0.470955951\times10^{+04}$	$0.763181298\times10^{+06}$	$-0.430728638\times10^{+09}$	$0.118093902\times10^{+12}$	$-0.170152283\times10^{+14}$	$0.123756313\times10^{+16}$	$-0.358125341\times10^{+17}$
6	$0.248118257\times10^{-07}$	$-0.176014398\times10^{+00}$	$0.546411270\times10^{+04}$	$0.382709019\times10^{+07}$	$-0.321036040\times10^{+10}$	$0.136543056\times10^{+13}$	$-0.304649186\times10^{+15}$	$0.339316565\times10^{+17}$	$-0.148556841\times10^{+19}$
7	$0.552823502\times10^{-08}$	$-0.853326551\times10^{-01}$	$0.738084609\times10^{+04}$	$0.509493799\times10^{+07}$	$-0.452686316\times10^{+10}$	$0.206014096\times10^{+13}$	$-0.507840858\times10^{+15}$	$0.645396348\times10^{+17}$	$-0.331488284\times10^{+19}$
8	$0.556027551\times10^{-14}$	$0.128059714\times10^{+01}$	$-0.722375766\times10^{+03}$	$0.293329373\times10^{+08}$	$-0.332551286\times10^{+11}$	$0.200746468\times10^{+14}$	$-0.655614154\times10^{+16}$	$0.108966954\times10^{+19}$	$-0.720949572\times10^{+20}$
9	$0.105702929\times10^{-22}$	$0.332530499\times10^{+01}$	$-0.159170215\times10^{+05}$	$0.877638068\times10^{+08}$	$-0.123120535\times10^{+12}$	$0.891243930\times10^{+14}$	$-0.342208461\times10^{+17}$	$0.661104891\times10^{+19}$	$-0.505031277\times10^{+21}$
10	$0.873323579\times10^{-29}$	$0.471156180\times10^{+01}$	$-0.309840419\times10^{+05}$	$0.157565181\times10^{+09}$	$-0.245838938\times10^{+12}$	$0.197254981\times10^{+15}$	$-0.842530563\times10^{+17}$	$0.181826211\times10^{+20}$	$-0.155700530\times10^{+22}$
11	$0.161691663\times10^{-11}$	$0.625264862\times10^{+00}$	$0.443656063\times10^{+04}$	$0.571317412\times10^{+08}$	$-0.111444978\times10^{+12}$	$0.108422852\times10^{+15}$	$-0.553083542\times10^{+17}$	$0.141285569\times10^{+20}$	$-0.142346921\times10^{+22}$
12	$0.996158345\times10^{-32}$	$0.526593701\times10^{+01}$	$-0.410318341\times10^{+05}$	$0.225769847\times10^{+09}$	$-0.403565597\times10^{+12}$	$0.377112648\times10^{+15}$	$-0.189442841\times10^{+18}$	$0.483813236\times10^{+20}$	$-0.492309385\times10^{+22}$
13	$0.137490891\times10^{-39}$	$0.701036896\times10^{+01}$	$-0.611600724\times10^{+05}$	$0.314708823\times10^{+09}$	$-0.582135637\times10^{+12}$	$0.568238103\times10^{+15}$	$-0.300018662\times10^{+18}$	$0.808463515\times10^{+20}$	$-0.870338374\times10^{+22}$
14	$0.137970987\times10^{-46}$	$0.855396426\times10^{+01}$	$-0.797231356\times10^{+05}$	$0.398979415\times10^{+09}$	$-0.761023171\times10^{+12}$	$0.771557989\times10^{+15}$	$-0.424818758\times10^{+18}$	$0.119669980\times10^{+21}$	$-0.134890370\times10^{+23}$
15	$0.188321221\times10^{-48}$	$0.891102185\times10^{+01}$	$-0.858227282\times10^{+05}$	$0.435553030\times10^{+09}$	$-0.854746186\times10^{+12}$	$0.895329760\times10^{+15}$	$-0.510613712\times10^{+18}$	$0.149231738\times10^{+21}$	$-0.174715509\times10^{+23}$
16	$0.253259645\times10^{-45}$	$0.814294465\times10^{+01}$	$-0.788662325\times10^{+05}$	$0.416268987\times10^{+09}$	$-0.832962953\times10^{+12}$	$0.891831242\times10^{+15}$	$-0.520838344\times10^{+18}$	$0.156096398\times10^{+21}$	$-0.187612211\times10^{+23}$
17	$0.880763483\times10^{-59}$	$0.111500449\times10^{+02}$	$-0.112805620\times10^{+06}$	$0.563857769\times10^{+09}$	$-0.114647825\times10^{+13}$	$0.125089015\times10^{+16}$	$-0.745700701\times10^{+18}$	$0.228371168\times10^{+21}$	$-0.280672906\times10^{+23}$

Journal of Physics: Conference Series **576** (2015) 012005 doi:10.1088/1742-6596/576/1/012005

References

[1] M. Larsson and A. E. Orel, *Dissociative recombination of molecular ions*, (Cambridge Univ. Press), (2008).
[2] A. Wolf, L. Lammich and P. Schmelcher (editors), *J. Phys. Conf. Series* **4**, Proceedings of the Sixth International Conference on Dissociative Recombination: Theory, Experiments and Applications, 12-16 July 2004, Mosbach, Germany (2005).
[3] W. J. van der Zande (editor), *J. Phys. Conf. Series* **192**, Proceedings of the Seventh International Conference on Dissociative Recombination: Theory, Experiments and Applications, 18-23 July 2007, Ameland, The Netherlands (2009).
[4] A. E. Orel and S. L. Guberman (editors), *J. Phys. Conf. Series* **300**, Proceedings of the Seventh International Conference on Dissociative Recombination: Theory, Experiments and Applications, August 16-20 2010, Lake Tahoe, USA (2011).
[5] I. F. Schneider, O. Dulieu and J. Robert (editors), *E. Phys. J. Conf. Series* Proceedings of the Eighth International Conference on Dissociative Recombination: Theory, Experiments and Applications, July 7-12 2013, Paris, France, in preparation (2013).
[6] G. D. Dickenson, T. I. Ivanov, M. Roudjane, N. de Oliveira, D. Joyeux, L. Nahon, W.-Ü. L. Tchang-Brillet, M. Glass-Maujean, I. Haas, A. Ehresmann and W. Ubachs, *J. Chem. Phys.* **133**, 144317 (2010).
[7] M. Glass-Maujean, Ch. Jungen, H. Schmoranzer, A. Knie, I. Haar, R. Hentges, W. Kielich, K. Jänkälä and A. Ehresmann, *Phys. Rev. Lett* **104**, 183002 (2010).
[8] M. Glass-Maujean, Ch. Jungen, G. Reichardt, A. Balzer, H. Schmoranzer, A. Ehresmann, I. Haar and P. Reiss, *Phys. Rev. A* **82**, 062511 (2010).
[9] J. Zs. Mezei, I. F. Schneider, E. Roueff, and Ch. Jungen, *Phys. Rev. A* **85**, 043411 (2012).
[10] P. Billaud, M. Geéléoc, Y. J. Picard, K. Veyrinas, J. F. Hergott, S. Marggi Poullain, P. Breger, T. Ruchon, M. Roulliay, F. Delmotte, F. Lepetit, A. Huetz, B. Carré and D. Dowek, *J. Phys. B. At. Molec. Opt. Phys.* **45**, 194013, (2012).
[11] A. Medina, G. Rahmat, C. R. de Carvalho, G. Jalbert, F. Zappa, R. F. Nascimento, R. Cireasa, N. Vanhaecke, I. F. Schneider, N. V. de Castro Faria et J. Robert, *J. Phys. B. At. Molec. Opt. Phys.* **44**, 215203, (2011).
[12] H. P. Summers, Badnell N. R., O'Mullane M. G. Whiteford A. D., Bingham R., Kellett B. J., Lang J. Behringer K. H, Fantz U. Zastrow K.-D., Loch S. D. Pindzola M. S., Griffin D. C. Ballance C. P., *Plasma Physics and Controlled Fusion* **44**, B323 (2002).
[13] U. Fantz, and P. T. Greenland, *Contrib. Plasm. Phys.* **42**, 694 (2002).
[14] R. E. H. Clark and D. Reiter (Eds.), *Nuclear Fusion Research: Understanding Plasma-Surface Interactions*, Springer series in Chemical Physics **78** (Springer Berlin Heidelberg), (2005).
[15] R. Celiberto, R. K. Janev and D. Reiter, *Plasma Phys. Control. Fusion* **54**, 035012 (2012).
[16] C. M. Coppola, S. Longo, M. Capitelli, F. Palla, and D. Galli, *ApJ. Suppl. Ser.* **193**, pp. 7 (2011).
[17] P. Gnacinski, *A&A* **532**, pp. 122 (2011).
[18] J. H. Black, *Phil. Trans. R. Soc. A* **370**, 5130 (2012).
[19] C. D. Gay, N. P. Abel, R. L. Porter, P. C. Stancil, G. J. Ferland, G. Shaw, P. A. M. van Hoof and R. J. R. Williams, *ApJ* **746**, pp. 78 (2012).
[20] A. Bultel, B. G. Chéron, A. Bourdon, O. Motapon and I. F. Schneider, *Phys. Plasmas* **13**, 043502 (2006).
[21] U. Fano, *Phys. Rev.* **124**, 1866 (1961).
[22] N. Bardsley, *J. Phys. B* **1**, 349, 365 (1968).
[23] A. Giusti-Suzor, *J. Phys. B: At. Mol. Phys.* **13**, 3867 (1980).
[24] A. Giusti-Suzor, J. N. Bardsley and C. Derkits, *Phys. Rev. A* **28**, 682(1983).
[25] O. Motapon, F. O. Waffeu Tamo, X. Urbain and I. F. Schneider, *Phys. Rev. A* **77**, 052711 (2008).
[26] M. J. Seaton, *Rept. Prog. Phys.* **46**, 167 (1983).
[27] C. H. Greene and Ch. Jungen, *Adv. At. Mol. Phys.* **21**, 51 (1985).
[28] Ch. Jungen (editor), *Molecular Applications of Quantum Defect Theory*, (IoP Publishing, Bristol), (1996).
[29] Ch. Jungen, in *Handbook of High resolution Spectroscopy*, edited by M. Quack and F. Merkt (Wiley, Chichester/New York), (2011).
[30] V. Sidis and H. Lefèbvre-Brion, *J. Phys. B: At. Mol. Phys.* **4**, 1040 (1971).
[31] Ch. Jungen and O. Atabek, *J. Chem. Phys.* **66**, 5584 (1977).
[32] V. Ngassam, A. Florescu, L. Pichl, I. F. Schneider, O. Motapon and A. Suzor-Weiner, *Euro. Phys. J. D* **26**, 165 (2003).
[33] U. Fano, *Phys. Rev. A* **2**, 353 (1970).
[34] E. S. Chang and U. Fano, *Phys. Rev. A* **6**, 173 (1972).
[35] B. Vâlcu, I. F. Schneider, M. Raoult, C. Strömholm, M. Larsson and A. Suzor-Weiner, *Euro. Phys. J. D* **1**, 71 (1998).
[36] H. Takagi, Physica Scripta **T96**, 52-60 (2002).
[37] T. Tanabe I. Katayama, H. Kamegaya, K. Chida, Y. Arakaki, T. Watanabe, M. Yoshizawa, M. Saito, Y.

Haruyama, K. Hosono, K. Hatanaka, T. Honma, K. Noda, S. Ohtani, and H. Takagi, *Phys. Rev. Lett.* **75**, 1066 (1995).
[38] A. Al-Khalili *et al.*, *Phys. Rev. A* **68**, 042702 (2003).
[39] K. Chakrabarti, D. R. Backodissa-Kiminou, N. Pop, J. Zs. Mezei, O. Motapon, F. Lique, O. Dulieu, A. Wolf, and I. F. Schneider, *Phys. Rev. A* **87**, 022702 (2013).
[40] J. Tennyson, *At. Data Nucl. Data Tables* **64**, 253 (1996).
[41] V. Ngassam, O. Motapon, A. Florescu, L. Pichl, I. F. Schneider and A. Suzor-Weiner, *Phys. Rev. A* **68**, 032704 (2003).
[42] F. O. Waffeu Tamo, Buhr, H., O. Motapon, S. Altevogt, V. M. Andrianarijaona, M. Grieser, L. Lammich, M. Lestinsky, M. Motsch, I. Nevo, S. Novotny, D. A. Orlov, H. B. Pedersen, D. Schwalm, F. Sprenger, X. Urbain, U. Weigel, A. Wolf, and I. F.Schneider, *Phys. Rev. A* **84**, 022710 (2011).
[43] R. von Hahn *et al.* , *Nucl. Instrum. and Meth. in Phys. Res. B* **269**, 2871 (2011).
[44] Roos, J. B. Roos, M. Larsson, Å. Larson and A. E. Orel, *Phys. Rev. A* **80**, 012501 (2009).
[45] S. Niyonzima, F. Lique, K. Chakrabarti, Å. Larson, A. E. Orel, I. F. Schneider, *Phys. Rev. A* **87**, 022713 (2013).
[46] L. Wolniewicz, *J. Chem. Phys.* **103**, 1792, (1995).
[47] L. Wolniewicz and G. Staszewska, *J. Mol. Spectrosc.* **220**, 45 (2003)
[48] A compilation of the numerical data published in Refs. [46, 47] can be found in `http://www.fizyka.umk.pl/publications/ifiz/luwo/`
[49] G. W. F. Drake, in *High Precision Calculations for Helium*, Springer Handbook of Atomic, Molecular, and Optical Physics, edited by G. W. F. Drake(Springer,New York, 2006), p.199.
[50] M. Glass-Maujean and Ch. Jungen, *J. Phys. Chem. A* **113**, 13124 (2009).
[51] N. Y. Du and C. H. Greene, *J. Chem. Phys.* **85**, 5430 (1986).
[52] S. C. Ross and Ch. Jungen, *Phys. Rev. A* **49**, 4364 (1994).
[53] J. J. Spezeski, O. F. Kalman and Jr. L. C. McIntyre, *Phys. Rev. A* **22** 1906 (1980).
[54] M. Leventhal, R. T. Robiscoe and K. R. Lea, *Phys. Rev.* **158** 49 (1967).
[55] R. Campargue, *J. Phys. Chem.* **88** 4466 (1984).
[56] R. Campargue, A. Lebehot, J. C. Lemonnier and D. Marette, *Rarefied Gas Dynamics* (New York: S. S. Fisher) (1981).
[57] V. Bocvarski, J. Robert, I. Colomb de Daunant, J. Reinhardt and J. Baudon, *J. Physique Lett.* **46** L13 (1985).
[58] J. Robert, V. Bocvarski, B. Stern, J. Reinhardt and J. Baudon, *Z. Phys. D.* **6** 279 (1987).
[59] J. E. Mental and E. P. Gentieu, *J. Chem. Phys.* **52** 5641 (1970).
[60] M. Misakian and C. Zorn, *Phys. Rev. A* **6** 2180 (1972).
[61] B. L. Carnahan and E. C. Zipf, *Phys. Rev. A* **16** 991(1977).
[62] A. U. Hazi and K. Wiemers, *J. Chem. Phys.* **66** 5296 (1977).
[63] A. Medina, G. Rahmat, G. Jalbert, R. Cireasa, F. Zappa, C. R. de Carvalho, N. V. de Castro Faria and J. Robert, Eur. Phys. J. D **66**, 134 (2012).
[64] O. Motapon, M. Fifirig, A. Florescu, F. O. Waffeu Tamo, O. Crumeyrolle, G. Varin-Bréant, A. Bultel, P. Vervisch, J. Tennyson and I. F. Schneider, *Plasma Sources Sci. Technol.* **15**, 23 (2006).
[65] H. Tawara, Y. Itikawa, H. Nishimura and M. Yoshino, *J. Phys. Chem. Ref. Data* **19**, 617 (1990).

Journal of Physics: Conference Series **576** (2015) 012006 doi:10.1088/1742-6596/576/1/012006

Polyatomic ions, branching ratios and hot molecules

J. Brian A. Mitchell

Institut de Physique de Rennes, U.M.R. N°. 6251 du CNRS, Université de Rennes, I35042 Rennes, France

Abstract. A discussion is given of the reason for the sharp fall-off observed in Dissociative Recombination (DR) cross sections above about 0.1 eV and of the need for accurate branching ratios being used in complex models of molecular ion chemistry. New measurements from TSR have shown that stored ions are not as cold as they were once thought to be and a new experiment facility is presented.

1. Introduction

Dissociative recombination is a process whereby molecular ions can be rapidly neutralized in collision with free electrons, with the recombination energy being converted into product kinetic energy. Rate coefficients for this process are generally large and for polyatomic ions, a wide variety of exit channels can occur leading to many atomic and radical species as well as stable neutral molecules. The process is normally important below 1 eV and much work has been devoted to studies relevant to astrochemistry where temperatures in the range of 10 – 100K are encountered. In particular, in this application, molecular ions are usually in there ro-vibrational ground state and there has been much interest over the last 20 years in using heavy ion storage rings to cool ions vibrationally and then to study their recombination down to sub-milli-electronvolt energies. The high signal count rates found in these experiments have allowed measurements to be performed over a wide energy range and in several cases, these experiments have revealed the presence of peaks in the cross sections above a few eV due to direct transitions from the ion ground state to a dissociative neutral excited state, a process not dissimilar to dissociative excitation where in that case both ions and neutral products are formed. These observations have particular significance to thermonuclear plasma research where electrons with tens and hundreds of electron-volts are to be found in the near edge and diverter regions of tokamaks. Recent general reviews of DR measurements and of the theoretical background to the subject have been given by Florescu-Mitchell and Mitchell [1], Larsson and Orel [2] and a detailed review of afterglow based measurements by Adams et al. [3]. Another source of information concerning DR is the series of books published from the conference series Dissociative Recombination: Theory, Experiment and Applications. (See list in the Appendix.)

2. Medium energy DR

The first systematic investigations of DR over a wide electron temperature (energy) range were performed using inclined beam [4] and merged beam experiments [5]. These initially concentrated on hydrogenic species (H_2^+, H_3^+ and their isotopomers) and it was found that the measured cross sections decreased linearly for diatomic species but not for polyatomic ions which generally displayed a sharp change of regime in the vicinity of 0.1 eV [6]. This is illustrated in Figure 1a and 1b that shows an example of this phenomenon with DR cross sections for H_3O^+ [7-9] measured using the single pass and multiple pass techniques respectively. Other examples are given graphically in [10]. The reason for this abrupt regime change was not however obvious. With the advent of heavy ion storage rings and the application of the merged beams technique at these facilities, the improved signal to noise thus achievable allowed measurements to be made for up to tens of electron volts of electron energy. This yielded further surprises with the appearance of peaks in some case that could be attributable to direct transitions from the ion ground state to repulsive excited, autoionising, neutral states that could dissociate through neutral daughter channels. Again in these studies, a sharp drop-off was observed in the vicinity of 0.1 eV.

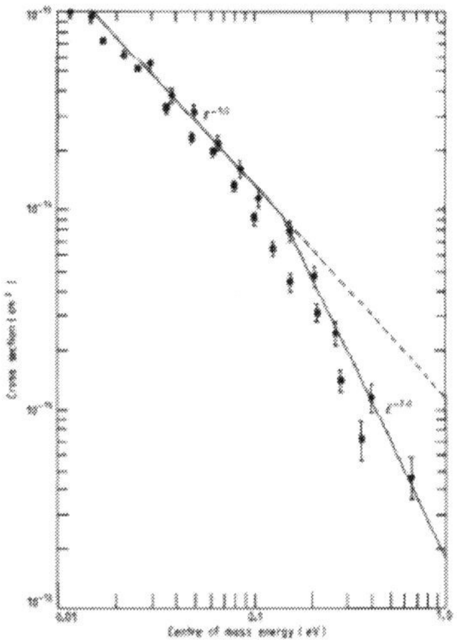

Fig. 2a. Single pass merged beam measurements of the DR of H_3O^+ (solid circles) and D_3O^+ (solid squares) (From [7] with permission)

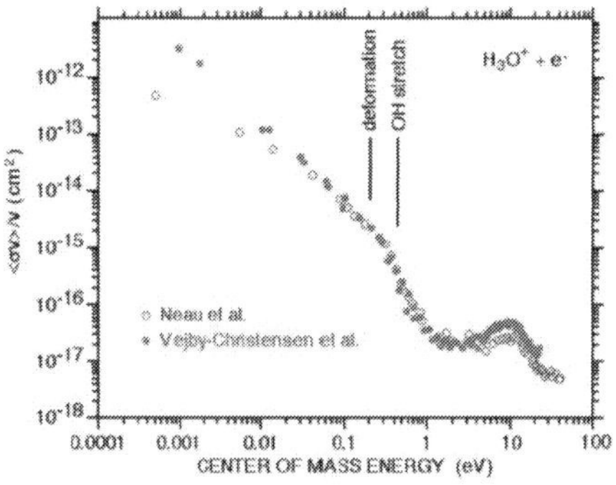

Fig. 2b. DR cross sections for H_3O^+ measured at ASTRID [8] and CRYRING [9] (From [10] with permission)

While theoretical studies of simple ions have addressed the low energy and "high energy" (few eV) portions of the cross section spectrum, and have shown sharp fall-offs in the 0.1 eV region, a clear reason for this was not given. Recently though, a new article by Jungen and Pratt [10] has addressed this particular issue directly. These authors have been very active in the experimental and theoretical study of the competition between different decay channels (ionization, dissociation, radiation) for autoionising neutral Rydberg states formed by excitation from ground state neutral molecules. In this process, it has been well recognized that there is a propensity rule whereby autoionisation can occur via a transfer of vibrational energy from a bound state to the electronic energy (thus liberating the electron) that is most likely to occur for a change of one vibrational quantum number i.e. $\Delta v= -1$.

Jungen and Pratt have combined their experience in vibrational autoionization with the concept of indirect DR and they have argued that indeed, for polyatomic systems, the process is dominated by the indirect process, proceeding via transitions involving $\Delta v=+1$. The energy required for a transition to the first vibrationally excited state is typically on the order of 0.1 eV and so when the electron energy exceeds this, DR would have to involve a transition of $\Delta v>+1$ which is much less likely than the $\Delta v=+1$ transition and the result is that the probability of indirect DR drops and hence the cross section drops dramatically. This is illustrated in Figure 3. This explanation presupposes that indirect DR is the main driving force for the recombination of such species and indeed this is supported by the observation that generally polyatomic ions have rate coefficients that do not differ one from another by much more than an order of magnitude [1]. This is to be contrasted with the similar process of dissociative electron attachment that can only proceed via a direct transition from a molecule to a negative ion state, there being no comparable indirect process. In this case it is observed that rates can vary by 5 orders of magnitude from one molecule to another [11]. This argues in favor of the dominance of the indirect process for DR.

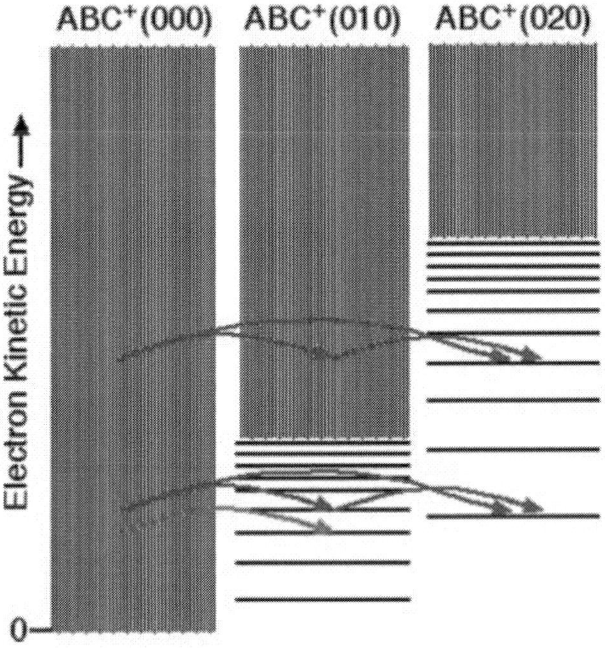

Fig. 3: Electron capture from the continuum of the ion into a bound neutral Rydberg state. At low electron energy the transition passes by $\Delta v=+1$, but at higher energy the transition would have to be by $\Delta v=+2$ (From [10, with permission].

3. Branching Ratios for DR

For many years, Branching Ratios, that is the ratio of the importance of respective dissociation channels for polyatomic ion recombination, was the Holy Grail of recombination research. These were of particular interest to astrophysicists since DR is one of the main mechanisms for the formation of interstellar molecules. While there were methods for determining the production of particular excited fragments using spectroscopic and chemical techniques, these methods were not sufficiently general to yield complete branching ratios and in particular were not able to determine the comparative likelihoods of the different channels that would lead to the formation of stable molecules. A notable advance in the measurement of branching ratios for polyatomic molecules was that of Mitchell et al. [12] for H_3^+ using a method that involved the merged beam technique coupled with a detection system in which some products were inhibited from reaching the energy sensitive surface barrier detector by placing a grid in front of it. This early measurement was fraught with background noise problems but with the development of high energy heavy particle storage rings, this limitation was largely removed and since then a wide variety of polyatomic ions have had their branching ratios determined at least with respect to fragment species of differing atomic masses [1,2,13]. However, complete branching ratios for hydrocarbon molecular ions with more than 2 or 3 carbon atoms are still beyond the reach of current experimental capabilities.

The importance of having accurate information concerning branching ratios has been highlighted recently by a modeling study by Pernot and associates [14] of the ionosphere of Titan where the atmosphere consists mainly of nitrogen and methane, yielding a rich chemistry. These authors have analysed the influence of the accuracy of known and speculated branching ratios on the final number densities of the products of polyatomic recombination reactions involving hydrocarbon species. In the astrophysical literature and other studies of Titan modeling [15], it has been generally assumed that

Journal of Physics: Conference Series **576** (2015) 012006 doi:10.1088/1742-6596/576/1/012006

when a polyatomic hydrocarbon molecular ion undergoes recombination, this leads to the ejection of a single hydrogen atom leaving behind the remaining dehydrogenated neutral partner:

$e + C_nH_m^+ \rightarrow C_nH_{m-1} + H$ ....(I)

While this is the main decay channel for CH_m^+ and $C_2H_m^+$ ions [1,2, 13], Mitchell et al. [16] showed using the Astrid storage ring in Aarhus, Denmark, that for $C_4H_m^+$ ions, this was no longer necessarily the case and indeed when m is large, scission of the C-C bonds dominates. This was subsequently confirmed by other experiments involving $C_3H_m^+$ and $C_4H_m^+$ ions [17,18] at Astrid and at CRYRING in Stockholm [19].

In their paper, Pernot and colleagues have determined the difference in the final predicted number densities of products if one considers only reactions of type I or if one uses actually measured branching ratios where C-C bond scission is included. The results are shown in Figure 4 and it can be seen that the differences can be dramatic, up to 5 orders of magnitude in some cases.

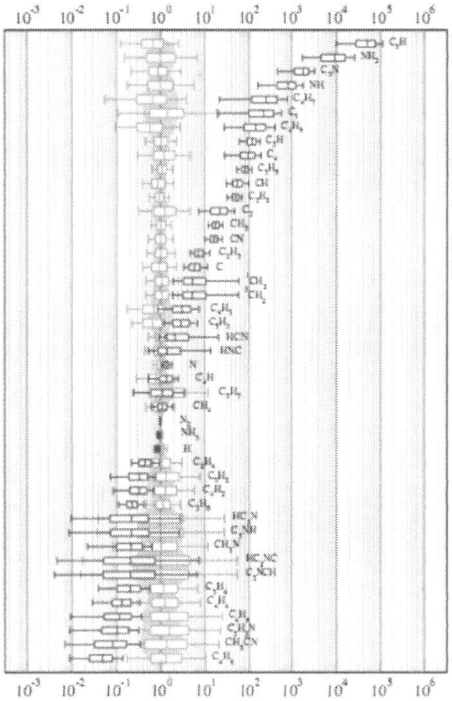

Fig. 4. Comparison of the ratio of the production rates for neutral species ($v^{Full}/v^{H\text{-}Loss}$) assuming that (a) the process passes by reaction I ($v^{H\text{-}Loss}$) (Brown boxes along the centre-line) or (b) by a full set of possible evaluated branching ratios (v^{Full}) (Blue boxes). (From [14] with permission).

This study highlights the need for continuing explorations of polyatomic ion recombination branching ratios. Given the problems related to the lack of experimental facilities capable of yielding such data, other approaches must be employed to address this problem. The work of Chabot and collaborators [19,20] perhaps points the way to new methods of determining branching ratios using sophisticated detector technology with the analysis of event timing. While this has not been directly used for DR research, the information concerning the dissociation of highly excited species can provide valuable information concerning fragmentation processes.

4. Ion cooling

Ion beams contain ions that are generally formed though violent processes such as electron impact and as such usually begin life in a variety of ro-vibrational and even electronically excited states. One of the major features of heavy ion storage rings is their ability to store ions for time longer than the typical relaxation times for vibrationally excited states (provided the ions possess a dipole moment) and so they were able to offer measurements involving vibrationally cold species. Rotational excitation is much longer lived and so it was perhaps not surprising that experimental measurements at the TSR ring in Heidelberg [21] found that H_3^+ ions used for their DR studies were indeed rotationally excited. What was surprising was how excited the ions really were (3000K). In other words they had as much internal energy as if they had been vibrationally excited. This observation prompted other techniques to be used to form the ions prior to their injection into the storage rings. In particular a supersonic expansion source was developed for this purpose [22]. It is well known that efficient rotational cooling can be achieved in the supersonic expansion of a gas through a nozzle, passing from a high to a low pressure environment. Supersonic expansion is much less efficient in producing vibrational cooling but this was not considered important given the radiative cooling achieved in the storage ring. In another approach, a cryogenically cooled (15K), multipole radiofrequency ion storage trap was used to produce the ions and this allowed them to be cooled via collisions with the background gas in the trap [23]. This approach [24,25] had originally been used in the single pass experiments of Mitchell et al. for H_3^+ [26], H_2^+ [27], HeH^+ [28] and N_2^+ [29] though in these experiments the trap was not cryogenically cooled. These different methods have been compared recently at TSR [30].

Again the results are quite surprising. Firstly it is found that the supersonic expansion source produced ions that were hotter (450 ± 100 K) that those in test-bed studies that did not involve ion storage in a ring. This points to a fundamental problem and that is that the electron cooler used to produce the high quality ion beam in the ring, can have an effect on the internal excitation state of the ions through both exciting and de-exciting collisions. Cross sections for electron induced rotational state change are in fact enormous at low electron energies [31]. Secondly it was found that the ion trap source was capable of producing relative cool ions provided the extraction voltage on the source was kept low. This in fact had been found in early merged beam studies of H_3^+ recombination performed using a trap source with the single pass experiment of Mitchell et al. [26]. When the ions are to leave the trap which is at a relatively high pressure (up to 1 Torr), they must be accelerated to extract them and then to form them into a beam and during this process, they pass through the gas cloud downstream of the extraction region where they can undergo exciting collisions. Thus the need to reduce this effect as much as possible by reducing the extraction voltage while sacrificing beam intensity. The final temperatures obtained in these experiments (150 ± 50K) are certainly not as low as originally thought but the high temperatures obtained with conventional sources are a cause for concern perhaps. This will probably not have a great influence on branching ratios but in some cases may play on the total rates measured with these machines.

5. New experimental developments

With the shutdown of the CRYRING and TSR facilities, the pursuit of heavy ion storage ring experiments has been curtailed. There are however, new developments on the horizon. There are two new electrostatic storage rings in development at Stockholm and Heidelberg that will be used for some recombination studies and these are discussed elsewhere in this issue [32] and in the proceedings of the 8th and 9th DR Conference. A third electrostatic storage ring is under construction in Saudi Arabia [33] and DR and DE measurements of light ions will be studied at this facility. This machine will have an injection energy of 20 keV and it is planned to build a merged beams setup into the ring that can be used for single and multiple pass experiments. Particular emphasis will be given to the examination of how these cross sections depend upon the electronic and ro-vibrational excitation of the ions.

A large heavy ion storage ring MIRFL has been designed in Langzhou China [34] and if built, this machine would offer a much higher magnetic rigidity that previously available in the three European rings. This would allow (a) much heavier ions to be studied and (b) higher ion energy to be used that would permit branching ratios to be measured for more complex species than previously possible. Some of the latter measurements use an energy sensitive surface barrier detector technology that when coupled with the grid technique mentioned earlier, allow products to be separated according to their mass. The higher the ion energy, the easier it is to perform this separation. At the present time, the schedule for the construction is uncertain [35].

While the heavy ion storage ring measurements have been designed to treat cold ions, those found in fusion wall and diverter plasmas are more likely to be excited and so future experiments will be needed to characterize the ions produced in low pressure ion sources and to perform single pass experiments to make prompt measurements of such excited ions. In previous studies, it has been found that excited ions can have cross sections that can differ by a factor of 3 or 4 from ground state species [36]. DR is such a complicated process and the position of the states implicated is so critical in some cases that it is not easy to predict a priori what the effects of an excitation will be. Hence the need for continuing experimental and theoretical efforts in a field that continues to surprise and for which there is a continual demand for data.

Acknowledgements
This work was carried out as a part of an IAEA Coordinated Research Project on "Light Element Atom, Molecule and Radical Behaviour in the Divertor and Edge Plasma Regions."

Appendix: The Dissociative Recombination Conferences

1988: Dissociative Recombination: Theory, Experiment and Applications, Lake Louise, Canada, (Editors: J.B.A. Mitchell and S. L. Guberman), World Scientific, Singapore, 1989.

1992: Dissociative Recombination: Theory, Experiment and Applications, St. Jacut de la Mer, France, (Editors: B. R. Rowe, J.B.A. Mitchell and A. Canosa), NATO ASI Series, Plenum Press, New York, World Scientific, 1993.

1995: Dissociative Recombination: Theory, Experiment and Applications, Ein Gedi, Israel, (Editors: D. Zajfman, J.B.A. Mitchell, D. Schwalm and B.R. Rowe), World Scientific, Singapore, 1996.

1999: Dissociative Recombination: Theory, Experiment and Applications IV, Stockholm, Sweden, (Editors: M. Larsson, J.B.A. Mitchell, I.F. Schneider), World Scientific, Singapore, 2000.

2001: Dissociative Recombination of Molecular Ions with Electrons (Chicago, USA) (Editor: S.L. Guberman) , Kluwer Academic/Plenum Publishers, New York, 2003.

2005: Sixth International Conference on Dissociative Recombination: Theory, Experiment and Applications (DR2004), Mosbach, Germany (Editors:A. Wolf, L. Lammich and P. Schmelcher), J. Phys Conference 4, 2005.

2007: Seventh International Conference on Dissociative Recombination: Theory, Experiment and Applications (DR2007), Ameland, The Netherlands (Editor: W.J. van de Zande), J. Phys Conference 192, 2009.

2010: Eighth International Conference on Dissociative Recombination: Theory, Experiment and Applications (DR2010), Lake Tahoe, USA (Editors: S.L. Guberman and A.E. Orel), J. Phys. Conference 300, 2011.

2013: Ninth International Conference on Dissociative Recombination: Theory, Experiment and Applications (DR2013), Paris, France (Editors: I. Schneider, O. Dulieu, J. Robert) EPJ Web of Conferences (to be published).

References

[1] A.I. Florescu-Mitchell and J.B.A. Mitchell, Phys. Reports, 430 (2006) 277.
[2] Dissociative Recombination of Molecular Ions, M. Larsson and A.E. Orel, Cambridge University Press, Cambridge, 2008.
[3] N.G. Adams, V. Poterya and L.M. Babcock, Mass Spec. Reviews 25 (2006) 798.

[4] B. Peart and K.T. Dolder J. Phys. B 7 (1974) 236.
[5] Auerbach, R. Cacak, R. Caudano, C.J. Keyser, T.D. Gaily, J.B.A. Mitchell, J. Wm. McGowan and S.F.J. Wilk, J. Phys. B 18 (1977) 3797.
[6] J. Wm McGowan, P.M. Mul, V.S. D'Angelo, J.B.A. Mitchell, and H.R. Froelich, Phys. Rev. Lett. 42 (1979) 373.
[7] P.M. Mul, J. Wm. McGowan and J.B.A. Mitchell, J. Phys. B 16 (1983) 3099.
[8] L. Vejby-Christensen, D. Kella, H.B. Pedersen and L.H. Andersen, Phys. Rev. A 57 (1998) 3267.
[9] A. Neau, A. Al Khalili, S. Rosen, A. Le Padellec, A.M. Derkatch, W. Shi, L. Vikor, and M. Larsson, J. Chem. Phys. 113 (2000) 1762.
[10] Ch. Jungen and S.T. Pratt, J. Chem Phys. 133 (2010) 214303.
[11] Electron Molecule Interactions and Their Applications. L.G. Christophorou (Ed.), Academic Press, N.Y. 1984.
[12] J.B.A. Mitchell, J.L. Forand, C.T. Ng, D.P. Levac, R.E. Mitchell, P.M. Mul, W. Claeys, A. Sen and J. Wm. McGowan, Phys. Rev. Lett. 51 (1983) 885.
[13] R.D. Thomas, Mass Spec. Reviews 27 (2008) 485.
[14] S. Plessis, N. Carrasco, and P. Pernot, J. Chem. Phys. 133 (2010) 134110.
[15] Chemistry of the Planets, Faraday Discussions, 147, 2010.
[16] J.B.A. Mitchell, C. Rebrion-Rowe, J.L. Le Garrec, G. Angelova, H. Bluhme, K. Seirsen and L.H. Andersen, Int. J. Mass Spec. 227 (2003) 273.
[17] G. Angelova, O. Novotny, J.B.A. Mitchell, C. Rebrion-Rowe, J.L. Le Garrec, H. Bluhme, K. Seiersen and L.H. Andersen, Int. J. Mass Spec. 232 (2004) 195.
[18] G. Angelova, O. Novotny, J.B.A. Mitchell, C. Rebrion-Rowe, J.L. Le Garrec, H. Bluhme, K. Seiersen and L.H. Andersen, Int. J. Mass Spec. 235 (2004) 7.
[19] M. Chabot, T. Tuna, K. Beroff, T. Pino, A. Le Padellec, P. Desquelles, G. Martinet, V.O. Nguyen-Thi, Y. Carpentier, F. Le Petit, E. Roueff and V. Wakelam, A&A 524 (2010) A39.
[20] T.Tuna, M.Chabot, T.Pino, P.Désesquelles, A.LePadellec, G.Martinet, M.Barat, B.Lucas, F.Mezdari, L.Montagnon, N.T.Van-Oanh, L.Lavergne, A.Lachaize, Y.Carpentier, K.Béroff J.Chem.Phys 128 (2008) 124312.
[21] D. Strasser, L. Lammich, S. Khrohn, L. Lange, H. Kreckel, J. Levin, D. Schwalm, Z. Vager, R. Wester, A. Wolf and D. Zajfman, Phys. Rev. Lett. 86 (2001) 779.
[22] B.J. McCall, A.J. Hunneycutt, R.J. Saykally, N. Djuric, G.H. Dunn, J. Semaniak, O. Novotny, A. Al-Khalili, A. Ehlerding, F. Hellberg, S. Kalhori, A. Neau, R.D. Thomas, A. Paal, F. Osterdahl and M. Larsson, Phys. Rev A 70 (2004) 052716.
[23] J. Mikosch, H. Kreckel, R. Wester, R. Plasil, J. Glosik, D. Gerlich, D. Schwalm and A. Wolf, J. Chem. Phys. 121 (2004) 11030.
[24] E. Teloy and D. Gerlich, Chem. Phys. 4 (1974) 417.
[25] A. Sen and J.B.A. Mitchell, J. Phys. B 19 (1986) L545.
[26] H. Hus, F.B. Yousif, A. Sen and J.B.A. Mitchell, Phys. Rev. A 38 (1988) 658.
[27] H. Hus, F.B. Yousif, C. Noren, A. Sen and J.B.A. Mitchell, Phys. Rev. Lett. 60 (1988) 1006
[28] F.B. Yousif and J.B.A. Mitchell, Phys. Rev. A 40 (1989) 4318.
[29] C. Noren, F.B. Yousif and J.B.A. Mitchell, J. Chem. Soc. Faraday Trans. II 85 (1989) 1697
[30] A. Petrignani et al., Phys. Rev. A 83(2011) 032711.
[31] A. Faure and J. Tennyson, M.N.R.A.S. 325 (2001) 443.
[32] A. Larson and M. Larsson, This issue.
[33] M.O.A. Ghazaly, S.M. Alshammari, C.P. Welch and H.H. Alharbi, Nucl. Instrum. Meth. A 709 (2013) 76
[34] W. Cai, F. Ruan, Y.L. Mao, R. Lu, C. Shao, M. Song, D. Yu, J. Phys Conference 300 (2011) 012009.
[35] X. Cai (Private communication, 2013).
[36] T. Mostefaoui, S. Laubé, G. Gauthier, C. Rebrion-Rowe, B.R. Rowe and J.B.A. Mitchell, J. Phys. B 32 (1999) 5247.

NIST program of spectroscopic data for light elements of fusion interest

Joseph Reader,[1] Alexander Kramida, Yuri Ralchenko, Wolfgang Wiese and Jeffrey Fuhr
National Institute of Standards and Technology, Gaithersburg, MD, USA

E-mail: joseph.reader@nist.gov

Abstract. This paper summarizes work at the National Institute of Standards and Technology (NIST) in support of the International Atomic Energy Agency (IAEA) Coordinated Research Project on "Light Element Atom, Molecule and Radical Behaviour in the Divertor and Edge Plasma Regions." It includes numerical data on radiative transition rates for ions of fluorine and neon critically compiled at NIST.

18 June 2013

Introduction

This paper summarizes work at the National Institute of Standards and Technology (NIST) in support of the International Atomic Energy Agency (IAEA) Coordinated Research Project on "Light Element Atom, Molecule and Radical Behaviour in the Divertor and Edge Plasma Regions." The work was carried out as part of the program of the NIST Atomic Spectroscopy Group, which operates as a part of the NIST Physical Measurement Laboratory. Members of the Group working in this area include the above authors, four Guest Researchers (C. Froese Fischer, L. Podobedova, E. Saloman, and J. Sansonetti), a Technical Information Specialist (R. Ibacache), and several student interns.

The work is comprised of several specific types of activities:

- Experimental investigations of atomic spectra
- Critical compilation of spectral data for light atoms
- Numerical databases available on the world wide web
- Bibliographic databases available on the world wide web
- Theoretical investigations of atomic spectra
- Development of methods for diagnosing plasmas by means of spectra of light atoms in tokamak plasmas

The following sections summarize results obtained during the last six years for each of these activities.

[1] To whom correspondence should be addressed.

1. Experiment

The following list enumerates the recent papers describing experimental work on light elements completed at NIST in the last six years:

1. Extension and new level optimization of the Ne IV spectrum [1].
2. A critical compilation of experimental data on spectral lines and energy levels of hydrogen, deuterium, and tritium [2].
3. Additions to the spectrum and energy levels and a critical compilation of helium-like and hydrogen-like boron, B IV and B V [3].
4. A critical compilation of energy levels and spectral lines of neutral boron [4].
5. Extended EUV and UV spectrum of Ne II [5].
6. The Ne II spectrum [6].
7. Energy levels and spectral lines of Ne VIII [7].
8. Energy levels and spectral lines of Ne VII [8].
9. New FTS measurements, optimized energy levels and refined VUV standards in the Ne III spectrum [9].
10. Transition probability measurements for some strong and weak lines of N I [10].
11. Experimental study of weak intersystem lines and related strong persistent lines of Ne II [11].
12. Absolute transition frequencies and quantum interference in a frequency comb based measurement of the ^{6,7}Li D lines [12].
13. Quantum interference and light polarization effects in unresolvable atomic lines: application to a precise measurement of the ^{6,7}Li D2 lines [13].
14. Progress at NIST in measuring the D-lines of Li isotopes using an optical frequency synthesizer [14].

2. Critical Compilations

In the past six years, the following critical compilations of transition probabilities for light elements have been published by the NIST team:

1. Tables of atomic transition probabilities for beryllium and boron [15].
2. Accurate atomic transition probabilities for hydrogen, helium, and lithium [16].
3. Improved critical compilations of selected atomic transition probabilities for neutral and singly ionized carbon and nitrogen [17].

The data on transition rates compiled in these publications are mostly theoretical. Scarcely available experimental laboratory data on transition rates and radiative lifetimes were used to verify the quality of theoretical calculations. In a few cases, when experimental data were proven to have better accuracy than theory, they were included in the compiled data sets. In all cases, whether the radiative rate is theoretical or experimental, the transition wavelength is always based on experimental measurements of energy levels.

The unpublished work on spectra of fluorine and neon ions recently completed at NIST is described in the following subsection.

2.1. Critical Compilation of Transition Probabilities for Ions of Fluorine and Neon

This section and the accompanying tables give transition probability data for ions of fluorine (F V-VIII) and neon (Ne VI-IX) that were recently critically compiled by W. Wiese and J. Fuhr.

Journal of Physics: Conference Series **576** (2015) 012007 doi:10.1088/1742-6596/576/1/012007

Throughout this section, the energy level values and classifications are quoted from the NIST Atomic Spectra Database (ASD) [18]. The wavelengths given in the tables, as well as in the text, are Ritz values derived from those energy levels. The data currently included in ASD for all spectra of fluorine, as well as Ne IX, have not been critically evaluated by our team. A comprehensive compilation of these spectra is currently underway. When this compilation is completed, the final adopted energy levels and wavelengths may differ from those given here. This may also affect the transition probability values (A), since all of them are obtained from calculated line strengths (S) with standard formulas (see, *e.g.*, Fuhr and Wiese [15]) using the Ritz wavelengths in vacuum.

In the tables, the wavelengths between 2000 Å and 20000 Å are given in standard air; outside this range they are given in vacuum. Conversion from vacuum to air and vice versa was made with the five-parameter formula for the refraction index of air from Peck and Reeder [19].

For some of the highly excited energy levels, the fine structure was not resolved in experiments. For transitions involving such levels, the A and S values given in the tables are sums of the values corresponding to all allowed transitions between the unresolved fine-structure component transitions, and all possible values of the total angular momentum J_i and J_k of the lower and upper level, respectively, are given as comma-delimited lists.

Uncertainties of theoretical A-values rely largely on comparisons between different sets of data. In such comparisons, we used logarithms of calculated line strengths, log(S), rather than the line strengths themselves. This is dictated by statistical properties of quantum-mechanical calculations. For log(S), statistical distribution of deviations between different calculations (or between calculations and measurements) is usually much closer to normal than for the A or S values, which allows one to use the term "uncertainty" with a well-defined meaning. Thus, uncertainties of the adopted A and S values are specified in the tables with a letter code instead of numerical values. This letter code is explained in Table 1.

Table 1. Transition probability uncertainty code.

Symbol	Uncertainty in S-value	Uncertainty in log(S)
AAA	≤ 0.3 %	≤ 0.0013
AA	≤ 1 %	≤ 0.004
A	≤ 3 %	≤ 0.013
B+	≤ 7 %	≤ 0.03
B	≤ 10 %	≤ 0.04
C+	≤ 18 %	≤ 0.08
C	≤ 25 %	≤ 0.11
D+	≤ 40 %	≤ 0.18
D	≤ 50 %	≤ 0.24
E	> 50 %	> 0.24

Many of the theoretical papers discussed in this section contain data on a large number of extremely weak parity-forbidden transitions. We included in our tabulation only those few that appear to be important in collisional-radiative modeling, *i.e.*, those that make a noticeable contribution to radiative lifetimes. However, in estimating the accuracy of these data, we used the complete sets of available data for extensive comparisons.

The procedure of data evaluation starts with selection of the most reliable data sets based on the available description of experimental or theoretical methods used to obtain the data, as well as the estimates of uncertainties sometimes available for the published data. Theoretical data often includes comparisons of results obtained in the length and velocity forms, which provides sufficient statistical basis for evaluating uncertainties, even if the authors did not provide them themselves. This initial critical selection, varying in size from several tens to several thousand transitions, depending on the spectrum, provides the basis for further comparisons with other data. These comparisons allow us to extend the evaluated data sets with reliably estimated uncertainties. As more uncertainties become estimated, the reference data set gradually extends until all available data can be evaluated.

The following subsections give details on the data evaluation and sources used for each spectrum.

2.1.1. B-like F V and Ne VI

The selected transition probability data are from advanced atomic structure calculations of Rynkun *et al.* [20], Tachiev and Froese Fischer [21] and Fernley *et al.* [22]. Rynkun *et al.* [20] made accurate and comprehensive calculations of transitions between configurations with principal quantum number $n = 2$. They used the fully-relativistic multiconfiguration Dirac–Hartree–Fock (MCDHF) and configuration interaction (CI) methods. For transitions from $n = 3$ configurations, Tachiev and Froese Fischer performed multiconfiguration Hartree-Fock (MCHF) calculations with extensive configuration mixing, involving thousands of possible interacting configurations, and included the relativistic Breit – Pauli terms. They determined the line strengths in the dipole length as well as dipole velocity representations and tabulated the presumably more accurate length data. Their tables also include the percentage differences with the velocity values. The differences, which are of the order of only one percent for strong lines, are good measures of the quality of the data.

The calculations by Fernley *et al.* [22] with the R-Matrix method have also produced very similar results, differing usually only by less than 10%. These *LS*-coupling-based calculations, made in frames of the Opacity Project (OP), produced transition probabilities for many highly excited levels not included in references [20, 21]. However, the results were obtained only for unresolved multiplets, *i.e.*, for transitions between *LS*-terms. To find the values corresponding to the fine-structure transitions, we decomposed the data of Fernley *et al.* [22] using well-known formulas for relative strengths of transitions within *LS* multiplets (see, for example, Cowan [23]). These formulas assume pure *LS* coupling. For Ne VI, Kramida *et al.* [24] showed that some of the highly excited levels are strongly mixed by interactions between series based on different ionic core configurations. Therefore, we excluded from our data tables transitions between levels having leading *LS* percentages less than 85%. Comparisons of the OP data with other theoretical results, mentioned above, included only levels with high *LS*-coupling purity.

For Ne VI, Koc [25] obtained results essentially identical to those of Tachiev and Froese Fischer [21] with a similar, slightly less detailed method. Since he covers far fewer lines, we have not utilized his results.

Further indication of the high quality of the theoretical data is good agreement between the calculated and experimental lifetime results available for a few levels. The lifetimes were measured in the 1970s with the beam-foil technique for F V by Knystautas *et al.* [26] (estimated uncertainties about 10%), and for Ne VI by Buchet and Druetta [27] (estimated uncertainties 10–25%) and by Kernahan *et al.* [28]. In table 2, we compare the lifetimes calculated by Tachiev and Froese Fischer [21] with these measurements. For F V, the agreement varies between 3% and 15% (we have not shown earlier lifetime measurements for two F V levels by Barrette and Drouin [29], since they do not discuss their uncertainties). For Ne VI, the agreement is also quite close, with the beam-foil values generally slightly larger, as would be expected due to cascading effects. Kernahan *et al.* have estimated that cascading effects should increase their measured lifetimes by only about 5 to 10%, as indicated by small replenishment ratios, *i.e.*, re-population/de-population ratios. The lifetime data for the Ne VI $1s^22s^23d$ level are a special case, insofar as the lifetime measured by Buchet and Druetta is almost at the stated time resolution of their experiment, while the calculated value is shorter than this limit.

Table 2. Comparison of calculated and measured lifetimes for some levels of F V and Ne VI.

Atomic Level	Lifetime (in 10^{-11} s)			
	MCHF Calculation [a]	Beam-Foil Measurements [b]	MCHF Calculation [a]	Beam-Foil Measurements [c,d]
	F V		Ne VI	
$1s^22s2p^2\ ^2P$	11.3	11.7 ± 1.1 [b]	9.6	10.2 [c]
$1s^22s2p^2\ ^2S$	22.6	21.5 ± 2 [b]	18.9	22.0 [c]
$1s^22p^3\ ^2P^\circ$	17.0	14.8 ± 1 [b]* 15.0 ± 1 [b]*	13.8	14.0 [c]
$1s^22p^3\ ^4S^\circ$	12.8	11.8 ± 1 [b]	10.7	10.8 [c]
$1s^22s2p3s\ ^4P^\circ$	4.9	5.4 ± 0.4 [b]	2.77	3.0 ± 0.4 [d]
$1s^22s^23d\ ^2D$			0.66	1.2 ± 0.3 [d]

[a] Tachiev and Froese Fischer [21].
[b] Knystautas *et al.* [26].
[c] Kernahan *et al.* [28].
[d] Buchet and Druetta [27].
*The lifetime of this level was measured utilizing two different transitions

We have tabulated the great majority of the results from references [20, 21, 22], but have omitted some of the very weak intersystem lines from Tachiev and Froese Fischer [21], since the "dipole length – dipole velocity" differences, computed in that work, turned out to be as large as 100%. For the strong, LS-allowed transitions, we estimate accuracies within 3% because of the excellent agreement of the dipole length and velocity values as well as the good agreement with the lifetime measurements (except for the special case of the Ne VI 1s22s23d level discussed above).

2.1.2. Be-like F VI and Ne VII

All selected transition probability data are from advanced MCDHF calculations made by Froese Fischer in 2004 for F VI [30] and in 2009 for Ne VII [31], as well as from the Opacity Project (OP) data calculated by Tully *et al.* [32]. Froese Fischer [30, 31] determined the line strengths in the Babushkin and Coulomb gauges, equivalent to the length and velocity forms in non-relativistic calculations. The differences between these two forms, which are good measures of the quality of the results, are of the order of a few percent for the stronger lines, but larger for the weaker ones. The OP calculations by Tully *et al.* [32] involved a much smaller number of interacting configurations and were based on the *LS* coupling approximation. However, they have produced very similar results, as Table 3 shows. We included the OP results for transitions from highly excited levels not covered in references [30, 31]. The OP database provides multiplet values only, so we had to utilize *LS*-coupling fractions (see, *e.g.*, Cowan [23]) to obtain values for the fine structure transitions.

Table 3. Comparison of F VI and Ne VII oscillator strengths from MCDHF and less extensive Opacity Project calculations.

Transition		MCDHF Calculation Froese Fischer [30]	Opacity Project Tully *et al.* [32]
$2s^2\ ^1S - 2s2p\ ^1P^\circ$	F VI	0.436	0.443
	Ne VII	0.388	0.389
$2s2p\ ^1P^\circ - 2p^2\ ^1D$	F VI	0.145	0.142
	Ne VII	0.132	0.130
$2s2p\ ^1P^\circ - 2p^2\ ^1S$	F VI	0.0989	0.101
	Ne VII	0.0883	0.0895
$2s2p\ ^3P^\circ - 2p^2\ ^3P$	F VI	0.167	0.169
	Ne VII	0.148	0.149

Another strong indication for the high quality of the theoretical data comes from the good agreement with experimental lifetime data, most of which were obtained with the beam-foil method in the 1970s. For F VI, these measurements were made by Knystautas *et al.* [26]. As Table 4 shows, their data agree with calculations close to, or within, the experimental uncertainty estimates. For Ne VII, many of the measured lifetimes [33–37] are somewhat longer, by amounts from 10% to 50%, than the calculated ones. This is most likely due to cascading effects in the beam-foil data that were insufficiently accounted for.

For parity-allowed transitions, we have tabulated the great majority of the results from references [30–32], but have omitted some very weak intercombination lines. Their "Babushkin – Coulomb" differences, as computed by Froese Fischer [30, 31], often turned out to be very large.

Table 4. Comparison of experimental and theoretical lifetimes (in ns) for F VI.

Atomic Level	Beam-foil experiment Knystautas *et al.* [26]	MCDHF Froese Fischer [30]	Opacity Project Tully *et al.* [32]
2s3s ^{3}S	0.031 ± 0.004	0.029	
2s2p ^{1}P°	0.31 ± 0.04	0.293	0.291
2p^2 ^{3}P	0.32 ± 0.04	0.375	0.370
2p^2 ^{1}S	0.21 ± 0.02	0.212	0.210
2p^2 ^{1}D	2.5 ± 0.1	2.248	2.28

We have included the data for only two of the many parity-forbidden transitions calculated by Froese Fischer [30, 31]. Their accuracy was estimated by extensive comparisons of all available data for magnetic dipole (M1), electric quadrupole (E2) and magnetic quadrupole (M2) transitions, which were also calculated by Kingston and Hibbert [38] using the CIV3 code. Similar to the MCDHF calculations [30, 31], the large-scale CIV3 calculations [38] included several thousand interacting configurations. However, they use a non-relativistic zero approximation with perturbative treatment of relativistic effects via the Breit-Pauli terms. Nevertheless, the agreement between the two very different calculations, CIV3 and MCDHF, is quite good, normally within 3%, 5%, and 20% for the M1, M2, and E2 transitions, respectively. For the M2 2s^2 $^1S_0$ – 2s2p $^3P^\circ_2$ transition, another recent result is available from the multi-configuration Dirac-Fock calculations of Majumder and Das [39] which agrees with the result of Froese Fischer within 1% for F VI and 0.5% for Ne VII.

2.1.3. Li-like F VII and Ne VIII

For E1 transitions, we selected two sophisticated multi-configuration calculations, the fully relativistic multi-configuration Dirac Hartree Fock (MCDHF) calculation by Froese Fischer [30] and the R-matrix method with Breit-Pauli relativistic terms (BPRM) by Nahar [40]. The MCDHF results are limited to relatively few transitions, but are based on a very extensive multi-configuration treatment, while the BPRM calculations, with smaller multi-configuration expansions, cover much larger parts of the F VII and Ne VIII spectra over wide wavelength ranges.

For a few lower-excited levels of F VII, lifetimes have been measured with the beam-foil method by Barrette *et al.* [41], Irwin and Drouin [42] and Engström and Bengtsson [43]. For Ne VIII, lifetimes were measured by Berkner *et al.* [44], Knystautas *et al.* [45], Barrette and Drouin [36], and Barrette *et al.* [46]. Table 5 shows the excellent agreement between these experimental results and the calculations. For the principal resonance level 2p, very accurate additional data are available from two other advanced theoretical treatments, the calculations of Yan *et al.* [47] with the Hylleraas-type variational method and the relativistic many body perturbation theory (RMBPT) calculations by Johnson *et al.* [48]; these data have been added to the comparison table. For Ne VIII, we included data from the 1971 study of systematic trends along the lithium isoelectronic sequence [49]. They agree closely with the recent advanced calculations. All data for the 2p level agree closely with each other, and for the other levels the differences between experimental and theoretical results do not exceed 15% for F VII and 23% for Ne VIII.

Table 5. Comparison of experimental and theoretical lifetimes (in ps) for F VII and Ne VIII.

F VII			Ne VIII			
Level	Experiment	Theory	Level	Experiment	Theory	Trend Study [49]
$1s^2 2p\ ^2P^\circ$	2040 ± 90[a]	2048[d] 2047[e] 2045[f] 2045[g]	$1s^2 2p\ ^2P^\circ_{1/2}$	1802 ± 62[h,l] 1880 ± 88[i]	1812[f] 1812[g] 1809[e] 1836[d,l]	
			$1s^2 2p\ ^2P^\circ_{3/2}$	1802 ± 62[h,l] 1859 ± 87[i]	1742[f] 1739[g] 1836[d,l]	
			$1s^2 3s\ ^2S_{1/2}$	21.3 ± 2.1[j]	21.49[f] 21.2[g]	
$1s^2 3p\ ^2P^\circ$	23.0 ± 1.0[b]	19.8[f] 20.02[g]	$1s^2 3p\ ^2P^\circ_{1/2}$	11.5 ± 1.7[j]	11.55[f] 11.55[g]	
$1s^2 3d\ ^2D$	6.8 ± 0.7[b] 6.3[c]	6.20[f] 6.20[g]	$1s^2 3d\ ^2D_{3/2}$	3.7 ± 0.6[j]	3.65[f] 3.63[g]	
			$1s^2 4s\ ^2S_{1/2}$	39.4 ± 1.2[k]	33.33[f] 32.13[g]	33.3
$1s^2 4p\ ^2P^\circ$	40.0 ± 1.0[b]	44.1[g]	$1s^2 4p\ ^2P^\circ$	22.0 ± 2.2[j] 21.4 ± 0.6[k]	20.15[g]	
			$1s^2 4d\ ^2D$	7.8 ± 0.8[j] 8.5 ± 2.4[k]	8.42[g]	8.4
			$1s^2 4f\ ^2F^\circ$	20.1 ± 0.6[k]	17.7[g]	17.9
			$1s^2 5d\ ^2D$	18.0 ± 4.5[j]	16.15[g]	
$1s2s3d\ ^4D$	2.4 ± 0.2[b]	—				

[a] Barrette *et al.* [41]
[b] Irwin and Drouin [42].
[c] Engström and Bengtsson [43].
[d] Yan *et al.* [47].
[e] Johnson *et al.* [48].
[f] Froese Fischer [30].
[g] Nahar [40].
[h] Berkner *et al.* [44].
[i] Knystautas *et al.* [45].
[j] Barrette and Drouin [36].
[k] Barrette *et al.* [46].
[l] The fine-structure splitting between $^2P^\circ_{1/2}$ and $^2P^\circ_{3/2}$ was not resolved in this study.

For forbidden transitions, we selected a small fraction of the data from Aggarwal *et al.* [50]. These authors computed the M1, E2, and M2 transition rates among the 24 lowest levels of several Li-like spectra, including F VII and Ne VIII. They used the General Purpose Relativistic Atomic Structure Package (GRASP) [51]. They also calculated the allowed (E1) transitions among these levels, and their results are in close agreement with, or often even identical to, the MCDHF data of Froese Fischer [30] and the relativistic R-Matrix data of Nahar [40] which we utilized for the allowed transitions. To evaluate the uncertainties of the data of Aggarwal *et al.* [50], we compared the full body of their results with all other available independent calculations. The selected transitions are those for which the radiative rates make a noticeable contribution to the lifetimes of their upper levels. All of them turned out to be electric-quadrupole (E2).

2.1.4. He-like F VIII and Ne IX

We selected the data for electric-dipole transitions from the elaborate non-relativistic variational calculation of Cann and Thakkar [52]. For Ne IX, most of the data were taken from the relativistic configuration-interaction (CI) calculations of Johnson *et al.* [53], but some data for transitions not covered in that work were also taken from Cann and Thakkar [52]. Both papers include data for several He-like ions. Although Johnson *et al.* [53] did not consider F VIII, the close agreement between these two essentially different calculations for other isoelectronic spectra indicates excellent quality of their results. Their high accuracy is also confirmed by the close agreement of the calculated lifetimes with those measured in beam-foil experiments made for F VIII by Knystautas *et al.* [26] and Engström *et al.* [54], and also by the close agreement with the non-relativistic MCHF calculations made for Ne VIII by Biémont [55]. Both Cann and Thakkar [52] and Johnson *et al.* [53] published only the multiplet-averaged values of transition rates. Therefore, to obtain the fine-structure data, we decomposed their original values using the standard formulas for relative line strengths within *LS*-multiplets (see, *e.g.*, Cowan [23]). Uncertainties of all selected data are estimated to be below 1%.

The critically evaluated radiative transition rates for the spectra of fluorine (F V–VIII) and neon (Ne VI–IX) are given in Table 6 in the Appendix.

3. Numerical Databases

NIST maintains a number of numerical databases that are available on the internet: http://www.nist.gov/pml. The most comprehensive of these databases is the Atomic Spectra Database (ASD) [18]. The latest version of this database, Ver. 5, was introduced in September 2012. In the past few years a number of modifications have been made to this database and a large amount of data have been added. A new database has been added within ASD that provides ionization energies for all atoms and ions.

Several other NIST databases that contain material of interest for fusion energy research. The Handbook of Basic Atomic Spectroscopic Data v.1.1.2 (online since 2005) contains data for prominent spectral lines of neutral and singly-ionized atoms of all elements hydrogen through einsteinium. In addition to the wavelengths, this database provides energy levels and transition probabilities for persistent lines. Other databases on the NIST Physics Measurement Laboratory website are also routinely used by the energy research community: Fundamental Physical Constants, precision energy levels for hydrogen and deuterium, x-ray wavelengths, molecular data, and electron-impact cross sections for excitation and ionization for atoms and molecules.

4. Bibliographic Databases

In addition to the numerical databases, NIST maintains the following bibliographic databases:

- Bibliographic Database on Atomic Transition Probabilities
- Bibliographic Database on Atomic Spectral Line Broadening and Shifts
- Bibliographic Database on Atomic Energy Levels and Spectra

These databases have the advantage that they are maintained to be constantly up-to-date; they are updated through literature searches every two or four weeks. They allow users to search for papers relevant to their specific needs, most notably papers on specific atoms and ions that contain observed or calculated wavelengths, energy levels, radiative rates, and line shapes. Unlike ASD, which contains data sets that are updated only very infrequently, the bibliographic databases are maintained current by adding all relevant papers within a few months after they are published.

5. Theoretical investigations of atomic spectra

The following theoretical papers related to spectra of light elements have been published in the last few years:

1. Energies and E1, M1, E2, M2 transition rates for states of the $2s^22p$, $2s2p^2$, and $2p^3$ configurations in boron-like ions between N III and Zn XXVI [20].
2. Multiconfiguration Dirac–Hartree–Fock energies and transition probabilities for $2p^4(^3P)3d$-$2p^4(^3P)4f$ transitions in Ne II [56].
3. Relativistic all-order and multiconfiguration Hartree–Fock calculations of the 4d-4f energy separation in Li I [57].
4. Some improved transition probabilities for neutral carbon [58].
5. Analysis of Breit–Pauli transition probabilities for lines in O III [59].
6. The effect of hot coronal electrons on extreme-ultraviolet spectral lines of He II emitted by solar transition region plasmas [60].
7. Spectroscopic diagnostics of magnetic fusion plasmas [61].
8. Online databases and computational tools for non-LTE spectroscopy [62].
9. Effect of radiative cascades on intensities of dielectronic satellites to He_α [63].
10. Comment on "Electron impact ionization of helium isoelectronic systems" [64].
11. Electron-impact excitation and ionization cross sections for ground state and excited helium atoms [65].
12. High-electron-temperature diagnostics of transient ionizing plasma using near-UV transitions [66].
13. Relativistic all-order and multiconfiguration Hartree–Fock calculations of the 4d-4f energy separation in Li I [67].
14. Atomic data for dielectronic recombination into C-like oxygen [68].

6. Development of methods for diagnosing plasmas by means of spectra of light atoms in tokamak plasmas

Energetic beams of hydrogen and deuterium are extensively used for heating and diagnostics of magnetic fusion plasmas. Various diagnostic techniques, such as charge exchange recombination spectroscopy (CXRS) and motional Stark effect (MSE) diagnostics, make use of spectroscopic signatures of interaction between the beam atoms and plasma particles to infer important plasma parameters, e.g., density, temperature, field strength and pitch angle. Needless to say, the output of diagnostic analysis directly depends on the quality of models used to simulate beam emission in hot fusion plasmas.

One of the key issues in simulations of beam emission is the proper choice of atomic model. The plasma in a magnetic fusion device is confined by a very strong magnetic field, typically 1–5 T. The fast neutrals moving with the velocity $\boldsymbol{v}$ cross the lines of the magnetic field $\boldsymbol{B}$ and therefore experience the induced electric field $\boldsymbol{E} = \boldsymbol{v} \times \boldsymbol{B}$. One of the primary consequences of this interaction is the modification of the atomic structure, which has to be described in terms of the parabolic eigenstates (n, k, m) with n being the principal quantum number, k the electric quantum number, and m the projection of the orbital angular momentum. While the basic properties of parabolic states and radiative transitions between them are well known, collisions of these states with plasma particles, of which protons are the most important, have not been adequately examined so far. This shortcoming impeded development of detailed collisional-radiative (CR) models capable of reliably explaining numerous measurements from tokamaks and other devices. In reference [69] we introduced a new general method for accurate calculation of collisional cross sections between parabolic states and developed an extensive CR model for analysis of beam emission. It was shown that the cross sections can be expressed as the linear combinations of on- and off-diagonal elements of the density matrix for the ordinary spherical (n, l, m) states. Large sets of collisional data were calculated by using the Glauber (eikonal) approximation and used as input to the modified code NOMAD [70]. The detailed comparisons of simulations with the experimental data from the JET tokamak [71] showed very good agreement and confirmed the predictive power of this model.

Most atomic models for neutral H/D beams assume the Boltzmann (statistical) distribution for excited states with the same n. The validity of this assumption, however, has not been thoroughly examined, and simple arguments indicate that it may fail for parabolic beam states under typical conditions of fusion plasmas. To analyze this problem in detail, we extended the previously developed CR model up to $n = 10$ (total of 210 states) and performed large-scale calculation of proton-impact collisional data using the atomic-orbital–close-coupling (AOCC) method and the Glauber approximation [72]. The model also included charge exchange, electron-impact collisions (both (de-)excitation and ionization), and electric field ionization.

A detailed comparison of excitation cross sections calculated with different techniques shows that generally all methods agree in the high-energy region. However, there are differences, sometimes significant, for the mid- and especially low-energy range. It should be mentioned that the nonperturbative close-coupling methods seem to show a higher level of agreement although there are nonphysical oscillations in the calculated data due to either numerical difficulties or limitations of the basis set size. It is worth noting that the total m-summed cross sections agree among all compared approximations.

Using the newly developed CR model, we calculated the populations of parabolic states in order to analyze the applicability of the statistical approximation for the m-sublevels under typical conditions of fusion plasmas (for instance, plasma temperature $T = 3$ keV, plasma density $n_i = 3\times10^{13}$ cm^{-3}, beam energy $E_B = 50$ keV/u). Without the field-induced ionization (FII), the statistical distribution is established only for the highest $n \geq 7$ states. The lower m-states within $n = 3$ and 4 show the most significant deviations, reaching 30 %; this is due to the different magnitude of excitation from the ground state into the final states differing by m. The account of FII reduces the absolute populations of the high-n states ($n \geq 8$), which are most susceptible to the FII, and greatly increases deviations from the statistical limit. This also affects the populations of the collisionally-coupled lower states with $n = 5$ to 7. We also performed simulations for plasma conditions similar to those in the international tokamak ITER, *i.e.*, $T = 20$ keV, $n_i = 10^{14}$ cm^{-3}, and $E_B = 100$ keV/u and $E_B = 500$ keV/u. It was found that deviations increase with the beam energy, and a strong asymmetry in populations of the m components is observed for lower values of n.

For quantitative characterization of non-statistical deviations for the m-resolved states within the same n, a new parameter ζ_n was introduced [72]:

$$\zeta_n = \sqrt{\frac{1}{M}\cdot\sum_{i=1}^{M}\left(\frac{n_i}{\bar{n}}-1\right)^2}, \quad \bar{n} = \frac{\sum_j N_j}{\sum_j g_j}, \tag{1}$$

where $M = n\,(n + 1)/2$ is the number of parabolic levels within the n state and $\bar{n}$ is the averaged population for the states with a principal quantum number n. The parameter ζ_n is the standard deviation from the Boltzmann values, and the limit $\zeta_n \to 0$ corresponds to the statistical distribution. This parameter was calculated over a large range of densities and beam energies and provided a very clear description of deviations from the statistical limit. Figure 1 presents the calculated values of ζ_n at $E_B = 100$ keV/u.

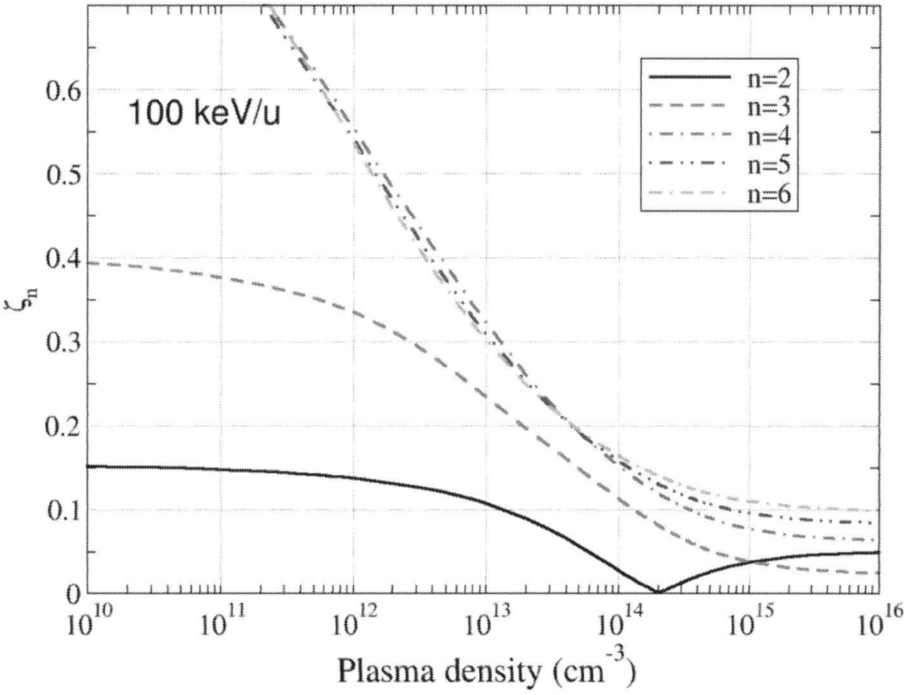

Figure 1 (From reference [72]). Deviations from the statistical limit for the beam energy of 100 keV/u, plasma temperature of 3 keV, and magnetic field of 3 T.

Finally, the most sensitive ratios of H_α components, σ_1/σ_0 and π_4/π_3, were calculated for typical conditions of ITER diagnostic and heating beams. A substantial deviation from the expected statistical values at the level of 20 % to 50 % was observed. We also found that the total σ- and π- emissions are sensitive to non-statistical effects. These results can be used for analysis of neutral beam spectra and kinetics in ITER and other existing and future fusion devices.

Our recent studies related to diagnostics of tokamak plasmas by means of spectra of light atoms have been published in the following papers:

1. Collisional excitation and emission of H_α Stark multiplet in fusion plasmas [69].
2. Non-statistical population distributions for hydrogen beams in fusion plasmas [72].
3. Non-statistical simulations for neutral beam spectroscopy in fusion plasmas [73].
4. A non-statistical atomic model for beam emission and motional Stark effect diagnostics in fusion plasmas [74].
5. Non-statistical populations of magnetic sublevels of hydrogen beam atoms in fusion plasmas [75].

Acknowledgment

This work was carried out as a part of an IAEA Coordinated Research Project on "Light Element Atom, Molecule and Radical Behaviour in the Divertor and Edge Plasma Regions." It was supported in part by the Office of Fusion Energy Sciences of the U.S. Department of Energy.

References

[1] Kramida A, Brown C M, Feldman U and Reader J 2012 *Physica Scripta* **85**, 025303
[2] Kramida A E 2010 *At. Data Nucl. Data Tables* **96** 586
[3] Kramida A E, Ryabtsev A N, Ekberg J O, Kink I, Mannervik S and Martinson I 2008 *Physica Scripta* **78** 025302
[4] Kramida A E and Ryabtsev A N 2007 *Physica Scripta* **76** 544
[5] Kramida A E, Brown C M, Feldman U and Reader J 2006 *Physica Scripta* **74** 156
[6] Kramida A E and Nave G 2006 *Eur. Phys. J.* D **39** 331
[7] Kramida A E and Buchet-Poulizac M-C 2006 *Eur. Phys. J.* D **39** 173
[8] Kramida A E and Buchet-Poulizac M-C 2006 *Eur. Phys. J.* D **38** 265
[9] Kramida A E and Nave G 2006 *Eur. Phys. J.* D **37** 1
[10] Bridges J M and Wiese W L 2010 *Phys. Rev.* A **82** 024502
[11] Bridges J M and Wiese W L 2007 *Phys. Rev.* A **76** 022513
[12] Sansonetti C J, Simien C E, Gillaspy J D, Tan J N, Brewer S M, Brown R C, Wu S and Porto J V 2011 *Phys. Rev. Lett.* **107** 023001; Erratum: 2012 *Phys. Rev. Lett.* **109**, 259901
[13] Brown R C, Wu S, Porto J V, Sansonetti C J, Simien C E, Brewer S M, Tan J N and Gillaspy J D 2013 *Phys. Rev.* A **87** 032504
[14] Simien C E, Brewer S M, Tan J N, Gillaspy J D and Sansonetti C J 2011 *Can. J. Phys.* **89** 59
[15] Fuhr J R and Wiese W L 2010 *J. Phys. Chem. Ref. Data* **39** 013101
[16] Wiese W L and Fuhr J R 2009 *J. Phys. Chem. Ref. Data* **38** 565; Erratum: 2009 *J. Phys. Chem. Ref. Data* **38** 1129
[17] Wiese W L and Fuhr J R 2007 *J. Phys. Chem. Ref. Data* **36** 1287; Erratum: 2007 *J. Phys. Chem. Ref. Data* **36** 1737
[18] Kramida A, Ralchenko Yu, Reader J. and NIST ASD Team 2012, NIST Atomic Spectra Database (version 5.0), [Online]. Available: http://physics.nist.gov/asd. National Institute of Standards and Technology, Gaithersburg, MD
[19] Peck E R and Reeder K 1972 *J. Opt. Soc. Am.* **62** 958
[20] Rynkun P, Jönsson P, Gaigalas G and Froese Fischer C 2012 *At. Data Nucl. Data Tables* **98** 481
[21] Tachiev G and Froese Fischer C 2000 *J. Phys. B* **33** 2419; Numerical results were taken from *The MCHF/MCDHF Collection*, http://nlte.nist.gov/MCHF/ (*ab initio* MCHF calculations, downloaded on April 28, 2006)

[22] Fernley J A, Hibbert A, Kingston A E and Seaton M J 1999 *J. Phys.* B **32** 5507; Numerical results were taken from *TOPBASE (Opacity Project)*, http://cdsweb.u-strasbg.fr/topbase/topbase.html (downloaded on April 1, 2012)
[23] Cowan R D 1981 *The Theory of Atomic Structure and Spectra* (Berkeley: University of California Press)
[24] Kramida A E, Bastin T, Biémont E, Dumont P-D and Garnir H-P 1999 *J. Opt. Soc. Am.* B **16** 1966
[25] Koc K 2003 *Physica Scripta* **67** 491
[26] Knystautas E J, Buchet-Poulicac M-C, Buchet J P and Druetta M 1979 *J. Opt. Soc. Am.* **69** 474
[27] Buchet J P and Druetta M 1975 *J. Opt. Soc. Am.* **65** 991
[28] Kernahan J A, Donnelly K E and Pinnington E H 1977 *Can. J. Phys.* **55** 1310
[29] Barrette L and Drouin R 1973 *Can. J. Spectrosc.* **18** 50
[30] Froese Fischer C 2004 *unpublished ab initio MCDHF calculations*; downloaded from *The MCHF/MCDHF Collection*, http://nlte.nist.gov/MCHF/ on December 21, 2005
[31] Froese Fischer C 2009 *unpublished ab initio MCDHF calculations*; downloaded from *The MCHF/MCDHF Collection*, http://nlte.nist.gov/MCHF/ on April 1, 2012
[32] Tully J A, Seaton M J and Berrington K A 1990 *J. Phys.* B **23** 3811; Numerical results were taken from *TOPBASE (Opacity Project)*, http://cdsweb.u-strasbg.fr/topbase/topbase.html (downloaded on April 1, 2012)
[33] Irwin D J G, Livingston A E and Kernahan J A 1973 *Nucl. Instrum. Methods* **110** 105
[34] Beauchemin G, Kernahan J A, Knystautas E, Irwin D J G and Drouin R 1972 *Phys. Lett.* A **40** 194
[35] Hardis J E, Curtis L J, Ramanujam P S, Livingston A E and Brooks R L 1983 *Phys. Rev.* A **27** 257
[36] Barrette L and Drouin R 1974 *Physica Scripta* **10** 213
[37] Buchet J P and Druetta M 1975 *J. Opt. Soc. Am.* **65** 991
[38] Kingston A E and Hibbert A 2001 *J. Phys.* B **34** 81
[39] Majumder S and Das B P 2000 *Phys. Rev.* A **62** 042508
[40] Nahar S N 2002 *Astron. Astrophys.* **389** 716
[41] Barrette L, Knystautas E J, Neveu B and Drouin R 1970 *Phys. Lett.* A **32** 435
[42] Irwin D J G and Drouin R 1976 *Beam-Foil Spectroscopy* vol 1, ed I A Sellin and D J Pegg (New York: Plenum) pp 347–354
[43] Engström L and Bengtsson P 1991*Physica Scripta* **43** 480
[44] Berkner K, Cooper III W S, Kaplan S N and Pyle R V 1965 *Phys. Lett.* **16** 35
[45] Knystautas E J, Barrette L, Neveu B and Drouin R 1971 *J. Quant. Spectrosc. Radiat. Transfer* **11** 75
[46] Barrette L, Irwin D J G and Drouin R 1975 *Physica Scripta* **12** 113
[47] Yan Z-C, Tambasco M and Drake G W F 1998 *Phys. Rev.* A **57** 1652
[48] Johnson W R, Liu Z W and Sapirstein J 1996 *At. Data Nucl. Data Tables* **64** 279
[49] Smith M W and Wiese W L 1971 *Astrophys. J., Suppl. Ser.* **23** 103
[50] Aggarwal K M, Keenan F P and Heeter R F 2010 *Physica Scripta* **81** 015303
[51] Grant I P, McKenzie B J, Norrington P H, Mayers D F and Pyper N C 1980 *Comput. Phys. Commun.* **21** 207
[52] Cann N M and Thakkar A J 1992 *Phys. Rev.* A **46** 5397
[53] Johnson W R, Savukov I M, Safronova U I and Dalgarno A 2002 *Astrophys. J., Suppl. Ser.* **141** 543
[54] Engström L, Jupén C, Denne B, Huldt S, Weng T M, Kaijser P, Ekberg J O, Litzén U and Martinson I 1981 *Physica Scripta* **22** 570
[55] Biémont E 1982 *Nucl. Instrum. Methods Phys. Res.* **202** 283
[56] Froese Fischer C and Ralchenko Yu 2008 *Int. J. Mass Spectrom.* **271** 85
[57] Safronova M S, Froese Fischer C and Ralchenko Yu 2007 *Phys. Rev.* A **76** 054502

[58] Froese Fischer C 2006 *J. Phys.* B **39** 2159
[59] Froese Fischer C, Tachiev G, Rubin R H and Rodríguez M 2009 *Astrophys. J.* **703** 500
[60] Feldman U, Ralchenko Yu and Doschek G A 2010 *Astrophys. J.* **708** 244
[61] Neu R, Summers H P and Ralchenko Yu 2010 *J. Phys.* B **43** 140201
[62] Ralchenko Yu 2009 *Physica Scripta* **T134** 014025
[63] Bernshtam V A, Zarnitsky Yu, Ralchenko Yu, Vainshtein L A, Weingarten L and Maron Y 2009 *Physica Scripta* **79** 035303
[64] Bernshtam V A, Ralchenko Yu and Maron Y 2009 *Eur. Phys. J.* D **51** 319
[65] Ralchenko Yu, Janev R K, Kato T, Fursa D V, Bray I and de Heer F J 2008 *At. Data Nucl. Data Tables* **94** 603
[66] Doron R, Arad R, Bernshtam V, Maron Y and Ralchenko Yu 2008 *Phys. Rev.* E **78** 036410
[67] Safronova M S, Froese Fischer C and Ralchenko Yu 2007 *Phys. Rev.* A **76** 054502
[68] Safronova U I, Ralchenko Yu, Murakami I, Kato T and Kato D 2006 *Physica Scripta* **73** 143
[69] Marchuk O, Ralchenko Yu, Janev R K, Biel W, Delabie E and Urnov A M 2010 *J. Phys.* B **43** 011002
[70] Ralchenko Yu V and Maron Y 2001 *J. Quant. Spectr. Rad. Transf.* **71** 609
[71] Delabie E *et al* 2010 *Plasma Phys. Control. Fusion* **52** 125008
[72] Marchuk O, Ralchenko Yu and Schultz D R 2012 *Plasma Phys. Controlled Fusion* **54** 095010
[73] Marchuk O, Ralchenko Yu, Schultz D R, Delabie E, Urnov A M, Biel W, Janev R K and Schlummer T 2012 *AIP Conf. Proc.* **1438** 169
[74] Ralchenko Yu, Marchuk O, Biel W, Schlummer T, Schultz D R and Stambulchik E 2012 *Rev. Sci. Instrum.* **83** 10D504
[75] Marchuk O, Ralchenko Yu, Janev R K, Delabie E, Biel W and Urnov A 2010 *Nucl. Instrum. Methods Phys. Res.* A **623** 738

APPENDIX

Table 6. Critically evaluated radiative transition rates for the spectra of fluorine (F V–VIII) and neon (Ne VI–IX).

λ_{Ritz} [a] (Å)	Configurations		Terms		Type	J_i [b]	J_k [b]	E_i (cm^{-1})	E_k (cm^{-1})	A_{ki} [c] (s^{-1})	S_{ik} [c]	Acc. [d]	Ref. [e]
F V (B-like)													
158.535	$2s2p^2$	$2s2p(^1P^o)3d$	^2D	$^2F^o$		5/2	5/2	152874.6	783650.	2.9+09	3.4−02	B	FH99LS
158.535	$2s2p^2$	$2s2p(^1P^o)3d$	^2D	$^2F^o$		5/2	7/2	152874.6	783650.	4.3+10	6.8−01	B	FH99LS
158.541	$2s2p^2$	$2s2p(^1P^o)3d$	^2D	$^2F^o$		3/2	5/2	152898.7	783650.	4.0+10	4.8−01	B	FH99LS
162.014	$2s2p^2$	$2s2p(^3P^o)3d$	^4P	$^4P^o$		1/2	1/2	85790.2	703020.0	1.07+10	4.5−02	B	FH99LS
162.055	$2s2p^2$	$2s2p(^3P^o)3d$	^4P	$^4P^o$		1/2	3/2	85790.2	702866.0	2.7+10	2.24−01	B	FH99LS
162.081	$2s2p^2$	$2s2p(^3P^o)3d$	^4P	$^4P^o$		3/2	1/2	86043.5	703020.0	5.3+10	2.24−01	B	FH99LS
162.121	$2s2p^2$	$2s2p(^3P^o)3d$	^4P	$^4P^o$		3/2	3/2	86043.5	702866.0	8.5+09	7.2−02	B	FH99LS
162.174	$2s2p^2$	$2s2p(^3P^o)3d$	^4P	$^4P^o$		3/2	5/2	86043.5	702664.0	1.92+10	2.42−01	B	FH99LS
162.217	$2s2p^2$	$2s2p(^3P^o)3d$	^4P	$^4P^o$		5/2	3/2	86407.0	702866.0	2.9+10	2.42−01	B	FH99LS
162.270	$2s2p^2$	$2s2p(^3P^o)3d$	^4P	$^4P^o$		5/2	5/2	86407.0	702664.0	4.5+10	5.6−01	B	FH99LS
163.456	$2s2p^2$	$2s2p(^3P^o)3d$	^4P	$^4D^o$		1/2	3/2	85790.2	697575.2	4.9+10	4.2−01	B	FH99LS
163.457	$2s2p^2$	$2s2p(^3P^o)3d$	^4P	$^4D^o$		1/2	1/2	85790.2	697573.	9.9+10	4.2−01	B	FH99LS
163.497	$2s2p^2$	$2s2p(^3P^o)3d$	^4P	$^4D^o$		3/2	5/2	86043.5	697675.0	8.3+10	1.07+00	B	FH99LS
163.524	$2s2p^2$	$2s2p(^3P^o)3d$	^4P	$^4D^o$		3/2	3/2	86043.5	697575.2	6.3+10	5.4−01	B	FH99LS
163.524	$2s2p^2$	$2s2p(^3P^o)3d$	^4P	$^4D^o$		3/2	1/2	86043.5	697573.	1.97+10	8.5−02	B	FH99LS
163.558	$2s2p^2$	$2s2p(^3P^o)3d$	^4P	$^4D^o$		5/2	7/2	86407.0	697811.0	1.18+11	2.04+00	B	FH99LS
163.594	$2s2p^2$	$2s2p(^3P^o)3d$	^4P	$^4D^o$		5/2	5/2	86407.0	697675.0	3.5+10	4.6−01	B	FH99LS
163.621	$2s2p^2$	$2s2p(^3P^o)3d$	^4P	$^4D^o$		5/2	3/2	86407.0	697575.2	5.9+09	5.1−02	B	FH99LS
165.982	$2s^22p$	$2s^23d$	$^2P^o$	^2D		1/2	3/2	0.0	602476.	6.54+10	5.91−01	A	TF00
166.176	$2s^22p$	$2s^23d$	$^2P^o$	^2D		3/2	5/2	744.5	602516.	7.83+10	1.06+00	A	TF00
166.187	$2s^22p$	$2s^23d$	$^2P^o$	^2D		3/2	3/2	744.5	602476.	1.31+10	1.18−01	A	TF00
171.214	$2p^3$	$2p^2(^3P)3d$	$^4S^o$	^4P		3/2	1/2	276413.8	860477.8	1.03+11	5.1−01	B	FH99LS
171.241	$2p^3$	$2p^2(^3P)3d$	$^4S^o$	^4P		3/2	3/2	276413.8	860385.8	1.03+11	1.02+00	B	FH99LS
171.302	$2p^3$	$2p^2(^3P)3d$	$^4S^o$	^4P		3/2	5/2	276413.8	860177.8	1.03+11	1.54+00	B	FH99LS
174.494	$2p^3$	$2p^2(^1D)3d$	$^2D^o$	^2F		5/2	5/2	307227.3	880312.	7.6+09	1.19−01	B	FH99LS
174.494	$2p^3$	$2p^2(^1D)3d$	$^2D^o$	^2F		5/2	7/2	307227.3	880312.	1.13+11	2.38+00	B	FH99LS

λ_{Ritz} [a] (Å)	Configurations		Terms		Type	J_i [b]	J_k [b]	E_i (cm^{-1})	E_k (cm^{-1})	A_{ki} [c] (s^{-1})	S_{ik} [c]	Acc. [d]	Ref. [e]
174.509	$2p^3$	$2p^2(^1D)3d$	$^2D^\circ$	^2F		3/2	5/2	307275.1	880312.	1.06+11	1.67+00	B	FH99LS
174.567	$2s2p^2$	$2s2p(^1P^\circ)3d$	^2P	$^2D^\circ$		1/2	3/2	214880.4	787725.	6.4+10	6.7−01	B	FH99LS
174.698	$2s2p^2$	$2s2p(^1P^\circ)3d$	^2P	$^2D^\circ$		3/2	5/2	215348.1	787764.	7.7+10	1.21+00	B	FH99LS
174.710	$2s2p^2$	$2s2p(^1P^\circ)3d$	^2P	$^2D^\circ$		3/2	3/2	215348.1	787725.	1.28+10	1.35−01	B	FH99LS
178.434	$2s2p^2$	$2s2p(^3P^\circ)3d$	^2D	$^2F^\circ$		5/2	7/2	152874.6	713306.	7.3+10	1.64+00	B	FH99LS
178.582	$2s2p^2$	$2s2p(^3P^\circ)3d$	^2D	$^2F^\circ$		5/2	5/2	152874.6	712840.	4.9+09	8.2−02	B	FH99LS
178.590	$2s2p^2$	$2s2p(^3P^\circ)3d$	^2D	$^2F^\circ$		3/2	5/2	152898.7	712840.	6.8+10	1.15+00	B	FH99LS
186.715	$2s2p^2$	$2s2p(^3P^\circ)3s$	^4P	$^4P^\circ$		3/2	5/2	86043.5	621619.0	6.10+09	1.18−01	A	TF00
186.790	$2s2p^2$	$2s2p(^3P^\circ)3s$	^4P	$^4P^\circ$		1/2	3/2	85790.2	621151.0	8.4+09	1.09−01	A	TF00
186.842	$2s2p^2$	$2s2p(^3P^\circ)3s$	^4P	$^4P^\circ$		5/2	5/2	86407.0	621619.0	1.42+10	2.75−01	A	TF00
186.878	$2s2p^2$	$2s2p(^3P^\circ)3s$	^4P	$^4P^\circ$		3/2	3/2	86043.5	621151.0	2.70+09	3.48−02	A	TF00
186.879	$2s2p^2$	$2s2p(^3P^\circ)3s$	^4P	$^4P^\circ$		1/2	1/2	85790.2	620894.5	3.37+09	2.17−02	A	TF00
186.968	$2s2p^2$	$2s2p(^3P^\circ)3s$	^4P	$^4P^\circ$		3/2	1/2	86043.5	620894.5	1.68+10	1.09−01	A	TF00
187.005	$2s2p^2$	$2s2p(^3P^\circ)3s$	^4P	$^4P^\circ$		5/2	3/2	86407.0	621151.0	9.1+09	1.18−01	A	TF00
189.940	$2p^3$	$2p^2(^1D)3d$	$^2P^\circ$	^2D		1/2	3/2	347420.6	873904.	2.7+10	3.7−01	B	FH99LS
189.947	$2p^3$	$2p^2(^1D)3d$	$^2P^\circ$	^2D		3/2	3/2	347442.5	873904.	5.4+09	7.4−02	B	FH99LS
189.947	$2p^3$	$2p^2(^1D)3d$	$^2P^\circ$	^2D		3/2	5/2	347442.5	873904.	3.3+10	6.6−01	B	FH99LS
190.567	$2s^22p$	$2s^23s$	$^2P^\circ$	^2S		1/2	1/2	0.0	524751.	5.49+09	3.75−02	A	TF00
190.837	$2s^22p$	$2s^23s$	$^2P^\circ$	^2S		3/2	1/2	744.5	524751.	1.10+10	7.56−02	A	TF00
208.440	$2s2p^2$	$2s^23p$	^4P	$^2P^\circ$		1/2	3/2	85790.2	565544.	2.9+03	5.1−08	C	TF00
208.517	$2s2p^2$	$2s^23p$	^4P	$^2P^\circ$		1/2	1/2	85790.2	565367.	1.1+04	1.0−07	C	TF00
208.550	$2s2p^2$	$2s^23p$	^4P	$^2P^\circ$		3/2	3/2	86043.5	565544.	1.1+03	2.0−08	D	TF00
208.627	$2s2p^2$	$2s^23p$	^4P	$^2P^\circ$		3/2	1/2	86043.5	565367.	1.5+04	1.3−07	C	TF00
208.709	$2s2p^2$	$2s^23p$	^4P	$^2P^\circ$		5/2	3/2	86407.0	565544.	5.9+04	1.1−06	C	TF00
213.336	$2s2p^2$	$2s2p(^3P^\circ)3s$	^2D	$^4P^\circ$		5/2	5/2	152874.6	621619.0	4.6+05	1.33−05	B	TF00
213.347	$2s2p^2$	$2s2p(^3P^\circ)3s$	^2D	$^4P^\circ$		3/2	5/2	152898.7	621619.0	2.9+04	8.2−07	B	TF00
213.549	$2s2p^2$	$2s2p(^3P^\circ)3s$	^2D	$^4P^\circ$		5/2	3/2	152874.6	621151.0	8.5+05	1.64−05	A	TF00
213.560	$2s2p^2$	$2s2p(^3P^\circ)3s$	^2D	$^4P^\circ$		3/2	3/2	152898.7	621151.0	3.6+05	6.9−06	B	TF00
213.677	$2s2p^2$	$2s2p(^3P^\circ)3s$	^2D	$^4P^\circ$		3/2	1/2	152898.7	620894.5	4.20+05	4.05−06	A	TF00

λ_{Ritz} [a] (Å)	Configurations		Terms		Type	J_i [b]	J_k [b]	E_i (cm^{-1})	E_k (cm^{-1})	A_{ki} [c] (s^{-1})	S_{ik} [c]	Acc. [d]	Ref. [e]
236.080	$2s2p^2$	$2s2p(^3P°)3s$	^2S	$^4P°$		1/2	3/2	197566.4	621151.0	1.71+05	4.44−06	A	TF00
236.223	$2s2p^2$	$2s2p(^3P°)3s$	^2S	$^4P°$		1/2	1/2	197566.4	620894.5	5.27+04	6.86−07	A	TF00
242.325	$2s2p^2$	$2s^23p$	^2D	$^2P°$		5/2	3/2	152874.6	565544.	1.00+09	2.82−02	A	TF00
242.339	$2s2p^2$	$2s^23p$	^2D	$^2P°$		3/2	3/2	152898.7	565544.	1.11+08	3.12−03	A	TF00
242.443	$2s2p^2$	$2s^23p$	^2D	$^2P°$		3/2	1/2	152898.7	565367.	1.12+09	1.57−02	A	TF00
271.756	$2s2p^2$	$2s^23p$	^2S	$^2P°$		1/2	3/2	197566.4	565544.	1.80+08	7.15−03	A	TF00
271.886	$2s2p^2$	$2s^23p$	^2S	$^2P°$		1/2	1/2	197566.4	565367.	1.76+08	3.50−03	A	TF00
285.174	$2s2p^2$	$2s^23p$	^2P	$^2P°$		1/2	3/2	214880.4	565544.	2.6+06	1.19−04	B	TF00
285.318	$2s2p^2$	$2s^23p$	^2P	$^2P°$		1/2	1/2	214880.4	565367.	1.67+07	3.8−04	B	TF00
285.554	$2s2p^2$	$2s^23p$	^2P	$^2P°$		3/2	3/2	215348.1	565544.	1.83+07	8.4−04	B	TF00
285.699	$2s2p^2$	$2s^23p$	^2P	$^2P°$		3/2	1/2	215348.1	565367.	6.4+06	1.47−04	B	TF00
338.652	$2p^3$	$2s^23d$	$^2D°$	^2D		5/2	5/2	307227.3	602516.	2.35+06	2.7−04	B	TF00
338.698	$2p^3$	$2s^23d$	$^2D°$	^2D		5/2	3/2	307227.3	602476.	2.35+05	1.80−05	B	TF00
338.706	$2p^3$	$2s^23d$	$^2D°$	^2D		3/2	5/2	307275.1	602516.	1.8+05	2.1−05	C	TF00
338.752	$2p^3$	$2s^23d$	$^2D°$	^2D		3/2	3/2	307275.1	602476.	2.30+06	1.77−04	B	TF00
382.187	$2s2p^2$	$2p^3$	^4P	$^2P°$		1/2	3/2	85790.2	347442.5	1.4+03	1.5−07	D+	R12
382.219	$2s2p^2$	$2p^3$	^4P	$^2P°$		1/2	1/2	85790.2	347420.6	5.2+04	2.9−06	B	R12
382.557	$2s2p^2$	$2p^3$	^4P	$^2P°$		3/2	3/2	86043.5	347442.5	1.27+05	1.41−05	B	R12
382.589	$2s2p^2$	$2p^3$	^4P	$^2P°$		3/2	1/2	86043.5	347420.6	9.+03	5.0−07	D+	R12
383.090	$2s2p^2$	$2p^3$	^4P	$^2P°$		5/2	3/2	86407.0	347442.5	4.8+04	5.3−06	B	R12
392.044	$2p^3$	$2s^23d$	$^2P°$	^2D		3/2	5/2	347442.5	602516.	2.3+05	4.1−05	C	TF00
392.072	$2p^3$	$2s^23d$	$^2P°$	^2D		1/2	3/2	347420.6	602476.	1.6+05	1.9−05	C	TF00
392.105	$2p^3$	$2s^23d$	$^2P°$	^2D		3/2	3/2	347442.5	602476.	3.0+04	3.6−06	C	TF00
402.678	$2p^3$	$2s^23s$	$^4S°$	^2S		3/2	1/2	276413.8	524751.	1.6+02	1.0−08	C	TF00
451.498	$2s2p^2$	$2p^3$	^4P	$^2D°$		1/2	3/2	85790.2	307275.1	2.7+03	4.9−07	D+	R12
452.015	$2s2p^2$	$2p^3$	^4P	$^2D°$		3/2	3/2	86043.5	307275.1	5.5+04	1.01−05	B	R12
452.113	$2s2p^2$	$2p^3$	^4P	$^2D°$		3/2	5/2	86043.5	307227.3	5.0+03	1.4−06	D+	R12
452.759	$2s2p^2$	$2p^3$	^4P	$^2D°$		5/2	3/2	86407.0	307275.1	1.07+04	1.96−06	B	R12
452.857	$2s2p^2$	$2p^3$	^4P	$^2D°$		5/2	5/2	86407.0	307227.3	1.81+05	5.0−05	B	R12

λ_{Ritz} [a] (Å)	Configurations		Terms		Type	J_i [b]	J_k [b]	E_i (cm^{-1})	E_k (cm^{-1})	A_{ki} [c] (s^{-1})	S_{ik} [c]	Acc. [d]	Ref. [e]
459.821	$2p^3$	$2s^23s$	$^2D°$	^2S		3/2	1/2	307275.1	524751.	1.3+02	1.3−08	D	TF00
464.364	$2s^22p$	$2s2p^2$	$^2P°$	^2P		1/2	3/2	0.0	215348.1	1.46+09	2.89−01	A	R12
465.375	$2s^22p$	$2s2p^2$	$^2P°$	^2P		1/2	1/2	0.0	214880.4	5.72+09	5.69−01	A	R12
465.975	$2s^22p$	$2s2p^2$	$^2P°$	^2P		3/2	3/2	744.5	215348.1	7.34+09	1.47+00	A	R12
466.993	$2s^22p$	$2s2p^2$	$^2P°$	^2P		3/2	1/2	744.5	214880.4	3.06+09	3.08−01	A	R12
506.159	$2s^22p$	$2s2p^2$	$^2P°$	^2S		1/2	1/2	0.0	197566.4	1.60+09	2.04−01	A	R12
508.074	$2s^22p$	$2s2p^2$	$^2P°$	^2S		3/2	1/2	744.5	197566.4	2.79+09	3.61−01	A	R12
513.959	$2s2p^2$	$2p^3$	^2D	$^2P°$		5/2	3/2	152874.6	347442.5	3.13+09	8.4−01	A	R12
514.023	$2s2p^2$	$2p^3$	^2D	$^2P°$		3/2	3/2	152898.7	347442.5	3.67+08	9.8−02	A	R12
514.081	$2s2p^2$	$2p^3$	^2D	$^2P°$		3/2	1/2	152898.7	347420.6	3.54+09	4.75−01	A	R12
524.021	$2s^23d$	$2s2p(^1P°)3d$	^2D	$^2P°$		3/2	1/2	602476.	793308.	3.8+09	5.3−01	B	FH99LS
524.021	$2s^23d$	$2s2p(^1P°)3d$	^2D	$^2P°$		3/2	3/2	602476.	793308.	3.8+08	1.07−01	B	FH99LS
524.131	$2s^23d$	$2s2p(^1P°)3d$	^2D	$^2P°$		5/2	3/2	602516.	793308.	3.4+09	9.6−01	B	FH99LS
524.594	$2s2p^2$	$2p^3$	^4P	$^4S°$		1/2	3/2	85790.2	276413.8	1.31+09	3.73−01	A	R12
525.292	$2s2p^2$	$2p^3$	^4P	$^4S°$		3/2	3/2	86043.5	276413.8	2.61+09	7.46−01	A	R12
526.297	$2s2p^2$	$2p^3$	^4P	$^4S°$		5/2	3/2	86407.0	276413.8	3.88+09	1.12+00	A	R12
531.098	$2s^23p$	$2s2p(^1P°)3p$	$^2P°$	^2P		1/2	3/2	565367.	753656.	6.3+08	1.87−01	B	FH99LS
531.457	$2s^23p$	$2s2p(^1P°)3p$	$^2P°$	^2P		1/2	1/2	565367.	753529.	2.5+09	3.7−01	B	FH99LS
531.598	$2s^23p$	$2s2p(^1P°)3p$	$^2P°$	^2P		3/2	3/2	565544.	753656.	3.2+09	9.4−01	B	FH99LS
531.904	$2s^23s$	$2s2p(^1P°)3s$	^2S	$^2P°$		1/2	1/2	524751.	712755.	3.2+09	4.8−01	B	FH99LS
531.904	$2s^23s$	$2s2p(^1P°)3s$	^2S	$^2P°$		1/2	3/2	524751.	712755.	3.2+09	9.6−01	B	FH99LS
531.957	$2s^23p$	$2s2p(^1P°)3p$	$^2P°$	^2P		3/2	1/2	565544.	753529.	1.26+09	1.87−01	B	FH99LS
537.522	$2s^23p$	$2s2p(^1P°)3p$	$^2P°$	^2D		1/2	3/2	565367.	751406.	1.74+09	5.3−01	B	FH99LS
537.900	$2s^23p$	$2s2p(^1P°)3p$	$^2P°$	^2D		3/2	5/2	565544.	751452.	2.08+09	9.6−01	B	FH99LS
538.034	$2s^23p$	$2s2p(^1P°)3p$	$^2P°$	^2D		3/2	3/2	565544.	751406.	3.5+08	1.07−01	B	FH99LS
539.403	$2s2p(^3P°)3p$	$2s2p(^3P°)4d$	^2P	$^2D°$		1/2	3/2	656208.	841598.	2.26+09	7.0−01	B	FH99LS
539.700	$2s^23d$	$2s2p(^1P°)3d$	^2D	$^2D°$		3/2	5/2	602476.	787764.	2.34+08	1.09−01	B	FH99LS
539.785	$2s2p(^3P°)3p$	$2s2p(^3P°)4d$	^2P	$^2D°$		3/2	5/2	656436.	841695.	2.7+09	1.26+00	B	FH99LS
539.814	$2s^23d$	$2s2p(^1P°)3d$	^2D	$^2D°$		3/2	3/2	602476.	787725.	3.2+09	9.8−01	B	FH99LS

λ_{Ritz} [a] (Å)	Configurations		Terms		Type	J_i [b]	J_k [b]	E_i (cm^{-1})	E_k (cm^{-1})	A_{ki} [c] (s^{-1})	S_{ik} [c]	Acc. [d]	Ref. [e]
539.817	$2s^23d$	$2s2p(^1P^\circ)3d$	^2D	$^2D^\circ$		5/2	5/2	602516.	787764.	3.3+09	1.52+00	B	FH99LS
539.931	$2s^23d$	$2s2p(^1P^\circ)3d$	^2D	$^2D^\circ$		5/2	3/2	602516.	787725.	3.5+08	1.09−01	B	FH99LS
540.068	$2s2p(^3P^\circ)3p$	$2s2p(^3P^\circ)4d$	^2P	$^2D^\circ$		3/2	3/2	656436.	841598.	4.5+08	1.40−01	B	FH99LS
546.304	$2s2p(^3P^\circ)3p$	$2s2p(^3P^\circ)4d$	^4D	$^4D^\circ$		1/2	3/2	657744.8	840793.0	1.83+08	5.9−02	B	FH99LS
546.358	$2s2p(^3P^\circ)3p$	$2s2p(^3P^\circ)4d$	^4D	$^4D^\circ$		1/2	1/2	657744.8	840775.0	3.7+08	5.9−02	B	FH99LS
546.563	$2s2p(^3P^\circ)3p$	$2s2p(^3P^\circ)4d$	^4D	$^4D^\circ$		3/2	5/2	657890.0	840851.5	1.70+08	8.2−02	B	FH99LS
546.701	$2s2p(^3P^\circ)3p$	$2s2p(^3P^\circ)4d$	^4D	$^4D^\circ$		5/2	7/2	658145.7	841061.0	1.04+08	6.7−02	B	FH99LS
546.738	$2s2p(^3P^\circ)3p$	$2s2p(^3P^\circ)4d$	^4D	$^4D^\circ$		3/2	3/2	657890.0	840793.0	2.9+08	9.4−02	B	FH99LS
546.792	$2s2p(^3P^\circ)3p$	$2s2p(^3P^\circ)4d$	^4D	$^4D^\circ$		3/2	1/2	657890.0	840775.0	3.6+08	5.9−02	B	FH99LS
547.328	$2s2p(^3P^\circ)3p$	$2s2p(^3P^\circ)4d$	^4D	$^4D^\circ$		5/2	5/2	658145.7	840851.5	4.2+08	2.03−01	B	FH99LS
547.503	$2s2p(^3P^\circ)3p$	$2s2p(^3P^\circ)4d$	^4D	$^4D^\circ$		5/2	3/2	658145.7	840793.0	2.5+08	8.2−02	B	FH99LS
547.903	$2s2p(^3P^\circ)3p$	$2s2p(^3P^\circ)4d$	^4D	$^4D^\circ$		7/2	7/2	658547.0	841061.0	6.2+08	4.0−01	B	FH99LS
548.533	$2s2p(^3P^\circ)3p$	$2s2p(^3P^\circ)4d$	^4D	$^4D^\circ$		7/2	5/2	658547.0	840851.5	1.37+08	6.7−02	B	FH99LS
551.956	$2s^23d$	$2s2p(^1P^\circ)3d$	^2D	$^2F^\circ$		3/2	5/2	602476.	783650.	3.2+09	1.61+00	B	FH99LS
552.077	$2s^23d$	$2s2p(^1P^\circ)3d$	^2D	$^2F^\circ$		5/2	5/2	602516.	783650.	2.31+08	1.15−01	B	FH99LS
552.077	$2s^23d$	$2s2p(^1P^\circ)3d$	^2D	$^2F^\circ$		5/2	7/2	602516.	783650.	3.5+09	2.30+00	B	FH99LS
559.776	$2s^23p$	$2s^24d$	$^2P^\circ$	^2D		1/2	3/2	565367.	744010.	3.6+09	1.26+00	B	FH99LS
560.249	$2s^23p$	$2s^24d$	$^2P^\circ$	^2D		3/2	5/2	565544.	744036.	4.3+09	2.26+00	B	FH99LS
560.331	$2s^23p$	$2s^24d$	$^2P^\circ$	^2D		3/2	3/2	565544.	744010.	7.2+08	2.5−01	B	FH99LS
563.919	$2p^3$	$2s^23s$	$^2P^\circ$	^2S		1/2	1/2	347420.6	524751.	8.8+04	1.6−05	C	TF00
563.989	$2p^3$	$2s^23s$	$^2P^\circ$	^2S		3/2	1/2	347442.5	524751.	1.8+05	3.2−05	C	TF00
567.499	$2s2p(^3P^\circ)3p$	$2s2p(^3P^\circ)4d$	^4S	$^4P^\circ$		3/2	5/2	665996.2	842208.0	2.01+09	1.09+00	B	FH99LS
568.557	$2s2p(^3P^\circ)3p$	$2s2p(^3P^\circ)4d$	^4S	$^4P^\circ$		3/2	3/2	665996.2	841880.	2.00+09	7.2−01	B	FH99LS
568.881	$2s2p(^3P^\circ)3p$	$2s2p(^3P^\circ)4d$	^4S	$^4P^\circ$		3/2	1/2	665996.2	841780.	1.99+09	3.6−01	B	FH99LS
582.839	$2s2p(^3P^\circ)3p$	$2s2p(^3P^\circ)4d$	^2D	$^2F^\circ$		3/2	5/2	675932.	847506.	1.80+09	1.06+00	B	FH99LS
583.448	$2s2p(^3P^\circ)3p$	$2s2p(^3P^\circ)4d$	^2D	$^2F^\circ$		5/2	7/2	676422.	847817.	1.92+09	1.51+00	B	FH99LS
584.508	$2s2p(^3P^\circ)3p$	$2s2p(^3P^\circ)4d$	^2D	$^2F^\circ$		5/2	5/2	676422.	847506.	1.28+08	7.5−02	B	FH99LS
594.601	$2s2p(^3P^\circ)3p$	$2p^2(^1D)3p$	^2D	$^2F^\circ$		3/2	5/2	675932.	844112.	1.34+09	8.3−01	B	FH99LS
595.79	$2s2p(^3P^\circ)3p$	$2p^2(^1D)3p$	^2D	$^2F^\circ$		5/2	7/2	676422.	844266.	1.42+09	1.19+00	B	FH99LS

λ_{Ritz} [a] (Å)	Configurations		Terms		Type	J_i [b]	J_k [b]	E_i (cm^{-1})	E_k (cm^{-1})	A_{ki} [c] (s^{-1})	S_{ik} [c]	Acc. [d]	Ref. [e]
596.34	2s2p(^{3}P°)3p	2p^2(^{1}D)3p	^{2}D	^{2}F°		5/2	5/2	676422.	844112.	9.5+07	5.9−02	B	FH99LS
597.11	2s2p(^{3}P°)3d	2p^2(^{1}D)3d	^{2}F°	^{2}F		5/2	5/2	712840.	880312.	7.0+08	4.4−01	B	FH99LS
597.11	2s2p(^{3}P°)3d	2p^2(^{1}D)3d	^{2}F°	^{2}F		5/2	7/2	712840.	880312.	2.6+07	2.21−02	B	FH99LS
598.78	2s2p(^{3}P°)3d	2p^2(^{1}D)3d	^{2}F°	^{2}F		7/2	5/2	713306.	880312.	3.5+07	2.21−02	B	FH99LS
598.78	2s2p(^{3}P°)3d	2p^2(^{1}D)3d	^{2}F°	^{2}F		7/2	7/2	713306.	880312.	7.1+08	6.0−01	B	FH99LS
603.27	2s2p(^{3}P°)3p	2s2p(^{3}P°)4d	^{2}D	^{2}D°		3/2	5/2	675932.	841695.	6.3+07	4.1−02	B	FH99LS
603.62	2s2p(^{3}P°)3p	2s2p(^{3}P°)4d	^{2}D	^{2}D°		3/2	3/2	675932.	841598.	8.5+08	3.7−01	B	FH99LS
605.06	2s2p(^{3}P°)3p	2s2p(^{3}P°)4d	^{2}D	^{2}D°		5/2	5/2	676422.	841695.	8.7+08	5.7−01	B	FH99LS
605.41	2s2p(^{3}P°)3p	2s2p(^{3}P°)4d	^{2}D	^{2}D°		5/2	3/2	676422.	841598.	9.4+07	4.1−02	B	FH99LS
611.180	2s2p(^{3}P°)3s	2p^2(^{3}P)3s	^{4}P°	^{4}P		3/2	5/2	621151.0	784768.8	9.2+08	6.2−01	B	FH99LS
611.739	2s2p(^{3}P°)3s	2p^2(^{3}P)3s	^{4}P°	^{4}P		1/2	3/2	620894.5	784362.8	1.27+09	5.7−01	B	FH99LS
612.701	2s2p(^{3}P°)3s	2p^2(^{3}P)3s	^{4}P°	^{4}P		3/2	3/2	621151.0	784362.8	4.1+08	1.84−01	B	FH99LS
612.728	2s2p(^{3}P°)3s	2p^2(^{3}P)3s	^{4}P°	^{4}P		1/2	1/2	620894.5	784099.	5.1+08	1.15−01	B	FH99LS
612.934	2s2p(^{3}P°)3s	2p^2(^{3}P)3s	^{4}P°	^{4}P		5/2	5/2	621619.0	784768.8	2.12+09	1.45+00	B	FH99LS
613.693	2s2p(^{3}P°)3s	2p^2(^{3}P)3s	^{4}P°	^{4}P		3/2	1/2	621151.0	784099.	2.5+09	5.7−01	B	FH99LS
613.855	2s2p(^{3}P°)3d	2p^2(^{3}P)3d	^{4}D°	^{4}P		1/2	1/2	697573.	860477.8	1.16+09	2.7−01	B	FH99LS
613.864	2s2p(^{3}P°)3d	2p^2(^{3}P)3d	^{4}D°	^{4}P		3/2	1/2	697575.2	860477.8	1.16+09	2.7−01	B	FH99LS
614.202	2s2p(^{3}P°)3d	2p^2(^{3}P)3d	^{4}D°	^{4}P		1/2	3/2	697573.	860385.8	1.16+08	5.3−02	B	FH99LS
614.211	2s2p(^{3}P°)3d	2p^2(^{3}P)3d	^{4}D°	^{4}P		3/2	3/2	697575.2	860385.8	7.4+08	3.4−01	B	FH99LS
614.463	2s2p(^{3}P°)3s	2p^2(^{3}P)3s	^{4}P°	^{4}P		5/2	3/2	621619.0	784362.8	1.36+09	6.2−01	B	FH99LS
614.587	2s2p(^{3}P°)3d	2p^2(^{3}P)3d	^{4}D°	^{4}P		5/2	3/2	697675.0	860385.8	1.46+09	6.7−01	B	FH99LS
614.996	2s2p(^{3}P°)3d	2p^2(^{3}P)3d	^{4}D°	^{4}P		3/2	5/2	697575.2	860177.8	4.6+07	3.2−02	B	FH99LS
615.374	2s2p(^{3}P°)3d	2p^2(^{3}P)3d	^{4}D°	^{4}P		5/2	5/2	697675.0	860177.8	4.2+08	2.9−01	B	FH99LS
615.889	2s2p(^{3}P°)3d	2p^2(^{3}P)3d	^{4}D°	^{4}P		7/2	5/2	697811.0	860177.8	1.84+09	1.28+00	B	FH99LS
630.27	2s2p(^{3}P°)3s	2p^2(^{3}P)3s	^{2}P°	^{2}P		1/2	3/2	638856.	797519.	8.8+08	4.3−01	B	FH99LS
632.10	2s2p(^{3}P°)3s	2p^2(^{3}P)3s	^{2}P°	^{2}P		1/2	1/2	638856.	797059.	3.5+09	8.7−01	B	FH99LS
632.30	2s2p(^{3}P°)3s	2p^2(^{3}P)3s	^{2}P°	^{2}P		3/2	3/2	639365.	797519.	4.3+09	2.17+00	B	FH99LS
634.028	2s2p(^{3}P°)3d	2p^2(^{3}P)3d	^{4}P°	^{4}P		5/2	3/2	702664.0	860385.8	4.0+08	2.02−01	B	FH99LS
634.14	2s2p(^{3}P°)3s	2p^2(^{3}P)3s	^{2}P°	^{2}P		3/2	1/2	639365.	797059.	1.72+09	4.3−01	B	FH99LS

λ_{Ritz} [a] (Å)	Configurations		Terms		Type	J_i [b]	J_k [b]	E_i (cm^{-1})	E_k (cm^{-1})	A_{ki} [c] (s^{-1})	S_{ik} [c]	Acc. [d]	Ref. [e]
634.470	2s2p(^{3}P°)3d	2p^2(^{3}P)3d	^{4}P°	^{4}P		3/2	1/2	702866.0	860477.8	7.4+08	1.87−01	B	FH99LS
634.841	2s2p(^{3}P°)3d	2p^2(^{3}P)3d	^{4}P°	^{4}P		3/2	3/2	702866.0	860385.8	1.18+08	6.0−02	B	FH99LS
634.865	2s2p(^{3}P°)3d	2p^2(^{3}P)3d	^{4}P°	^{4}P		5/2	5/2	702664.0	860177.8	6.2+08	4.7−01	B	FH99LS
635.091	2s2p(^{3}P°)3d	2p^2(^{3}P)3d	^{4}P°	^{4}P		1/2	1/2	703020.0	860477.8	1.48+08	3.7−02	B	FH99LS
635.385	2s2p(^{3}P°)3p	2p^2(^{3}P)3p	^{4}S	^{4}P°		3/2	5/2	665996.2	823381.	1.34+09	1.02+00	B	FH99LS
635.462	2s2p(^{3}P°)3d	2p^2(^{3}P)3d	^{4}P°	^{4}P		1/2	3/2	703020.0	860385.8	3.7+08	1.87−01	B	FH99LS
635.680	2s2p(^{3}P°)3d	2p^2(^{3}P)3d	^{4}P°	^{4}P		3/2	5/2	702866.0	860177.8	2.7+08	2.02−01	B	FH99LS
636.396	2s2p(^{3}P°)3p	2p^2(^{3}P)3p	^{4}S	^{4}P°		3/2	3/2	665996.2	823131.	1.34+09	6.8−01	B	FH99LS
636.554	2s2p(^{3}P°)3p	2p^2(^{3}P)3p	^{4}S	^{4}P°		3/2	1/2	665996.2	823092.	1.34+09	3.4−01	B	FH99LS
647.666	2s2p^2	2p^3	^{2}D	^{2}D°		5/2	3/2	152874.6	307275.1	2.07+08	1.11−01	A	R12
647.767	2s2p^2	2p^3	^{2}D	^{2}D°		3/2	3/2	152898.7	307275.1	1.70+09	9.1−01	A	R12
647.867	2s2p^2	2p^3	^{2}D	^{2}D°		5/2	5/2	152874.6	307227.3	1.79+09	1.44+00	A	R12
647.968	2s2p^2	2p^3	^{2}D	^{2}D°		3/2	5/2	152898.7	307227.3	1.33+08	1.07−01	A	R12
648.72	2s2p(^{3}P°)3d	2p^2(^{3}P)3d	^{2}D°	^{2}P		3/2	1/2	699293.	853442.	1.71+09	4.6−01	B	FH99LS
650.44	2s2p(^{3}P°)3d	2p^2(^{3}P)3d	^{2}D°	^{2}P		3/2	3/2	699293.	853035.	1.69+08	9.2−02	B	FH99LS
650.85	2s2p(^{3}P°)3d	2p^2(^{3}P)3d	^{2}D°	^{2}P		5/2	3/2	699389.	853035.	1.52+09	8.3−01	B	FH99LS
654.028	2s^{2}2p	2s2p^2	^{2}P°	^{2}D		1/2	3/2	0.0	152898.7	7.88+08	4.35−01	A	R12
657.009	2s2p(^{3}P°)3p	2s2p(^{3}P°)4s	^{4}D	^{4}P°		3/2	5/2	657890.0	810095.0	5.6+07	4.7−02	B	FH99LS
657.228	2s^{2}2p	2s2p^2	^{2}P°	^{2}D		3/2	3/2	744.5	152898.7	1.48+08	8.27−02	A	R12
657.332	2s^{2}2p	2s2p^2	^{2}P°	^{2}D		3/2	5/2	744.5	152874.6	9.2+08	7.75−01	A	R12
658.076	2s2p(^{3}P°)3p	2s2p(^{3}P°)4s	^{4}D	^{4}P°		1/2	3/2	657744.8	809703.0	1.38+08	7.8−02	B	FH99LS
658.114	2s2p(^{3}P°)3p	2s2p(^{3}P°)4s	^{4}D	^{4}P°		5/2	5/2	658145.7	810095.0	5.0+08	4.2−01	B	FH99LS
658.705	2s2p(^{3}P°)3p	2s2p(^{3}P°)4s	^{4}D	^{4}P°		3/2	3/2	657890.0	809703.0	8.8+08	5.0−01	B	FH99LS
659.654	2s2p(^{3}P°)3p	2s2p(^{3}P°)4s	^{4}D	^{4}P°		1/2	1/2	657744.8	809339.5	1.37+09	3.9−01	B	FH99LS
659.816	2s2p(^{3}P°)3p	2s2p(^{3}P°)4s	^{4}D	^{4}P°		5/2	3/2	658145.7	809703.0	1.73+09	9.8−01	B	FH99LS
659.857	2s2p(^{3}P°)3p	2s2p(^{3}P°)4s	^{4}D	^{4}P°		7/2	5/2	658547.0	810095.0	2.20+09	1.87+00	B	FH99LS
660.286	2s2p(^{3}P°)3p	2s2p(^{3}P°)4s	^{4}D	^{4}P°		3/2	1/2	657890.0	809339.5	1.37+09	3.9−01	B	FH99LS
667.218	2s2p^2	2p^3	^{2}S	^{2}P°		1/2	3/2	197566.4	347442.5	5.82+08	3.41−01	A	R12
667.315	2s2p^2	2p^3	^{2}S	^{2}P°		1/2	1/2	197566.4	347420.6	5.12+08	1.50−01	A	R12

λ_{Ritz} [a] (Å)	Configurations		Terms		Type	J_i [b]	J_k [b]	E_i (cm^{-1})	E_k (cm^{-1})	A_{ki} [c] (s^{-1})	S_{ik} [c]	Acc. [d]	Ref. [e]
677.65	$2s^2 3p$	$2s^2 4s$	$^2P^{\circ}$	^2S		1/2	1/2	565367.	712936.	1.64+09	5.0−01	B	FH99LS
678.46	$2s^2 3p$	$2s^2 4s$	$^2P^{\circ}$	^2S		3/2	1/2	565544.	712936.	3.3+09	1.01+00	B	FH99LS
694.21	$2s^2 3d$	$2s^2 4f$	^2D	$^2F^{\circ}$		3/2	5/2	602476.	746525.	6.7+09	6.6+00	B	FH99LS
694.40	$2s^2 3d$	$2s^2 4f$	^2D	$^2F^{\circ}$		5/2	5/2	602516.	746525.	4.8+08	4.7−01	B	FH99LS
694.40	$2s^2 3d$	$2s^2 4f$	^2D	$^2F^{\circ}$		5/2	7/2	602516.	746525.	7.1+09	9.4+00	B	FH99LS
705.95	$2s2p(^3P^{\circ})3d$	$2s2p(^3P^{\circ})4f$	$^2D^{\circ}$	^2F		3/2	5/2	699293.	840946.	2.37+09	2.47+00	B	FH99LS
706.43	$2s2p(^3P^{\circ})3d$	$2s2p(^3P^{\circ})4f$	$^2D^{\circ}$	^2F		5/2	5/2	699389.	840946.	1.69+08	1.76−01	B	FH99LS
706.43	$2s2p(^3P^{\circ})3d$	$2s2p(^3P^{\circ})4f$	$^2D^{\circ}$	^2F		5/2	7/2	699389.	840946.	2.5+09	3.5+00	B	FH99LS
742.37	$2s2p(^3P^{\circ})3d$	$2s2p(^3P^{\circ})4f$	$^2F^{\circ}$	^2G		5/2	7/2	712840.	847543.	8.3+09	1.34+01	B	FH99LS
744.95	$2s2p(^3P^{\circ})3d$	$2s2p(^3P^{\circ})4f$	$^2F^{\circ}$	^2G		7/2	7/2	713306.	847543.	3.0+08	5.0−01	B	FH99LS
744.95	$2s2p(^3P^{\circ})3d$	$2s2p(^3P^{\circ})4f$	$^2F^{\circ}$	^2G		7/2	9/2	713306.	847543.	8.5+09	1.73+01	B	FH99LS
754.363	$2s2p^2$	$2p^3$	^2P	$^2P^{\circ}$		1/2	3/2	214880.4	347442.5	2.69+08	2.28−01	A	R12
754.488	$2s2p^2$	$2p^3$	^2P	$^2P^{\circ}$		1/2	1/2	214880.4	347420.6	1.22+09	5.19−01	A	R12
757.034	$2s2p^2$	$2p^3$	^2P	$^2P^{\circ}$		3/2	3/2	215348.1	347442.5	1.49+09	1.28+00	A	R12
757.160	$2s2p^2$	$2p^3$	^2P	$^2P^{\circ}$		3/2	1/2	215348.1	347420.6	5.82+08	2.49−01	A	R12
780.60	$2s2p(^3P^{\circ})3d$	$2s2p(^3P^{\circ})4f$	$^2F^{\circ}$	^2F		5/2	5/2	712840.	840946.	1.05+09	1.48+00	B	FH99LS
780.60	$2s2p(^3P^{\circ})3d$	$2s2p(^3P^{\circ})4f$	$^2F^{\circ}$	^2F		5/2	7/2	712840.	840946.	3.9+07	7.4−02	B	FH99LS
783.45	$2s2p(^3P^{\circ})3d$	$2s2p(^3P^{\circ})4f$	$^2F^{\circ}$	^2F		7/2	5/2	713306.	840946.	5.2+07	7.4−02	B	FH99LS
783.45	$2s2p(^3P^{\circ})3d$	$2s2p(^3P^{\circ})4f$	$^2F^{\circ}$	^2F		7/2	7/2	713306.	840946.	1.05+09	1.99+00	B	FH99LS
809.460	$2s2p^2$	$2p^3$	^2D	$^4S^{\circ}$		5/2	3/2	152874.6	276413.8	7.+02	7.−07	D+	R12
809.618	$2s2p^2$	$2p^3$	^2D	$^4S^{\circ}$		3/2	3/2	152898.7	276413.8	1.9+00	2.0−09	D+	R12
825.90	$2s2p(^3P^{\circ})3d$	$2s2p(^3P^{\circ})4p$	$^2F^{\circ}$	^2D		5/2	5/2	712840.	833920.	3.2+07	5.3−02	B	FH99LS
828.77	$2s2p(^3P^{\circ})3d$	$2s2p(^3P^{\circ})4p$	$^2F^{\circ}$	^2D		5/2	3/2	712840.	833501.	6.6+08	7.4−01	B	FH99LS
829.09	$2s2p(^3P^{\circ})3d$	$2s2p(^3P^{\circ})4p$	$^2F^{\circ}$	^2D		7/2	5/2	713306.	833920.	6.3+08	1.06+00	B	FH99LS
911.505	$2s2p^2$	$2p^3$	^2S	$^2D^{\circ}$		1/2	3/2	197566.4	307275.1	5.64+05	8.4−04	A	R12
965.58	$2s2p(^1P^{\circ})3p$	$2p^2(^1D)3p$	^2D	$^2D^{\circ}$		3/2	3/2	751406.	854971.	3.2+08	5.7−01	B	FH99LS
965.58	$2s2p(^1P^{\circ})3p$	$2p^2(^1D)3p$	^2D	$^2D^{\circ}$		3/2	5/2	751406.	854971.	2.38+07	6.3−02	B	FH99LS
966.01	$2s2p(^1P^{\circ})3p$	$2p^2(^1D)3p$	^2D	$^2D^{\circ}$		5/2	3/2	751452.	854971.	3.6+07	6.3−02	B	FH99LS
966.01	$2s2p(^1P^{\circ})3p$	$2p^2(^1D)3p$	^2D	$^2D^{\circ}$		5/2	5/2	751452.	854971.	3.3+08	8.9−01	B	FH99LS

λ_{Ritz} [a] (Å)	Configurations		Terms		Type	J_i [b]	J_k [b]	E_i (cm^{-1})	E_k (cm^{-1})	A_{ki} [c] (s^{-1})	S_{ik} [c]	Acc. [d]	Ref. [e]
1017.09	$2s2p(^1P^o)3s$	$2p^2(^1D)3s$	$^2P^o$	^2D		3/2	3/2	712755.	811075.	1.15+08	2.38−01	B	FH99LS
1017.09	$2s2p(^1P^o)3s$	$2p^2(^1D)3s$	$^2P^o$	^2D		3/2	5/2	712755.	811075.	6.9+08	2.14+00	B	FH99LS
1017.09	$2s2p(^1P^o)3s$	$2p^2(^1D)3s$	$^2P^o$	^2D		1/2	3/2	712755.	811075.	5.7+08	1.19+00	B	FH99LS
1034.53	$2s2p(^1P^o)3d$	$2p^2(^1D)3d$	$^2F^o$	^2F		7/2	5/2	783650.	880312.	1.44+07	4.7−02	B	FH99LS
1034.53	$2s2p(^1P^o)3d$	$2p^2(^1D)3d$	$^2F^o$	^2F		7/2	7/2	783650.	880312.	2.9+08	1.28+00	B	FH99LS
1034.53	$2s2p(^1P^o)3d$	$2p^2(^1D)3d$	$^2F^o$	^2F		5/2	5/2	783650.	880312.	2.9+08	9.5−01	B	FH99LS
1034.53	$2s2p(^1P^o)3d$	$2p^2(^1D)3d$	$^2F^o$	^2F		5/2	7/2	783650.	880312.	1.08+07	4.7−02	B	FH99LS
1037.34	$2s^23s$	$2s2p(^3P^o)3s$	^2S	$^4P^o$		1/2	3/2	524751.	621151.0	4.35+04	9.6−05	A	TF00
1037.72	$2s2p(^1P^o)3p$	$2s2p(^3P^o)4d$	^2D	$^2F^o$		5/2	7/2	751452.	847817.	4.3+08	1.89+00	B	FH99LS
1040.11	$2s^23s$	$2s2p(^3P^o)3s$	^2S	$^4P^o$		1/2	1/2	524751.	620894.5	1.50+04	1.67−05	A	TF00
1040.58	$2s2p(^1P^o)3p$	$2s2p(^3P^o)4d$	^2D	$^2F^o$		3/2	5/2	751406.	847506.	4.0+08	1.32+00	B	FH99LS
1041.08	$2s2p(^1P^o)3p$	$2s2p(^3P^o)4d$	^2D	$^2F^o$		5/2	5/2	751452.	847506.	2.8+07	9.5−02	B	FH99LS
1077.42	$2s2p(^1P^o)3p$	$2p^2(^1D)3p$	^2D	$^2F^o$		5/2	7/2	751452.	844266.	1.26+08	6.2−01	B	FH99LS
1078.68	$2s2p(^1P^o)3p$	$2p^2(^1D)3p$	^2D	$^2F^o$		3/2	5/2	751406.	844112.	1.17+08	4.4−01	B	FH99LS
1079.21	$2s2p(^1P^o)3p$	$2p^2(^1D)3p$	^2D	$^2F^o$		5/2	5/2	751452.	844112.	8.4+06	3.1−02	B	FH99LS
1079.88	$2s2p(^3P^o)3d$	$2p^2(^1D)3s$	$^2P^o$	^2D		3/2	3/2	718472.	811075.	3.0+07	7.5−02	B	FH99LS
1079.88	$2s2p(^3P^o)3d$	$2p^2(^1D)3s$	$^2P^o$	^2D		3/2	5/2	718472.	811075.	1.80+08	6.7−01	B	FH99LS
1082.313	$2s2p^2$	$2p^3$	^2P	$^2D^o$		1/2	3/2	214880.4	307275.1	3.21+08	8.04−01	A	R12
1082.44	$2s2p(^3P^o)3d$	$2p^2(^1D)3s$	$^2P^o$	^2D		1/2	3/2	718691.	811075.	1.49+08	3.7−01	B	FH99LS
1087.820	$2s2p^2$	$2p^3$	^2P	$^2D^o$		3/2	3/2	215348.1	307275.1	5.92+07	1.50−01	A	R12
1088.386	$2s2p^2$	$2p^3$	^2P	$^2D^o$		3/2	5/2	215348.1	307227.3	3.72+08	1.42+00	A	R12
1113.90	$2s2p(^1P^o)3d$	$2p^2(^1D)3d$	$^2P^o$	^2P		3/2	3/2	793308.	883083.	3.2+08	8.7−01	B	FH99LS
1113.90	$2s2p(^1P^o)3d$	$2p^2(^1D)3d$	$^2P^o$	^2P		1/2	3/2	793308.	883083.	6.4+07	1.74−01	B	FH99LS
1115.80	$2s2p(^1P^o)3d$	$2p^2(^1D)3d$	$^2P^o$	^2P		3/2	1/2	793308.	882930.	1.27+08	1.74−01	B	FH99LS
1115.80	$2s2p(^1P^o)3d$	$2p^2(^1D)3d$	$^2P^o$	^2P		1/2	1/2	793308.	882930.	2.5+08	3.5−01	B	FH99LS
1160.38	$2s2p(^1P^o)3d$	$2p^2(^1D)3d$	$^2D^o$	^2D		3/2	3/2	787725.	873904.	3.8+08	1.18+00	B	FH99LS
1160.38	$2s2p(^1P^o)3d$	$2p^2(^1D)3d$	$^2D^o$	^2D		3/2	5/2	787725.	873904.	2.8+07	1.32−01	B	FH99LS
1160.90	$2s2p(^1P^o)3d$	$2p^2(^1D)3d$	$^2D^o$	^2D		5/2	3/2	787764.	873904.	4.3+07	1.32−01	B	FH99LS
1160.90	$2s2p(^1P^o)3d$	$2p^2(^1D)3d$	$^2D^o$	^2D		5/2	5/2	787764.	873904.	4.0+08	1.84+00	B	FH99LS

λ_{Ritz} [a] (Å)	Configurations		Terms		Type	J_i [b]	J_k [b]	E_i (cm^{-1})	E_k (cm^{-1})	A_{ki} [c] (s^{-1})	S_{ik} [c]	Acc. [d]	Ref. [e]
1162.203	$2s^22p$	$2s2p^2$	$^2P°$	^4P		1/2	3/2	0.0	86043.5	1.3+02	4.1−07	D+	R12
1165.634	$2s^22p$	$2s2p^2$	$^2P°$	^4P		1/2	1/2	0.0	85790.2	4.9+03	7.6−06	B	R12
1167.372	$2s^22p$	$2s2p^2$	$^2P°$	^4P		3/2	5/2	744.5	86407.0	4.0+03	1.87−05	B	R12
1172.347	$2s^22p$	$2s2p^2$	$^2P°$	^4P		3/2	3/2	744.5	86043.5	1.00+03	3.2−06	B	R12
1175.838	$2s^22p$	$2s2p^2$	$^2P°$	^4P		3/2	1/2	744.5	85790.2	4.6+03	7.4−06	B	R12
1268.273	$2s2p^2$	$2p^3$	^2S	$^4S°$		1/2	3/2	197566.4	276413.8	1.1+02	4.3−07	D+	R12
1600.44	$2s2p(^3P°)3p$	$2s2p(^3P°)3d$	^2P	$^2P°$		1/2	1/2	656208.	718691.	1.54+08	6.2−01	B	FH99LS
1606.06	$2s2p(^3P°)3p$	$2s2p(^3P°)3d$	^2P	$^2P°$		1/2	3/2	656208.	718472.	3.8+07	3.1−01	B	FH99LS
1606.30	$2s2p(^3P°)3p$	$2s2p(^3P°)3d$	^2P	$^2P°$		3/2	1/2	656436.	718691.	7.6+07	3.1−01	B	FH99LS
1611.97	$2s2p(^3P°)3p$	$2s2p(^3P°)3d$	^2P	$^2P°$		3/2	3/2	656436.	718472.	1.88+08	1.55+00	B	FH99LS
1625.134	$2s2p^2$	$2p^3$	^2P	$^4S°$		1/2	3/2	214880.4	276413.8	1.23+03	1.04−05	B	R12
1637.581	$2s2p^2$	$2p^3$	^2P	$^4S°$		3/2	3/2	215348.1	276413.8	4.6+03	4.0−05	B	R12
1726.76	$2p^2(^3P)3s$	$2p^2(^1D)3p$	^2P	$^2D°$		1/2	3/2	797059.	854971.	2.18+08	2.22+00	B	FH99LS
1740.58	$2p^2(^3P)3s$	$2p^2(^1D)3p$	^2P	$^2D°$		3/2	3/2	797519.	854971.	4.3+07	4.4−01	B	FH99LS
1740.58	$2p^2(^3P)3s$	$2p^2(^1D)3p$	^2P	$^2D°$		3/2	5/2	797519.	854971.	2.6+08	4.0+00	B	FH99LS
1981.76	$2p^2(^3P)3s$	$2p^2(^3P)3p$	^4P	$^4S°$		1/2	3/2	784099.	834559.1	7.6+07	1.17+00	B	FH99LS
1992.18	$2p^2(^3P)3s$	$2p^2(^3P)3p$	^4P	$^4S°$		3/2	3/2	784362.8	834559.1	1.50+08	2.35+00	B	FH99LS
2007.77	$2p^2(^3P)3s$	$2p^2(^3P)3p$	^4P	$^4S°$		5/2	3/2	784768.8	834559.1	2.20+08	3.5+00	B	FH99LS
2027.8	$2s^24d$	$2s2p(^1P°)3d$	^2D	$^2P°$		3/2	1/2	744010.	793308.	1.04+08	8.6−01	B	FH99LS
2027.8	$2s^24d$	$2s2p(^1P°)3d$	^2D	$^2P°$		3/2	3/2	744010.	793308.	1.04+07	1.72−01	B	FH99LS
2028.9	$2s^24d$	$2s2p(^1P°)3d$	^2D	$^2P°$		5/2	3/2	744036.	793308.	9.4+07	1.55+00	B	FH99LS
2042.2	$2s2p(^3P°)3s$	$2s2p(^3P°)3p$	$^2P°$	^2S		1/2	1/2	638856.	687806.	1.22+08	1.02+00	B	FH99LS
2063.7	$2s2p(^3P°)3s$	$2s2p(^3P°)3p$	$^2P°$	^2S		3/2	1/2	639365.	687806.	2.36+08	2.05+00	B	FH99LS
2100.7	$2s2p(^1P°)3s$	$2s2p(^1P°)3p$	$^2P°$	^2S		3/2	1/2	712755.	760342.	1.01+08	9.3−01	B	FH99LS
2100.7	$2s2p(^1P°)3s$	$2s2p(^1P°)3p$	$^2P°$	^2S		1/2	1/2	712755.	760342.	5.1+07	4.6−01	B	FH99LS
2208.03	$2s2p(^3P°)3p$	$2s2p(^3P°)3d$	^4D	$^4P°$		1/2	1/2	657744.8	703020.0	1.41+07	1.50−01	B	FH99LS
2215.13	$2s2p(^3P°)3p$	$2s2p(^3P°)3d$	^4D	$^4P°$		3/2	1/2	657890.0	703020.0	1.40+07	1.50−01	B	FH99LS
2215.56	$2s2p(^3P°)3p$	$2s2p(^3P°)3d$	^4D	$^4P°$		1/2	3/2	657744.8	702866.0	1.40+06	3.0−02	B	FH99LS
2216.52	$2s2p(^3P°)3s$	$2s2p(^3P°)3p$	$^4P°$	^4S		1/2	3/2	620894.5	665996.2	4.1+07	8.8−01	B	FH99LS

λ_{Ritz} [a] (Å)	Configurations		Terms		Type	J_i [b]	J_k [b]	E_i (cm^{-1})	E_k (cm^{-1})	A_{ki} [c] (s^{-1})	S_{ik} [c]	Acc. [d]	Ref. [e]
2222.72	2s2p(^{3}P°)3p	2s2p(^{3}P°)3d	^{4}D	^{4}P°		3/2	3/2	657890.0	702866.0	8.9+06	1.92−01	B	FH99LS
2229.20	2s2p(^{3}P°)3s	2s2p(^{3}P°)3p	^{4}P°	^{4}S		3/2	3/2	621151.0	665996.2	8.1+07	1.76+00	B	FH99LS
2232.75	2s2p(^{3}P°)3p	2s2p(^{3}P°)3d	^{4}D	^{4}P°		3/2	5/2	657890.0	702664.0	5.5+05	1.80−02	B	FH99LS
2235.43	2s2p(^{3}P°)3p	2s2p(^{3}P°)3d	^{4}D	^{4}P°		5/2	3/2	658145.7	702866.0	1.72+07	3.8−01	B	FH99LS
2244.5	2p^2(^{3}P)3s	2s2p(^{3}P°)4d	^{2}P	^{2}D°		1/2	3/2	797059.	841598.	8.2+07	1.83+00	B	FH99LS
2245.57	2s2p(^{3}P°)3p	2s2p(^{3}P°)3d	^{4}D	^{4}P°		5/2	5/2	658145.7	702664.0	4.8+06	1.62−01	B	FH99LS
2252.71	2s2p(^{3}P°)3s	2s2p(^{3}P°)3p	^{4}P°	^{4}S		5/2	3/2	621619.0	665996.2	1.17+08	2.6+00	B	FH99LS
2263.0	2p^2(^{3}P)3s	2s2p(^{3}P°)4d	^{2}P	^{2}D°		3/2	5/2	797519.	841695.	9.6+07	3.3+00	B	FH99LS
2266.00	2s2p(^{3}P°)3p	2s2p(^{3}P°)3d	^{4}D	^{4}P°		7/2	5/2	658547.0	702664.0	2.09+07	7.2−01	B	FH99LS
2268.0	2p^2(^{3}P)3s	2s2p(^{3}P°)4d	^{2}P	^{2}D°		3/2	3/2	797519.	841598.	1.59+07	3.7−01	B	FH99LS
2277.4	2p^2(^{1}D)3s	2p^2(^{1}D)3p	^{2}D	^{2}D°		5/2	3/2	811075.	854971.	2.05+07	4.8−01	B	FH99LS
2277.4	2p^2(^{1}D)3s	2p^2(^{1}D)3p	^{2}D	^{2}D°		5/2	5/2	811075.	854971.	1.91+08	6.7+00	B	FH99LS
2277.4	2p^2(^{1}D)3s	2p^2(^{1}D)3p	^{2}D	^{2}D°		3/2	3/2	811075.	854971.	1.85+08	4.3+00	B	FH99LS
2277.4	2p^2(^{1}D)3s	2p^2(^{1}D)3p	^{2}D	^{2}D°		3/2	5/2	811075.	854971.	1.37+07	4.8−01	B	FH99LS
2320.3	2s2p(^{3}P°)3p	2s2p(^{3}P°)3d	^{2}P	^{2}D°		1/2	3/2	656208.	699293.	1.36+08	3.4+00	B	FH99LS
2327.4	2s2p(^{3}P°)3p	2s2p(^{3}P°)3d	^{2}P	^{2}D°		3/2	5/2	656436.	699389.	1.62+08	6.1+00	B	FH99LS
2332.6	2s2p(^{3}P°)3p	2s2p(^{3}P°)3d	^{2}P	^{2}D°		3/2	3/2	656436.	699293.	2.7+07	6.7−01	B	FH99LS
2444.2	2s2p(^{1}P°)3s	2s2p(^{1}P°)3p	^{2}P°	^{2}P		3/2	3/2	712755.	753656.	1.16+08	3.4+00	B	FH99LS
2444.2	2s2p(^{1}P°)3s	2s2p(^{1}P°)3p	^{2}P°	^{2}P		1/2	3/2	712755.	753656.	2.33+07	6.7−01	B	FH99LS
2450.7	2s^{2}3s	2s^{2}3p	^{2}S	^{2}P°		1/2	3/2	524751.	565544.	1.72+08	5.01+00	A	TF00
2451.8	2s2p(^{1}P°)3s	2s2p(^{1}P°)3p	^{2}P°	^{2}P		3/2	1/2	712755.	753529.	4.6+07	6.7−01	B	FH99LS
2451.8	2s2p(^{1}P°)3s	2s2p(^{1}P°)3p	^{2}P°	^{2}P		1/2	1/2	712755.	753529.	9.2+07	1.34+00	B	FH99LS
2461.3	2s^{2}3s	2s^{2}3p	^{2}S	^{2}P°		1/2	1/2	524751.	565367.	1.70+08	2.50+00	A	TF00
2509.89	2s2p(^{3}P°)3p	2s2p(^{3}P°)3d	^{4}D	^{4}D°		1/2	3/2	657744.8	697575.2	9.8+06	3.1−01	B	FH99LS
2510.0	2s2p(^{3}P°)3p	2s2p(^{3}P°)3d	^{4}D	^{4}D°		1/2	1/2	657744.8	697573.	1.95+07	3.1−01	B	FH99LS
2512.75	2s2p(^{3}P°)3p	2s2p(^{3}P°)3d	^{4}D	^{4}D°		3/2	5/2	657890.0	697675.0	9.1+06	4.3−01	B	FH99LS
2513.1	2s2p(^{1}P°)3p	2s2p(^{1}P°)3d	^{2}P	^{2}P°		1/2	1/2	753529.	793308.	2.8+07	4.4−01	B	FH99LS
2513.1	2s2p(^{1}P°)3p	2s2p(^{1}P°)3d	^{2}P	^{2}P°		1/2	3/2	753529.	793308.	7.0+06	2.21−01	B	FH99LS
2519.07	2s2p(^{3}P°)3p	2s2p(^{3}P°)3d	^{4}D	^{4}D°		3/2	3/2	657890.0	697575.2	1.55+07	4.9−01	B	FH99LS

λ_{Ritz} [a] (Å)	Configurations		Terms		Type	J_i [b]	J_k [b]	E_i (cm^{-1})	E_k (cm^{-1})	A_{ki} [c] (s^{-1})	S_{ik} [c]	Acc. [d]	Ref. [e]
2519.2	$2s2p(^3P^\circ)3p$	$2s2p(^3P^\circ)3d$	^4D	$^4D^\circ$		3/2	1/2	657890.0	697573.	1.93+07	3.1−01	B	FH99LS
2520.34	$2s2p(^3P^\circ)3p$	$2s2p(^3P^\circ)3d$	^4D	$^4D^\circ$		5/2	7/2	658145.7	697811.0	5.5+06	3.5−01	B	FH99LS
2521.2	$2s2p(^1P^\circ)3p$	$2s2p(^1P^\circ)3d$	^2P	$^2P^\circ$		3/2	1/2	753656.	793308.	1.39+07	2.21−01	B	FH99LS
2521.2	$2s2p(^1P^\circ)3p$	$2s2p(^1P^\circ)3d$	^2P	$^2P^\circ$		3/2	3/2	753656.	793308.	3.5+07	1.10+00	B	FH99LS
2521.9	$2s^24d$	$2s2p(^1P^\circ)3d$	^2D	$^2F^\circ$		3/2	5/2	744010.	783650.	1.44+08	6.9+00	B	FH99LS
2523.6	$2s^24d$	$2s2p(^1P^\circ)3d$	^2D	$^2F^\circ$		5/2	5/2	744036.	783650.	1.03+07	4.9−01	B	FH99LS
2523.6	$2s^24d$	$2s2p(^1P^\circ)3d$	^2D	$^2F^\circ$		5/2	7/2	744036.	783650.	1.54+08	9.8+00	B	FH99LS
2529.01	$2s2p(^3P^\circ)3p$	$2s2p(^3P^\circ)3d$	^4D	$^4D^\circ$		5/2	5/2	658145.7	697675.0	2.19+07	1.05+00	B	FH99LS
2535.41	$2s2p(^3P^\circ)3p$	$2s2p(^3P^\circ)3d$	^4D	$^4D^\circ$		5/2	3/2	658145.7	697575.2	1.33+07	4.3−01	B	FH99LS
2546.10	$2s2p(^3P^\circ)3p$	$2s2p(^3P^\circ)3d$	^4D	$^4D^\circ$		7/2	7/2	658547.0	697811.0	3.2+07	2.09+00	B	FH99LS
2554.95	$2s2p(^3P^\circ)3p$	$2s2p(^3P^\circ)3d$	^4D	$^4D^\circ$		7/2	5/2	658547.0	697675.0	7.0+06	3.5−01	B	FH99LS
2561.2	$2p^2(^3P)3s$	$2p^2(^3P)3p$	^4P	$^4P^\circ$		1/2	3/2	784099.	823131.	6.4+07	2.12+00	B	FH99LS
2562.1	$2p^2(^3P)3s$	$2p^2(^3P)3p$	^4P	$^4P^\circ$		3/2	5/2	784362.8	823381.	4.6+07	2.29+00	B	FH99LS
2563.8	$2p^2(^3P)3s$	$2p^2(^3P)3p$	^4P	$^4P^\circ$		1/2	1/2	784099.	823092.	2.5+07	4.2−01	B	FH99LS
2578.7	$2p^2(^3P)3s$	$2p^2(^3P)3p$	^4P	$^4P^\circ$		3/2	3/2	784362.8	823131.	2.00+07	6.8−01	B	FH99LS
2581.3	$2p^2(^3P)3s$	$2p^2(^3P)3p$	^4P	$^4P^\circ$		3/2	1/2	784362.8	823092.	1.25+08	2.12+00	B	FH99LS
2582.3	$2s2p(^3P^\circ)4d$	$2p^2(^1D)3d$	$^2D^\circ$	^2F		3/2	5/2	841598.	880312.	3.1+07	1.60+00	B	FH99LS
2583.4	$2s2p(^1P^\circ)3s$	$2s2p(^1P^\circ)3p$	$^2P^\circ$	^2D		3/2	5/2	712755.	751452.	9.5+07	4.8+00	B	FH99LS
2586.5	$2s2p(^1P^\circ)3s$	$2s2p(^1P^\circ)3p$	$^2P^\circ$	^2D		3/2	3/2	712755.	751406.	1.57+07	5.4−01	B	FH99LS
2586.5	$2s2p(^1P^\circ)3s$	$2s2p(^1P^\circ)3p$	$^2P^\circ$	^2D		1/2	3/2	712755.	751406.	7.9+07	2.7+00	B	FH99LS
2588.8	$2s2p(^3P^\circ)4d$	$2p^2(^1D)3d$	$^2D^\circ$	^2F		5/2	5/2	841695.	880312.	2.23+06	1.15−01	B	FH99LS
2588.8	$2s2p(^3P^\circ)4d$	$2p^2(^1D)3d$	$^2D^\circ$	^2F		5/2	7/2	841695.	880312.	3.3+07	2.29+00	B	FH99LS
2589.1	$2p^2(^3P)3s$	$2p^2(^3P)3p$	^4P	$^4P^\circ$		5/2	5/2	784768.8	823381.	1.04+08	5.3+00	B	FH99LS
2606.0	$2p^2(^3P)3s$	$2p^2(^3P)3p$	^4P	$^4P^\circ$		5/2	3/2	784768.8	823131.	6.5+07	2.29+00	B	FH99LS
2674.0	$2p^2(^3P)3p$	$2p^2(^3P)3d$	$^4P^\circ$	^4P		1/2	1/2	823092.	860477.8	1.61+07	3.0−01	B	FH99LS
2676.8	$2p^2(^3P)3p$	$2p^2(^3P)3d$	$^4P^\circ$	^4P		3/2	1/2	823131.	860477.8	8.0+07	1.52+00	B	FH99LS
2680.6	$2p^2(^3P)3p$	$2p^2(^3P)3d$	$^4P^\circ$	^4P		1/2	3/2	823092.	860385.8	4.0+07	1.52+00	B	FH99LS
2683.4	$2p^2(^3P)3p$	$2p^2(^3P)3d$	$^4P^\circ$	^4P		3/2	3/2	823131.	860385.8	1.27+07	4.9−01	B	FH99LS
2694.0	$2s^23p$	$2s^23d$	$^2P^\circ$	^2D		1/2	3/2	565367.	602476.	1.09+08	4.20+00	A	TF00

λ_{Ritz} [a] (Å)	Configurations		Terms		Type	J_i [b]	J_k [b]	E_i (cm^{-1})	E_k (cm^{-1})	A_{ki} [c] (s^{-1})	S_{ik} [c]	Acc. [d]	Ref. [e]
2696.4	2s2p(^{3}P°)3s	2s2p(^{3}P°)3p	^{2}P°	^{2}D		1/2	3/2	638856.	675932.	1.14+08	4.4+00	B	FH99LS
2697.7	2s2p(^{3}P°)3s	2s2p(^{3}P°)3p	^{2}P°	^{2}D		3/2	5/2	639365.	676422.	1.37+08	8.0+00	B	FH99LS
2698.5	2p^2(^{3}P)3p	2p^2(^{3}P)3d	^{4}P°	^{4}P		3/2	5/2	823131.	860177.8	2.8+07	1.64+00	B	FH99LS
2700.16	2s2p(^{3}P°)3p	2s2p(^{3}P°)3d	^{4}S	^{4}P°		3/2	1/2	665996.2	703020.0	9.5+07	1.84+00	B	FH99LS
2701.6	2p^2(^{3}P)3p	2p^2(^{3}P)3d	^{4}P°	^{4}P		5/2	3/2	823381.	860385.8	4.2+07	1.64+00	B	FH99LS
2702.23	2s2p(^{3}P°)3s	2s2p(^{3}P°)3p	^{4}P°	^{4}D		1/2	3/2	620894.5	657890.0	5.8+07	2.26+00	B	FH99LS
2702.29	2s2p(^{3}P°)3s	2s2p(^{3}P°)3p	^{4}P°	^{4}D		3/2	5/2	621151.0	658145.7	9.7+07	5.7+00	B	FH99LS
2703.9	2s^{2}3p	2s^{2}3d	^{2}P°	^{2}D		3/2	5/2	565544.	602516.	1.29+08	7.56+00	A	TF00
2706.9	2s^{2}3p	2s^{2}3d	^{2}P°	^{2}D		3/2	3/2	565544.	602476.	2.14+07	8.4−01	A	TF00
2707.17	2s2p(^{3}P°)3s	2s2p(^{3}P°)3p	^{4}P°	^{4}D		5/2	7/2	621619.0	658547.0	1.38+08	1.08+01	B	FH99LS
2708.6	2s2p(^{3}P°)3p	2s2p(^{3}P°)3d	^{2}D	^{2}F°		3/2	5/2	675932.	712840.	1.51+08	8.9+00	B	FH99LS
2710.4	2s2p(^{3}P°)3p	2s2p(^{3}P°)3d	^{2}D	^{2}F°		5/2	7/2	676422.	713306.	1.61+08	1.27+01	B	FH99LS
2711.44	2s2p(^{3}P°)3p	2s2p(^{3}P°)3d	^{4}S	^{4}P°		3/2	3/2	665996.2	702866.0	9.4+07	3.7+00	B	FH99LS
2712.88	2s2p(^{3}P°)3s	2s2p(^{3}P°)3p	^{4}P°	^{4}D		1/2	1/2	620894.5	657744.8	1.14+08	2.26+00	B	FH99LS
2716.8	2p^2(^{3}P)3p	2p^2(^{3}P)3d	^{4}P°	^{4}P		5/2	5/2	823381.	860177.8	6.4+07	3.8+00	B	FH99LS
2721.10	2s2p(^{3}P°)3s	2s2p(^{3}P°)3p	^{4}P°	^{4}D		3/2	3/2	621151.0	657890.0	7.3+07	2.9+00	B	FH99LS
2726.38	2s2p(^{3}P°)3p	2s2p(^{3}P°)3d	^{4}S	^{4}P°		3/2	5/2	665996.2	702664.0	9.2+07	5.5+00	B	FH99LS
2731.89	2s2p(^{3}P°)3s	2s2p(^{3}P°)3p	^{4}P°	^{4}D		3/2	1/2	621151.0	657744.8	2.24+07	4.5−01	B	FH99LS
2733.9	2s2p(^{3}P°)3s	2s2p(^{3}P°)3p	^{2}P°	^{2}D		3/2	3/2	639365.	675932.	2.19+07	8.9−01	B	FH99LS
2736.91	2s2p(^{3}P°)3s	2s2p(^{3}P°)3p	^{4}P°	^{4}D		5/2	5/2	621619.0	658145.7	4.0+07	2.44+00	B	FH99LS
2745.1	2s2p(^{3}P°)3p	2s2p(^{3}P°)3d	^{2}D	^{2}F°		5/2	5/2	676422.	712840.	1.03+07	6.3−01	B	FH99LS
2749.6	2s2p(^{1}P°)3p	2s2p(^{1}P°)3d	^{2}D	^{2}D°		3/2	5/2	751406.	787764.	2.07+06	1.28−01	B	FH99LS
2752.6	2s2p(^{1}P°)3p	2s2p(^{1}P°)3d	^{2}D	^{2}D°		3/2	3/2	751406.	787725.	2.8+07	1.15+00	B	FH99LS
2753.1	2s2p(^{1}P°)3p	2s2p(^{1}P°)3d	^{2}D	^{2}D°		5/2	5/2	751452.	787764.	2.9+07	1.79+00	B	FH99LS
2756.1	2s2p(^{1}P°)3p	2s2p(^{1}P°)3d	^{2}D	^{2}D°		5/2	3/2	751452.	787725.	3.1+06	1.28−01	B	FH99LS
2756.21	2s2p(^{3}P°)3s	2s2p(^{3}P°)3p	^{4}P°	^{4}D		5/2	3/2	621619.0	657890.0	6.5+06	2.7−01	B	FH99LS
2761.6	2p^2(^{1}D)3p	2p^2(^{1}D)3d	^{2}F°	^{2}F		5/2	5/2	844112.	880312.	1.91+07	1.19+00	B	FH99LS
2761.6	2p^2(^{1}D)3p	2p^2(^{1}D)3d	^{2}F°	^{2}F		5/2	7/2	844112.	880312.	7.2+05	6.0−02	B	FH99LS
2773.4	2p^2(^{1}D)3p	2p^2(^{1}D)3d	^{2}F°	^{2}F		7/2	5/2	844266.	880312.	9.4+05	6.0−02	B	FH99LS

λ_{Ritz} [a] (Å)	Configurations		Terms		Type	J_i [b]	J_k [b]	E_i (cm^{-1})	E_k (cm^{-1})	A_{ki} [c] (s^{-1})	S_{ik} [c]	Acc. [d]	Ref. [e]
2773.4	$2p^2(^1D)3p$	$2p^2(^1D)3d$	$^2F°$	^2F		7/2	7/2	844266.	880312.	1.91+07	1.61+00	B	FH99LS
2841.4	$2s2p(^3P°)3d$	$2s2p(^1P°)3p$	$^2P°$	^2P		3/2	3/2	718472.	753656.	1.39+07	6.3−01	B	FH99LS
2851.7	$2s2p(^3P°)3d$	$2s2p(^1P°)3p$	$^2P°$	^2P		3/2	1/2	718472.	753529.	5.5+06	1.26−01	B	FH99LS
2859.2	$2s2p(^3P°)3d$	$2s2p(^1P°)3p$	$^2P°$	^2P		1/2	3/2	718691.	753656.	2.7+06	1.26−01	B	FH99LS
2869.6	$2s2p(^3P°)3d$	$2s2p(^1P°)3p$	$^2P°$	^2P		1/2	1/2	718691.	753529.	1.08+07	2.5−01	B	FH99LS
2923.5	$2s2p(^1P°)3p$	$2s2p(^1P°)3d$	^2P	$^2D°$		1/2	3/2	753529.	787725.	5.7+07	2.8+00	B	FH99LS
2931.0	$2s2p(^1P°)3p$	$2s2p(^1P°)3d$	^2P	$^2D°$		3/2	5/2	753656.	787764.	6.8+07	5.1+00	B	FH99LS
2934.4	$2s2p(^1P°)3p$	$2s2p(^1P°)3d$	^2P	$^2D°$		3/2	3/2	753656.	787725.	1.13+07	5.6−01	B	FH99LS
3012.0	$2p^2(^1D)3s$	$2p^2(^1D)3p$	^2D	$^2F°$		5/2	7/2	811075.	844266.	1.19+08	1.29+01	B	FH99LS
3026.0	$2p^2(^1D)3s$	$2p^2(^1D)3p$	^2D	$^2F°$		5/2	5/2	811075.	844112.	7.8+06	6.4−01	B	FH99LS
3026.0	$2p^2(^1D)3s$	$2p^2(^1D)3p$	^2D	$^2F°$		3/2	5/2	811075.	844112.	1.10+08	9.0+00	B	FH99LS
3031.3	$2s2p(^3P°)3d$	$2s2p(^1P°)3p$	$^2P°$	^2D		3/2	5/2	718472.	751452.	1.23+07	1.01+00	B	FH99LS
3032.5	$2s2p(^1P°)3p$	$2s2p(^1P°)3d$	^2S	$^2P°$		1/2	1/2	760342.	793308.	8.5+07	2.34+00	B	FH99LS
3032.5	$2s2p(^1P°)3p$	$2s2p(^1P°)3d$	^2S	$^2P°$		1/2	3/2	760342.	793308.	8.5+07	4.7+00	B	FH99LS
3035.5	$2s2p(^3P°)3d$	$2s2p(^1P°)3p$	$^2P°$	^2D		3/2	3/2	718472.	751406.	2.03+06	1.12−01	B	FH99LS
3055.8	$2s2p(^3P°)3d$	$2s2p(^1P°)3p$	$^2P°$	^2D		1/2	3/2	718691.	751406.	1.0+07	5.6−01	B	FH99LS
3100.5	$2s2p(^1P°)3p$	$2s2p(^1P°)3d$	^2D	$^2F°$		3/2	5/2	751406.	783650.	5.5+07	4.9+00	B	FH99LS
3104.9	$2s2p(^1P°)3p$	$2s2p(^1P°)3d$	^2D	$^2F°$		5/2	5/2	751452.	783650.	3.9+06	3.5−01	B	FH99LS
3104.9	$2s2p(^1P°)3p$	$2s2p(^1P°)3d$	^2D	$^2F°$		5/2	7/2	751452.	783650.	5.9+07	6.9+00	B	FH99LS
3195.9	$2s2p(^1P°)3s$	$2s^24d$	$^2P°$	^2D		3/2	5/2	712755.	744036.	1.83+07	1.77+00	B	FH99LS
3198.6	$2s2p(^1P°)3s$	$2s^24d$	$^2P°$	^2D		3/2	3/2	712755.	744010.	3.0+06	1.97−01	B	FH99LS
3198.6	$2s2p(^1P°)3s$	$2s^24d$	$^2P°$	^2D		1/2	3/2	712755.	744010.	1.52+07	9.8−01	B	FH99LS
3236.9	$2s2p(^3P°)3p$	$2s2p(^3P°)3d$	^2S	$^2P°$		1/2	1/2	687806.	718691.	4.9+07	1.64+00	B	FH99LS
3260.0	$2s2p(^3P°)3p$	$2s2p(^3P°)3d$	^2S	$^2P°$		1/2	3/2	687806.	718472.	4.8+07	3.3+00	B	FH99LS
3556.2	$2p^2(^1D)3p$	$2p^2(^1D)3d$	$^2D°$	^2P		5/2	3/2	854971.	883083.	1.03+07	9.1−01	B	FH99LS
3556.2	$2p^2(^1D)3p$	$2p^2(^1D)3d$	$^2D°$	^2P		3/2	3/2	854971.	883083.	1.14+06	1.02−01	B	FH99LS
3575.6	$2p^2(^1D)3p$	$2p^2(^1D)3d$	$^2D°$	^2P		3/2	1/2	854971.	882930.	1.12+07	5.1−01	B	FH99LS
3787.1	$2s2p(^3P°)4d$	$2p^2(^1D)3d$	$^2F°$	^2D		5/2	3/2	847506.	873904.	5.5+06	5.9−01	B	FH99LS
3787.1	$2s2p(^3P°)4d$	$2p^2(^1D)3d$	$^2F°$	^2D		5/2	5/2	847506.	873904.	2.6+05	4.2−02	B	FH99LS

λ_{Ritz}[a] (Å)	Configurations		Terms		Type	J_i[b]	J_k[b]	E_i ($\mathrm{cm^{-1}}$)	E_k ($\mathrm{cm^{-1}}$)	A_{ki}[c] ($\mathrm{s^{-1}}$)	S_{ik}[c]	Acc.[d]	Ref.[e]
3832.2	2s2p(^{3}P°)4d	2p^2(^{1}D)3d	^{2}F°	^{2}D		7/2	5/2	847817.	873904.	5.0+06	8.4−01	B	FH99LS
3857.1	2p^2(^{3}P)3p	2p^2(^{3}P)3d	^{4}S°	^{4}P		3/2	1/2	834559.1	860477.8	2.7+07	1.51+00	B	FH99LS
3870.9	2p^2(^{3}P)3p	2p^2(^{3}P)3d	^{4}S°	^{4}P		3/2	3/2	834559.1	860385.8	2.6+07	3.0+00	B	FH99LS
3885.1	2p^2(^{3}P)3s	2s2p(^{3}P°)4s	^{4}P	^{4}P°		3/2	5/2	784362.8	810095.0	2.9+06	5.0−01	B	FH99LS
3902.3	2p^2(^{3}P)3p	2p^2(^{3}P)3d	^{4}S°	^{4}P		3/2	5/2	834559.1	860177.8	2.6+07	4.5+00	B	FH99LS
3904.5	2p^2(^{3}P)3s	2s2p(^{3}P°)4s	^{4}P	^{4}P°		1/2	3/2	784099.	809703.0	4.0+06	4.7−01	B	FH99LS
3945.1	2p^2(^{1}D)3p	2p^2(^{1}D)3d	^{2}D°	^{2}F		5/2	5/2	854971.	880312.	3.0+06	5.4−01	B	FH99LS
3945.1	2p^2(^{1}D)3p	2p^2(^{1}D)3d	^{2}D°	^{2}F		5/2	7/2	854971.	880312.	4.5+07	1.09+01	B	FH99LS
3945.1	2p^2(^{1}D)3p	2p^2(^{1}D)3d	^{2}D°	^{2}F		3/2	5/2	854971.	880312.	4.2+07	7.6+00	B	FH99LS
3945.2	2p^2(^{3}P)3s	2s2p(^{3}P°)4s	^{4}P	^{4}P°		3/2	3/2	784362.8	809703.0	1.23+06	1.49−01	B	FH99LS
3947.4	2p^2(^{3}P)3s	2s2p(^{3}P°)4s	^{4}P	^{4}P°		5/2	5/2	784768.8	810095.0	6.4+06	1.17+00	B	FH99LS
3960.8	2p^2(^{3}P)3s	2s2p(^{3}P°)4s	^{4}P	^{4}P°		1/2	1/2	784099.	809339.5	1.52+06	9.3−02	B	FH99LS
4002.6	2p^2(^{3}P)3s	2s2p(^{3}P°)4s	^{4}P	^{4}P°		3/2	1/2	784362.8	809339.5	7.3+06	4.7−01	B	FH99LS
4009.4	2p^2(^{3}P)3s	2s2p(^{3}P°)4s	^{4}P	^{4}P°		5/2	3/2	784768.8	809703.0	3.9+06	5.0−01	B	FH99LS
4261.9	2s2p(^{3}P°)3p	2s2p(^{3}P°)3d	^{2}D	^{2}D°		3/2	5/2	675932.	699389.	5.4+05	1.24−01	B	FH99LS
4279.4	2s2p(^{3}P°)3p	2s2p(^{3}P°)3d	^{2}D	^{2}D°		3/2	3/2	675932.	699293.	7.2+06	1.12+00	B	FH99LS
4352.8	2s2p(^{3}P°)3p	2s2p(^{3}P°)3d	^{2}D	^{2}D°		5/2	5/2	676422.	699389.	7.1+06	1.74+00	B	FH99LS
4371.1	2s2p(^{3}P°)3p	2s2p(^{3}P°)3d	^{2}D	^{2}D°		5/2	3/2	676422.	699293.	7.5+05	1.24−01	B	FH99LS
5074.0	2s2p(^{3}P°)4d	2p^2(^{3}P)3d	^{4}D°	^{4}P		1/2	1/2	840775.0	860477.8	6.5+05	8.4−02	B	FH99LS
5078.6	2s2p(^{3}P°)4d	2p^2(^{3}P)3d	^{4}D°	^{4}P		3/2	1/2	840793.0	860477.8	6.5+05	8.4−02	B	FH99LS
5097.8	2s2p(^{3}P°)4d	2p^2(^{3}P)3d	^{4}D°	^{4}P		1/2	3/2	840775.0	860385.8	6.4+04	1.68−02	B	FH99LS
5102.5	2s2p(^{3}P°)4d	2p^2(^{3}P)3d	^{4}D°	^{4}P		3/2	3/2	840793.0	860385.8	4.1+05	1.08−01	B	FH99LS
5117.8	2s2p(^{3}P°)4d	2p^2(^{3}P)3d	^{4}D°	^{4}P		5/2	3/2	840851.5	860385.8	8.0+05	2.12−01	B	FH99LS
5157.2	2s2p(^{3}P°)4d	2p^2(^{3}P)3d	^{4}D°	^{4}P		3/2	5/2	840793.0	860177.8	2.48+04	1.01−02	B	FH99LS
5172.9	2s2p(^{3}P°)4d	2p^2(^{3}P)3d	^{4}D°	^{4}P		5/2	5/2	840851.5	860177.8	2.21+05	9.1−02	B	FH99LS
5229.5	2s2p(^{3}P°)4d	2p^2(^{3}P)3d	^{4}D°	^{4}P		7/2	5/2	841061.0	860177.8	9.5+05	4.0−01	B	FH99LS
5280.3	2p^2(^{1}D)3p	2p^2(^{1}D)3d	^{2}D°	^{2}D		5/2	3/2	854971.	873904.	5.1+05	1.49−01	B	FH99LS
5280.3	2p^2(^{1}D)3p	2p^2(^{1}D)3d	^{2}D°	^{2}D		5/2	5/2	854971.	873904.	4.8+06	2.09+00	B	FH99LS
5280.3	2p^2(^{1}D)3p	2p^2(^{1}D)3d	^{2}D°	^{2}D		3/2	3/2	854971.	873904.	4.6+06	1.34+00	B	FH99LS

λ_{Ritz} [a] (Å)	Configurations		Terms		Type	J_i [b]	J_k [b]	E_i (cm^{-1})	E_k (cm^{-1})	A_{ki} [c] (s^{-1})	S_{ik} [c]	Acc. [d]	Ref. [e]
5280.3	2p^2(^{1}D)3p	2p^2(^{1}D)3d	^{2}D°	^{2}D		3/2	5/2	854971.	873904.	3.4+05	1.49−01	B	FH99LS
5686.7	2s2p(^{3}P°)3s	2s2p(^{3}P°)3p	^{2}P°	^{2}P		1/2	3/2	638856.	656436.	2.22+06	8.0−01	B	FH99LS
5761.4	2s2p(^{3}P°)3s	2s2p(^{3}P°)3p	^{2}P°	^{2}P		1/2	1/2	638856.	656208.	8.5+06	1.61+00	B	FH99LS
5856.3	2s2p(^{3}P°)3s	2s2p(^{3}P°)3p	^{2}P°	^{2}P		3/2	3/2	639365.	656436.	1.01+07	4.0+00	B	FH99LS
5935.5	2s2p(^{3}P°)3s	2s2p(^{3}P°)3p	^{2}P°	^{2}P		3/2	1/2	639365.	656208.	3.9+06	8.0−01	B	FH99LS
7138	2s2p(^{3}P°)4p	2s2p(^{3}P°)4d	^{2}D	^{2}F°		3/2	5/2	833501.	847506.	2.19+07	2.36+01	B	FH99LS
7194	2s2p(^{3}P°)4p	2s2p(^{3}P°)4d	^{2}D	^{2}F°		5/2	7/2	833920.	847817.	2.29+07	3.4+01	B	FH99LS
7358	2s2p(^{3}P°)4p	2s2p(^{3}P°)4d	^{2}D	^{2}F°		5/2	5/2	833920.	847506.	1.43+06	1.68+00	B	FH99LS
8220	2s2p(^{3}P°)4p	2s2p(^{3}P°)4d	^{2}P	^{2}D°		1/2	3/2	829436.	841598.	1.25+07	1.37+01	B	FH99LS
8339	2s2p(^{3}P°)4p	2s2p(^{3}P°)4d	^{2}P	^{2}D°		3/2	5/2	829707.	841695.	1.43+07	2.46+01	B	FH99LS
8407	2s2p(^{3}P°)4p	2s2p(^{3}P°)4d	^{2}P	^{2}D°		3/2	3/2	829707.	841598.	2.33+06	2.7+00	B	FH99LS
9422	2s2p(^{3}P°)4p	2p^2(^{1}D)3p	^{2}D	^{2}F°		3/2	5/2	833501.	844112.	4.8+06	1.20+01	B	FH99LS
9663	2s2p(^{3}P°)4p	2p^2(^{1}D)3p	^{2}D	^{2}F°		5/2	7/2	833920.	844266.	4.8+06	1.72+01	B	FH99LS
9809	2s2p(^{3}P°)4p	2p^2(^{1}D)3p	^{2}D	^{2}F°		5/2	5/2	833920.	844112.	3.1+05	8.6−01	B	FH99LS
10208	2s2p(^{1}P°)3d	2p^2(^{3}P)3s	^{2}D°	^{2}P		3/2	3/2	787725.	797519.	4.8+04	1.00−01	B	FH99LS
10248	2s2p(^{1}P°)3d	2p^2(^{3}P)3s	^{2}D°	^{2}P		5/2	3/2	787764.	797519.	4.2+05	9.0−01	B	FH99LS
10711	2s2p(^{1}P°)3d	2p^2(^{3}P)3s	^{2}D°	^{2}P		3/2	1/2	787725.	797059.	4.1+05	5.0−01	B	FH99LS
12201	2s2p(^{3}P°)4p	2s2p(^{3}P°)4d	^{2}D	^{2}D°		3/2	5/2	833501.	841695.	9.6+04	5.2−01	B	FH99LS
12347	2s2p(^{3}P°)4p	2s2p(^{3}P°)4d	^{2}D	^{2}D°		3/2	3/2	833501.	841598.	1.25+06	4.6+00	B	FH99LS
12858	2s2p(^{3}P°)4p	2s2p(^{3}P°)4d	^{2}D	^{2}D°		5/2	5/2	833920.	841695.	1.15+06	7.2+00	B	FH99LS
13021	2s2p(^{3}P°)4p	2s2p(^{3}P°)4d	^{2}D	^{2}D°		5/2	3/2	833920.	841598.	1.18+05	5.2−01	B	FH99LS
14550	2s2p(^{3}P°)4f	2s2p(^{3}P°)4d	^{2}F	^{2}F°		7/2	7/2	840946.	847817.	1.40+05	1.71+00	B	FH99LS
14550	2s2p(^{3}P°)4f	2s2p(^{3}P°)4d	^{2}F	^{2}F°		5/2	7/2	840946.	847817.	5.2+03	6.3−02	B	FH99LS
15240	2s2p(^{3}P°)4f	2s2p(^{3}P°)4d	^{2}F	^{2}F°		7/2	5/2	840946.	847506.	6.0+03	6.3−02	B	FH99LS
15240	2s2p(^{3}P°)4f	2s2p(^{3}P°)4d	^{2}F	^{2}F°		5/2	5/2	840946.	847506.	1.21+05	1.27+00	B	FH99LS
20300	2s^{2}4f	2s2p(^{1}P°)3p	^{2}F°	^{2}D		7/2	5/2	746525.	751452.	4.4+04	1.09+00	B	FH99LS
20300	2s^{2}4f	2s2p(^{1}P°)3p	^{2}F°	^{2}D		5/2	5/2	746525.	751452.	2.20+03	5.4−02	B	FH99LS
20490	2s^{2}4f	2s2p(^{1}P°)3p	^{2}F°	^{2}D		5/2	3/2	746525.	751406.	4.5+04	7.6−01	B	FH99LS
29150	2p^2(^{1}D)3p	2s2p(^{3}P°)4f	^{2}F°	^{2}G		5/2	7/2	844112.	847543.	1.53+05	1.49+01	B	FH99LS

λ_{Ritz} [a] (Å)	Configurations		Terms		Type	J_i [b]	J_k [b]	E_i (cm^{-1})	E_k (cm^{-1})	A_{ki} [c] (s^{-1})	S_{ik} [c]	Acc. [d]	Ref. [e]
30520	$2p^2(^1D)3p$	$2s2p(^3P^o)4f$	$^2F^o$	^2G		7/2	7/2	844266.	847543.	4.9+03	5.5−01	B	FH99LS
30520	$2p^2(^1D)3p$	$2s2p(^3P^o)4f$	$^2F^o$	^2G		7/2	9/2	844266.	847543.	1.38+05	1.93+01	B	FH99LS
39760	$2s^24d$	$2s^24f$	^2D	$^2F^o$		3/2	5/2	744010.	746525.	1.14+05	2.13+01	B	FH99LS
40180	$2s^24d$	$2s^24f$	^2D	$^2F^o$		5/2	5/2	744036.	746525.	7.9+03	1.52+00	B	FH99LS
40180	$2s^24d$	$2s^24f$	^2D	$^2F^o$		5/2	7/2	744036.	746525.	1.19+05	3.0+01	B	FH99LS
133500	$2s2p(^3P^o)4f$	$2s2p(^3P^o)4d$	^2F	$^2D^o$		7/2	5/2	840946.	841695.	1.50+03	1.06+01	B	FH99LS
133500	$2s2p(^3P^o)4f$	$2s2p(^3P^o)4d$	^2F	$^2D^o$		5/2	5/2	840946.	841695.	7.5+01	5.3−01	B	FH99LS
134320	$2s^22p$	$2s^22p$	$^2P^o$	$^2P^o$	M1	1/2	3/2	0.0	744.5	3.7−03	1.33+00	B	R12
153400	$2s2p(^3P^o)4f$	$2s2p(^3P^o)4d$	^2F	$^2D^o$		5/2	3/2	840946.	841598.	1.04+03	7.4+00	B	FH99LS
552000	$2s2p(^1P^o)3s$	$2s^24s$	$^2P^o$	^2S		3/2	1/2	712755.	712936.	5.7+00	9.5−01	B	FH99LS
552000	$2s2p(^1P^o)3s$	$2s^24s$	$^2P^o$	^2S		1/2	1/2	712755.	712936.	2.9+00	4.8−01	B	FH99LS
F VI (Be-like)													
108.955	2s2p	2s4d	$^3P^o$	^3D		0	1	96590	1014404	2.46+10	4.7−02	B	TS90LS
108.984	2s2p	2s4d	$^3P^o$	^3D		1	2	96850	1014413	3.3+10	1.06−01	B	TS90LS
108.985	2s2p	2s4d	$^3P^o$	^3D		1	1	96850	1014404	1.84+10	3.5−02	B	TS90LS
109.051	2s2p	2s4d	$^3P^o$	^3D		2	3	97427	1014428	4.4+10	1.98−01	B	TS90LS
109.053	2s2p	2s4d	$^3P^o$	^3D		2	2	97427	1014413	1.10+10	3.5−02	B	TS90LS
109.054	2s2p	2s4d	$^3P^o$	^3D		2	1	97427	1014404	1.2+09	2.4−03	C	TS90LS
112.830	$2s^2$	2p3s	^1S	$^1P^o$		0	1	0	886288	1.6+09	3.5−03	D	F04
114.753	$2s^2$	2p3s	^1S	$^3P^o$		0	1	0	871437	1.4+06	3.1−06	C	F04
126.532	$2s^2$	2s3p	^1S	$^3P^o$		0	1	0	790312	1.45+08	4.35−04	A	F04
126.929	$2s^2$	2s3p	^1S	$^1P^o$		0	1	0	787844	6.19+10	1.88−01	A	F04
136.988	2s2p	2s3d	$^3P^o$	^1D		1	2	96850	826843	4.6+05	2.9−06	C	F04
137.096	2s2p	2s3d	$^3P^o$	^1D		2	2	97427	826843	8.+04	5.−07	D	F04
139.752	2s2p	2s3d	$^3P^o$	^3D		0	1	96590	812142	7.66+10	3.10−01	A	F04
139.798	2s2p	2s3d	$^3P^o$	^3D		1	2	96850	812166	1.03+11	6.96−01	A	F04
139.803	2s2p	2s3d	$^3P^o$	^3D		1	1	96850	812142	5.74+10	2.32−01	A	F04
139.904	2s2p	2s3d	$^3P^o$	^3D		2	3	97427	812202	1.38+11	1.30+00	A	F04

λ_{Ritz} [a] (Å)	Configurations		Terms		Type	J_i [b]	J_k [b]	E_i (cm^{-1})	E_k (cm^{-1})	A_{ki} [c] (s^{-1})	S_{ik} [c]	Acc. [d]	Ref. [e]
139.911	2s2p	2s3d	$^3P^\circ$	^3D		2	2	97427	812166	3.44+10	2.33−01	A	F04
139.916	2s2p	2s3d	$^3P^\circ$	^3D		2	1	97427	812142	3.83+09	1.55−02	A	F04
145.463	$2p^2$	2p3d	^3P	$^3P^\circ$		0	1	251336	938798	2.8+10	1.29−01	B	TS90LS
145.495	$2p^2$	2p3d	^3P	$^3P^\circ$		1	0	251629	938939	8.5+10	1.29−01	B	TS90LS
145.525	$2p^2$	2p3d	^3P	$^3P^\circ$		1	1	251629	938798	2.13+10	9.7−02	B	TS90LS
145.581	$2p^2$	2p3d	^3P	$^3P^\circ$		1	2	251629	938530	2.12+10	1.62−01	B	TS90LS
145.633	$2p^2$	2p3d	^3P	$^3P^\circ$		2	1	252139	938798	3.5+10	1.62−01	B	TS90LS
145.690	$2p^2$	2p3d	^3P	$^3P^\circ$		2	2	252139	938530	6.4+10	4.8−01	B	TS90LS
146.574	$2p^2$	2p3d	^3P	$^3D^\circ$		0	1	251336	933585	9.0+10	4.2−01	B	TS90LS
146.609	$2p^2$	2p3d	^3P	$^3D^\circ$		1	2	251629	933716	1.22+11	9.5−01	B	TS90LS
146.637	$2p^2$	2p3d	^3P	$^3D^\circ$		1	1	251629	933585	6.7+10	3.2−01	B	TS90LS
146.671	$2p^2$	2p3d	^3P	$^3D^\circ$		2	3	252139	933937	1.62+11	1.76+00	B	TS90LS
146.719	$2p^2$	2p3d	^3P	$^3D^\circ$		2	2	252139	933716	4.0+10	3.2−01	B	TS90LS
146.747	$2p^2$	2p3d	^3P	$^3D^\circ$		2	1	252139	933585	4.5+09	2.10−02	B	TS90LS
148.652	$2p^2$	2p3d	^1D	$^1F^\circ$		2	3	274602	947313	1.80+11	2.04+00	B	TS90
153.682	2s2p	2s3s	$^3P^\circ$	^3S		0	1	96590	747284	3.81+09	2.05−02	A	F04
153.744	2s2p	2s3s	$^3P^\circ$	^3S		1	1	96850	747284	1.15+10	6.16−02	A	F04
153.880	2s2p	2s3s	$^3P^\circ$	^3S		2	1	97427	747284	1.91+10	1.03−01	A	F04
154.506	$2p^2$	2p3d	^1D	$^1D^\circ$		2	2	274602	921824	6.0+10	5.5−01	B	TS90
156.250	2s2p	2s3d	$^1P^\circ$	^1D		1	2	186844	826843	8.25+10	7.77−01	A	F04
157.492	$2p^2$	2p3s	^3P	$^1P^\circ$		0	1	251336	886288	1.75+06	1.02−05	B	F04
157.565	$2p^2$	2p3s	^3P	$^1P^\circ$		1	1	251629	886288	4.4+06	2.6−05	B	F04
157.692	$2p^2$	2p3s	^3P	$^1P^\circ$		2	1	252139	886288	2.0+04	1.2−07	D	F04
159.918	2s2p	2s3d	$^1P^\circ$	^3D		1	2	186844	812166	4.6+05	5.−06	D	F04
159.924	2s2p	2s3d	$^1P^\circ$	^3D		1	1	186844	812142	9.2+05	5.6−06	C	F04
161.172	$2p^2$	2p3s	^3P	$^3P^\circ$		1	2	251629	872084	6.05+09	6.25−02	A	F04
161.264	$2p^2$	2p3s	^3P	$^3P^\circ$		0	1	251336	871437	8.12+09	5.04−02	A	F04
161.305	$2p^2$	2p3s	^3P	$^3P^\circ$		2	2	252139	872084	1.83+10	1.89−01	A	F04
161.340	$2p^2$	2p3s	^3P	$^3P^\circ$		1	1	251629	871437	6.11+09	3.80−02	A	F04

λ_{Ritz} [a] (Å)	Configurations		Terms		Type	J_i [b]	J_k [b]	E_i (cm^{-1})	E_k (cm^{-1})	A_{ki} [c] (s^{-1})	S_{ik} [c]	Acc. [d]	Ref. [e]
161.418	$2p^2$	2p3s	^3P	$^3P°$		1	0	251629	871140	2.45+10	5.09−02	A	F04
161.473	$2p^2$	2p3s	^3P	$^3P°$		2	1	252139	871437	1.02+10	6.36−02	A	F04
163.140	$2p^2$	2p3d	^1S	$^1P°$		0	1	340434	953404	9.0+10	5.8−01	B	TS90
163.483	$2p^2$	2p3s	^1D	$^1P°$		2	1	274602	886288	1.95+10	1.26−01	A	F04
167.369	$2p^2$	2p3s	^1D	$^3P°$		2	2	274602	872084	4.3+06	5.0−05	C	F04
167.550	$2p^2$	2p3s	^1D	$^3P°$		2	1	274602	871437	3.5+06	2.47−05	B	F04
173.146	2s2p	2s3s	$^1P°$	^1S		1	0	186844	764392	8.1+09	2.1−02	C	F04
178.431	2s2p	2s3s	$^1P°$	^3S		1	1	186844	747284	2.1+05	1.8−06	D	F04
183.199	$2p^2$	2p3s	^1S	$^1P°$		0	1	340434	886288	5.+09	4.2−02	D	F04
185.537	$2p^2$	2s3p	^3P	$^3P°$		0	1	251336	790312	5.+07	5.−04	D	F04
185.587	$2p^2$	2s3p	^3P	$^3P°$		1	2	251629	790460	3.7+07	6.−04	D	F04
185.638	$2p^2$	2s3p	^3P	$^3P°$		1	1	251629	790312	4.0+07	3.8−04	D	F04
185.663	$2p^2$	2s3p	^3P	$^3P°$		1	0	251629	790240	1.6+08	5.−04	D	F04
185.763	$2p^2$	2s3p	^3P	$^3P°$		2	2	252139	790460	1.2+08	1.8−03	D	F04
185.814	$2p^2$	2s3p	^3P	$^3P°$		2	1	252139	790312	7.+07	6.−04	D	F04
186.391	$2p^2$	2s3p	^3P	$^1P°$		0	1	251336	787844	2.9+05	2.8−06	D	F04
186.492	$2p^2$	2s3p	^3P	$^1P°$		1	1	251629	787844	1.9+05	1.8−06	C	F04
186.670	$2p^2$	2s3p	^3P	$^1P°$		2	1	252139	787844	2.5+06	2.4−05	D	F04
188.323	$2p^2$	2p3s	^1S	$^3P°$		0	1	340434	871437	2.1+06	2.0−05	D	F04
193.852	$2p^2$	2s3p	^1D	$^3P°$		2	2	274602	790460	7.+02	1.3−08	E	F04
193.907	$2p^2$	2s3p	^1D	$^3P°$		2	1	274602	790312	7.+06	8.−05	D	F04
194.840	$2p^2$	2s3p	^1D	$^1P°$		2	1	274602	787844	2.7+09	2.9−02	D	F04
222.282	$2p^2$	2s3p	^1S	$^3P°$		0	1	340434	790312	2.3+05	3.7−06	E	F04
223.509	$2p^2$	2s3p	^1S	$^1P°$		0	1	340434	787844	5.+07	8.−04	E	F04
405.930	2p3s	2p4p	$^3P°$	^3D		0	1	871140	1117488	1.92+09	1.90−01	B	TS90LS
405.999	2p3s	2p4p	$^3P°$	^3D		1	2	871437	1117743	2.6+09	4.3−01	B	TS90LS
406.228	2p3s	2p4p	$^3P°$	^3D		2	3	872084	1118251	3.5+09	8.0−01	B	TS90LS
406.420	2p3s	2p4p	$^3P°$	^3D		1	1	871437	1117488	1.44+09	1.43−01	B	TS90LS
407.068	2p3s	2p4p	$^3P°$	^3D		2	2	872084	1117743	8.6+08	1.43−01	B	TS90LS

λ_{Ritz} [a] (Å)	Configurations		Terms		Type	J_i [b]	J_k [b]	E_i (cm^{-1})	E_k (cm^{-1})	A_{ki} [c] (s^{-1})	S_{ik} [c]	Acc. [d]	Ref. [e]
407.491	2p3s	2p4p	$^3P^\circ$	^3D		2	1	872084	1117488	9.5+07	9.5−03	C	TS90LS
410.536	2s2p	$2p^2$	$^3P^\circ$	^1S		1	0	96850	340434	6.8+04	2.31−06	B	F04
416.898	2p3s	2p4p	$^1P^\circ$	^1D		1	2	886288	1126155	3.1+09	5.5−01	B	TS90
431.928	2s3p	2s4d	$^1P^\circ$	^1D		1	2	787844	1019364	7.7+09	1.52+00	B	TS90
433.112	2p3p	2p4d	^1P	$^1D^\circ$		1	2	895286	1126173	7.2+09	1.44+00	B	TS90
435.385	2p3s	2p4p	$^1P^\circ$	^1P		1	1	886288	1115970	3.3+09	4.1−01	B	TS90
446.102	2s3p	2s4d	$^3P^\circ$	^3D		0	1	790240	1014404	5.2+09	6.9−01	B	TS90LS
446.227	2s3p	2s4d	$^3P^\circ$	^3D		1	2	790312	1014413	7.0+09	1.54+00	B	TS90LS
446.245	2s3p	2s4d	$^3P^\circ$	^3D		1	1	790312	1014404	3.9+09	5.1−01	B	TS90LS
446.492	2s3p	2s4d	$^3P^\circ$	^3D		2	3	790460	1014428	9.4+09	2.9+00	B	TS90LS
446.522	2s3p	2s4d	$^3P^\circ$	^3D		2	2	790460	1014413	2.34+09	5.1−01	B	TS90LS
446.540	2s3p	2s4d	$^3P^\circ$	^3D		2	1	790460	1014404	2.6+08	3.4−02	B	TS90LS
474.904	2p3p	2p4d	^1D	$^1F^\circ$		2	3	925389	1135958	7.3+09	2.7+00	B	TS90
479.750	2s3d	2s4f	^3D	$^3F^\circ$		1	2	812142	1020584	1.68+10	4.6+00	B	TS90LS
479.805	2s3d	2s4f	^3D	$^3F^\circ$		2	2	812166	1020584	3.1+09	8.5−01	B	TS90LS
479.805	2s3d	2s4f	^3D	$^3F^\circ$		2	3	812166	1020584	1.78+10	6.8+00	B	TS90LS
479.888	2s3d	2s4f	^3D	$^3F^\circ$		3	2	812202	1020584	8.8+07	2.40−02	B	TS90LS
479.888	2s3d	2s4f	^3D	$^3F^\circ$		3	3	812202	1020584	2.23+09	8.5−01	B	TS90LS
479.888	2s3d	2s4f	^3D	$^3F^\circ$		3	4	812202	1020584	2.00+10	9.8+00	B	TS90LS
492.871	2p3p	2p4d	^1S	$^1P^\circ$		0	1	934650	1137543	4.6+09	8.1−01	B	TS90
498.048	2p3p	2p4d	^1D	$^1D^\circ$		2	2	925389	1126173	2.7+09	8.1−01	B	TS90
500.774	2s3p	2s4s	$^3P^\circ$	^3S		0	1	790240	989931	5.9+08	1.09−01	B	TS90LS
500.954	2s3p	2s4s	$^3P^\circ$	^3S		1	1	790312	989931	1.75+09	3.3−01	B	TS90LS
501.326	2s3p	2s4s	$^3P^\circ$	^3S		2	1	790460	989931	2.9+09	5.4−01	B	TS90LS
511.310	2s3d	2s4f	^1D	$^1F^\circ$		2	3	826843	1022419	1.91+10	8.8+00	B	TS90
534.911	2p3p	2p4s	^1D	$^1P^\circ$		2	1	925389	1112336	1.96+09	4.5−01	B	TS90
535.206	$2s^2$	2s2p	^1S	$^1P^\circ$		0	1	0	186844	3.37+09	7.66−01	A	F04
546.983	2p3d	2p4p	$^3P^\circ$	^3S		2	1	938530	1121351	9.9+08	2.40−01	B	TS90LS
547.786	2p3d	2p4p	$^3P^\circ$	^3S		1	1	938798	1121351	5.9+08	1.44−01	B	TS90LS

λ_{Ritz} [a] (Å)	Configurations		Terms		Type	J_i [b]	J_k [b]	E_i (cm^{-1})	E_k (cm^{-1})	A_{ki} [c] (s^{-1})	S_{ik} [c]	Acc. [d]	Ref. [e]
548.210	2p3d	2p4p	$^3P^o$	^3S		0	1	938939	1121351	1.96+08	4.8−02	B	TS90LS
562.582	2s2p	$2p^2$	$^3P^o$	^1D		1	2	96850	274602	6.6+04	2.9−05	B	F04
564.414	2s2p	$2p^2$	$^3P^o$	^1D		2	2	97427	274602	9.7+05	4.33−04	A	F04
643.96	2s2p	$2p^2$	$^3P^o$	^3P		1	2	96850	252139	6.73+08	4.44−01	A	F04
645.00	2s2p	$2p^2$	$^3P^o$	^3P		0	1	96590	251629	8.9+08	3.55−01	A	F04
646.08	2s2p	$2p^2$	$^3P^o$	^3P		1	1	96850	251629	6.66+08	2.66−01	A	F04
646.36	2s2p	$2p^2$	$^3P^o$	^3P		2	2	97427	252139	1.99+09	1.33+00	A	F04
647.31	2s2p	$2p^2$	$^3P^o$	^3P		1	0	96850	251336	2.65+09	3.55−01	A	F04
648.50	2s2p	$2p^2$	$^3P^o$	^3P		2	1	97427	251629	1.10+09	4.43−01	A	F04
651.08	2s2p	$2p^2$	$^1P^o$	^1S		1	0	186844	340434	4.64+09	6.32−01	A	F04
719.40	2s3s	2p3s	^3S	$^1P^o$		1	1	747284	886288	5.3+05	2.9−04	B	F04
751.45	2p3s	2s4d	$^1P^o$	^1D		1	2	886288	1019364	6.1+08	6.3−01	B	TS90
788.66	2s3d	2p3d	^3D	$^3P^o$		1	0	812142	938939	9.1+08	2.20−01	B	TS90LS
789.54	2s3d	2p3d	^3D	$^3P^o$		1	1	812142	938798	2.26+08	1.65−01	B	TS90LS
789.69	2s3d	2p3d	^3D	$^3P^o$		2	1	812166	938798	6.8+08	4.9−01	B	TS90LS
790.13	2s3d	2p3d	^1D	$^1P^o$		2	1	826843	953404	9.7+08	7.1−01	B	TS90
791.21	2s3d	2p3d	^3D	$^3P^o$		1	2	812142	938530	9.0+06	1.10−02	B	TS90LS
791.36	2s3d	2p3d	^3D	$^3P^o$		2	2	812166	938530	1.35+08	1.65−01	B	TS90LS
791.59	2s3d	2p3d	^3D	$^3P^o$		3	2	812202	938530	7.5+08	9.2−01	B	TS90LS
797.14	2s3p	2p3p	$^3P^o$	^3P		1	2	790312	915760	1.74+08	2.18−01	B	TS90LS
798.08	2s3p	2p3p	$^3P^o$	^3P		2	2	790460	915760	5.2+08	6.5−01	B	TS90LS
798.96	2s3p	2p3p	$^3P^o$	^3P		0	1	790240	915403	2.30+08	1.74−01	B	TS90LS
799.42	2s3p	2p3p	$^3P^o$	^3P		1	1	790312	915403	1.73+08	1.31−01	B	TS90LS
800.36	2s3p	2p3p	$^3P^o$	^3P		2	1	790460	915403	2.9+08	2.18−01	B	TS90LS
800.84	2s3p	2p3p	$^3P^o$	^3P		1	0	790312	915181	6.9+08	1.74−01	B	TS90LS
801.28	2s3s	2p3s	^3S	$^3P^o$		1	2	747284	872084	5.92+08	7.51−01	A	F04
805.46	2s3s	2p3s	^3S	$^3P^o$		1	1	747284	871437	5.81+08	4.50−01	A	F04
807.39	2s3s	2p3s	^3S	$^3P^o$		1	0	747284	871140	5.77+08	1.50−01	A	F04
820.37	2s3s	2p3s	^1S	$^1P^o$		0	1	764392	886288	1.2+09	1.0+00	D	F04

λ_{Ritz} [a] (Å)	Configurations		Terms		Type	J_i [b]	J_k [b]	E_i (cm^{-1})	E_k (cm^{-1})	A_{ki} [c] (s^{-1})	S_{ik} [c]	Acc. [d]	Ref. [e]
821.21	2s3d	2p3d	^3D	$^3D^o$		2	3	812166	933937	8.5+07	1.63−01	B	TS90LS
821.46	2s3d	2p3d	^3D	$^3D^o$		3	3	812202	933937	6.8+08	1.30+00	B	TS90LS
822.54	2s3d	2p3d	^3D	$^3D^o$		1	2	812142	933716	1.14+08	1.57−01	B	TS90LS
822.71	2s3d	2p3d	^3D	$^3D^o$		2	2	812166	933716	5.3+08	7.3−01	B	TS90LS
822.95	2s3d	2p3d	^3D	$^3D^o$		3	2	812202	933716	1.19+08	1.63−01	B	TS90LS
823.43	2s3d	2p3d	^3D	$^3D^o$		1	1	812142	933585	5.7+08	4.7−01	B	TS90LS
823.59	2s3d	2p3d	^3D	$^3D^o$		2	1	812166	933585	1.90+08	1.57−01	B	TS90LS
832.31	2p3p	2s4f	^3D	$^3F^o$		1	2	900436	1020584	1.47+08	2.09−01	B	TS90LS
834.77	2p3p	2s4f	^3D	$^3F^o$		2	2	900791	1020584	2.7+07	3.9−02	B	TS90LS
834.77	2p3p	2s4f	^3D	$^3F^o$		2	3	900791	1020584	1.54+08	3.1−01	B	TS90LS
839.00	2p3p	2s4f	^3D	$^3F^o$		3	2	901395	1020584	7.5+05	1.1−03	C	TS90LS
839.00	2p3p	2s4f	^3D	$^3F^o$		3	3	901395	1020584	1.90+07	3.9−02	B	TS90LS
839.00	2p3p	2s4f	^3D	$^3F^o$		3	4	901395	1020584	1.71+08	4.5−01	B	TS90LS
839.87	2s3p	2p3p	$^3P^o$	^3S		0	1	790240	909306	1.07+08	9.4−02	B	TS90LS
840.38	2s3p	2p3p	$^3P^o$	^3S		1	1	790312	909306	3.2+08	2.8−01	B	TS90LS
841.43	2s3p	2p3p	$^3P^o$	^3S		2	1	790460	909306	5.3+08	4.7−01	B	TS90LS
846.17	2s4d	2p4d	^1D	$^1P^o$		2	1	1019364	1137543	6.8+08	6.1−01	B	TS90
857.68	2s4d	2p4d	^1D	$^1F^o$		2	3	1019364	1135958	5.3+08	1.16+00	B	TS90
901.43	2s3p	2p3p	$^3P^o$	^3D		2	3	790460	901395	5.6+08	1.41+00	B	TS90LS
905.15	2s3p	2p3p	$^3P^o$	^3D		1	2	790312	900791	4.1+08	7.5−01	B	TS90LS
906.36	2s3p	2p3p	$^3P^o$	^3D		2	2	790460	900791	1.37+08	2.5−01	B	TS90LS
907.47	2s3p	2p3p	$^3P^o$	^3D		0	1	790240	900436	3.0+08	3.3−01	B	TS90LS
908.07	2s3p	2p3p	$^3P^o$	^3D		1	1	790312	900436	2.26+08	2.5−01	B	TS90LS
909.29	2s3p	2p3p	$^3P^o$	^3D		2	1	790460	900436	1.50+07	1.67−02	B	TS90LS
924.98	2s4p	2p4p	$^1P^o$	^1P		1	1	1007860	1115970	7.6+08	8.9−01	B	TS90
930.73	2s3p	2p3p	$^1P^o$	^1P		1	1	787844	895286	1.04+09	1.24+00	B	TS90
934.19	2s3s	2p3s	^1S	$^3P^o$		0	1	764392	871437	6.1+05	7.3−04	C	F04
936.25	2s4d	2p4d	^1D	$^1D^o$		2	2	1019364	1126173	4.3+08	8.8−01	B	TS90
1026.41	$2s^2$	2s2p	^1S	$^3P^o$	M2	0	2	0	97427	3.66−02	1.40+01	A	F04

λ_{Ritz} [a] (Å)	Configurations		Terms		Type	J_i [b]	J_k [b]	E_i (cm^{-1})	E_k (cm^{-1})	A_{ki} [c] (s^{-1})	S_{ik} [c]	Acc. [d]	Ref. [e]
1030.61	2p3p	2s4f	^1D	$^1F^o$		2	3	925389	1022419	1.29+08	4.9−01	B	TS90
1032.52	$2s^2$	2s2p	^1S	$^3P^o$		0	1	0	96850	7.+03	1.1−05	D	F04
1052.84	2s3d	2p3d	^1D	$^1D^o$		2	2	826843	921824	3.1+08	8.8−01	B	TS90
1139.50	2s2p	$2p^2$	$^1P^o$	^1D		1	2	186844	274602	4.5+08	1.63+00	B	F04
1317.56	2p3d	2s4d	$^3P^o$	^3D		2	3	938530	1014428	4.6+07	3.6−01	B	TS90LS
1317.82	2p3d	2s4d	$^3P^o$	^3D		2	2	938530	1014413	1.14+07	6.4−02	B	TS90LS
1317.97	2p3d	2s4d	$^3P^o$	^3D		2	1	938530	1014404	1.3+06	4.3−03	C	TS90LS
1322.49	2p3d	2s4d	$^3P^o$	^3D		1	2	938798	1014413	3.4+07	1.93−01	B	TS90LS
1322.65	2p3d	2s4d	$^3P^o$	^3D		1	1	938798	1014404	1.88+07	6.4−02	B	TS90LS
1325.12	2p3d	2s4d	$^3P^o$	^3D		0	1	938939	1014404	2.49+07	8.6−02	B	TS90LS
1531.51	2s2p	$2p^2$	$^1P^o$	^3P		1	2	186844	252139	4.3+04	3.8−04	B	F04
1550.58	2s2p	$2p^2$	$^1P^o$	^3P		1	0	186844	251336	1.3+04	2.3−05	E	F04
1668.28	2s3d	2p3s	^3D	$^3P^o$		1	2	812142	872084	7.9+04	9.0−04	B	F04
1668.95	2s3d	2p3s	^3D	$^3P^o$		2	2	812166	872084	1.16+06	1.33−02	B	F04
1669.95	2s3d	2p3s	^3D	$^3P^o$		3	2	812202	872084	6.3+06	7.2−02	B	F04
1682.23	2s3d	2p3s	^1D	$^1P^o$		2	1	826843	886288	2.2+07	1.6−01	D	F04
1686.48	2s3d	2p3s	^3D	$^3P^o$		1	1	812142	871437	1.89+06	1.34−02	B	F04
1687.17	2s3d	2p3s	^3D	$^3P^o$		2	1	812166	871437	5.6+06	3.9−02	B	F04
1694.97	2s3d	2p3s	^3D	$^3P^o$		1	0	812142	871140	7.4+06	1.79−02	B	F04
1720.64	2p3p	2p3d	^1P	$^1P^o$		1	1	895286	953404	2.29+08	1.73+00	B	TS90
2067.1	2p3s	2p3p	$^1P^o$	^1S		1	0	886288	934650	4.9+08	2.13+00	B	TS90
2255.5	2p3s	2p3p	$^3P^o$	^3P		1	2	871437	915760	5.9+07	1.68+00	B	TS90LS
2258.5	2p3s	2p3p	$^3P^o$	^3P		0	1	871140	915403	7.9+07	1.35+00	B	TS90LS
2273.8	2p3s	2p3p	$^3P^o$	^3P		1	1	871437	915403	5.8+07	1.01+00	B	TS90LS
2285.3	2p3s	2p3p	$^3P^o$	^3P		1	0	871437	915181	2.28+08	1.35+00	B	TS90LS
2288.9	2p3s	2p3p	$^3P^o$	^3P		2	2	872084	915760	1.71+08	5.1+00	B	TS90LS
2307.7	2p3s	2p3p	$^3P^o$	^3P		2	1	872084	915403	9.2+07	1.68+00	B	TS90LS
2315.4	2s3s	2s3p	^3S	$^3P^o$		1	2	747284	790460	1.75+08	5.35+00	A	F04
2323.4	2s3s	2s3p	^3S	$^3P^o$		1	1	747284	790312	1.72+08	3.20+00	A	F04

λ_{Ritz} [a] (Å)	Configurations		Terms		Type	J_i [b]	J_k [b]	E_i (cm^{-1})	E_k (cm^{-1})	A_{ki} [c] (s^{-1})	S_{ik} [c]	Acc.[d]	Ref.[e]
2327.2	2s3s	2s3p	^3S	$^3P^\circ$		1	0	747284	790240	1.72+08	1.07+00	A	F04
2464.7	2s3s	2s3p	^3S	$^1P^\circ$		1	1	747284	787844	3.58+05	7.94−03	A	F04
2556.7	2p3s	2p3p	$^1P^\circ$	^1D		1	2	886288	925389	1.78+08	7.3+00	B	TS90
2563.4	2s3p	2s3d	$^1P^\circ$	^1D		1	2	787844	826843	1.14+08	4.75+00	A	F04
2596.4	2p3p	2p3d	^3D	$^3P^\circ$		1	0	900436	938939	2.24+07	1.94−01	B	TS90LS
2606.0	2p3p	2p3d	^3D	$^3P^\circ$		1	1	900436	938798	5.5+06	1.45−01	B	TS90LS
2619.4	2p3s	2p3p	$^3P^\circ$	^3S		0	1	871140	909306	1.45+07	3.9−01	B	TS90LS
2624.3	2p3p	2p3d	^3D	$^3P^\circ$		1	2	900436	938530	2.2+05	9.7−03	C	TS90LS
2630.3	2p3p	2p3d	^3D	$^3P^\circ$		2	1	900791	938798	1.62+07	4.4−01	B	TS90LS
2639.9	2p3s	2p3p	$^3P^\circ$	^3S		1	1	871437	909306	4.2+07	1.16+00	B	TS90LS
2649.0	2p3p	2p3d	^3D	$^3P^\circ$		2	2	900791	938530	3.2+06	1.45−01	B	TS90LS
2685.8	2p3s	2p3p	$^3P^\circ$	^3S		2	1	872084	909306	6.7+07	1.93+00	B	TS90LS
2692.1	2p3p	2p3d	^3D	$^3P^\circ$		3	2	901395	938530	1.69+07	8.1−01	B	TS90LS
2736.6	2s3p	2s3d	$^3P^\circ$	^1D		1	2	790312	826843	2.4+05	1.2−02	C	F04
2747.7	2s3p	2s3d	$^3P^\circ$	^1D		2	2	790460	826843	4.1+01	2.1−06	D	F04
3003.9	2p3p	2p3d	^3D	$^3D^\circ$		1	2	900436	933716	2.6+06	1.74−01	B	TS90LS
3015.8	2p3p	2p3d	^3D	$^3D^\circ$		1	1	900436	933585	1.29+07	5.2−01	B	TS90LS
3016.1	2p3p	2p3d	^3D	$^3D^\circ$		2	3	900791	933937	1.91+06	1.81−01	B	TS90LS
3036.3	2p3p	2p3d	^3D	$^3D^\circ$		2	2	900791	933716	1.17+07	8.1−01	B	TS90LS
3048.5	2p3p	2p3d	^3D	$^3D^\circ$		2	1	900791	933585	4.1+06	1.74−01	B	TS90LS
3072.1	2p3p	2p3d	^3D	$^3D^\circ$		3	3	901395	933937	1.44+07	1.44+00	B	TS90LS
3093.1	2p3p	2p3d	^3D	$^3D^\circ$		3	2	901395	933716	2.48+06	1.81−01	B	TS90LS
3373.6	2p3p	2p3d	^3S	$^3P^\circ$		1	0	909306	938939	4.4+07	8.3−01	B	TS90LS
3389.8	2p3p	2p3d	^3S	$^3P^\circ$		1	1	909306	938798	4.3+07	2.49+00	B	TS90LS
3405.7	2p3s	2p3p	$^3P^\circ$	^3D		1	2	871437	900791	4.6+07	4.5+00	B	TS90LS
3410.7	2p3s	2p3p	$^3P^\circ$	^3D		2	3	872084	901395	6.1+07	8.4+00	B	TS90LS
3412.5	2p3s	2p3p	$^3P^\circ$	^3D		0	1	871140	900436	3.4+07	2.00+00	B	TS90LS
3420.9	2p3p	2p3d	^3S	$^3P^\circ$		1	2	909306	938530	4.2+07	4.1+00	B	TS90LS
3447.4	2p3s	2p3p	$^3P^\circ$	^3D		1	1	871437	900436	2.47+07	1.50+00	B	TS90LS

λ_{Ritz} [a] (Å)	Configurations		Terms		Type	J_i [b]	J_k [b]	E_i (cm^{-1})	E_k (cm^{-1})	A_{ki} [c] (s^{-1})	S_{ik} [c]	Acc. [d]	Ref. [e]
3482.5	2p3s	2p3p	^{3}P°	^{3}D		2	2	872084	900791	1.44+07	1.50+00	B	TS90LS
3526.1	2p3s	2p3p	^{3}P°	^{3}D		2	1	872084	900436	1.54+06	1.00−01	B	TS90LS
3767.1	2p3p	2p3d	^{1}P	^{1}D°		1	2	895286	921824	3.0+07	3.9+00	B	TS90
3856.9	2s3s	2s3p	^{1}S	^{3}P°		0	1	764392	790312	7.37+04	6.27−03	A	F04
4110.3	2s3p	2s3d	^{1}P°	^{3}D		1	2	787844	812166	5.1+04	8.8−03	C	F04
4114.4	2s3p	2s3d	^{1}P°	^{3}D		1	1	787844	812142	2.55+04	2.64−03	A	F04
4233.0	2p3p	2p3d	^{3}P	^{3}P°		0	1	915181	938798	2.9+06	3.3−01	B	TS90LS
4247.6	2p3p	2p3d	^{3}P	^{3}P°		1	0	915403	938939	8.7+06	3.3−01	B	TS90LS
4262.8	2s3s	2s3p	^{1}S	^{1}P°		0	1	764392	787844	2.20+07	2.5+00	B	F04
4273.2	2p3p	2p3d	^{3}P	^{3}P°		1	1	915403	938798	2.13+06	2.46−01	B	TS90LS
4322.7	2p3p	2p3d	^{3}P	^{3}P°		1	2	915403	938530	2.06+06	4.1−01	B	TS90LS
4339.4	2p3p	2p3d	^{3}P	^{3}P°		2	1	915760	938798	3.4+06	4.1−01	B	TS90LS
4390.5	2p3p	2p3d	^{3}P	^{3}P°		2	2	915760	938530	5.9+06	1.23+00	B	TS90LS
4559.9	2p3p	2p3d	^{1}D	^{1}F°		2	3	925389	947313	2.5+07	8.2+00	B	TS90
4564.5	2s3p	2s3d	^{3}P°	^{3}D		0	1	790240	812142	1.05+07	1.48+00	B	F04
4574.5	2s3p	2s3d	^{3}P°	^{3}D		1	2	790312	812166	1.41+07	3.3+00	B	F04
4579.6	2s3p	2s3d	^{3}P°	^{3}D		1	1	790312	812142	7.8+06	1.11+00	B	F04
4598.1	2s3p	2s3d	^{3}P°	^{3}D		2	3	790460	812202	1.86+07	6.2+00	B	F04
4605.7	2s3p	2s3d	^{3}P°	^{3}D		2	2	790460	812166	4.6+06	1.11+00	B	F04
4610.8	2s3p	2s3d	^{3}P°	^{3}D		2	1	790460	812142	5.1+05	7.4−02	B	F04
4634.1	2p4p	2p4d	^{1}P	^{1}P°		1	1	1115970	1137543	4.6+07	6.8+00	B	TS90
5330.7	2p3p	2p3d	^{1}S	^{1}P°		0	1	934650	953404	9.5+06	2.13+00	B	TS90
5432.1	2p3p	2p3d	^{3}P	^{3}D°		0	1	915181	933585	5.7+06	1.37+00	B	TS90LS
5459.1	2p3p	2p3d	^{3}P	^{3}D°		1	2	915403	933716	7.6+06	3.1+00	B	TS90LS
5498.4	2p3p	2p3d	^{3}P	^{3}D°		1	1	915403	933585	4.2+06	1.02+00	B	TS90LS
5499.9	2p3p	2p3d	^{3}P	^{3}D°		2	3	915760	933937	1.0+07	5.7+00	B	TS90LS
5567.6	2p3p	2p3d	^{3}P	^{3}D°		2	2	915760	933716	2.40+06	1.02+00	B	TS90LS
5608.5	2p3p	2p3d	^{3}P	^{3}D°		2	1	915760	933585	2.6+05	6.8−02	B	TS90LS
7234	2p4s	2p4p	^{1}P°	^{1}D		1	2	1112336	1126155	1.86+07	1.74+01	B	TS90

λ_{Ritz} [a] (Å)	Configurations		Terms		Type	J_i [b]	J_k [b]	E_i (cm^{-1})	E_k (cm^{-1})	A_{ki} [c] (s^{-1})	S_{ik} [c]	Acc.[d]	Ref.[e]
8690	2s4p	2s4d	$^1P°$	^1D		1	2	1007860	1019364	1.39+07	2.25+01	B	TS90
8779	2p4p	2p4d	^1D	$^1P°$		2	1	1126155	1137543	7.8+05	7.9−01	B	TS90
9798	2p4p	2p4d	^1P	$^1D°$		1	2	1115970	1126173	7.4+06	1.72+01	B	TS90
9811	2s4s	2s4p	^1S	$^1P°$		0	1	997670	1007860	8.4+06	1.18+01	B	TS90
10198	2p4p	2p4d	^1D	$^1F°$		2	3	1126155	1135958	8.8+06	3.2+01	B	TS90
11111	2p3s	2p3p	$^1P°$	^1P		1	1	886288	895286	1.35+06	2.7+00	B	TS90
16177	2s4d	2s4f	^3D	$^3F°$		1	2	1014404	1020584	1.22+06	1.28+01	B	TS90LS
16200	2s4d	2s4f	^3D	$^3F°$		2	2	1014413	1020584	2.26+05	2.37+00	B	TS90LS
16200	2s4d	2s4f	^3D	$^3F°$		2	3	1014413	1020584	1.29+06	1.89+01	B	TS90LS
16240	2s4d	2s4f	^3D	$^3F°$		3	2	1014428	1020584	6.3+03	6.7−02	B	TS90LS
16240	2s4d	2s4f	^3D	$^3F°$		3	3	1014428	1020584	1.60+05	2.37+00	B	TS90LS
16240	2s4d	2s4f	^3D	$^3F°$		3	4	1014428	1020584	1.44+06	2.7+01	B	TS90LS
27520	2p4s	2p4p	$^1P°$	^1P		1	1	1112336	1115970	3.1+05	9.6+00	B	TS90
28050	2p3d	2p3p	$^1D°$	^1D		2	2	921824	925389	2.06+04	1.12+00	B	TS90
32730	2s4d	2s4f	^1D	$^1F°$		2	3	1019364	1022419	1.75+05	2.12+01	B	TS90
173300	2s2p	2s2p	$^3P°$	$^3P°$	M1	1	2	96850	97427	2.6−03	2.50+00	B	F04
5600000	2p4p	2p4d	^1D	$^1D°$		2	2	1126155	1126173	1.12−02	4.7+00	B	TS90
F VII (Li-like)													
69.4478	$1s^22s$	$1s^210p$	^2S	$^2P°$		1/2	1/2	0	1439930	1.51+09	5.00−04	A	N02
69.4478	$1s^22s$	$1s^210p$	^2S	$^2P°$		1/2	3/2	0	1439930	1.51+09	1.0−03	A	N02
70.1198	$1s^22s$	$1s^29p$	^2S	$^2P°$		1/2	1/2	0	1426130	2.07+09	7.04−04	A	N02
70.1198	$1s^22s$	$1s^29p$	^2S	$^2P°$		1/2	3/2	0	1426130	2.06+09	1.41−03	A	N02
70.9800	$1s^22s$	$1s^28p$	^2S	$^2P°$		1/2	1/2	0	1408848	2.94+09	1.04−03	A	N02
70.9800	$1s^22s$	$1s^28p$	^2S	$^2P°$		1/2	3/2	0	1408848	2.94+09	2.07−03	A	N02
72.3140	$1s^22s$	$1s^27p$	^2S	$^2P°$		1/2	1/2	0	1382858	4.38+09	1.64−03	A	N02
72.3140	$1s^22s$	$1s^27p$	^2S	$^2P°$		1/2	3/2	0	1382858	4.37+09	3.26−03	A	N02
74.4654	$1s^22s$	$1s^26p$	^2S	$^2P°$		1/2	1/2	0	1342906	6.93+09	2.83−03	A	N02
74.4654	$1s^22s$	$1s^26p$	^2S	$^2P°$		1/2	3/2	0	1342906	6.91+09	5.63−03	A	N02

λ_{Ritz} [a] (Å)	Configurations		Terms		Type	J_i [b]	J_k [b]	E_i (cm^{-1})	E_k (cm^{-1})	A_{ki} [c] (s^{-1})	S_{ik} [c]	Acc.[d]	Ref.[e]
75.2729	$1s^22p$	$1s^210d$	$^2P°$	^2D		1/2	3/2	112260	1440760	2.27+09	1.92−03	A	N02
75.3282	$1s^22p$	$1s^210d$	$^2P°$	^2D		3/2	3/2	113236	1440760	4.53+08	3.83−04	A	N02
75.3282	$1s^22p$	$1s^210d$	$^2P°$	^2D		3/2	5/2	113236	1440760	2.72+09	3.45−03	A	N02
76.0479	$1s^22p$	$1s^29d$	$^2P°$	^2D		1/2	3/2	112260	1427220	3.14+09	2.73−03	A	N02
76.1044	$1s^22p$	$1s^29d$	$^2P°$	^2D		3/2	3/2	113236	1427220	6.27+08	5.45−04	A	N02
76.1044	$1s^22p$	$1s^29d$	$^2P°$	^2D		3/2	5/2	113236	1427220	3.76+09	4.91−03	A	N02
77.0826	$1s^22p$	$1s^28d$	$^2P°$	^2D		1/2	3/2	112260	1409569	4.54+09	4.10−03	A	N02
77.1407	$1s^22p$	$1s^28d$	$^2P°$	^2D		3/2	3/2	113236	1409569	9.0+08	8.20−04	A	N02
77.1407	$1s^22p$	$1s^28d$	$^2P°$	^2D		3/2	5/2	113236	1409569	5.43+09	7.39−03	A	N02
78.3566	$1s^22s$	$1s^25p$	^2S	$^2P°$		1/2	3/2	0	1276216	1.18+10	1.13−02	A	N02
78.3606	$1s^22s$	$1s^25p$	^2S	$^2P°$		1/2	1/2	0	1276152	1.19+10	5.64−03	A	N02
78.6430	$1s^22p$	$1s^27d$	$^2P°$	^2D		1/2	3/2	112260	1383829	6.91+09	6.63−03	A	N02
78.7034	$1s^22p$	$1s^27d$	$^2P°$	^2D		3/2	3/2	113236	1383829	1.38+09	1.33−03	A	N02
78.7034	$1s^22p$	$1s^27d$	$^2P°$	^2D		3/2	5/2	113236	1383829	8.28+09	1.20−02	A	N02
78.8323	$1s^22p$	$1s^27s$	$^2P°$	^2S		1/2	1/2	112260	1380775	6.49+08	3.14−04	A	N02
78.8930	$1s^22p$	$1s^27s$	$^2P°$	^2S		3/2	1/2	113236	1380775	1.31+09	6.33−04	A	N02
81.1755	$1s^22p$	$1s^26d$	$^2P°$	^2D		1/2	3/2	112260	1344159	1.13+10	1.20−02	A	N02
81.2391	$1s^22p$	$1s^26d$	$^2P°$	^2D		3/2	5/2	113236	1344170	1.36+10	2.15−02	A	N02
81.2399	$1s^22p$	$1s^26d$	$^2P°$	^2D		3/2	3/2	113236	1344159	2.26+09	2.39−03	A	N02
81.4977	$1s^22p$	$1s^26s$	$^2P°$	^2S		1/2	1/2	112260	1339289	1.05+09	5.59−04	A	N02
81.5625	$1s^22p$	$1s^26s$	$^2P°$	^2S		3/2	1/2	113236	1339289	2.10+09	1.13−03	A	N02
85.7444	$1s^22p$	$1s^25f$	$^2P°$	$^2F°$	E2	1/2	5/2	112260	1278517	3.8+06	9.5−02	B	A10
85.7553	$1s^22p$	$1s^25d$	$^2P°$	^2D		1/2	3/2	112260	1278369	2.07+10	2.57−02	A	N02
85.8162	$1s^22p$	$1s^25f$	$^2P°$	$^2F°$	E2	3/2	5/2	113236	1278517	1.09+06	2.7−02	B	A10
85.8162	$1s^22p$	$1s^25f$	$^2P°$	$^2F°$	E2	3/2	7/2	113236	1278517	4.9+06	1.63−01	B	A10
85.8257	$1s^22p$	$1s^25d$	$^2P°$	^2D		3/2	5/2	113236	1278388	2.47+10	4.63−02	A	N02
85.8271	$1s^22p$	$1s^25d$	$^2P°$	^2D		3/2	3/2	113236	1278369	4.12+09	5.14−03	A	N02
86.3841	$1s^22p$	$1s^25s$	$^2P°$	^2S		1/2	1/2	112260	1269881	1.85+09	1.18−03	A	N02
86.4091	$1s^22s$	$1s^24d$	^2S	^2D	E2	1/2	5/2	0	1157285	1.82+06	4.7−02	B	A10

λ_{Ritz} [a] (Å)	Configurations		Terms		Type	J_i [b]	J_k [b]	E_i (cm^{-1})	E_k (cm^{-1})	A_{ki} [c] (s^{-1})	S_{ik} [c]	Acc. [d]	Ref. [e]
86.4118	$1s^22s$	$1s^24d$	^2S	^2D	E2	1/2	3/2	0	1157249	1.82+06	3.1−02	B	A10
86.4570	$1s^22p$	$1s^25s$	$^2P°$	^2S		3/2	1/2	113236	1269881	3.72+09	2.37−03	A	N02
86.7295	$1s^22s$	$1s^24p$	^2S	$^2P°$		1/2	3/2	0	1153010	2.27+10	2.92−02	A	N02
86.7383	$1s^22s$	$1s^24p$	^2S	$^2P°$		1/2	1/2	0	1152893	2.27+10	1.47−02	A	N02
95.6697	$1s^22p$	$1s^24f$	$^2P°$	$^2F°$	E2	1/2	5/2	112260	1157523	5.7+06	2.43−01	B	A10
95.6948	$1s^22p$	$1s^24d$	$^2P°$	^2D		1/2	3/2	112260	1157249	4.46+10	7.71−02	A	N02
95.7591	$1s^22p$	$1s^24f$	$^2P°$	$^2F°$	E2	3/2	7/2	113236	1157523	7.3+06	4.2−01	B	A10
95.7809	$1s^22p$	$1s^24d$	$^2P°$	^2D		3/2	5/2	113236	1157285	5.34+10	1.39−01	A	N02
95.7842	$1s^22p$	$1s^24d$	$^2P°$	^2D		3/2	3/2	113236	1157249	8.9+09	1.54−02	A	N02
96.1856	$1s^22p$	$1s^24p$	$^2P°$	$^2P°$	E2	3/2	1/2	113236	1152893	1.28+06	1.88−02	B	A10
97.2600	$1s^22p$	$1s^24s$	$^2P°$	^2S		1/2	1/2	112260	1140432	3.81+09	3.46−03	A	F04
97.3524	$1s^22p$	$1s^24s$	$^2P°$	^2S		3/2	1/2	113236	1140432	7.69+09	7.00−03	A	F04
111.644	$1s^22s$	$1s^23d$	^2S	^2D	E2	1/2	5/2	0	895706	9.8+06	9.1−01	B	A10
111.655	$1s^22s$	$1s^23d$	^2S	^2D	E2	1/2	3/2	0	895619	9.8+06	6.1−01	B	A10
112.942	$1s^22s$	$1s^23p$	^2S	$^2P°$		1/2	3/2	0	885414	5.03+10	1.430−01	AA	F04
112.978	$1s^22s$	$1s^23p$	^2S	$^2P°$		1/2	1/2	0	885128	5.05+10	7.18−02	AA	F04
127.655	$1s^22p$	$1s^23d$	$^2P°$	^2D		1/2	3/2	112260	895619	1.346+11	5.53−01	AA	F04
127.800	$1s^22p$	$1s^23d$	$^2P°$	^2D		3/2	5/2	113236	895706	1.612+11	9.97−01	AA	F04
127.815	$1s^22p$	$1s^23d$	$^2P°$	^2D		3/2	3/2	113236	895619	2.69+10	1.108−01	AA	F04
129.340	$1s^22p$	$1s^23p$	$^2P°$	$^2P°$	E2	1/2	3/2	112260	885414	1.50+06	1.94−01	B	A10
129.504	$1s^22p$	$1s^23p$	$^2P°$	$^2P°$	E2	3/2	3/2	113236	885414	1.50+06	1.95−01	B	A10
129.552	$1s^22p$	$1s^23p$	$^2P°$	$^2P°$	E2	3/2	1/2	113236	885128	3.0+06	1.96−01	B	A10
134.705	$1s^22p$	$1s^23s$	$^2P°$	^2S		1/2	1/2	112260	854621	9.6+09	2.33−02	A	F04
134.883	$1s^22p$	$1s^23s$	$^2P°$	^2S		3/2	1/2	113236	854621	1.93+10	4.67−02	A	F04
170.850	$1s^23s$	$1s^210p$	^2S	$^2P°$		1/2	1/2	854621	1439930	4.47+08	2.20−03	A	N02
170.850	$1s^23s$	$1s^210p$	^2S	$^2P°$		1/2	3/2	854621	1439930	4.45+08	4.38−03	A	N02
174.975	$1s^23s$	$1s^29p$	^2S	$^2P°$		1/2	1/2	854621	1426130	6.10+08	3.23−03	A	N02
174.975	$1s^23s$	$1s^29p$	^2S	$^2P°$		1/2	3/2	854621	1426130	6.08+08	6.43−03	A	N02
179.975	$1s^23p$	$1s^210d$	$^2P°$	^2D		1/2	3/2	885128	1440760	7.73+08	8.9−03	A	N02

λ_{Ritz} [a] (Å)	Configurations		Terms		Type	J_i [b]	J_k [b]	E_i (cm^{-1})	E_k (cm^{-1})	A_{ki} [c] (s^{-1})	S_{ik} [c]	Acc. [d]	Ref. [e]
180.068	$1s^23p$	$1s^210d$	$^2P°$	^2D		3/2	3/2	885414	1440760	1.55+08	1.78−03	A	N02
180.068	$1s^23p$	$1s^210d$	$^2P°$	^2D		3/2	5/2	885414	1440760	9.3+08	1.60−02	A	N02
180.431	$1s^23s$	$1s^28p$	^2S	$^2P°$		1/2	1/2	854621	1408848	8.7+08	5.07−03	A	N02
180.431	$1s^23s$	$1s^28p$	^2S	$^2P°$		1/2	3/2	854621	1408848	8.7+08	1.01−02	A	N02
183.718	$1s^23d$	$1s^210p$	^2D	$^2P°$		3/2	1/2	895619	1439930	4.34+07	2.66−04	A	N02
183.718	$1s^23d$	$1s^210p$	^2D	$^2P°$		3/2	3/2	895619	1439930	4.30+06	5.27−05	A	N02
183.748	$1s^23d$	$1s^210p$	^2D	$^2P°$		5/2	3/2	895706	1439930	3.88+07	4.76−04	A	N02
184.471	$1s^23p$	$1s^29d$	$^2P°$	^2D		1/2	3/2	885128	1427220	1.07+09	1.32−02	A	N02
184.568	$1s^23p$	$1s^29d$	$^2P°$	^2D		3/2	3/2	885414	1427220	2.13+08	2.65−03	A	N02
184.568	$1s^23p$	$1s^29d$	$^2P°$	^2D		3/2	5/2	885414	1427220	1.28+09	2.38−02	A	N02
188.498	$1s^23d$	$1s^29p$	^2D	$^2P°$		3/2	1/2	895619	1426130	6.09+07	4.03−04	A	N02
188.498	$1s^23d$	$1s^29p$	^2D	$^2P°$		3/2	3/2	895619	1426130	6.03+06	7.98−05	A	N02
188.528	$1s^23d$	$1s^29p$	^2D	$^2P°$		5/2	3/2	895706	1426130	5.45+07	7.20−04	A	N02
189.309	$1s^23s$	$1s^27p$	^2S	$^2P°$		1/2	1/2	854621	1382858	1.31+09	8.8−03	A	N02
189.309	$1s^23s$	$1s^27p$	^2S	$^2P°$		1/2	3/2	854621	1382858	1.31+09	1.75−02	A	N02
190.679	$1s^23p$	$1s^28d$	$^2P°$	^2D		1/2	3/2	885128	1409569	1.54+09	2.11−02	A	N02
190.783	$1s^23p$	$1s^28d$	$^2P°$	^2D		3/2	3/2	885414	1409569	3.08+08	4.23−03	A	N02
190.783	$1s^23p$	$1s^28d$	$^2P°$	^2D		3/2	5/2	885414	1409569	1.85+09	3.81−02	A	N02
194.845	$1s^23d$	$1s^28p$	^2D	$^2P°$		3/2	1/2	895619	1408848	9.1+07	6.61−04	A	N02
194.845	$1s^23d$	$1s^28p$	^2D	$^2P°$		3/2	3/2	895619	1408848	9.0+06	1.31−04	A	N02
194.878	$1s^23d$	$1s^28p$	^2D	$^2P°$		5/2	3/2	895706	1408848	8.10+07	1.18−03	A	N02
200.521	$1s^23p$	$1s^27d$	$^2P°$	^2D		1/2	3/2	885128	1383829	2.36+09	3.76−02	A	N02
200.636	$1s^23p$	$1s^27d$	$^2P°$	^2D		3/2	3/2	885414	1383829	4.72+08	7.53−03	A	N02
200.636	$1s^23p$	$1s^27d$	$^2P°$	^2D		3/2	5/2	885414	1383829	2.83+09	6.78−02	A	N02
201.756	$1s^23p$	$1s^27s$	$^2P°$	^2S		1/2	1/2	885128	1380775	3.96+08	3.21−03	A	N02
201.873	$1s^23p$	$1s^27s$	$^2P°$	^2S		3/2	1/2	885414	1380775	7.97+08	6.47−03	A	N02
204.798	$1s^23s$	$1s^26p$	^2S	$^2P°$		1/2	1/2	854621	1342906	2.10+09	1.78−02	A	N02
204.798	$1s^23s$	$1s^26p$	^2S	$^2P°$		1/2	3/2	854621	1342906	2.09+09	3.54−02	A	N02
204.808	$1s^23d$	$1s^27f$	^2D	$^2F°$		3/2	5/2	895619	1383881	2.66+09	6.77−02	A	N02

λ_{Ritz} [a] (Å)	Configurations		Terms		Type	J_i [b]	J_k [b]	E_i (cm^{-1})	E_k (cm^{-1})	A_{ki} [c] (s^{-1})	S_{ik} [c]	Acc. [d]	Ref. [e]
204.845	$1s^23d$	$1s^27f$	^2D	$^2F^\circ$		5/2	5/2	895706	1383881	1.90+08	4.83−03	A	N02
204.845	$1s^23d$	$1s^27f$	^2D	$^2F^\circ$		5/2	7/2	895706	1383881	2.85+09	9.7−02	A	N02
205.238	$1s^23d$	$1s^27p$	^2D	$^2P^\circ$		3/2	1/2	895619	1382858	1.44+08	1.23−03	A	N02
205.238	$1s^23d$	$1s^27p$	^2D	$^2P^\circ$		3/2	3/2	895619	1382858	1.42+07	2.43−04	A	N02
205.275	$1s^23d$	$1s^27p$	^2D	$^2P^\circ$		5/2	3/2	895706	1382858	1.28+08	2.19−03	A	N02
217.850	$1s^23p$	$1s^26d$	$^2P^\circ$	^2D		1/2	3/2	885128	1344159	3.91+09	7.97−02	A	N02
217.981	$1s^23p$	$1s^26d$	$^2P^\circ$	^2D		3/2	5/2	885414	1344170	4.69+09	1.44−01	A	N02
217.986	$1s^23p$	$1s^26d$	$^2P^\circ$	^2D		3/2	3/2	885414	1344159	7.80+08	1.60−02	A	N02
220.186	$1s^23p$	$1s^26s$	$^2P^\circ$	^2S		1/2	1/2	885128	1339289	6.72+08	7.08−03	A	N02
220.325	$1s^23p$	$1s^26s$	$^2P^\circ$	^2S		3/2	1/2	885414	1339289	1.35+09	1.43−02	A	N02
222.902	$1s^23d$	$1s^26f$	^2D	$^2F^\circ$		3/2	5/2	895619	1344247	4.79+09	1.57−01	A	N02
222.945	$1s^23d$	$1s^26f$	^2D	$^2F^\circ$		5/2	5/2	895706	1344247	3.42+08	1.12−02	A	N02
222.945	$1s^23d$	$1s^26f$	^2D	$^2F^\circ$		5/2	7/2	895706	1344247	5.13+09	2.25−01	A	N02
223.570	$1s^23d$	$1s^26p$	^2D	$^2P^\circ$		3/2	1/2	895619	1342906	2.51+08	2.77−03	A	N02
223.570	$1s^23d$	$1s^26p$	^2D	$^2P^\circ$		3/2	3/2	895619	1342906	2.49+07	5.49−04	A	N02
223.614	$1s^23d$	$1s^26p$	^2D	$^2P^\circ$		5/2	3/2	895706	1342906	2.24+08	4.96−03	A	N02
237.194	$1s^23s$	$1s^25p$	^2S	$^2P^\circ$		1/2	3/2	854621	1276216	3.58+09	9.4−02	A	N02
237.230	$1s^23s$	$1s^25p$	^2S	$^2P^\circ$		1/2	1/2	854621	1276152	3.60+09	4.74−02	A	N02
254.297	$1s^23p$	$1s^25d$	$^2P^\circ$	^2D		1/2	3/2	885128	1278369	7.16+09	2.32−01	A	N02
254.470	$1s^23p$	$1s^25d$	$^2P^\circ$	^2D		3/2	5/2	885414	1278388	8.6+09	4.19−01	A	N02
254.482	$1s^23p$	$1s^25d$	$^2P^\circ$	^2D		3/2	3/2	885414	1278369	1.43+09	4.66−02	A	N02
259.907	$1s^23p$	$1s^25s$	$^2P^\circ$	^2S		1/2	1/2	885128	1269881	1.27+09	2.21−02	A	N02
260.100	$1s^23p$	$1s^25s$	$^2P^\circ$	^2S		3/2	1/2	885414	1269881	2.56+09	4.45−02	A	N02
261.153	$1s^23d$	$1s^25g$	^2D	^2G	E2	3/2	7/2	895619	1278537	1.23+06	1.07+01	B	A10
261.166	$1s^23d$	$1s^25f$	^2D	$^2F^\circ$		3/2	5/2	895619	1278517	1.04+10	5.50−01	A	N02
261.212	$1s^23d$	$1s^25g$	^2D	^2G	E2	5/2	9/2	895706	1278537	1.37+06	1.48+01	B	A10
261.226	$1s^23d$	$1s^25f$	^2D	$^2F^\circ$		5/2	5/2	895706	1278517	7.43+08	3.92−02	A	N02
261.226	$1s^23d$	$1s^25f$	^2D	$^2F^\circ$		5/2	7/2	895706	1278517	1.12+10	7.85−01	A	N02
262.745	$1s^23d$	$1s^25p$	^2D	$^2P^\circ$		3/2	3/2	895619	1276216	4.96+07	1.78−03	A	N02

λ_{Ritz} [a] (Å)	Configurations		Terms		Type	J_i [b]	J_k [b]	E_i (cm^{-1})	E_k (cm^{-1})	A_{ki} [c] (s^{-1})	S_{ik} [c]	Acc. [d]	Ref. [e]
262.789	$1s^23d$	$1s^25p$	^2D	$^2P^\circ$		3/2	1/2	895619	1276152	5.00+08	9.0−03	A	N02
262.805	$1s^23d$	$1s^25p$	^2D	$^2P^\circ$		5/2	3/2	895706	1276216	4.47+08	1.60−02	A	N02
333.892	$1s^24s$	$1s^210p$	^2S	$^2P^\circ$		1/2	1/2	1140432	1439930	2.01+08	7.38−03	A	N02
333.892	$1s^24s$	$1s^210p$	^2S	$^2P^\circ$		1/2	3/2	1140432	1439930	2.00+08	1.47−02	A	N02
335.133	$1s^23s$	$1s^24p$	^2S	$^2P^\circ$		1/2	3/2	854621	1153010	5.85+09	4.35−01	A	N02
335.264	$1s^23s$	$1s^24p$	^2S	$^2P^\circ$		1/2	1/2	854621	1152893	5.89+09	2.19−01	A	N02
347.383	$1s^24p$	$1s^210d$	$^2P^\circ$	^2D		1/2	3/2	1152893	1440760	3.61+08	2.99−02	A	N02
347.524	$1s^24p$	$1s^210d$	$^2P^\circ$	^2D		3/2	3/2	1153010	1440760	7.23+07	5.99−03	A	N02
347.524	$1s^24p$	$1s^210d$	$^2P^\circ$	^2D		3/2	5/2	1153010	1440760	4.33+08	5.39−02	A	N02
350.02	$1s^24s$	$1s^29p$	^2S	$^2P^\circ$		1/2	1/2	1140432	1426130	2.71+08	1.15−02	A	N02
350.02	$1s^24s$	$1s^29p$	^2S	$^2P^\circ$		1/2	3/2	1140432	1426130	2.70+08	2.29−02	A	N02
353.06	$1s^24f$	$1s^210d$	$^2F^\circ$	^2D		7/2	5/2	1157523	1440760	6.96+06	9.1−04	A	N02
353.06	$1s^24f$	$1s^210d$	$^2F^\circ$	^2D		5/2	3/2	1157523	1440760	7.32+06	6.36−04	A	N02
353.06	$1s^24f$	$1s^210d$	$^2F^\circ$	^2D		5/2	5/2	1157523	1440760	3.48+05	4.53−05	A	N02
353.76	$1s^24d$	$1s^210p$	^2D	$^2P^\circ$		3/2	1/2	1157249	1439930	4.92+07	2.15−03	A	N02
353.76	$1s^24d$	$1s^210p$	^2D	$^2P^\circ$		3/2	3/2	1157249	1439930	4.88+06	4.27−04	A	N02
353.80	$1s^24d$	$1s^210p$	^2D	$^2P^\circ$		5/2	3/2	1157285	1439930	4.40+07	3.85−03	A	N02
364.53	$1s^24p$	$1s^29d$	$^2P^\circ$	^2D		1/2	3/2	1152893	1427220	4.93+08	4.72−02	A	N02
364.68	$1s^24p$	$1s^29d$	$^2P^\circ$	^2D		3/2	3/2	1153010	1427220	9.9+07	9.5−03	A	N02
364.68	$1s^24p$	$1s^29d$	$^2P^\circ$	^2D		3/2	5/2	1153010	1427220	5.92+08	8.5−02	A	N02
367.484	$1s^23p$	$1s^24d$	$^2P^\circ$	^2D		1/2	3/2	885128	1157249	1.38+10	1.35+00	A	N02
367.822	$1s^23p$	$1s^24d$	$^2P^\circ$	^2D		3/2	5/2	885414	1157285	1.66+10	2.44+00	A	N02
367.870	$1s^23p$	$1s^24d$	$^2P^\circ$	^2D		3/2	3/2	885414	1157249	2.77+09	2.72−01	A	N02
370.79	$1s^24f$	$1s^29d$	$^2F^\circ$	^2D		7/2	5/2	1157523	1427220	1.00+07	1.51−03	A	N02
370.79	$1s^24f$	$1s^29d$	$^2F^\circ$	^2D		5/2	3/2	1157523	1427220	1.05+07	1.06−03	A	N02
370.79	$1s^24f$	$1s^29d$	$^2F^\circ$	^2D		5/2	5/2	1157523	1427220	5.00+05	7.55−05	A	N02
371.91	$1s^24d$	$1s^29p$	^2D	$^2P^\circ$		3/2	1/2	1157249	1426130	6.89+07	3.50−03	A	N02
371.91	$1s^24d$	$1s^29p$	^2D	$^2P^\circ$		3/2	3/2	1157249	1426130	6.83+06	6.94−04	A	N02
371.96	$1s^24d$	$1s^29p$	^2D	$^2P^\circ$		5/2	3/2	1157285	1426130	6.16+07	6.26−03	A	N02

λ_{Ritz} [a] (Å)	Configurations		Terms		Type	J_i [b]	J_k [b]	E_i (cm^{-1})	E_k (cm^{-1})	A_{ki} [c] (s^{-1})	S_{ik} [c]	Acc. [d]	Ref. [e]
372.556	$1s^24s$	$1s^28p$	^2S	$^2P°$		1/2	1/2	1140432	1408848	3.84+08	1.96−02	A	N02
372.556	$1s^24s$	$1s^28p$	^2S	$^2P°$		1/2	3/2	1140432	1408848	3.83+08	3.91−02	A	N02
381.819	$1s^23d$	$1s^24f$	^2D	$^2F°$		3/2	5/2	895619	1157523	3.09+10	5.09+00	A	N02
381.946	$1s^23d$	$1s^24f$	^2D	$^2F°$		5/2	5/2	895706	1157523	2.21+09	3.64−01	A	N02
381.946	$1s^23d$	$1s^24f$	^2D	$^2F°$		5/2	7/2	895706	1157523	3.31+10	7.28+00	A	N02
388.514	$1s^23d$	$1s^24p$	^2D	$^2P°$		3/2	3/2	895619	1153010	1.07+08	1.24−02	A	N02
388.645	$1s^23d$	$1s^24p$	^2D	$^2P°$		5/2	3/2	895706	1153010	9.7+08	1.12−01	A	N02
388.691	$1s^23d$	$1s^24p$	^2D	$^2P°$		3/2	1/2	895619	1152893	1.08+09	6.29−02	A	N02
389.596	$1s^24p$	$1s^28d$	$^2P°$	^2D		1/2	3/2	1152893	1409569	7.07+08	8.26−02	A	N02
389.774	$1s^24p$	$1s^28d$	$^2P°$	^2D		3/2	3/2	1153010	1409569	1.42+08	1.66−02	A	N02
389.774	$1s^24p$	$1s^28d$	$^2P°$	^2D		3/2	5/2	1153010	1409569	8.5+08	1.49−01	A	N02
391.690	$1s^23p$	$1s^24s$	$^2P°$	^2S		1/2	1/2	885128	1140432	2.19+09	1.30−01	B	F04
392.129	$1s^23p$	$1s^24s$	$^2P°$	^2S		3/2	1/2	885414	1140432	4.4+09	2.6−01	B	F04
396.69	$1s^24f$	$1s^28g$	$^2F°$	^2G		7/2	7/2	1157523	1409610	3.12+07	7.70−03	A	N02
396.69	$1s^24f$	$1s^28g$	$^2F°$	^2G		7/2	9/2	1157523	1409610	8.8+08	2.70−01	A	N02
396.69	$1s^24f$	$1s^28g$	$^2F°$	^2G		5/2	7/2	1157523	1409610	8.4+08	2.08−01	A	N02
396.753	$1s^24f$	$1s^28d$	$^2F°$	^2D		7/2	5/2	1157523	1409569	1.55+07	2.87−03	A	N02
396.753	$1s^24f$	$1s^28d$	$^2F°$	^2D		5/2	3/2	1157523	1409569	1.63+07	2.01−03	A	N02
396.753	$1s^24f$	$1s^28d$	$^2F°$	^2D		5/2	5/2	1157523	1409569	7.73+05	1.43−04	A	N02
397.458	$1s^24d$	$1s^28p$	^2D	$^2P°$		3/2	1/2	1157249	1408848	1.03+08	6.39−03	A	N02
397.458	$1s^24d$	$1s^28p$	^2D	$^2P°$		3/2	3/2	1157249	1408848	1.02+07	1.27−03	A	N02
397.515	$1s^24d$	$1s^28p$	^2D	$^2P°$		5/2	3/2	1157285	1408848	9.2+07	1.14−02	A	N02
412.497	$1s^24s$	$1s^27p$	^2S	$^2P°$		1/2	1/2	1140432	1382858	5.64+08	3.91−02	A	N02
412.497	$1s^24s$	$1s^27p$	^2S	$^2P°$		1/2	3/2	1140432	1382858	5.61+08	7.78−02	A	N02
433.020	$1s^24p$	$1s^27d$	$^2P°$	^2D		1/2	3/2	1152893	1383829	1.06+09	1.71−01	A	N02
433.240	$1s^24p$	$1s^27d$	$^2P°$	^2D		3/2	3/2	1153010	1383829	2.13+08	3.42−02	A	N02
433.240	$1s^24p$	$1s^27d$	$^2P°$	^2D		3/2	5/2	1153010	1383829	1.28+09	3.08−01	A	N02
438.824	$1s^24p$	$1s^27s$	$^2P°$	^2S		1/2	1/2	1152893	1380775	2.77+08	2.31−02	A	N02
439.049	$1s^24p$	$1s^27s$	$^2P°$	^2S		3/2	1/2	1153010	1380775	5.56+08	4.64−02	A	N02

λ_{Ritz} [a] (Å)	Configurations		Terms		Type	J_i [b]	J_k [b]	E_i (cm^{-1})	E_k (cm^{-1})	A_{ki} [c] (s^{-1})	S_{ik} [c]	Acc. [d]	Ref. [e]
441.244	$1s^24d$	$1s^27f$	^2D	$^2F^o$		3/2	5/2	1157249	1383881	1.65+09	4.19−01	A	N02
441.314	$1s^24d$	$1s^27f$	^2D	$^2F^o$		5/2	5/2	1157285	1383881	1.17+08	2.99−02	A	N02
441.314	$1s^24d$	$1s^27f$	^2D	$^2F^o$		5/2	7/2	1157285	1383881	1.76+09	5.98−01	A	N02
441.880	$1s^24f$	$1s^27d$	$^2F^o$	^2D		7/2	5/2	1157523	1383829	2.61+07	6.67−03	A	N02
441.880	$1s^24f$	$1s^27d$	$^2F^o$	^2D		5/2	3/2	1157523	1383829	2.75+07	4.68−03	A	N02
441.880	$1s^24f$	$1s^27d$	$^2F^o$	^2D		5/2	5/2	1157523	1383829	1.30+06	3.33−04	A	N02
443.245	$1s^24d$	$1s^27p$	^2D	$^2P^o$		3/2	1/2	1157249	1382858	1.64+08	1.41−02	A	N02
443.245	$1s^24d$	$1s^27p$	^2D	$^2P^o$		3/2	3/2	1157249	1382858	1.63+07	2.80−03	A	N02
443.315	$1s^24d$	$1s^27p$	^2D	$^2P^o$		5/2	3/2	1157285	1382858	1.47+08	2.53−02	A	N02
443.760	$1s^24f$	$1s^27g$	$^2F^o$	^2G		7/2	7/2	1157523	1382870	5.46+07	1.89−02	A	N02
443.760	$1s^24f$	$1s^27g$	$^2F^o$	^2G		7/2	9/2	1157523	1382870	1.53+09	6.60−01	A	N02
443.760	$1s^24f$	$1s^27g$	$^2F^o$	^2G		5/2	7/2	1157523	1382870	1.48+09	5.10−01	A	N02
493.89	$1s^24s$	$1s^26p$	^2S	$^2P^o$		1/2	1/2	1140432	1342906	8.6+08	1.03−01	A	N02
493.89	$1s^24s$	$1s^26p$	^2S	$^2P^o$		1/2	3/2	1140432	1342906	8.6+08	2.05−01	A	N02
522.83	$1s^24p$	$1s^26d$	$^2P^o$	^2D		1/2	3/2	1152893	1344159	1.70+09	4.79−01	A	N02
523.12	$1s^24p$	$1s^26d$	$^2P^o$	^2D		3/2	5/2	1153010	1344170	2.04+09	8.6−01	A	N02
523.15	$1s^24p$	$1s^26d$	$^2P^o$	^2D		3/2	3/2	1153010	1344159	3.40+08	9.6−02	A	N02
534.77	$1s^24d$	$1s^26f$	^2D	$^2F^o$		3/2	5/2	1157249	1344247	2.89+09	1.31+00	A	N02
534.87	$1s^24d$	$1s^26f$	^2D	$^2F^o$		5/2	5/2	1157285	1344247	2.06+08	9.3−02	A	N02
534.87	$1s^24d$	$1s^26f$	^2D	$^2F^o$		5/2	7/2	1157285	1344247	3.09+09	1.87+00	A	N02
535.52	$1s^24f$	$1s^26g$	$^2F^o$	^2G		7/2	7/2	1157523	1344259	1.18+08	7.14−02	A	N02
535.52	$1s^24f$	$1s^26g$	$^2F^o$	^2G		7/2	9/2	1157523	1344259	3.30+09	2.50+00	A	N02
535.52	$1s^24f$	$1s^26g$	$^2F^o$	^2G		5/2	7/2	1157523	1344259	3.18+09	1.93+00	A	N02
535.77	$1s^24f$	$1s^26d$	$^2F^o$	^2D		7/2	5/2	1157523	1344170	5.02+07	2.28−02	A	N02
535.77	$1s^24f$	$1s^26d$	$^2F^o$	^2D		5/2	5/2	1157523	1344170	2.50+06	1.14−03	A	N02
535.80	$1s^24f$	$1s^26d$	$^2F^o$	^2D		5/2	3/2	1157523	1344159	5.28+07	1.60−02	A	N02
536.49	$1s^24p$	$1s^26s$	$^2P^o$	^2S		1/2	1/2	1152893	1339289	4.65+08	7.09−02	A	N02
536.83	$1s^24p$	$1s^26s$	$^2P^o$	^2S		3/2	1/2	1153010	1339289	9.3+08	1.43−01	A	N02
538.63	$1s^24d$	$1s^26p$	^2D	$^2P^o$		3/2	1/2	1157249	1342906	2.90+08	4.47−02	A	N02

λ_{Ritz} [a] (Å)	Configurations		Terms		Type	J_i [b]	J_k [b]	E_i (cm^{-1})	E_k (cm^{-1})	A_{ki} [c] (s^{-1})	S_{ik} [c]	Acc. [d]	Ref. [e]
538.63	$1s^24d$	$1s^26p$	^2D	$^2P^o$		3/2	3/2	1157249	1342906	2.87+07	8.9−03	A	N02
538.73	$1s^24d$	$1s^26p$	^2D	$^2P^o$		5/2	3/2	1157285	1342906	2.59+08	7.99−02	A	N02
588.07	$1s^25s$	$1s^210p$	^2S	$^2P^o$		1/2	1/2	1269881	1439930	1.08+08	2.18−02	A	N02
588.07	$1s^25s$	$1s^210p$	^2S	$^2P^o$		1/2	3/2	1269881	1439930	1.08+08	4.33−02	A	N02
607.50	$1s^25p$	$1s^210d$	$^2P^o$	^2D		1/2	3/2	1276152	1440760	1.97+08	8.7−02	A	N02
607.74	$1s^25p$	$1s^210d$	$^2P^o$	^2D		3/2	3/2	1276216	1440760	3.94+07	1.75−02	A	N02
607.74	$1s^25p$	$1s^210d$	$^2P^o$	^2D		3/2	5/2	1276216	1440760	2.36+08	1.57−01	A	N02
616.36	$1s^25f$	$1s^210d$	$^2F^o$	^2D		7/2	5/2	1278517	1440760	1.09+07	7.56−03	A	N02
616.36	$1s^25f$	$1s^210d$	$^2F^o$	^2D		5/2	3/2	1278517	1440760	1.15+07	5.30−03	A	N02
616.36	$1s^25f$	$1s^210d$	$^2F^o$	^2D		5/2	5/2	1278517	1440760	5.45+05	3.78−04	A	N02
618.96	$1s^25d$	$1s^210p$	^2D	$^2P^o$		3/2	1/2	1278369	1439930	4.39+07	1.03−02	A	N02
618.96	$1s^25d$	$1s^210p$	^2D	$^2P^o$		3/2	3/2	1278369	1439930	4.36+06	2.04−03	A	N02
619.03	$1s^25d$	$1s^210p$	^2D	$^2P^o$		5/2	3/2	1278388	1439930	3.93+07	1.84−02	A	N02
640.00	$1s^25s$	$1s^29p$	^2S	$^2P^o$		1/2	1/2	1269881	1426130	1.45+08	3.74−02	A	N02
640.00	$1s^25s$	$1s^29p$	^2S	$^2P^o$		1/2	3/2	1269881	1426130	1.44+08	7.44−02	A	N02
661.95	$1s^25p$	$1s^29d$	$^2P^o$	^2D		1/2	3/2	1276152	1427220	2.66+08	1.52−01	A	N02
662.23	$1s^25p$	$1s^29d$	$^2P^o$	^2D		3/2	3/2	1276216	1427220	5.33+07	3.05−02	A	N02
662.23	$1s^25p$	$1s^29d$	$^2P^o$	^2D		3/2	5/2	1276216	1427220	3.19+08	2.75−01	A	N02
672.48	$1s^25f$	$1s^29d$	$^2F^o$	^2D		7/2	5/2	1278517	1427220	1.59+07	1.43−02	A	N02
672.48	$1s^25f$	$1s^29d$	$^2F^o$	^2D		5/2	3/2	1278517	1427220	1.67+07	1.00−02	A	N02
672.48	$1s^25f$	$1s^29d$	$^2F^o$	^2D		5/2	5/2	1278517	1427220	7.92+05	7.13−04	A	N02
676.77	$1s^25d$	$1s^29p$	^2D	$^2P^o$		3/2	1/2	1278369	1426130	6.18+07	1.89−02	A	N02
676.77	$1s^25d$	$1s^29p$	^2D	$^2P^o$		3/2	3/2	1278369	1426130	6.13+06	3.75−03	A	N02
676.86	$1s^25d$	$1s^29p$	^2D	$^2P^o$		5/2	3/2	1278388	1426130	5.53+07	3.39−02	A	N02
719.60	$1s^25s$	$1s^28p$	^2S	$^2P^o$		1/2	1/2	1269881	1408848	2.04+08	7.52−02	A	N02
719.60	$1s^25s$	$1s^28p$	^2S	$^2P^o$		1/2	3/2	1269881	1408848	2.03+08	1.50−01	A	N02
724.87	$1s^24s$	$1s^25d$	^2S	^2D	E2	1/2	5/2	1140432	1278388	7.4+08	7.9+05	B	A10
724.97	$1s^24s$	$1s^25d$	^2S	^2D	E2	1/2	3/2	1140432	1278369	7.4+08	5.3+05	B	A10
736.46	$1s^24s$	$1s^25p$	^2S	$^2P^o$		1/2	3/2	1140432	1276216	1.31+09	1.03+00	A	N02

λ_{Ritz} [a] (Å)	Configurations		Terms		Type	J_i [b]	J_k [b]	E_i (cm^{-1})	E_k (cm^{-1})	A_{ki} [c] (s^{-1})	S_{ik} [c]	Acc. [d]	Ref. [e]
736.81	$1s^24s$	$1s^25p$	^2S	$^2P^o$		1/2	1/2	1140432	1276152	1.31+09	5.19−01	A	N02
749.53	$1s^25p$	$1s^28d$	$^2P^o$	^2D		1/2	3/2	1276152	1409569	3.79+08	3.15−01	A	N02
749.89	$1s^25p$	$1s^28d$	$^2P^o$	^2D		3/2	3/2	1276216	1409569	7.59+07	6.32−02	A	N02
749.89	$1s^25p$	$1s^28d$	$^2P^o$	^2D		3/2	5/2	1276216	1409569	4.55+08	5.68−01	A	N02
762.82	$1s^25f$	$1s^28g$	$^2F^o$	^2G		7/2	7/2	1278517	1409610	2.69+07	4.71−02	A	N02
762.82	$1s^25f$	$1s^28g$	$^2F^o$	^2G		7/2	9/2	1278517	1409610	7.53+08	1.65+00	A	N02
762.82	$1s^25f$	$1s^28g$	$^2F^o$	^2G		5/2	7/2	1278517	1409610	7.26+08	1.27+00	A	N02
763.06	$1s^25f$	$1s^28d$	$^2F^o$	^2D		7/2	5/2	1278517	1409569	2.52+07	3.31−02	A	N02
763.06	$1s^25f$	$1s^28d$	$^2F^o$	^2D		5/2	3/2	1278517	1409569	2.65+07	2.32−02	A	N02
763.06	$1s^25f$	$1s^28d$	$^2F^o$	^2D		5/2	5/2	1278517	1409569	1.26+06	1.65−03	A	N02
766.41	$1s^25d$	$1s^28p$	^2D	$^2P^o$		3/2	1/2	1278369	1408848	9.5+07	4.22−02	A	N02
766.41	$1s^25d$	$1s^28p$	^2D	$^2P^o$		3/2	3/2	1278369	1408848	9.4+06	8.4−03	A	N02
766.52	$1s^25d$	$1s^28p$	^2D	$^2P^o$		5/2	3/2	1278388	1408848	8.5+07	7.56−02	A	N02
796.97	$1s^24p$	$1s^25d$	$^2P^o$	^2D		1/2	3/2	1152893	1278369	2.83+09	2.83+00	A	N02
797.59	$1s^24p$	$1s^25d$	$^2P^o$	^2D		3/2	5/2	1153010	1278388	3.40+09	5.11+00	A	N02
797.71	$1s^24p$	$1s^25d$	$^2P^o$	^2D		3/2	3/2	1153010	1278369	5.68+08	5.69−01	A	N02
824.62	$1s^24d$	$1s^25f$	^2D	$^2F^o$		3/2	5/2	1157249	1278517	5.80+09	9.6+00	A	N02
824.86	$1s^24d$	$1s^25f$	^2D	$^2F^o$		5/2	5/2	1157285	1278517	4.15+08	6.89−01	A	N02
824.86	$1s^24d$	$1s^25f$	^2D	$^2F^o$		5/2	7/2	1157285	1278517	6.22+09	1.38+01	A	N02
826.35	$1s^24f$	$1s^25g$	$^2F^o$	^2G		7/2	7/2	1157523	1278537	3.65+08	8.14−01	A	N02
826.35	$1s^24f$	$1s^25g$	$^2F^o$	^2G		7/2	9/2	1157523	1278537	1.02+10	2.85+01	A	N02
826.35	$1s^24f$	$1s^25g$	$^2F^o$	^2G		5/2	7/2	1157523	1278537	9.9+09	2.20+01	A	N02
827.37	$1s^24f$	$1s^25d$	$^2F^o$	^2D		7/2	5/2	1157523	1278388	1.18+08	1.98−01	A	N02
827.37	$1s^24f$	$1s^25d$	$^2F^o$	^2D		5/2	5/2	1157523	1278388	5.88+06	9.9−03	A	N02
827.50	$1s^24f$	$1s^25d$	$^2F^o$	^2D		5/2	3/2	1157523	1278369	1.24+08	1.39−01	A	N02
840.57	$1s^24d$	$1s^25p$	^2D	$^2P^o$		3/2	3/2	1157249	1276216	5.78+07	6.78−02	A	N02
840.82	$1s^24d$	$1s^25p$	^2D	$^2P^o$		5/2	3/2	1157285	1276216	5.22+08	6.13−01	A	N02
841.02	$1s^24d$	$1s^25p$	^2D	$^2P^o$		3/2	1/2	1157249	1276152	5.83+08	3.43−01	A	N02
854.79	$1s^24p$	$1s^25s$	$^2P^o$	^2S		1/2	1/2	1152893	1269881	8.8+08	5.43−01	A	N02

λ_{Ritz} [a] (Å)	Configurations		Terms		Type	J_i [b]	J_k [b]	E_i (cm^{-1})	E_k (cm^{-1})	A_{ki} [c] (s^{-1})	S_{ik} [c]	Acc. [d]	Ref. [e]
855.64	$1s^24p$	$1s^25s$	$^2P°$	^2S		3/2	1/2	1153010	1269881	1.77+09	1.09+00	A	N02
883.11	$1s^22s$	$1s^22p$	^2S	$^2P°$		1/2	3/2	0	113236	4.93+08	6.71−01	AA	F04
885.14	$1s^25s$	$1s^27p$	^2S	$^2P°$		1/2	1/2	1269881	1382858	2.94+08	2.01−01	A	N02
885.14	$1s^25s$	$1s^27p$	^2S	$^2P°$		1/2	3/2	1269881	1382858	2.92+08	4.00−01	A	N02
890.79	$1s^22s$	$1s^22p$	^2S	$^2P°$		1/2	1/2	0	112260	4.80+08	3.35−01	AA	F04
928.70	$1s^25p$	$1s^27d$	$^2P°$	^2D		1/2	3/2	1276152	1383829	5.61+08	8.9−01	A	N02
929.26	$1s^25p$	$1s^27d$	$^2P°$	^2D		3/2	3/2	1276216	1383829	1.12+08	1.78−01	A	N02
929.26	$1s^25p$	$1s^27d$	$^2P°$	^2D		3/2	5/2	1276216	1383829	6.74+08	1.60+00	A	N02
947.76	$1s^25d$	$1s^27f$	^2D	$^2F°$		3/2	5/2	1278369	1383881	9.7+08	2.45+00	A	N02
947.93	$1s^25d$	$1s^27f$	^2D	$^2F°$		5/2	5/2	1278388	1383881	6.94+07	1.75−01	A	N02
947.93	$1s^25d$	$1s^27f$	^2D	$^2F°$		5/2	7/2	1278388	1383881	1.04+09	3.50+00	A	N02
949.27	$1s^25g$	$1s^27f$	^2G	$^2F°$		9/2	7/2	1278537	1383881	1.09+07	3.68−02	A	N02
949.27	$1s^25g$	$1s^27f$	^2G	$^2F°$		7/2	5/2	1278537	1383881	1.12+07	2.84−02	A	N02
949.27	$1s^25g$	$1s^27f$	^2G	$^2F°$		7/2	7/2	1278537	1383881	3.11+05	1.05−03	A	N02
949.56	$1s^25f$	$1s^27d$	$^2F°$	^2D		7/2	5/2	1278517	1383829	4.43+07	1.12−01	A	N02
949.56	$1s^25f$	$1s^27d$	$^2F°$	^2D		5/2	3/2	1278517	1383829	4.66+07	7.88−02	A	N02
949.56	$1s^25f$	$1s^27d$	$^2F°$	^2D		5/2	5/2	1278517	1383829	2.21+06	5.61−03	A	N02
955.81	$1s^25p$	$1s^27s$	$^2P°$	^2S		1/2	1/2	1276152	1380775	2.02+08	1.74−01	A	N02
956.40	$1s^25p$	$1s^27s$	$^2P°$	^2S		3/2	1/2	1276216	1380775	4.05+08	3.50−01	A	N02
957.04	$1s^25d$	$1s^27p$	^2D	$^2P°$		3/2	1/2	1278369	1382858	1.56+08	1.35−01	A	N02
957.04	$1s^25d$	$1s^27p$	^2D	$^2P°$		3/2	3/2	1278369	1382858	1.55+07	2.68−02	A	N02
957.21	$1s^25d$	$1s^27p$	^2D	$^2P°$		5/2	3/2	1278388	1382858	1.40+08	2.42−01	A	N02
958.29	$1s^25f$	$1s^27g$	$^2F°$	^2G		7/2	7/2	1278517	1382870	4.57+07	1.59−01	A	N02
958.29	$1s^25f$	$1s^27g$	$^2F°$	^2G		7/2	9/2	1278517	1382870	1.28+09	5.56+00	A	N02
958.29	$1s^25f$	$1s^27g$	$^2F°$	^2G		5/2	7/2	1278517	1382870	1.23+09	4.29+00	A	N02
993.63	$1s^26s$	$1s^210p$	^2S	$^2P°$		1/2	1/2	1339289	1439930	6.66+07	6.45−02	A	N02
993.63	$1s^26s$	$1s^210p$	^2S	$^2P°$		1/2	3/2	1339289	1439930	6.62+07	1.28−01	A	N02
1021.93	$1s^26p$	$1s^210d$	$^2P°$	^2D		3/2	3/2	1342906	1440760	2.41+07	5.1−02	B	N02
1021.93	$1s^26p$	$1s^210d$	$^2P°$	^2D		3/2	5/2	1342906	1440760	1.44+08	4.6−01	B	N02

λ_{Ritz} [a] (Å)	Configurations		Terms		Type	J_i [b]	J_k [b]	E_i (cm^{-1})	E_k (cm^{-1})	A_{ki} [c] (s^{-1})	S_{ik} [c]	Acc. [d]	Ref. [e]
1021.93	$1s^26p$	$1s^210d$	$^2P°$	^2D		1/2	3/2	1342906	1440760	1.20+08	2.5−01	B	N02
1036.13	$1s^26f$	$1s^210d$	$^2F°$	^2D		7/2	5/2	1344247	1440760	1.30+07	4.29−02	A	N02
1036.13	$1s^26f$	$1s^210d$	$^2F°$	^2D		5/2	3/2	1344247	1440760	1.37+07	3.00−02	A	N02
1036.13	$1s^26f$	$1s^210d$	$^2F°$	^2D		5/2	5/2	1344247	1440760	6.49+05	2.14−03	A	N02
1044.16	$1s^26d$	$1s^210p$	^2D	$^2P°$		3/2	1/2	1344159	1439930	3.95+07	4.44−02	A	N02
1044.16	$1s^26d$	$1s^210p$	^2D	$^2P°$		3/2	3/2	1344159	1439930	3.92+06	8.8−03	A	N02
1044.28	$1s^26d$	$1s^210p$	^2D	$^2P°$		5/2	3/2	1344170	1439930	3.53+07	7.95−02	A	N02
1151.5	$1s^26s$	$1s^29p$	^2S	$^2P°$		1/2	1/2	1339289	1426130	8.7+07	1.31−01	B	N02
1151.5	$1s^26s$	$1s^29p$	^2S	$^2P°$		1/2	3/2	1339289	1426130	8.6+07	2.6−01	B	N02
1186.0	$1s^26p$	$1s^29d$	$^2P°$	^2D		3/2	3/2	1342906	1427220	3.2+07	1.06−01	B	N02
1186.0	$1s^26p$	$1s^29d$	$^2P°$	^2D		3/2	5/2	1342906	1427220	1.92+08	9.5−01	B	N02
1186.0	$1s^26p$	$1s^29d$	$^2P°$	^2D		1/2	3/2	1342906	1427220	1.60+08	5.3−01	B	N02
1205.2	$1s^26f$	$1s^29d$	$^2F°$	^2D		7/2	5/2	1344247	1427220	1.92+07	9.9−02	A	N02
1205.2	$1s^26f$	$1s^29d$	$^2F°$	^2D		5/2	3/2	1344247	1427220	2.02+07	6.97−02	A	N02
1205.2	$1s^26f$	$1s^29d$	$^2F°$	^2D		5/2	5/2	1344247	1427220	9.6+05	4.97−03	A	N02
1219.9	$1s^26d$	$1s^29p$	^2D	$^2P°$		3/2	1/2	1344159	1426130	5.56+07	1.0−01	A	N02
1219.9	$1s^26d$	$1s^29p$	^2D	$^2P°$		3/2	3/2	1344159	1426130	5.52+06	1.98−02	A	N02
1220.1	$1s^26d$	$1s^29p$	^2D	$^2P°$		5/2	3/2	1344170	1426130	4.98+07	1.78−01	A	N02
1369.39	$1s^25s$	$1s^26p$	^2S	$^2P°$		1/2	1/2	1269881	1342906	4.14+08	1.05+00	A	N02
1369.39	$1s^25s$	$1s^26p$	^2S	$^2P°$		1/2	3/2	1269881	1342906	4.10+08	2.08+00	A	N02
1437.6	$1s^26s$	$1s^28p$	^2S	$^2P°$		1/2	1/2	1339289	1408848	1.21+08	3.6−01	B	N02
1437.6	$1s^26s$	$1s^28p$	^2S	$^2P°$		1/2	3/2	1339289	1408848	1.20+08	7.1−01	B	N02
1470.44	$1s^25p$	$1s^26d$	$^2P°$	^2D		1/2	3/2	1276152	1344159	8.4+08	5.3+00	B	N02
1471.58	$1s^25p$	$1s^26d$	$^2P°$	^2D		3/2	5/2	1276216	1344170	1.01+09	9.5+00	B	N02
1471.82	$1s^25p$	$1s^26d$	$^2P°$	^2D		3/2	3/2	1276216	1344159	1.69+08	1.06+00	B	N02
1500.1	$1s^26p$	$1s^28d$	$^2P°$	^2D		3/2	3/2	1342906	1409569	4.5+07	3.0−01	B	N02
1500.1	$1s^26p$	$1s^28d$	$^2P°$	^2D		3/2	5/2	1342906	1409569	2.7+08	2.7+00	B	N02
1500.1	$1s^26p$	$1s^28d$	$^2P°$	^2D		1/2	3/2	1342906	1409569	2.24+08	1.50+00	B	N02
1518.0	$1s^25d$	$1s^26f$	^2D	$^2F°$		3/2	5/2	1278369	1344247	1.62+09	1.68+01	B	N02

λ_{Ritz} [a] (Å)	Configurations		Terms		Type	J_i [b]	J_k [b]	E_i (cm^{-1})	E_k (cm^{-1})	A_{ki} [c] (s^{-1})	S_{ik} [c]	Acc. [d]	Ref. [e]
1518.4	$1s^25d$	$1s^26f$	^2D	$^2F^o$		5/2	5/2	1278388	1344247	1.16+08	1.20+00	B	N02
1518.4	$1s^25d$	$1s^26f$	^2D	$^2F^o$		5/2	7/2	1278388	1344247	1.74+09	2.40+01	B	N02
1521.1	$1s^25f$	$1s^26g$	$^2F^o$	^2G		7/2	7/2	1278517	1344259	9.5+07	1.32+00	B	N02
1521.1	$1s^25f$	$1s^26g$	$^2F^o$	^2G		7/2	9/2	1278517	1344259	2.7+09	4.6+01	B	N02
1521.1	$1s^25f$	$1s^26g$	$^2F^o$	^2G		5/2	7/2	1278517	1344259	2.6+09	3.6+01	B	N02
1521.8	$1s^25g$	$1s^26f$	^2G	$^2F^o$		9/2	7/2	1278537	1344247	2.7+07	3.7−01	B	N02
1521.8	$1s^25g$	$1s^26f$	^2G	$^2F^o$		7/2	5/2	1278537	1344247	2.7+07	2.9−01	B	N02
1521.8	$1s^25g$	$1s^26f$	^2G	$^2F^o$		7/2	7/2	1278537	1344247	7.6+05	1.06−02	B	N02
1523.2	$1s^25f$	$1s^26d$	$^2F^o$	^2D		7/2	5/2	1278517	1344170	9.1+07	9.5−01	B	N02
1523.2	$1s^25f$	$1s^26d$	$^2F^o$	^2D		5/2	5/2	1278517	1344170	4.5+06	4.7−02	B	N02
1523.4	$1s^25f$	$1s^26d$	$^2F^o$	^2D		5/2	3/2	1278517	1344159	9.5+07	6.6−01	B	N02
1529.9	$1s^26f$	$1s^28g$	$^2F^o$	^2G		7/2	7/2	1344247	1409610	1.94+07	2.75−01	A	N02
1529.9	$1s^26f$	$1s^28g$	$^2F^o$	^2G		7/2	9/2	1344247	1409610	5.44+08	9.6+00	A	N02
1529.9	$1s^26f$	$1s^28g$	$^2F^o$	^2G		5/2	7/2	1344247	1409610	5.24+08	7.41+00	A	N02
1530.9	$1s^26f$	$1s^28d$	$^2F^o$	^2D		7/2	5/2	1344247	1409569	3.17+07	3.37−01	A	N02
1530.9	$1s^26f$	$1s^28d$	$^2F^o$	^2D		5/2	3/2	1344247	1409569	3.33+07	2.36−01	A	N02
1530.9	$1s^26f$	$1s^28d$	$^2F^o$	^2D		5/2	5/2	1344247	1409569	1.58+06	1.68−02	A	N02
1545.9	$1s^26d$	$1s^28p$	^2D	$^2P^o$		3/2	1/2	1344159	1408848	8.8+07	3.22−01	A	N02
1545.9	$1s^26d$	$1s^28p$	^2D	$^2P^o$		3/2	3/2	1344159	1408848	8.8+06	6.38−02	A	N02
1546.1	$1s^26d$	$1s^28p$	^2D	$^2P^o$		5/2	3/2	1344170	1408848	7.90+07	5.76−01	A	N02
1549.5	$1s^25d$	$1s^26p$	^2D	$^2P^o$		3/2	1/2	1278369	1342906	2.9+08	1.06+00	B	N02
1549.5	$1s^25d$	$1s^26p$	^2D	$^2P^o$		3/2	3/2	1278369	1342906	2.9+07	2.10−01	B	N02
1550.0	$1s^25d$	$1s^26p$	^2D	$^2P^o$		5/2	3/2	1278388	1342906	2.6+08	1.89+00	B	N02
1583.9	$1s^25p$	$1s^26s$	$^2P^o$	^2S		1/2	1/2	1276152	1339289	3.5+08	1.39+00	B	N02
1585.5	$1s^25p$	$1s^26s$	$^2P^o$	^2S		3/2	1/2	1276216	1339289	7.1+08	2.8+00	B	N02
1690.5	$1s^27s$	$1s^210p$	^2S	$^2P^o$		1/2	1/2	1380775	1439930	4.42+07	2.11−01	A	N02
1690.5	$1s^27s$	$1s^210p$	^2S	$^2P^o$		1/2	3/2	1380775	1439930	4.40+07	4.20−01	A	N02
1727.1	$1s^27p$	$1s^210d$	$^2P^o$	^2D		3/2	3/2	1382858	1440760	1.62+07	1.65−01	A	N02
1727.1	$1s^27p$	$1s^210d$	$^2P^o$	^2D		3/2	5/2	1382858	1440760	9.7+07	1.48+00	A	N02

λ_{Ritz} [a] (Å)	Configurations		Terms		Type	J_i [b]	J_k [b]	E_i (cm^{-1})	E_k (cm^{-1})	A_{ki} [c] (s^{-1})	S_{ik} [c]	Acc. [d]	Ref. [e]
1727.1	$1s^27p$	$1s^210d$	$^2P^\circ$	^2D		1/2	3/2	1382858	1440760	8.07+07	8.21−01	A	N02
1758.1	$1s^27f$	$1s^210d$	$^2F^\circ$	^2D		7/2	5/2	1383881	1440760	1.43+07	2.31−01	A	N02
1758.1	$1s^27f$	$1s^210d$	$^2F^\circ$	^2D		5/2	3/2	1383881	1440760	1.51+07	1.62−01	A	N02
1758.1	$1s^27f$	$1s^210d$	$^2F^\circ$	^2D		5/2	5/2	1383881	1440760	7.16+05	1.15−02	A	N02
1782.5	$1s^27d$	$1s^210p$	^2D	$^2P^\circ$		5/2	3/2	1383829	1439930	3.20+07	3.58−01	A	N02
1782.5	$1s^27d$	$1s^210p$	^2D	$^2P^\circ$		3/2	1/2	1383829	1439930	3.57+07	2.00−01	A	N02
1782.5	$1s^27d$	$1s^210p$	^2D	$^2P^\circ$		3/2	3/2	1383829	1439930	3.55+06	3.97−02	A	N02
2204.1	$1s^27s$	$1s^29p$	^2S	$^2P^\circ$		1/2	1/2	1380775	1426130	5.50+07	5.82−01	A	N02
2204.1	$1s^27s$	$1s^29p$	^2S	$^2P^\circ$		1/2	3/2	1380775	1426130	5.46+07	1.16+00	A	N02
2253.5	$1s^27p$	$1s^29d$	$^2P^\circ$	^2D		3/2	3/2	1382858	1427220	2.10+07	4.74−01	A	N02
2253.5	$1s^27p$	$1s^29d$	$^2P^\circ$	^2D		3/2	5/2	1382858	1427220	1.26+08	4.26+00	A	N02
2253.5	$1s^27p$	$1s^29d$	$^2P^\circ$	^2D		1/2	3/2	1382858	1427220	1.04+08	2.36+00	A	N02
2294.5	$1s^26s$	$1s^27p$	^2S	$^2P^\circ$		1/2	1/2	1339289	1382858	1.59+08	1.90+00	B	N02
2294.5	$1s^26s$	$1s^27p$	^2S	$^2P^\circ$		1/2	3/2	1339289	1382858	1.58+08	3.8+00	B	N02
2306.7	$1s^27f$	$1s^29d$	$^2F^\circ$	^2D		7/2	5/2	1383881	1427220	2.15+07	7.83−01	A	N02
2306.7	$1s^27f$	$1s^29d$	$^2F^\circ$	^2D		5/2	3/2	1383881	1427220	2.26+07	5.48−01	A	N02
2306.7	$1s^27f$	$1s^29d$	$^2F^\circ$	^2D		5/2	5/2	1383881	1427220	1.07+06	3.91−02	A	N02
2363.3	$1s^27d$	$1s^29p$	^2D	$^2P^\circ$		5/2	3/2	1383829	1426130	4.47+07	1.17+00	A	N02
2363.3	$1s^27d$	$1s^29p$	^2D	$^2P^\circ$		3/2	1/2	1383829	1426130	4.99+07	6.51−01	A	N02
2363.3	$1s^27d$	$1s^29p$	^2D	$^2P^\circ$		3/2	3/2	1383829	1426130	4.96+06	1.29−01	A	N02
2442.9	$1s^26p$	$1s^27d$	$^2P^\circ$	^2D		3/2	3/2	1342906	1383829	6.3+07	1.82+00	B	N02
2442.9	$1s^26p$	$1s^27d$	$^2P^\circ$	^2D		3/2	5/2	1342906	1383829	3.8+08	1.64+01	B	N02
2442.9	$1s^26p$	$1s^27d$	$^2P^\circ$	^2D		1/2	3/2	1342906	1383829	3.1+08	9.0+00	B	N02
2516.7	$1s^26d$	$1s^27f$	^2D	$^2F^\circ$		3/2	5/2	1344159	1383881	5.8+08	2.7+01	B	N02
2517.4	$1s^26d$	$1s^27f$	^2D	$^2F^\circ$		5/2	5/2	1344170	1383881	4.1+07	1.96+00	B	N02
2517.4	$1s^26d$	$1s^27f$	^2D	$^2F^\circ$		5/2	7/2	1344170	1383881	6.2+08	3.9+01	B	N02
2523.1	$1s^26g$	$1s^27f$	^2G	$^2F^\circ$		9/2	7/2	1344259	1383881	2.60+07	1.65+00	A	N02
2523.1	$1s^26g$	$1s^27f$	^2G	$^2F^\circ$		7/2	5/2	1344259	1383881	2.67+07	1.27+00	A	N02
2523.1	$1s^26g$	$1s^27f$	^2G	$^2F^\circ$		7/2	7/2	1344259	1383881	7.41+05	4.71−02	A	N02

λ_{Ritz} [a] (Å)	Configurations		Terms		Type	J_i [b]	J_k [b]	E_i (cm^{-1})	E_k (cm^{-1})	A_{ki} [c] (s^{-1})	S_{ik} [c]	Acc. [d]	Ref. [e]
2525.6	$1s^26f$	$1s^27d$	$^2F°$	^2D		7/2	5/2	1344247	1383829	5.87+07	2.80+00	A	N02
2525.6	$1s^26f$	$1s^27d$	$^2F°$	^2D		5/2	3/2	1344247	1383829	6.17+07	1.97+00	A	N02
2525.6	$1s^26f$	$1s^27d$	$^2F°$	^2D		5/2	5/2	1344247	1383829	2.93+06	1.40−01	A	N02
2583.3	$1s^26d$	$1s^27p$	^2D	$^2P°$		3/2	1/2	1344159	1382858	1.49+08	2.5+00	B	N02
2583.3	$1s^26d$	$1s^27p$	^2D	$^2P°$		3/2	3/2	1344159	1382858	1.48+07	5.0−01	B	N02
2584.0	$1s^26d$	$1s^27p$	^2D	$^2P°$		5/2	3/2	1344170	1382858	1.34+08	4.6+00	B	N02
2588.4	$1s^26f$	$1s^27g$	$^2F°$	^2G		7/2	7/2	1344247	1382870	2.99+07	2.05+00	A	N02
2588.4	$1s^26f$	$1s^27g$	$^2F°$	^2G		7/2	9/2	1344247	1382870	8.4+08	7.17+01	A	N02
2588.4	$1s^26f$	$1s^27g$	$^2F°$	^2G		5/2	7/2	1344247	1382870	8.06+08	5.53+01	A	N02
2639.9	$1s^26p$	$1s^27s$	$^2P°$	^2S		3/2	1/2	1342906	1380775	3.3+08	6.0+00	B	N02
2639.9	$1s^26p$	$1s^27s$	$^2P°$	^2S		1/2	1/2	1342906	1380775	1.64+08	3.0+00	B	N02
3132.7	$1s^28p$	$1s^210d$	$^2P°$	^2D		3/2	3/2	1408848	1440760	1.17+07	7.12−01	A	N02
3132.7	$1s^28p$	$1s^210d$	$^2P°$	^2D		3/2	5/2	1408848	1440760	7.02+07	6.40+00	A	N02
3132.7	$1s^28p$	$1s^210d$	$^2P°$	^2D		1/2	3/2	1408848	1440760	5.83+07	3.54+00	A	N02
3246.6	$1s^23s$	$1s^23p$	^2S	$^2P°$		1/2	3/2	854621	885414	6.1+07	4.1+00	B	F04
3277.0	$1s^23s$	$1s^23p$	^2S	$^2P°$		1/2	1/2	854621	885128	5.9+07	2.06+00	B	F04
3292.8	$1s^28d$	$1s^210p$	^2D	$^2P°$		5/2	3/2	1409569	1439930	3.00+07	2.12+00	A	N02
3292.8	$1s^28d$	$1s^210p$	^2D	$^2P°$		3/2	1/2	1409569	1439930	3.35+07	1.18+00	A	N02
3292.8	$1s^28d$	$1s^210p$	^2D	$^2P°$		3/2	3/2	1409569	1439930	3.33+06	2.35−01	A	N02
3561.1	$1s^27s$	$1s^28p$	^2S	$^2P°$		1/2	1/2	1380775	1408848	7.14+07	3.19+00	A	N02
3561.1	$1s^27s$	$1s^28p$	^2S	$^2P°$		1/2	3/2	1380775	1408848	7.08+07	6.31+00	A	N02
3742.7	$1s^27p$	$1s^28d$	$^2P°$	^2D		3/2	3/2	1382858	1409569	2.82+07	2.92+00	A	N02
3742.7	$1s^27p$	$1s^28d$	$^2P°$	^2D		3/2	5/2	1382858	1409569	1.69+08	2.63+01	A	N02
3742.7	$1s^27p$	$1s^28d$	$^2P°$	^2D		1/2	3/2	1382858	1409569	1.40+08	1.45+01	A	N02
3886	$1s^27f$	$1s^28g$	$^2F°$	^2G		7/2	7/2	1383881	1409610	1.32+07	3.06+00	A	N02
3886	$1s^27f$	$1s^28g$	$^2F°$	^2G		7/2	9/2	1383881	1409610	3.69+08	1.07+02	A	N02
3886	$1s^27f$	$1s^28g$	$^2F°$	^2G		5/2	7/2	1383881	1409610	3.56+08	8.25+01	A	N02
3891.8	$1s^27f$	$1s^28d$	$^2F°$	^2D		7/2	5/2	1383881	1409569	3.71+07	6.47+00	A	N02
3891.8	$1s^27f$	$1s^28d$	$^2F°$	^2D		5/2	3/2	1383881	1409569	3.90+07	4.54+00	A	N02

λ_{Ritz} [a] (Å)	Configurations		Terms		Type	J_i [b]	J_k [b]	E_i (cm^{-1})	E_k (cm^{-1})	A_{ki} [c] (s^{-1})	S_{ik} [c]	Acc. [d]	Ref. [e]
3891.8	$1s^2 7f$	$1s^2 8d$	$^2F^o$	^2D		5/2	5/2	1383881	1409569	1.85+06	3.23−01	A	N02
3995.8	$1s^2 7d$	$1s^2 8p$	^2D	$^2P^o$		5/2	3/2	1383829	1408848	7.39+07	9.3+00	A	N02
3995.8	$1s^2 7d$	$1s^2 8p$	^2D	$^2P^o$		3/2	1/2	1383829	1408848	8.26+07	5.21+00	A	N02
3995.8	$1s^2 7d$	$1s^2 8p$	^2D	$^2P^o$		3/2	3/2	1383829	1408848	8.19+06	1.03+00	A	N02
5442	$1s^2 8p$	$1s^2 9d$	$^2P^o$	^2D		3/2	3/2	1408848	1427220	1.40+07	4.46+00	A	N02
5442	$1s^2 8p$	$1s^2 9d$	$^2P^o$	^2D		3/2	5/2	1408848	1427220	8.4+07	4.01+01	A	N02
5442	$1s^2 8p$	$1s^2 9d$	$^2P^o$	^2D		1/2	3/2	1408848	1427220	6.95+07	2.21+01	A	N02
6037	$1s^2 8d$	$1s^2 9p$	^2D	$^2P^o$		5/2	3/2	1409569	1426130	3.93+07	1.71+01	A	N02
6037	$1s^2 8d$	$1s^2 9p$	^2D	$^2P^o$		3/2	1/2	1409569	1426130	4.39+07	9.5+00	A	N02
6037	$1s^2 8d$	$1s^2 9p$	^2D	$^2P^o$		3/2	3/2	1409569	1426130	4.35+06	1.89+00	A	N02
6833	$1s^2 9p$	$1s^2 10d$	$^2P^o$	^2D		3/2	3/2	1426130	1440760	1.04+07	6.54+00	A	N02
6833	$1s^2 9p$	$1s^2 10d$	$^2P^o$	^2D		3/2	5/2	1426130	1440760	6.21+07	5.88+01	A	N02
6833	$1s^2 9p$	$1s^2 10d$	$^2P^o$	^2D		1/2	3/2	1426130	1440760	5.15+07	3.25+01	A	N02
7866	$1s^2 9d$	$1s^2 10p$	^2D	$^2P^o$		5/2	3/2	1427220	1439930	3.00+07	2.89+01	A	N02
7866	$1s^2 9d$	$1s^2 10p$	^2D	$^2P^o$		3/2	1/2	1427220	1439930	3.35+07	1.61+01	A	N02
7866	$1s^2 9d$	$1s^2 10p$	^2D	$^2P^o$		3/2	3/2	1427220	1439930	3.32+06	3.20+00	A	N02
7948	$1s^2 4s$	$1s^2 4p$	^2S	$^2P^o$		1/2	3/2	1140432	1153010	1.41+07	1.40+01	B	N02
8023	$1s^2 4s$	$1s^2 4p$	^2S	$^2P^o$		1/2	1/2	1140432	1152893	1.37+07	7.0+00	B	N02
9529	$1s^2 3p$	$1s^2 3d$	$^2P^o$	^2D		1/2	3/2	885128	895619	1.62+06	2.8+00	B	F04
9714	$1s^2 3p$	$1s^2 3d$	$^2P^o$	^2D		3/2	5/2	885414	895706	1.83+06	5.0+00	B	F04
9796	$1s^2 3p$	$1s^2 3d$	$^2P^o$	^2D		3/2	3/2	885414	895619	3.0+05	5.5−01	B	F04
15781	$1s^2 5s$	$1s^2 5p$	^2S	$^2P^o$		1/2	3/2	1269881	1276216	4.5+06	3.5+01	B	N02
15942	$1s^2 5s$	$1s^2 5p$	^2S	$^2P^o$		1/2	1/2	1269881	1276152	4.4+06	1.77+01	B	N02
F VIII (He-like)													
13.334	$1s^2$	1s6p	^1S	$^1P^o$		0	1	0	7499630.	1.904+11	6.68−04	AA	CT92
13.488	$1s^2$	1s5p	^1S	$^1P^o$		0	1	0	7414000.	3.31+11	1.201−03	AA	CT92
13.781	$1s^2$	1s4p	^1S	$^1P^o$		0	1	0	7256370.	6.51+11	2.52−03	AA	CT92
14.458	$1s^2$	1s3p	^1S	$^1P^o$		0	1	0	6916590.	1.573+12	7.04−03	AA	CT92

λ_{Ritz} [a] (Å)	Configurations		Terms		Type	J_i [b]	J_k [b]	E_i (cm^{-1})	E_k (cm^{-1})	A_{ki} [c] (s^{-1})	S_{ik} [c]	Acc. [d]	Ref. [e]
16.807	$1s^2$	1s2p	^{1}S	^{1}P°		0	1	0	5949900.	5.59+12	3.93−02	AA	CT92
59.979	1s2s	1s6p	^{3}S	^{3}P°		1	1	5830040.	7497300.	1.222+10	3.90−03	AA	CT92LS
59.979	1s2s	1s6p	^{3}S	^{3}P°		1	2	5830040.	7497300.	1.222+10	6.51−03	AA	CT92LS
62.532	1s2p	1s6d	^{3}P°	^{3}D		1	2	5900750.	7499940.	1.716+10	1.035−02	AA	CT92LS
62.569	1s2p	1s6d	^{3}P°	^{3}D		2	2	5901700.	7499940.	5.71+09	3.45−03	AA	CT92LS
62.569	1s2p	1s6d	^{3}P°	^{3}D		2	3	5901700.	7499940.	2.284+10	1.933−02	AA	CT92LS
62.636	1s2s	1s6p	^{1}S	^{1}P°		0	1	5903100.	7499630.	1.159+10	4.22−03	AA	CT92
63.282	1s2s	1s5p	^{3}S	^{3}P°		1	1	5830040.	7410270.	2.105+10	7.90−03	AA	CT92LS
63.282	1s2s	1s5p	^{3}S	^{3}P°		1	2	5830040.	7410270.	2.105+10	1.316−02	AA	CT92LS
64.581	1s2p	1s6d	^{1}P°	^{1}D		1	2	5949900.	7498340.	2.017+10	1.341−02	AA	CT92
66.086	1s2p	1s5d	^{3}P°	^{3}D		1	2	5900750.	7413940.	3.13+10	2.227−02	AA	CT92LS
66.127	1s2p	1s5d	^{3}P°	^{3}D		2	2	5901700.	7413940.	1.040+10	7.42−03	AA	CT92LS
66.127	1s2p	1s5d	^{3}P°	^{3}D		2	3	5901700.	7413940.	4.16+10	4.16−02	AA	CT92LS
66.186	1s2s	1s5p	^{1}S	^{1}P°		0	1	5903100.	7414000.	1.995+10	8.57−03	AA	CT92
68.381	1s2p	1s5d	^{1}P°	^{1}D		1	2	5949900.	7412290.	3.71+10	2.93−02	AA	CT92
70.405	1s2s	1s4p	^{3}S	^{3}P°		1	1	5830040.	7250390.	4.07+10	2.104−02	AA	CT92LS
70.405	1s2s	1s4p	^{3}S	^{3}P°		1	2	5830040.	7250390.	4.07+10	3.51−02	AA	CT92LS
73.789	1s2p	1s4d	^{3}P°	^{3}D		1	2	5900750.	7255960.	6.77+10	6.72−02	AA	CT92LS
73.841	1s2p	1s4d	^{3}P°	^{3}D		2	2	5901700.	7255960.	2.253+10	2.239−02	AA	CT92LS
73.841	1s2p	1s4d	^{3}P°	^{3}D		2	3	5901700.	7255960.	9.01+10	1.254−01	AA	CT92LS
73.895	1s2s	1s4p	^{1}S	^{1}P°		0	1	5903100.	7256370.	3.86+10	2.304−02	AA	CT92
74.423	1s2p	1s4s	^{3}P°	^{3}S		0	1	5900600.	7244270.	1.620+09	9.89−04	AA	CT92LS
74.431	1s2p	1s4s	^{3}P°	^{3}S		1	1	5900750.	7244270.	4.86+09	2.97−03	AA	CT92LS
74.484	1s2p	1s4s	^{3}P°	^{3}S		2	1	5901700.	7244270.	8.08+09	4.95−03	AA	CT92LS
76.667	1s2p	1s4d	^{1}P°	^{1}D		1	2	5949900.	7254240.	8.17+10	9.09−02	AA	CT92
93.238	1s2s	1s3p	^{3}S	^{3}P°		1	1	5830040.	6902560.	9.14+10	1.097−01	AA	CT92LS
93.238	1s2s	1s3p	^{3}S	^{3}P°		1	2	5830040.	6902560.	9.14+10	1.829−01	AA	CT92LS
98.669	1s2s	1s3p	^{1}S	^{1}P°		0	1	5903100.	6916590.	8.69+10	1.236−01	AA	CT92
98.698	1s2p	1s3d	^{3}P°	^{3}D		2	3	5901700.	6914890.	2.75+11	9.14−01	AA	CT92LS

λ_{Ritz} [a] (Å)	Configurations		Terms		Type	J_i [b]	J_k [b]	E_i (cm^{-1})	E_k (cm^{-1})	A_{ki} [c] (s^{-1})	S_{ik} [c]	Acc. [d]	Ref. [e]
98.797	1s2p	1s3d	$^3P^o$	^3D		1	2	5900750.	6912930.	2.057+11	4.89−01	AA	CT92LS
98.889	1s2p	1s3d	$^3P^o$	^3D		2	2	5901700.	6912930.	6.84+10	1.631−01	AA	CT92LS
101.575	1s2p	1s3s	$^3P^o$	^3S		0	1	5900600.	6885090.	4.10+09	6.36−03	AA	CT92LS
101.591	1s2p	1s3s	$^3P^o$	^3S		1	1	5900750.	6885090.	1.229+10	1.909−02	AA	CT92LS
101.689	1s2p	1s3s	$^3P^o$	^3S		2	1	5901700.	6885090.	2.043+10	3.18−02	AA	CT92LS
103.520	1s2p	1s3d	$^1P^o$	^1D		1	2	5949900.	6915900.	2.62+11	7.17−01	AA	CT92
104.891	1s2p	1s3s	$^1P^o$	^1S		1	0	5949900.	6903270.	3.33+10	1.897−02	AA	CT92
163.34	1s3s	1s6p	^3S	$^3P^o$		1	1	6885090.	7497300.	3.80+09	2.450−02	AA	CT92LS
163.34	1s3s	1s6p	^3S	$^3P^o$		1	2	6885090.	7497300.	3.80+09	4.08−02	AA	CT92LS
167.40	1s3p	1s6d	$^3P^o$	^3D		2	2	6902560.	7499940.	1.984+09	2.297−02	AA	CT92LS
167.40	1s3p	1s6d	$^3P^o$	^3D		2	3	6902560.	7499940.	7.94+09	1.286−01	AA	CT92LS
167.40	1s3p	1s6d	$^3P^o$	^3D		1	2	6902560.	7499940.	5.95+09	6.89−02	AA	CT92LS
167.68	1s3s	1s6p	^1S	$^1P^o$		0	1	6903270.	7499630.	3.75+09	2.61−02	AA	CT92
171.12	1s3d	1s6p	^3D	$^3P^o$		2	1	6912930.	7497300.	2.94+08	2.180−03	AA	CT92LS
171.12	1s3d	1s6p	^3D	$^3P^o$		2	2	6912930.	7497300.	5.88+07	7.27−04	AA	CT92LS
171.31	1s3d	1s6p	^1D	$^1P^o$		2	1	6915900.	7499630.	2.98+08	2.220−03	AA	CT92
171.70	1s3d	1s6p	^3D	$^3P^o$		3	2	6914890.	7497300.	3.26+08	4.07−03	AA	CT92LS
171.90	1s3p	1s6d	$^1P^o$	^1D		1	2	6916590.	7498340.	7.61+09	9.53−02	AA	CT92
190.41	1s3s	1s5p	^3S	$^3P^o$		1	1	6885090.	7410270.	6.42+09	6.56−02	AA	CT92LS
190.41	1s3s	1s5p	^3S	$^3P^o$		1	2	6885090.	7410270.	6.42+09	1.093−01	AA	CT92LS
195.55	1s3p	1s5d	$^3P^o$	^3D		2	2	6902560.	7413940.	3.55+09	6.55−02	AA	CT92LS
195.55	1s3p	1s5d	$^3P^o$	^3D		2	3	6902560.	7413940.	1.419+10	3.67−01	AA	CT92LS
195.55	1s3p	1s5d	$^3P^o$	^3D		1	2	6902560.	7413940.	1.064+10	1.964−01	AA	CT92LS
195.80	1s3s	1s5p	^1S	$^1P^o$		0	1	6903270.	7414000.	6.35+09	7.05−02	AA	CT92
200.76	1s3d	1s5p	^1D	$^1P^o$		2	1	6915900.	7414000.	5.69+08	6.81−03	AA	CT92
201.07	1s3d	1s5p	^3D	$^3P^o$		2	1	6912930.	7410270.	5.62+08	6.76−03	AA	CT92LS
201.07	1s3d	1s5p	^3D	$^3P^o$		2	2	6912930.	7410270.	1.123+08	2.253−03	AA	CT92LS
201.73	1s3p	1s5d	$^1P^o$	^1D		1	2	6916590.	7412290.	1.377+10	2.79−01	AA	CT92
201.87	1s3d	1s5p	^3D	$^3P^o$		3	2	6914890.	7410270.	6.22+08	1.262−02	AA	CT92LS

λ_{Ritz} [a] (Å)	Configurations		Terms		Type	J_i [b]	J_k [b]	E_i (cm^{-1})	E_k (cm^{-1})	A_{ki} [c] (s^{-1})	S_{ik} [c]	Acc. [d]	Ref. [e]
273.75	1s3s	1s4p	^3S	$^3P°$		1	1	6885090.	7250390.	1.158+10	3.52−01	AA	CT92LS
273.75	1s3s	1s4p	^3S	$^3P°$		1	2	6885090.	7250390.	1.158+10	5.86−01	AA	CT92LS
282.97	1s3p	1s4d	$^3P°$	^3D		2	2	6902560.	7255960.	7.19+09	4.02−01	AA	CT92LS
282.97	1s3p	1s4d	$^3P°$	^3D		2	3	6902560.	7255960.	2.88+10	2.253+00	AA	CT92LS
282.97	1s3p	1s4d	$^3P°$	^3D		1	2	6902560.	7255960.	2.158+10	1.207+00	AA	CT92LS
283.21	1s3s	1s4p	^1S	$^1P°$		0	1	6903270.	7256370.	1.152+10	3.87−01	AA	CT92
292.65	1s3p	1s4s	$^3P°$	^3S		2	1	6902560.	7244270.	5.70+09	2.114−01	AA	CT92LS
292.65	1s3p	1s4s	$^3P°$	^3S		1	1	6902560.	7244270.	3.42+09	1.268−01	AA	CT92LS
293.71	1s3d	1s4p	^1D	$^1P°$		2	1	6915900.	7256370.	1.311+09	4.92−02	AA	CT92
296.16	1s3p	1s4d	$^1P°$	^1D		1	2	6916590.	7254240.	2.88+10	1.848+00	AA	CT92
296.33	1s3d	1s4p	^3D	$^3P°$		2	1	6912930.	7250390.	1.316+09	5.07−02	AA	CT92LS
296.33	1s3d	1s4p	^3D	$^3P°$		2	2	6912930.	7250390.	2.63+08	1.691−02	AA	CT92LS
298.06	1s3d	1s4p	^3D	$^3P°$		3	2	6914890.	7250390.	1.449+09	9.47−02	AA	CT92LS
395.2	1s4s	1s6p	^3S	$^3P°$		1	1	7244270.	7497300.	1.679+09	1.535−01	AA	CT92LS
395.2	1s4s	1s6p	^3S	$^3P°$		1	2	7244270.	7497300.	1.679+09	2.56−01	AA	CT92LS
400.7	1s4p	1s6d	$^3P°$	^3D		2	2	7250390.	7499940.	8.85+08	1.405−01	AA	CT92LS
400.7	1s4p	1s6d	$^3P°$	^3D		2	3	7250390.	7499940.	3.54+09	7.87−01	AA	CT92LS
400.7	1s4p	1s6d	$^3P°$	^3D		1	2	7250390.	7499940.	2.65+09	4.21−01	AA	CT92LS
407.5	1s4d	1s6p	^1D	$^1P°$		2	1	7254240.	7499630.	3.69+08	3.69−02	AA	CT92
413.3	1s4p	1s6d	$^1P°$	^1D		1	2	7256370.	7498340.	3.52+09	6.13−01	AA	CT92
414.4	1s4d	1s6p	^3D	$^3P°$		3	2	7255960.	7497300.	3.80+08	6.68−02	AA	CT92LS
414.4	1s4d	1s6p	^3D	$^3P°$		2	1	7255960.	7497300.	3.40+08	3.58−02	AA	CT92LS
414.4	1s4d	1s6p	^3D	$^3P°$		2	2	7255960.	7497300.	6.79+07	1.192−02	AA	CT92LS
602.4	1s4s	1s5p	^3S	$^3P°$		1	1	7244270.	7410270.	2.65+09	8.58−01	AA	CT92LS
602.4	1s4s	1s5p	^3S	$^3P°$		1	2	7244270.	7410270.	2.65+09	1.430+00	AA	CT92LS
611.4	1s4p	1s5d	$^3P°$	^3D		2	2	7250390.	7413940.	1.493+09	8.42−01	AA	CT92LS
611.4	1s4p	1s5d	$^3P°$	^3D		2	3	7250390.	7413940.	5.97+09	4.72+00	AA	CT92LS
611.4	1s4p	1s5d	$^3P°$	^3D		1	2	7250390.	7413940.	4.48+09	2.53+00	AA	CT92LS
625.9	1s4d	1s5p	^1D	$^1P°$		2	1	7254240.	7414000.	7.38+08	2.68−01	AA	CT92

λ_{Ritz} [a] (Å)	Configurations		Terms		Type	J_i [b]	J_k [b]	E_i (cm^{-1})	E_k (cm^{-1})	A_{ki} [c] (s^{-1})	S_{ik} [c]	Acc. [d]	Ref. [e]
641.4	1s4p	1s5d	$^1P^o$	^1D		1	2	7256370.	7412290.	6.09+09	3.96+00	AA	CT92
648.0	1s4d	1s5p	^3D	$^3P^o$		3	2	7255960.	7410270.	7.59+08	5.10−01	AA	CT92LS
648.0	1s4d	1s5p	^3D	$^3P^o$		2	1	7255960.	7410270.	6.78+08	2.73−01	AA	CT92LS
648.0	1s4d	1s5p	^3D	$^3P^o$		2	2	7255960.	7410270.	1.356+08	9.11−02	AA	CT92LS
1115	1s5p	1s6d	$^3P^o$	^3D		2	2	7410270.	7499940.	4.55+08	1.559+00	AA	CT92LS
1115	1s5p	1s6d	$^3P^o$	^3D		2	3	7410270.	7499940.	1.822+09	8.73+00	AA	CT92LS
1115	1s5p	1s6d	$^3P^o$	^3D		1	2	7410270.	7499940.	1.366+09	4.68+00	AA	CT92LS
1145	1s5d	1s6p	^1D	$^1P^o$		2	1	7412290.	7499630.	3.83+08	8.51−01	AA	CT92
1186	1s5p	1s6d	$^1P^o$	^1D		1	2	7414000.	7498340.	1.830+09	7.53+00	AA	CT92
1200	1s5d	1s6p	^3D	$^3P^o$		3	2	7413940.	7497300.	3.77+08	1.606+00	AA	CT92LS
1200	1s5d	1s6p	^3D	$^3P^o$		2	1	7413940.	7497300.	3.37+08	8.60−01	AA	CT92LS
1200	1s5d	1s6p	^3D	$^3P^o$		2	2	7413940.	7497300.	6.73+07	2.87−01	AA	CT92LS
1395	1s2s	1s2p	^3S	$^3P^o$		1	2	5830040.	5901700.	9.57+07	6.41−01	AA	CT92LS
1414	1s2s	1s2p	^3S	$^3P^o$		1	1	5830040.	5900750.	9.19+07	3.85−01	AA	CT92LS
1417	1s2s	1s2p	^3S	$^3P^o$		1	0	5830040.	5900600.	9.13+07	1.283−01	AA	CT92LS
2136	1s2s	1s2p	^1S	$^1P^o$		0	1	5903100.	5949900.	2.91+07	4.21−01	AA	CT92
5720	1s3s	1s3p	^3S	$^3P^o$		1	1	6885090.	6902560.	9.46+06	2.63+00	AA	CT92LS
5720	1s3s	1s3p	^3S	$^3P^o$		1	2	6885090.	6902560.	9.46+06	4.38+00	AA	CT92LS
7510	1s3s	1s3p	^1S	$^1P^o$		0	1	6903270.	6916590.	4.11+06	2.57+00	AA	CT92
8110	1s3p	1s3d	$^3P^o$	^3D		2	3	6902560.	6914890.	2.150+06	3.96+00	AA	CT92LS
9640	1s3p	1s3d	$^3P^o$	^3D		2	2	6902560.	6912930.	3.20+05	7.07−01	AA	CT92LS
9640	1s3p	1s3d	$^3P^o$	^3D		1	2	6902560.	6912930.	9.59+05	2.122+00	AA	CT92LS
16300	1s4s	1s4p	^3S	$^3P^o$		1	1	7244270.	7250390.	1.600+06	1.033+01	AA	CT92LS
16300	1s4s	1s4p	^3S	$^3P^o$		1	2	7244270.	7250390.	1.600+06	1.722+01	AA	CT92LS
17900	1s4p	1s4d	$^3P^o$	^3D		2	2	7250390.	7255960.	1.766+05	2.52+00	AA	CT92LS
17900	1s4p	1s4d	$^3P^o$	^3D		2	3	7250390.	7255960.	7.07+05	1.412+01	AA	CT92LS
17900	1s4p	1s4d	$^3P^o$	^3D		1	2	7250390.	7255960.	5.30+05	7.57+00	AA	CT92LS
27200	1s5p	1s5d	$^3P^o$	^3D		2	2	7410270.	7413940.	1.059+05	5.29+00	AA	CT92LS
27200	1s5p	1s5d	$^3P^o$	^3D		2	3	7410270.	7413940.	4.24+05	2.96+01	AA	CT92LS

λ_{Ritz} [a] (Å)	Configurations		Terms		Type	J_i [b]	J_k [b]	E_i (cm^{-1})	E_k (cm^{-1})	A_{ki} [c] (s^{-1})	S_{ik} [c]	Acc. [d]	Ref. [e]
27200	1s5p	1s5d	$^3P°$	^3D		1	2	7410270.	7413940.	3.18+05	1.586+01	AA	CT92LS
38000	1s6p	1s6d	$^3P°$	^3D		2	2	7497300.	7499940.	6.91+04	9.26+00	AA	CT92LS
38000	1s6p	1s6d	$^3P°$	^3D		2	3	7497300.	7499940.	2.76+05	5.19+01	AA	CT92LS
38000	1s6p	1s6d	$^3P°$	^3D		1	2	7497300.	7499940.	2.072+05	2.78+01	AA	CT92LS
47000	1s4d	1s4p	^1D	$^1P°$		2	1	7254240.	7256370.	5.31+04	8.14+00	AA	CT92
58000	1s5d	1s5p	^1D	$^1P°$		2	1	7412290.	7414000.	4.64+04	1.374+01	AA	CT92
78000	1s6d	1s6p	^1D	$^1P°$		2	1	7498340.	7499630.	3.29+04	2.266+01	AA	CT92
145000	1s3d	1s3p	^1D	$^1P°$		2	1	6915900.	6916590.	3.32+03	1.497+01	AA	CT92
Ne VI (B-like)													
86.115	$2s^22p$	$2s^26d$	$^2P°$	^2D		1/2	3/2	0.00	1161238	1.17+10	1.48−02	B	FH99LS
86.211	$2s^22p$	$2s^26d$	$^2P°$	^2D		3/2	5/2	1306.81	1161252	1.40+10	2.7−02	B	FH99LS
86.212	$2s^22p$	$2s^26d$	$^2P°$	^2D		3/2	3/2	1306.81	1161238	2.3+09	3.0−03	C	FH99LS
89.978	$2s^22p$	$2s^25d$	$^2P°$	^2D		1/2	3/2	0.00	1111379	2.30+10	3.3−02	B	FH99LS
90.084	$2s^22p$	$2s^25d$	$^2P°$	^2D		3/2	3/2	1306.81	1111379	4.6+09	6.6−03	C	FH99LS
90.084	$2s^22p$	$2s^25d$	$^2P°$	^2D		3/2	5/2	1306.81	1111379	2.8+10	6.0−02	B	FH99LS
96.85	$2s2p^2$	$2s2p(^3P°)4d$	^4P	$^4P°$		1/2	1/2	100297.1	1132800	7.3+09	6.6−03	C	FH99LS
96.89	$2s2p^2$	$2s2p(^3P°)4d$	^4P	$^4P°$		3/2	1/2	100737.1	1132800	3.7+10	3.3−02	B	FH99LS
97.05	$2s2p^2$	$2s2p(^3P°)4d$	^4P	$^4D°$		1/2	1/2	100297.1	1130700	6.7+10	6.0−02	B	FH99LS
97.09	$2s2p^2$	$2s2p(^3P°)4d$	^4P	$^4D°$		3/2	1/2	100737.1	1130700	1.33+10	1.20−02	B	FH99LS
97.11	$2s2p^2$	$2s2p(^3P°)4d$	^4P	$^4D°$		5/2	7/2	101383.6	1131100	8.0+10	2.9−01	B	FH99LS
98.126	$2s^22p$	$2s^24d$	$^2P°$	^2D		1/2	3/2	0.00	1019100	4.2+10	7.9−02	B	FH99LS
98.252	$2s^22p$	$2s^24d$	$^2P°$	^2D		3/2	3/2	1306.81	1019100	8.4+09	1.58−02	B	FH99LS
98.252	$2s^22p$	$2s^24d$	$^2P°$	^2D		3/2	5/2	1306.81	1019100	5.1+10	1.42−01	B	FH99LS
101.54	$2s^22p$	$2s2p(^1P°)3p$	$^2P°$	^2D		3/2	5/2	1306.81	986100	7.6+09	2.36−02	B	FH99LS
101.75	$2s2p^2$	$2p^2(^3P)3p$	^4P	$^4S°$		1/2	3/2	100297.1	1083100	4.8+09	1.0−02	C	FH99LS
101.80	$2s2p^2$	$2p^2(^3P)3p$	^4P	$^4S°$		3/2	3/2	100737.1	1083100	9.6+09	1.99−02	B	FH99LS
101.86	$2s2p^2$	$2p^2(^3P)3p$	^4P	$^4S°$		5/2	3/2	101383.6	1083100	1.43+10	3.0−02	B	FH99LS
103.548	$2s2p^2$	$2p^2(^3P)3p$	^4P	$^4P°$		3/2	5/2	100737.1	1066470	8.9+09	2.9−02	B	FH99LS

λ_{Ritz} [a] (Å)	Configurations		Terms		Type	J_i [b]	J_k [b]	E_i (cm^{-1})	E_k (cm^{-1})	A_{ki} [c] (s^{-1})	S_{ik} [c]	Acc. [d]	Ref. [e]
103.58	$2s2p^2$	$2p^2(^3P)3p$	^4P	$^4P^\circ$		1/2	3/2	100297.1	1065700	1.24+10	2.7−02	B	FH99LS
103.597	$2s2p^2$	$2p^2(^3P)3p$	^4P	$^4P^\circ$		1/2	1/2	100297.1	1065580	4.9+09	5.4−03	C	FH99LS
103.618	$2s2p^2$	$2p^2(^3P)3p$	^4P	$^4P^\circ$		5/2	5/2	101383.6	1066470	2.07+10	6.8−02	B	FH99LS
103.63	$2s2p^2$	$2p^2(^3P)3p$	^4P	$^4P^\circ$		3/2	3/2	100737.1	1065700	3.9+09	8.7−03	C	FH99LS
103.644	$2s2p^2$	$2p^2(^3P)3p$	^4P	$^4P^\circ$		3/2	1/2	100737.1	1065580	2.47+10	2.7−02	B	FH99LS
103.70	$2s2p^2$	$2p^2(^3P)3p$	^4P	$^4P^\circ$		5/2	3/2	101383.6	1065700	1.33+10	2.9−02	B	FH99LS
104.100	$2s2p^2$	$2p^2(^3P)3p$	^4P	$^4D^\circ$		3/2	5/2	100737.1	1061350	1.29+10	4.3−02	B	FH99LS
104.102	$2s2p^2$	$2p^2(^3P)3p$	^4P	$^4D^\circ$		5/2	7/2	101383.6	1061980	1.85+10	8.2−02	B	FH99LS
104.103	$2s2p^2$	$2p^2(^3P)3p$	^4P	$^4D^\circ$		1/2	3/2	100297.1	1060880	7.7+09	1.72−02	B	FH99LS
104.135	$2s2p^2$	$2p^2(^3P)3p$	^4P	$^4D^\circ$		1/2	1/2	100297.1	1060590	1.54+10	1.72−02	B	FH99LS
104.151	$2s2p^2$	$2p^2(^3P)3p$	^4P	$^4D^\circ$		3/2	3/2	100737.1	1060880	9.9+09	2.20−02	B	FH99LS
104.17	$2s2p^2$	$2s2p(^3P^\circ)4d$	^2D	$^2F^\circ$		5/2	7/2	178988.1	1139000	6.6+10	2.9−01	B	FH99LS
104.170	$2s2p^2$	$2p^2(^3P)3p$	^4P	$^4D^\circ$		5/2	5/2	101383.6	1061350	5.5+09	1.85−02	B	FH99LS
104.183	$2s2p^2$	$2p^2(^3P)3p$	^4P	$^4D^\circ$		3/2	1/2	100737.1	1060590	3.1+09	3.4−03	C	FH99LS
104.221	$2s2p^2$	$2p^2(^3P)3p$	^4P	$^4D^\circ$		5/2	3/2	101383.6	1060880	9.2+08	2.1−03	C	FH99LS
104.25	$2s2p^2$	$2s2p(^3P^\circ)4d$	^2D	$^2F^\circ$		5/2	5/2	178988.1	1138200	4.4+09	1.47−02	B	FH99LS
104.26	$2s2p^2$	$2s2p(^3P^\circ)4d$	^2D	$^2F^\circ$		3/2	5/2	179018.0	1138200	6.1+10	2.06−01	B	FH99LS
109.306	$2s^22p$	$2s2p(^3P^\circ)3p$	$^2P^\circ$	^2S		1/2	1/2	0.00	914860	1.63+10	2.10−02	B	FH99LS
109.463	$2s^22p$	$2s2p(^3P^\circ)3p$	$^2P^\circ$	^2S		3/2	1/2	1306.81	914860	3.2+10	4.2−02	B	FH99LS
111.099	$2s^22p$	$2s2p(^3P^\circ)3p$	$^2P^\circ$	^2D		1/2	3/2	0.00	900100	4.5+10	1.23−01	B	FH99LS
111.142	$2s^22p$	$2s2p(^3P^\circ)3p$	$^2P^\circ$	^2D		3/2	5/2	1306.81	901054	5.4+10	2.21−01	B	FH99LS
111.260	$2s^22p$	$2s2p(^3P^\circ)3p$	$^2P^\circ$	^2D		3/2	3/2	1306.81	900100	9.0+09	2.46−02	B	FH99LS
117.47	$2s2p^2$	$2s2p(^1P^\circ)3d$	^2D	$^2D^\circ$		5/2	5/2	178988.1	1030250.6	2.03+10	9.7−02	B	FH99LS
117.48	$2s2p^2$	$2s2p(^1P^\circ)3d$	^2D	$^2D^\circ$		3/2	5/2	179018.0	1030250.6	1.4+09	6.9−03	C	FH99LS
117.49	$2s2p^2$	$2s2p(^1P^\circ)3d$	^2D	$^2D^\circ$		5/2	3/2	178988.1	1030100	2.2+09	6.9−03	C	FH99LS
117.50	$2s2p^2$	$2s2p(^1P^\circ)3d$	^2D	$^2D^\circ$		3/2	3/2	179018.0	1030100	1.95+10	6.3−02	B	FH99LS
120.141	$2s2p^2$	$2s2p(^3P^\circ)3d$	^4P	$^4P^\circ$		1/2	1/2	100297.1	932649	2.07+10	3.5−02	B	FH99LS
120.175	$2s2p^2$	$2s2p(^3P^\circ)3d$	^4P	$^4P^\circ$		1/2	3/2	100297.1	932415	5.2+10	1.77−01	B	FH99LS
120.205	$2s2p^2$	$2s2p(^3P^\circ)3d$	^4P	$^4P^\circ$		3/2	1/2	100737.1	932649	1.03+11	1.77−01	B	FH99LS

λ_{Ritz} [a] (Å)	Configurations		Terms		Type	J_i [b]	J_k [b]	E_i (cm^{-1})	E_k (cm^{-1})	A_{ki} [c] (s^{-1})	S_{ik} [c]	Acc. [d]	Ref. [e]
120.239	$2s2p^2$	$2s2p(^3P°)3d$	^4P	$^4P°$		3/2	3/2	100737.1	932415	1.65+10	5.7−02	B	FH99LS
120.289	$2s2p^2$	$2s2p(^3P°)3d$	^4P	$^4P°$		3/2	5/2	100737.1	932070	3.7+10	1.91−01	B	FH99LS
120.332	$2s2p^2$	$2s2p(^3P°)3d$	^4P	$^4P°$		5/2	3/2	101383.6	932415	5.6+10	1.91−01	B	FH99LS
120.382	$2s2p^2$	$2s2p(^3P°)3d$	^4P	$^4P°$		5/2	5/2	101383.6	932070	8.6+10	4.5−01	B	FH99LS
121.05	$2s2p^2$	$2s2p(^3P°)3d$	^4P	$^4D°$		1/2	1/2	100297.1	926400	1.92+11	3.4−01	B	FH99LS
121.05	$2s2p^2$	$2s2p(^3P°)3d$	^4P	$^4D°$		1/2	3/2	100297.1	926400	9.6+10	3.4−01	B	FH99LS
121.099	$2s2p^2$	$2s2p(^3P°)3d$	^4P	$^4D°$		3/2	5/2	100737.1	926510	1.61+11	8.5−01	B	FH99LS
121.11	$2s2p^2$	$2s2p(^3P°)3d$	^4P	$^4D°$		3/2	1/2	100737.1	926400	3.8+10	6.7−02	B	FH99LS
121.11	$2s2p^2$	$2s2p(^3P°)3d$	^4P	$^4D°$		3/2	3/2	100737.1	926400	1.22+11	4.3−01	B	FH99LS
121.150	$2s2p^2$	$2s2p(^3P°)3d$	^4P	$^4D°$		5/2	7/2	101383.6	926810	2.29+11	1.61+00	B	FH99LS
121.194	$2s2p^2$	$2s2p(^3P°)3d$	^4P	$^4D°$		5/2	5/2	101383.6	926510	6.9+10	3.6−01	B	FH99LS
121.21	$2s2p^2$	$2s2p(^3P°)3d$	^4P	$^4D°$		5/2	3/2	101383.6	926400	1.14+10	4.0−02	B	FH99LS
122.5163	$2s^22p$	$2s^23d$	$^2P°$	^2D		1/2	3/2	0.00	816218	1.26+11	4.58−01	A	TF00
122.7013	$2s^22p$	$2s^23d$	$^2P°$	^2D		3/2	5/2	1306.81	816294	1.51+11	8.26−01	A	TF00
122.7128	$2s^22p$	$2s^23d$	$^2P°$	^2D		3/2	3/2	1306.81	816218	2.52+10	9.2−02	A	TF00
128.07	$2s2p^2$	$2s2p(^1P°)3d$	^2P	$^2D°$		1/2	3/2	249286.0	1030100	1.29+11	5.4−01	B	FH99LS
128.18	$2s2p^2$	$2s2p(^1P°)3d$	^2P	$^2D°$		3/2	5/2	250099.4	1030250.6	1.55+11	9.7−01	B	FH99LS
128.21	$2s2p^2$	$2s2p(^1P°)3d$	^2P	$^2D°$		3/2	3/2	250099.4	1030100	2.6+10	1.07−01	B	FH99LS
130.269	$2s2p^2$	$2s2p(^3P°)3d$	^2D	$^2F°$		5/2	7/2	178988.1	946633	1.45+11	1.26+00	B	FH99LS
130.39	$2s2p^2$	$2s2p(^3P°)3d$	^2D	$^2F°$		5/2	5/2	178988.1	945900	9.6+09	6.3−02	B	FH99LS
130.40	$2s2p^2$	$2s2p(^3P°)3d$	^2D	$^2F°$		3/2	5/2	179018.0	945900	1.35+11	8.8−01	B	FH99LS
133.48	$2s2p^2$	$2s2p(^3P°)3d$	^2D	$^2D°$		5/2	5/2	178988.1	928140	5.8+10	4.1−01	B	FH99LS
133.49	$2s2p^2$	$2s2p(^3P°)3d$	^2D	$^2D°$		3/2	5/2	179018.0	928140	4.2+09	2.9−02	B	FH99LS
133.50	$2s2p^2$	$2s2p(^3P°)3d$	^2D	$^2D°$		5/2	3/2	178988.1	928030	6.3+09	2.9−02	B	FH99LS
133.51	$2s2p^2$	$2s2p(^3P°)3d$	^2D	$^2D°$		3/2	3/2	179018.0	928030	5.6+10	2.6−01	B	FH99LS
136.215	$2s2p^2$	$2s2p(^3P°)3s$	^4P	$^4P°$		3/2	5/2	100737.1	834869.7	1.09+10	8.15−02	A	TF00
136.284	$2s2p^2$	$2s2p(^3P°)3s$	^4P	$^4P°$		1/2	3/2	100297.1	834061.1	1.50+10	7.51−02	A	TF00
136.335	$2s2p^2$	$2s2p(^3P°)3s$	^4P	$^4P°$		5/2	5/2	101383.6	834869.7	2.54+10	1.90−01	A	TF00
136.365	$2s2p^2$	$2s2p(^3P°)3s$	^4P	$^4P°$		3/2	3/2	100737.1	834061.1	4.80+09	2.41−02	A	TF00

λ_{Ritz} [a] (Å)	Configurations		Terms		Type	J_i [b]	J_k [b]	E_i (cm^{-1})	E_k (cm^{-1})	A_{ki} [c] (s^{-1})	S_{ik} [c]	Acc. [d]	Ref. [e]
136.366	$2s2p^2$	$2s2p(^3P^o)3s$	^4P	$^4P^o$		1/2	1/2	100297.1	833617.2	5.99+09	1.50−02	A	TF00
136.448	$2s2p^2$	$2s2p(^3P^o)3s$	^4P	$^4P^o$		3/2	1/2	100737.1	833617.2	2.99+10	7.51−02	A	TF00
136.486	$2s2p^2$	$2s2p(^3P^o)3s$	^4P	$^4P^o$		5/2	3/2	101383.6	834061.1	1.62+10	8.13−02	A	TF00
138.392	$2s^22p$	$2s^23s$	$^2P^o$	^2S		1/2	1/2	0.00	722582.53	9.9+09	2.59−02	A	TF00
138.643	$2s^22p$	$2s^23s$	$^2P^o$	^2S		3/2	1/2	1306.81	722582.53	1.99+10	5.23−02	A	TF00
142.520	$2p^3$	$2p^2(^3P)3s$	$^4S^o$	^4P		3/2	5/2	321611.8	1023270	1.90+10	1.63−01	B	FH99LS
142.71	$2p^3$	$2p^2(^3P)3s$	$^4S^o$	^4P		3/2	3/2	321611.8	1022320	1.90+10	1.09−01	B	FH99LS
142.82	$2p^3$	$2p^2(^3P)3s$	$^4S^o$	^4P		3/2	1/2	321611.8	1021770	1.89+10	5.4−02	B	FH99LS
143.52	$2p^3$	$2p^2(^1D)3s$	$^2D^o$	^2D		5/2	3/2	359537.1	1056300	2.47+09	1.44−02	B	FH99LS
143.52	$2p^3$	$2p^2(^1D)3s$	$^2D^o$	^2D		5/2	5/2	359537.1	1056300	2.31+10	2.02−01	B	FH99LS
143.53	$2p^3$	$2p^2(^1D)3s$	$^2D^o$	^2D		3/2	3/2	359605.0	1056300	2.23+10	1.30−01	B	FH99LS
143.53	$2p^3$	$2p^2(^1D)3s$	$^2D^o$	^2D		3/2	5/2	359605.0	1056300	1.65+09	1.44−02	B	FH99LS
147.33	$2s2p^2$	$2s2p(^3P^o)3d$	^2P	$^2D^o$		1/2	3/2	249286.0	928030	1.62+10	1.02−01	B	FH99LS
147.48	$2s2p^2$	$2s2p(^3P^o)3d$	^2P	$^2D^o$		3/2	5/2	250099.4	928140	1.94+10	1.84−01	B	FH99LS
147.51	$2s2p^2$	$2s2p(^3P^o)3d$	^2P	$^2D^o$		3/2	3/2	250099.4	928030	3.2+09	2.05−02	B	FH99LS
147.60	$2s2p^2$	$2s2p(^3P^o)3s$	^2D	$^2P^o$		5/2	3/2	178988.1	856510	1.72+10	1.09−01	B	FH99LS
147.60	$2s2p^2$	$2s2p(^3P^o)3s$	^2D	$^2P^o$		3/2	3/2	179018.0	856510	1.91+09	1.21−02	B	FH99LS
147.78	$2s2p^2$	$2s2p(^3P^o)3s$	^2D	$^2P^o$		3/2	1/2	179018.0	855720	1.90+10	6.1−02	B	FH99LS
152.467	$2s2p^2$	$2s2p(^3P^o)3s$	^2D	$^4P^o$		5/2	5/2	178988.1	834869.7	1.86+06	1.95−05	B	TF00
152.473	$2s2p^2$	$2s2p(^3P^o)3s$	^2D	$^4P^o$		3/2	5/2	179018.0	834869.7	1.15+05	1.21−06	B	TF00
152.655	$2s2p^2$	$2s2p(^3P^o)3s$	^2D	$^4P^o$		5/2	3/2	178988.1	834061.1	3.1+06	2.15−05	B	TF00
152.662	$2s2p^2$	$2s2p(^3P^o)3s$	^2D	$^4P^o$		3/2	3/2	179018.0	834061.1	1.37+06	9.6−06	B	TF00
152.765	$2s2p^2$	$2s2p(^3P^o)3s$	^2D	$^4P^o$		3/2	1/2	179018.0	833617.2	1.56+06	5.5−06	B	TF00
153.78	$2p^3$	$2p^2(^1D)3s$	$^2P^o$	^2D		1/2	3/2	406007.3	1056300	6.1+09	4.3−02	B	FH99LS
153.79	$2p^3$	$2p^2(^1D)3s$	$^2P^o$	^2D		3/2	3/2	406049.5	1056300	1.2+09	8.7−03	C	FH99LS
153.79	$2p^3$	$2p^2(^1D)3s$	$^2P^o$	^2D		3/2	5/2	406049.5	1056300	7.3+09	7.8−02	B	FH99LS
159.60	$2p^3$	$2s2p(^1P^o)3p$	$^2D^o$	^2D		5/2	5/2	359537.1	986100	1.7+08	2.1−03	C	FH99LS
159.62	$2p^3$	$2s2p(^1P^o)3p$	$^2D^o$	^2D		3/2	5/2	359605.0	986100	1.2+07	1.5−04	C	FH99LS
159.83	$2s2p^2$	$2s2p(^3P^o)3s$	^2S	$^2P^o$		1/2	3/2	230851.2	856510	5.7+09	4.6−02	B	FH99LS

λ_{Ritz} [a] (Å)	Configurations		Terms		Type	J_i [b]	J_k [b]	E_i (cm^{-1})	E_k (cm^{-1})	A_{ki} [c] (s^{-1})	S_{ik} [c]	Acc. [d]	Ref. [e]
160.03	$2s2p^2$	$2s2p(^3P°)3s$	^2S	$^2P°$		1/2	1/2	230851.2	855720	5.7+09	2.30−02	B	FH99LS
164.68	$2s2p^2$	$2s2p(^3P°)3s$	^2P	$^2P°$		1/2	3/2	249286.0	856510	3.0+08	2.6−03	C	FH99LS
164.90	$2s2p^2$	$2s2p(^3P°)3s$	^2P	$^2P°$		1/2	1/2	249286.0	855720	1.2+09	5.2−03	C	FH99LS
164.90	$2s2p^2$	$2s2p(^3P°)3s$	^2P	$^2P°$		3/2	3/2	250099.4	856510	1.47+09	1.31−02	B	FH99LS
165.12	$2s2p^2$	$2s2p(^3P°)3s$	^2P	$^2P°$		3/2	1/2	250099.4	855720	5.9+08	2.6−03	C	FH99LS
165.780	$2s2p^2$	$2s2p(^3P°)3s$	^2S	$^4P°$		1/2	3/2	230851.2	834061.1	6.0+05	5.4−06	B+	TF00
165.902	$2s2p^2$	$2s2p(^3P°)3s$	^2S	$^4P°$		1/2	1/2	230851.2	833617.2	1.88+05	8.5−07	B+	TF00
168.765	$2s2p^2$	$2s^23p$	^2D	$^2P°$		5/2	3/2	178988.1	771529.25	1.90+09	1.80−02	A	TF00
168.773	$2s2p^2$	$2s^23p$	^2D	$^2P°$		3/2	3/2	179018.0	771529.25	2.09+08	1.99−03	A	TF00
168.865	$2s2p^2$	$2s^23p$	^2D	$^2P°$		3/2	1/2	179018.0	771206.7	2.11+09	1.01−02	A	TF00
184.67	$2p^3$	$2s2p(^3P°)3p$	$^2D°$	^2D		5/2	5/2	359537.1	901054	6.5+08	1.21−02	B	FH99LS
184.69	$2p^3$	$2s2p(^3P°)3p$	$^2D°$	^2D		3/2	5/2	359605.0	901054	4.6+07	8.6−04	C	FH99LS
184.953	$2s2p^2$	$2s^23p$	^2S	$^2P°$		1/2	3/2	230851.2	771529.25	3.14+08	3.92−03	A	TF00
184.99	$2p^3$	$2s2p(^3P°)3p$	$^2D°$	^2D		5/2	3/2	359537.1	900100	6.9+07	8.6−04	C	FH99LS
185.02	$2p^3$	$2s2p(^3P°)3p$	$^2D°$	^2D		3/2	3/2	359605.0	900100	6.2+08	7.8−03	C	FH99LS
185.063	$2s2p^2$	$2s^23p$	^2S	$^2P°$		1/2	1/2	230851.2	771206.7	2.99+08	1.87−03	A	TF00
191.482	$2s2p^2$	$2s^23p$	^2P	$^2P°$		1/2	3/2	249286.0	771529.25	6.9+06	9.6−05	B	TF00
191.600	$2s2p^2$	$2s^23p$	^2P	$^2P°$		1/2	1/2	249286.0	771206.7	5.2+07	3.6−04	B	TF00
191.780	$2s2p^2$	$2s^23p$	^2P	$^2P°$		3/2	3/2	250099.4	771529.25	5.4+07	7.6−04	B	TF00
191.899	$2s2p^2$	$2s^23p$	^2P	$^2P°$		3/2	1/2	250099.4	771206.7	1.83+07	1.28−04	B	TF00
196.52	$2p^3$	$2s2p(^3P°)3p$	$^2P°$	^2S		1/2	1/2	406007.3	914860	1.2+09	8.8−03	C	FH99LS
196.54	$2p^3$	$2s2p(^3P°)3p$	$^2P°$	^2S		3/2	1/2	406049.5	914860	2.35+09	1.76−02	B	FH99LS
202.02	$2p^3$	$2s2p(^3P°)3p$	$^2P°$	^2D		3/2	5/2	406049.5	901054	5.1+08	1.24−02	B	FH99LS
202.39	$2p^3$	$2s2p(^3P°)3p$	$^2P°$	^2D		1/2	3/2	406007.3	900100	4.2+08	6.9−03	C	FH99LS
202.41	$2p^3$	$2s2p(^3P°)3p$	$^2P°$	^2D		3/2	3/2	406049.5	900100	8.4+07	1.4−03	C	FH99LS
218.935	$2p^3$	$2s^23d$	$^2D°$	^2D		5/2	5/2	359537.1	816294	5.8+06	1.8−04	C	TF00
218.967	$2p^3$	$2s^23d$	$^2D°$	^2D		3/2	5/2	359605.0	816294	4.8+05	1.5−05	C	TF00
218.971	$2p^3$	$2s^23d$	$^2D°$	^2D		5/2	3/2	359537.1	816218	5.6+05	1.2−05	C	TF00
219.004	$2p^3$	$2s^23d$	$^2D°$	^2D		3/2	3/2	359605.0	816218	5.7+06	1.2−04	C	TF00

λ_{Ritz} [a] (Å)	Configurations		Terms		Type	J_i [b]	J_k [b]	E_i (cm^{-1})	E_k (cm^{-1})	A_{ki} [c] (s^{-1})	S_{ik} [c]	Acc. [d]	Ref. [e]
243.757	$2p^3$	$2s^23d$	$^2P°$	^2D		3/2	5/2	406049.5	816294	4.5+05	1.9−05	C+	TF00
243.777	$2p^3$	$2s^23d$	$^2P°$	^2D		1/2	3/2	406007.3	816218	2.8+05	8.1−06	C+	TF00
243.802	$2p^3$	$2s^23d$	$^2P°$	^2D		3/2	3/2	406049.5	816218	5.0+04	1.4−06	C+	TF00
249.39	$2p^3$	$2s^23s$	$^4S°$	^2S		3/2	1/2	321611.8	722582.53	8.+02	1.2−08	D	TF00
256.39	$2s^23p$	$2s^26d$	$^2P°$	^2D		1/2	3/2	771206.7	1161238	2.4+09	7.9−02	C	FH99LS
256.59	$2s^23p$	$2s^26d$	$^2P°$	^2D		3/2	5/2	771529.25	1161252	2.9+09	1.4−01	C	FH99LS
256.60	$2s^23p$	$2s^26d$	$^2P°$	^2D		3/2	3/2	771529.25	1161238	4.8+08	1.6−02	C	FH99LS
293.97	$2s^23p$	$2s^25d$	$^2P°$	^2D		1/2	3/2	771206.7	1111379	4.0+09	2.0−01	C	FH99LS
294.25	$2s^23p$	$2s^25d$	$^2P°$	^2D		3/2	3/2	771529.25	1111379	7.9+08	4.0−02	C	FH99LS
294.25	$2s^23p$	$2s^25d$	$^2P°$	^2D		3/2	5/2	771529.25	1111379	4.7+09	3.6−01	C	FH99LS
309.9	$2s^23d$	$2s2p(^3P°)4d$	^2D	$^2F°$		5/2	7/2	816294	1139000	6.+07	7.−03	D	FH99LS
310.6	$2s^23d$	$2s2p(^3P°)4d$	^2D	$^2F°$		3/2	5/2	816218	1138200	5.+07	5.−03	D	FH99LS
310.6	$2s^23d$	$2s2p(^3P°)4d$	^2D	$^2F°$		5/2	5/2	816294	1138200	3.8+06	3.4−04	D	FH99LS
315.88	$2p^3$	$2s^23s$	$^2P°$	^2S		1/2	1/2	406007.3	722582.53	2.25+05	7.0−06	B+	TF00
315.92	$2p^3$	$2s^23s$	$^2P°$	^2S		3/2	1/2	406049.5	722582.53	4.8+05	1.49−05	B+	TF00
350.8	$2s^23p$	$2p^2(^1D)3s$	$^2P°$	^2D		1/2	3/2	771206.7	1056300	1.6+08	1.3−02	C	FH99LS
351.2	$2s^23p$	$2p^2(^1D)3s$	$^2P°$	^2D		3/2	3/2	771529.25	1056300	3.1+07	2.7−03	D	FH99LS
351.2	$2s^23p$	$2p^2(^1D)3s$	$^2P°$	^2D		3/2	5/2	771529.25	1056300	1.9+08	2.4−02	C	FH99LS
391.1	$2s2p(^3P°)3s$	$2s^25d$	$^2P°$	^2D		1/2	3/2	855720	1111379	1.51+08	1.79−02	B	FH99LS
392.36	$2s2p(^3P°)3s$	$2s^25d$	$^2P°$	^2D		3/2	3/2	856510	1111379	3.0+07	3.6−03	C	FH99LS
392.36	$2s2p(^3P°)3s$	$2s^25d$	$^2P°$	^2D		3/2	5/2	856510	1111379	1.80+08	3.2−02	B	FH99LS
399.841	$2s^22p$	$2s2p^2$	$^2P°$	^2P		1/2	3/2	0.00	250099.4	1.72+09	2.17−01	A	R12
401.146	$2s^22p$	$2s2p^2$	$^2P°$	^2P		1/2	1/2	0.00	249286.0	6.64+09	4.23−01	A	R12
401.941	$2s^22p$	$2s2p^2$	$^2P°$	^2P		3/2	3/2	1306.81	250099.4	8.7+09	1.11+00	A	R12
403.260	$2s^22p$	$2s2p^2$	$^2P°$	^2P		3/2	1/2	1306.81	249286.0	3.72+09	2.41−01	A	R12
403.40	$2s^23p$	$2s^24d$	$^2P°$	^2D		1/2	3/2	771206.7	1019100	6.8+09	8.8−01	C	FH99LS
403.92	$2s^23p$	$2s^24d$	$^2P°$	^2D		3/2	3/2	771529.25	1019100	1.4+09	1.8−01	C	FH99LS
403.92	$2s^23p$	$2s^24d$	$^2P°$	^2D		3/2	5/2	771529.25	1019100	8.2+09	1.6+00	C	FH99LS
420.0	$2s2p(^3P°)3p$	$2s2p(^3P°)4d$	^2D	$^2F°$		3/2	5/2	900100	1138200	6.9+09	1.50+00	B	FH99LS

λ_{Ritz} [a] (Å)	Configurations		Terms		Type	J_i [b]	J_k [b]	E_i (cm^{-1})	E_k (cm^{-1})	A_{ki} [c] (s^{-1})	S_{ik} [c]	Acc.[d]	Ref.[e]
420.3	2s2p(^{3}P°)3p	2s2p(^{3}P°)4d	^{2}D	^{2}F°		5/2	7/2	901054	1139000	7.3+09	2.15+00	B	FH99LS
421.7	2s2p(^{3}P°)3p	2s2p(^{3}P°)4d	^{2}D	^{2}F°		5/2	5/2	901054	1138200	4.8+08	1.07−01	B	FH99LS
428.8	2s2p(^{3}P°)3d	2s^{2}6d	^{2}D°	^{2}D		3/2	5/2	928030	1161252	3.6+05	8.4−05	C	FH99LS
428.8	2s2p(^{3}P°)3d	2s^{2}6d	^{2}D°	^{2}D		3/2	3/2	928030	1161238	4.9+06	7.6−04	C	FH99LS
429.0	2s2p(^{3}P°)3d	2s^{2}6d	^{2}D°	^{2}D		5/2	5/2	928140	1161252	5.0+06	1.2−03	C	FH99LS
429.0	2s2p(^{3}P°)3d	2s^{2}6d	^{2}D°	^{2}D		5/2	3/2	928140	1161238	5.4+05	8.4−05	C	FH99LS
433.179	2s^{2}2p	2s2p^2	^{2}P°	^{2}S		1/2	1/2	0.00	230851.2	2.02+09	1.62−01	A	R12
435.646	2s^{2}2p	2s2p^2	^{2}P°	^{2}S		3/2	1/2	1306.81	230851.2	3.24+09	2.64−01	A	R12
440.410	2s2p^2	2p^3	^{2}D	^{2}P°		5/2	3/2	178988.1	406049.5	3.74+09	6.30−01	A	R12
440.468	2s2p^2	2p^3	^{2}D	^{2}P°		3/2	3/2	179018.0	406049.5	4.51+08	7.61−02	A	R12
440.549	2s2p^2	2p^3	^{2}D	^{2}P°		3/2	1/2	179018.0	406007.3	4.26+09	3.60−01	A	R12
451.85	2s2p^2	2p^3	^{4}P	^{4}S°		1/2	3/2	100297.1	321611.8	1.56+09	2.85−01	A	R12
452.75	2s2p^2	2p^3	^{4}P	^{4}S°		3/2	3/2	100737.1	321611.8	3.11+09	5.70−01	A	R12
454.07	2s2p^2	2p^3	^{4}P	^{4}S°		5/2	3/2	101383.6	321611.8	4.62+09	8.5−01	A	R12
466.0	2s^{2}3p	2s2p(^{1}P°)3p	^{2}P°	^{2}D		3/2	5/2	771529.25	986100	3.2+09	9.6−01	C	FH99LS
467.2	2s^{2}3d	2s2p(^{1}P°)3d	^{2}D	^{2}D°		3/2	5/2	816218	1030250.6	2.8+08	8.5−02	C	FH99LS
467.4	2s^{2}3d	2s2p(^{1}P°)3d	^{2}D	^{2}D°		5/2	5/2	816294	1030250.6	3.9+09	1.2+00	C	FH99LS
467.5	2s^{2}3d	2s2p(^{1}P°)3d	^{2}D	^{2}D°		3/2	3/2	816218	1030100	3.8+09	7.6−01	C	FH99LS
467.7	2s^{2}3d	2s2p(^{1}P°)3d	^{2}D	^{2}D°		5/2	3/2	816294	1030100	4.2+08	8.5−02	C	FH99LS
478.24	2s2p(^{3}P°)3d	2s^{2}6d	^{2}P°	^{2}D		3/2	5/2	952150	1161252	9.+06	2.8−03	D	FH99LS
478.27	2s2p(^{3}P°)3d	2s^{2}6d	^{2}P°	^{2}D		3/2	3/2	952150	1161238	1.4+06	3.1−04	D	FH99LS
479.3	2s2p(^{3}P°)3d	2s^{2}6d	^{2}P°	^{2}D		1/2	3/2	952610	1161238	7.+06	1.5−03	D	FH99LS
498.6	2s2p(^{3}P°)3s	2p^2(^{1}D)3s	^{2}P°	^{2}D		1/2	3/2	855720	1056300	6.1+07	1.49−02	B	FH99LS
500.5	2s2p(^{3}P°)3s	2p^2(^{1}D)3s	^{2}P°	^{2}D		3/2	3/2	856510	1056300	1.2+07	3.0−03	C	FH99LS
500.5	2s2p(^{3}P°)3s	2p^2(^{1}D)3s	^{2}P°	^{2}D		3/2	5/2	856510	1056300	7.2+07	2.7−02	B	FH99LS
528.5	2s2p(^{3}P°)3s	2p^2(^{3}P)3s	^{4}P°	^{4}P		3/2	5/2	834061.1	1023270	1.08+09	4.7−01	B	FH99LS
529.9	2s2p(^{3}P°)3s	2p^2(^{3}P)3s	^{4}P°	^{4}P		1/2	3/2	833617.2	1022320	1.49+09	4.4−01	B	FH99LS
530.8	2s2p(^{3}P°)3s	2p^2(^{3}P)3s	^{4}P°	^{4}P		5/2	5/2	834869.7	1023270	2.48+09	1.10+00	B	FH99LS
531.2	2s2p(^{3}P°)3s	2p^2(^{3}P)3s	^{4}P°	^{4}P		3/2	3/2	834061.1	1022320	4.7+08	1.40−01	B	FH99LS

λ_{Ritz} [a] (Å)	Configurations		Terms		Type	J_i [b]	J_k [b]	E_i (cm^{-1})	E_k (cm^{-1})	A_{ki} [c] (s^{-1})	S_{ik} [c]	Acc. [d]	Ref. [e]
531.5	2s2p(^{3}P°)3s	2p^2(^{3}P)3s	^{4}P°	^{4}P		1/2	1/2	833617.2	1021770	5.9+08	8.7−02	B	FH99LS
532.7	2s2p(^{3}P°)3s	2p^2(^{3}P)3s	^{4}P°	^{4}P		3/2	1/2	834061.1	1021770	2.9+09	4.4−01	B	FH99LS
533.5	2s2p(^{3}P°)3s	2p^2(^{3}P)3s	^{4}P°	^{4}P		5/2	3/2	834869.7	1022320	1.57+09	4.7−01	B	FH99LS
553.658	2s2p^2	2p^3	^{2}D	^{2}D°		5/2	3/2	178988.1	359605.0	2.67+08	9.0−02	A	R12
553.750	2s2p^2	2p^3	^{2}D	^{2}D°		3/2	3/2	179018.0	359605.0	2.11+09	7.07−01	A	R12
553.866	2s2p^2	2p^3	^{2}D	^{2}D°		5/2	5/2	178988.1	359537.1	2.23+09	1.12+00	A	R12
553.958	2s2p^2	2p^3	^{2}D	^{2}D°		3/2	5/2	179018.0	359537.1	1.70+08	8.5−02	A	R12
558.603	2s^{2}2p	2s2p^2	^{2}P°	^{2}D		1/2	3/2	0.00	179018.0	9.8+08	3.39−01	A	R12
562.711	2s^{2}2p	2s2p^2	^{2}P°	^{2}D		3/2	3/2	1306.81	179018.0	1.78+08	6.26−02	A	R12
562.805	2s^{2}2p	2s2p^2	^{2}P°	^{2}D		3/2	5/2	1306.81	178988.1	1.14+09	5.99−01	A	R12
570.782	2s2p^2	2p^3	^{2}S	^{2}P°		1/2	3/2	230851.2	406049.5	7.54+08	2.77−01	A	R12
570.919	2s2p^2	2p^3	^{2}S	^{2}P°		1/2	1/2	230851.2	406007.3	6.12+08	1.13−01	A	R12
604.3	2s2p(^{3}P°)3d	2s^{2}5d	^{2}F°	^{2}D		5/2	3/2	945900	1111379	4.6+06	2.0−03	C	FH99LS
604.3	2s2p(^{3}P°)3d	2s^{2}5d	^{2}F°	^{2}D		5/2	5/2	945900	1111379	2.2+05	1.4−04	C	FH99LS
607.0	2s2p(^{3}P°)3d	2s^{2}5d	^{2}F°	^{2}D		7/2	5/2	946633	1111379	4.3+06	2.9−03	C	FH99LS
612.1	2s2p(^{3}P°)3s	2s^{2}4d	^{2}P°	^{2}D		1/2	3/2	855720	1019100	2.8+08	1.28−01	B	FH99LS
615.0	2s2p(^{3}P°)3s	2s^{2}4d	^{2}P°	^{2}D		3/2	3/2	856510	1019100	5.6+07	2.6−02	B	FH99LS
615.0	2s2p(^{3}P°)3s	2s^{2}4d	^{2}P°	^{2}D		3/2	5/2	856510	1019100	3.4+08	2.31−01	B	FH99LS
627.4	2s^{2}4p	2s^{2}6d	^{2}P°	^{2}D		1/2	3/2	1001840	1161238	8.3+08	4.0−01	C	FH99LS
627.8	2s^{2}4p	2s^{2}6d	^{2}P°	^{2}D		3/2	5/2	1001965	1161252	9.9+08	7.3−01	C	FH99LS
627.9	2s^{2}4p	2s^{2}6d	^{2}P°	^{2}D		3/2	3/2	1001965	1161238	1.7+08	8.1−02	C	FH99LS
628.0	2s2p(^{3}P°)3d	2s^{2}5d	^{2}P°	^{2}D		3/2	3/2	952150	1111379	1.0+06	5.−04	D	FH99LS
628.0	2s2p(^{3}P°)3d	2s^{2}5d	^{2}P°	^{2}D		3/2	5/2	952150	1111379	6.+06	4.4−03	D	FH99LS
629.8	2s2p(^{3}P°)3d	2s^{2}5d	^{2}P°	^{2}D		1/2	3/2	952610	1111379	5.+06	2.4−03	D	FH99LS
637.904	2s2p^2	2p^3	^{2}P	^{2}P°		1/2	3/2	249286.0	406049.5	3.22+08	1.65−01	A	R12
638.075	2s2p^2	2p^3	^{2}P	^{2}P°		1/2	1/2	249286.0	406007.3	1.59+09	4.09−01	A	R12
641.231	2s2p^2	2p^3	^{2}P	^{2}P°		3/2	3/2	250099.4	406049.5	1.92+09	1.0+00	A	R12
641.404	2s2p^2	2p^3	^{2}P	^{2}P°		3/2	1/2	250099.4	406007.3	7.35+08	1.92−01	A	R12
654	2s2p(^{1}P°)3p	2s2p(^{3}P°)4d	^{2}D	^{2}F°		5/2	7/2	986100	1139000	6.4+07	7.0−02	B	FH99LS

λ_{Ritz} [a] (Å)	Configurations		Terms		Type	J_i [b]	J_k [b]	E_i (cm^{-1})	E_k (cm^{-1})	A_{ki} [c] (s^{-1})	S_{ik} [c]	Acc. [d]	Ref. [e]
657	2s2p(^{1}P°)3p	2s2p(^{3}P°)4d	^{2}D	^{2}F°		3/2	5/2	986000	1138200	5.9+07	4.9−02	C	FH99LS
657	2s2p(^{1}P°)3p	2s2p(^{3}P°)4d	^{2}D	^{2}F°		5/2	5/2	986100	1138200	4.2+06	3.5−03	C	FH99LS
696.1	2s^{2}3p	2s2p(^{3}P°)3p	^{2}P°	^{2}S		1/2	1/2	771206.7	914860	4.8+07	1.6−02	C	FH99LS
697.7	2s^{2}3p	2s2p(^{3}P°)3p	^{2}P°	^{2}S		3/2	1/2	771529.25	914860	9.5+07	3.2−02	C	FH99LS
701.15	2s2p^2	2p^3	^{2}D	^{4}S°		5/2	3/2	178988.1	321611.8	1.34+03	9.1−07	B	R12
701.29	2s2p^2	2p^3	^{2}D	^{4}S°		3/2	3/2	179018.0	321611.8	3.2+01	2.17−08	B	R12
746.7	2s^{2}3s	2s2p(^{3}P°)3s	^{2}S	^{2}P°		1/2	3/2	722582.53	856510	1.4+08	1.2−01	C	FH99LS
751.1	2s^{2}3s	2s2p(^{3}P°)3s	^{2}S	^{2}P°		1/2	1/2	722582.53	855720	1.4+08	5.8−02	C	FH99LS
768.3	2s2p(^{3}P°)3p	2s2p(^{1}P°)3d	^{2}D	^{2}D°		3/2	5/2	900100	1030250.6	6.5+06	8.8−03	C	FH99LS
769.2	2s2p(^{3}P°)3p	2s2p(^{1}P°)3d	^{2}D	^{2}D°		3/2	3/2	900100	1030100	8.8+07	7.9−02	B	FH99LS
772	2s2p(^{3}P°)3s	2s2p(^{1}P°)3p	^{2}P°	^{2}D		3/2	5/2	856510	986100	2.9+08	3.9−01	B	FH99LS
772.1	2s^{2}3p	2s2p(^{3}P°)3p	^{2}P°	^{2}D		3/2	5/2	771529.25	901054	2.8+08	3.8−01	C	FH99LS
774.0	2s2p(^{3}P°)3p	2s2p(^{1}P°)3d	^{2}D	^{2}D°		5/2	5/2	901054	1030250.6	9.0+07	1.23−01	B	FH99LS
774.9	2s2p(^{3}P°)3p	2s2p(^{1}P°)3d	^{2}D	^{2}D°		5/2	3/2	901054	1030100	9.6+06	8.8−03	C	FH99LS
775.8	2s^{2}3p	2s2p(^{3}P°)3p	^{2}P°	^{2}D		1/2	3/2	771206.7	900100	2.3+08	2.1−01	C	FH99LS
776.68	2s2p^2	2p^3	^{2}S	^{2}D°		1/2	3/2	230851.2	359605.0	1.87+06	1.73−03	A	R12
777.8	2s^{2}3p	2s2p(^{3}P°)3p	^{2}P°	^{2}D		3/2	3/2	771529.25	900100	4.5+07	4.2−02	C	FH99LS
779.6	2s2p(^{3}P°)3d	2p^2(^{1}D)3s	^{2}D°	^{2}D		3/2	3/2	928030	1056300	1.89+07	1.76−02	B	FH99LS
779.6	2s2p(^{3}P°)3d	2p^2(^{1}D)3s	^{2}D°	^{2}D		3/2	5/2	928030	1056300	1.4+06	2.0−03	C	FH99LS
780.3	2s2p(^{3}P°)3d	2p^2(^{1}D)3s	^{2}D°	^{2}D		5/2	3/2	928140	1056300	2.1+06	2.0−03	C	FH99LS
780.3	2s2p(^{3}P°)3d	2p^2(^{1}D)3s	^{2}D°	^{2}D		5/2	5/2	928140	1056300	1.95+07	2.7−02	B	FH99LS
818.7	2s2p(^{1}P°)3d	2s^{2}6d	^{2}P°	^{2}D		1/2	3/2	1039100	1161238	3.9+07	4.2−02	C	FH99LS
819.3	2s2p(^{1}P°)3d	2s^{2}6d	^{2}P°	^{2}D		3/2	5/2	1039200	1161252	4.6+07	7.6−02	C	FH99LS
819.4	2s2p(^{1}P°)3d	2s^{2}6d	^{2}P°	^{2}D		3/2	3/2	1039200	1161238	8.+06	8.−03	D	FH99LS
834	2s^{2}4d	2s2p(^{3}P°)4d	^{2}D	^{2}F°		5/2	7/2	1019100	1139000	1.3+06	2.9−03	C	FH99LS
840	2s^{2}4d	2s2p(^{3}P°)4d	^{2}D	^{2}F°		5/2	5/2	1019100	1138200	8.3+04	1.5−04	C	FH99LS
840	2s^{2}4d	2s2p(^{3}P°)4d	^{2}D	^{2}F°		3/2	5/2	1019100	1138200	1.2+06	2.0−03	C	FH99LS
862.0	2s2p(^{1}P°)3s	2p^2(^{1}D)3s	^{2}P°	^{2}D		3/2	3/2	940290	1056300	1.8+08	2.2−01	C	FH99LS
862.0	2s2p(^{1}P°)3s	2p^2(^{1}D)3s	^{2}P°	^{2}D		3/2	5/2	940290	1056300	1.1+09	2.0+00	C	FH99LS

λ_{Ritz} [a] (Å)	Configurations		Terms		Type	J_i [b]	J_k [b]	E_i (cm^{-1})	E_k (cm^{-1})	A_{ki} [c] (s^{-1})	S_{ik} [c]	Acc. [d]	Ref. [e]
862.2	$2s2p(^1P^o)3s$	$2p^2(^1D)3s$	$^2P^o$	^2D		1/2	3/2	940320	1056300	8.7+08	1.1+00	C	FH99LS
893.5	$2s^23d$	$2s2p(^3P^o)3d$	^2D	$^2D^o$		3/2	5/2	816218	928140	1.3+05	2.7−04	D	FH99LS
894.1	$2s^23d$	$2s2p(^3P^o)3d$	^2D	$^2D^o$		5/2	5/2	816294	928140	1.8+06	3.8−03	D	FH99LS
894.4	$2s^23d$	$2s2p(^3P^o)3d$	^2D	$^2D^o$		3/2	3/2	816218	928030	1.7+06	2.4−03	D	FH99LS
895.0	$2s^23d$	$2s2p(^3P^o)3d$	^2D	$^2D^o$		5/2	3/2	816294	928030	1.9+05	2.7−04	D	FH99LS
897.0	$2s^23s$	$2s2p(^3P^o)3s$	^2S	$^4P^o$		1/2	3/2	722582.53	834061.1	1.13+05	1.62−04	B+	TF00
900.6	$2s^23s$	$2s2p(^3P^o)3s$	^2S	$^4P^o$		1/2	1/2	722582.53	833617.2	4.0+04	2.86−05	B+	TF00
906	$2s2p(^3P^o)3d$	$2p^2(^1D)3s$	$^2F^o$	^2D		5/2	3/2	945900	1056300	6.1+07	9.0−02	B	FH99LS
906	$2s2p(^3P^o)3d$	$2p^2(^1D)3s$	$^2F^o$	^2D		5/2	5/2	945900	1056300	2.9+06	6.4−03	C	FH99LS
906.46	$2s2p^2$	$2p^3$	^2P	$^2D^o$		1/2	3/2	249286.0	359605.0	4.33+08	6.37−01	A	R12
911.9	$2s2p(^3P^o)3d$	$2p^2(^1D)3s$	$^2F^o$	^2D		7/2	5/2	946633	1056300	5.7+07	1.29−01	B	FH99LS
912.9	$2s^24p$	$2s^25d$	$^2P^o$	^2D		1/2	3/2	1001840	1111379	1.2+09	1.9+00	C	FH99LS
913.20	$2s2p^2$	$2p^3$	^2P	$^2D^o$		3/2	3/2	250099.4	359605.0	7.68+07	1.15−01	A	R12
913.76	$2s2p^2$	$2p^3$	^2P	$^2D^o$		3/2	5/2	250099.4	359537.1	4.95+08	1.12+00	A	R12
914.0	$2s^24p$	$2s^25d$	$^2P^o$	^2D		3/2	3/2	1001965	1111379	2.5+08	3.7−01	C	FH99LS
914.0	$2s^24p$	$2s^25d$	$^2P^o$	^2D		3/2	5/2	1001965	1111379	1.5+09	3.3+00	C	FH99LS
960.2	$2s2p(^3P^o)3d$	$2p^2(^1D)3s$	$^2P^o$	^2D		3/2	3/2	952150	1056300	9.6+06	1.7−02	C	FH99LS
960.2	$2s2p(^3P^o)3d$	$2p^2(^1D)3s$	$^2P^o$	^2D		3/2	5/2	952150	1056300	5.8+07	1.5−01	C	FH99LS
964.4	$2s2p(^3P^o)3d$	$2p^2(^1D)3s$	$^2P^o$	^2D		1/2	3/2	952610	1056300	4.8+07	8.4−02	C	FH99LS
992.68	$2s^22p$	$2s2p^2$	$^2P^o$	^4P		1/2	3/2	0.00	100737.1	3.6+02	7.0−07	B	R12
997.04	$2s^22p$	$2s2p^2$	$^2P^o$	^4P		1/2	1/2	0.00	100297.1	1.37+04	1.34−05	B	R12
999.23	$2s^22p$	$2s2p^2$	$^2P^o$	^4P		3/2	5/2	1306.81	101383.6	1.11+04	3.27−05	A	R12
1005.73	$2s^22p$	$2s2p^2$	$^2P^o$	^4P		3/2	3/2	1306.81	100737.1	2.9+03	5.8−06	B	R12
1010.20	$2s^22p$	$2s2p^2$	$^2P^o$	^4P		3/2	1/2	1306.81	100297.1	1.25+04	1.27−05	B	R12
1032.3	$2s2p(^3P^o)3d$	$2p^2(^3P)3s$	$^4D^o$	^4P		3/2	5/2	926400	1023270	3.8+05	1.2−03	C	FH99LS
1033.5	$2s2p(^3P^o)3d$	$2p^2(^3P)3s$	$^4D^o$	^4P		5/2	5/2	926510	1023270	3.4+06	1.10−02	B	FH99LS
1036.7	$2s2p(^3P^o)3d$	$2p^2(^3P)3s$	$^4D^o$	^4P		7/2	5/2	926810	1023270	1.49+07	4.9−02	B	FH99LS
1043	$2s2p(^3P^o)3d$	$2p^2(^3P)3s$	$^4D^o$	^4P		3/2	3/2	926400	1022320	5.8+06	1.31−02	B	FH99LS
1043	$2s2p(^3P^o)3d$	$2p^2(^3P)3s$	$^4D^o$	^4P		1/2	3/2	926400	1022320	9.1+05	2.0−03	C	FH99LS

λ_{Ritz} [a] (Å)	Configurations		Terms		Type	J_i [b]	J_k [b]	E_i (cm^{-1})	E_k (cm^{-1})	A_{ki} [c] (s^{-1})	S_{ik} [c]	Acc. [d]	Ref. [e]
1044	$2s2p(^3P^\circ)3d$	$2p^2(^3P)3s$	$^4D^\circ$	^4P		5/2	3/2	926510	1022320	1.15+07	2.6−02	B	FH99LS
1049	$2s2p(^3P^\circ)3d$	$2p^2(^3P)3s$	$^4D^\circ$	^4P		3/2	1/2	926400	1021770	9.0+06	1.02−02	B	FH99LS
1049	$2s2p(^3P^\circ)3d$	$2p^2(^3P)3s$	$^4D^\circ$	^4P		1/2	1/2	926400	1021770	9.0+06	1.02−02	B	FH99LS
1096.5	$2s2p(^3P^\circ)3d$	$2p^2(^3P)3s$	$^4P^\circ$	^4P		5/2	5/2	932070	1023270	3.2+06	1.24−02	B	FH99LS
1100.7	$2s2p(^3P^\circ)3d$	$2p^2(^3P)3s$	$^4P^\circ$	^4P		3/2	5/2	932415	1023270	1.3+06	5.3−03	C	FH99LS
1101.80	$2s2p^2$	$2p^3$	^2S	$^4S^\circ$		1/2	3/2	230851.2	321611.8	3.2+02	8.3−07	B	R12
1108.0	$2s2p(^3P^\circ)3d$	$2p^2(^3P)3s$	$^4P^\circ$	^4P		5/2	3/2	932070	1022320	2.0+06	5.3−03	C	FH99LS
1112	$2s2p(^3P^\circ)3d$	$2p^2(^3P)3s$	$^4P^\circ$	^4P		3/2	3/2	932415	1022320	5.8+05	1.6−03	C	FH99LS
1115	$2s2p(^3P^\circ)3d$	$2p^2(^3P)3s$	$^4P^\circ$	^4P		1/2	3/2	932649	1022320	1.8+06	4.9−03	C	FH99LS
1119	$2s2p(^3P^\circ)3d$	$2p^2(^3P)3s$	$^4P^\circ$	^4P		3/2	1/2	932415	1021770	3.6+06	4.9−03	C	FH99LS
1122	$2s2p(^3P^\circ)3d$	$2p^2(^3P)3s$	$^4P^\circ$	^4P		1/2	1/2	932649	1021770	7.1+05	9.9−04	C	FH99LS
1209	$2p^2(^1D)3s$	$2s2p(^3P^\circ)4d$	^2D	$^2F^\circ$		5/2	7/2	1056300	1139000	9.4+07	6.6−01	B	FH99LS
1221	$2p^2(^1D)3s$	$2s2p(^3P^\circ)4d$	^2D	$^2F^\circ$		5/2	5/2	1056300	1138200	6.1+06	3.3−02	B	FH99LS
1221	$2p^2(^1D)3s$	$2s2p(^3P^\circ)4d$	^2D	$^2F^\circ$		3/2	5/2	1056300	1138200	8.5+07	4.6−01	B	FH99LS
1269	$2s2p(^1P^\circ)3s$	$2s^24d$	$^2P^\circ$	^2D		3/2	3/2	940290	1019100	2.9+06	1.2−02	C	FH99LS
1269	$2s2p(^1P^\circ)3s$	$2s^24d$	$^2P^\circ$	^2D		3/2	5/2	940290	1019100	1.7+07	1.1−01	C	FH99LS
1269	$2s2p(^1P^\circ)3s$	$2s^24d$	$^2P^\circ$	^2D		1/2	3/2	940320	1019100	1.4+07	5.8−02	C	FH99LS
1384	$2s2p(^1P^\circ)3d$	$2s^25d$	$^2P^\circ$	^2D		1/2	3/2	1039100	1111379	8.3+07	4.4−01	C	FH99LS
1385	$2s2p(^1P^\circ)3d$	$2s^25d$	$^2P^\circ$	^2D		3/2	3/2	1039200	1111379	1.7+07	8.7−02	C	FH99LS
1385	$2s2p(^1P^\circ)3d$	$2s^25d$	$^2P^\circ$	^2D		3/2	5/2	1039200	1111379	9.9+07	7.8−01	C	FH99LS
1631	$2p^2(^3P)3s$	$2p^2(^3P)3p$	^4P	$^4S^\circ$		1/2	3/2	1021770	1083100	1.01+08	8.6−01	B	FH99LS
1645	$2p^2(^3P)3s$	$2p^2(^3P)3p$	^4P	$^4S^\circ$		3/2	3/2	1022320	1083100	1.96+08	1.72+00	B	FH99LS
1671	$2p^2(^3P)3s$	$2p^2(^3P)3p$	^4P	$^4S^\circ$		5/2	3/2	1023270	1083100	2.8+08	2.6+00	B	FH99LS
1691	$2s2p(^3P^\circ)3s$	$2s2p(^3P^\circ)3p$	$^2P^\circ$	^2S		1/2	1/2	855720	914860	1.52+08	7.2−01	B	FH99LS
1714	$2s2p(^3P^\circ)3s$	$2s2p(^3P^\circ)3p$	$^2P^\circ$	^2S		3/2	1/2	856510	914860	2.9+08	1.45+00	B	FH99LS
1722	$2s2p(^3P^\circ)3d$	$2s2p(^1P^\circ)3p$	$^2D^\circ$	^2D		3/2	5/2	928030	986100	3.8+05	5.7−03	C	FH99LS
1725	$2s2p(^3P^\circ)3d$	$2s2p(^1P^\circ)3p$	$^2D^\circ$	^2D		5/2	5/2	928140	986100	5.2+06	8.0−02	B	FH99LS
1836	$2s^24p$	$2p^2(^1D)3s$	$^2P^\circ$	^2D		1/2	3/2	1001840	1056300	3.2+05	3.9−03	D	FH99LS
1840	$2s^24p$	$2p^2(^1D)3s$	$^2P^\circ$	^2D		3/2	3/2	1001965	1056300	6.+04	8.−04	D	FH99LS

λ_{Ritz} [a] (Å)	Configurations		Terms		Type	J_i [b]	J_k [b]	E_i (cm^{-1})	E_k (cm^{-1})	A_{ki} [c] (s^{-1})	S_{ik} [c]	Acc. [d]	Ref. [e]
1840	$2s^24p$	$2p^2(^1D)3s$	$^2P^o$	^2D		3/2	5/2	1001965	1056300	3.8+05	7.−03	D	FH99LS
2042	$2s^23s$	$2s^23p$	^2S	$^2P^o$		1/2	3/2	722582.53	771529.25	2.17+08	3.65+00	A	TF00
2056	$2s^23s$	$2s^23p$	^2S	$^2P^o$		1/2	1/2	722582.53	771206.7	2.12+08	1.82+00	A	TF00
2182	$2s2p(^1P^o)3s$	$2s2p(^1P^o)3p$	$^2P^o$	^2D		3/2	5/2	940290	986100	1.5+08	4.7+00	C	FH99LS
2183	$2s2p(^3P^o)3p$	$2s2p(^3P^o)3d$	^2D	$^2F^o$		3/2	5/2	900100	945900	1.99+08	6.1+00	B	FH99LS
2193	$2s2p(^3P^o)3p$	$2s2p(^3P^o)3d$	^2D	$^2F^o$		5/2	7/2	901054	946633	2.10+08	8.8+00	B	FH99LS
2221	$2s^23p$	$2s^23d$	$^2P^o$	^2D		1/2	3/2	771206.7	816218	1.34+08	2.9+00	B	TF00
2229	$2s2p(^3P^o)3p$	$2s2p(^3P^o)3d$	^2D	$^2F^o$		5/2	5/2	901054	945900	1.33+07	4.4−01	B	FH99LS
2233	$2s^23p$	$2s^23d$	$^2P^o$	^2D		3/2	5/2	771529.25	816294	1.59+08	5.2+00	B	TF00
2237	$2s^23p$	$2s^23d$	$^2P^o$	^2D		3/2	3/2	771529.25	816218	2.6+07	5.8−01	B	TF00
2244	$2s2p(^3P^o)3s$	$2s2p(^3P^o)3p$	$^2P^o$	^2D		3/2	5/2	856510	901054	1.73+08	5.8+00	B	FH99LS
2253	$2s2p(^3P^o)3s$	$2s2p(^3P^o)3p$	$^2P^o$	^2D		1/2	3/2	855720	900100	1.43+08	3.2+00	B	FH99LS
2259	$2s2p(^1P^o)3p$	$2s2p(^1P^o)3d$	^2D	$^2D^o$		3/2	5/2	986000	1030250.6	2.7+06	9.2−02	C	FH99LS
2264	$2s2p(^1P^o)3p$	$2s2p(^1P^o)3d$	^2D	$^2D^o$		5/2	5/2	986100	1030250.6	3.7+07	1.29+00	B	FH99LS
2264	$2p^2(^3P)3s$	$2p^2(^3P)3p$	^4P	$^4P^o$		3/2	5/2	1022320	1066470	5.8+07	1.98+00	B	FH99LS
2267	$2s2p(^1P^o)3p$	$2s2p(^1P^o)3d$	^2D	$^2D^o$		3/2	3/2	986000	1030100	3.6+07	8.3−01	C	FH99LS
2272	$2s2p(^1P^o)3p$	$2s2p(^1P^o)3d$	^2D	$^2D^o$		5/2	3/2	986100	1030100	4.0+06	9.2−02	B	FH99LS
2276	$2p^2(^3P)3s$	$2p^2(^3P)3p$	^4P	$^4P^o$		1/2	3/2	1021770	1065700	7.9+07	1.84+00	B	FH99LS
2282	$2p^2(^3P)3s$	$2p^2(^3P)3p$	^4P	$^4P^o$		1/2	1/2	1021770	1065580	3.1+07	3.7−01	B	FH99LS
2293	$2s2p(^3P^o)3s$	$2s2p(^3P^o)3p$	$^2P^o$	^2D		3/2	3/2	856510	900100	2.7+07	6.5−01	B	FH99LS
2305	$2p^2(^3P)3s$	$2p^2(^3P)3p$	^4P	$^4P^o$		3/2	3/2	1022320	1065700	2.43+07	5.9−01	B	FH99LS
2311	$2p^2(^3P)3s$	$2p^2(^3P)3p$	^4P	$^4P^o$		3/2	1/2	1022320	1065580	1.51+08	1.84+00	B	FH99LS
2314	$2p^2(^3P)3s$	$2p^2(^3P)3p$	^4P	$^4P^o$		5/2	5/2	1023270	1066470	1.26+08	4.6+00	B	FH99LS
2356	$2p^2(^3P)3s$	$2p^2(^3P)3p$	^4P	$^4P^o$		5/2	3/2	1023270	1065700	7.7+07	1.98+00	B	FH99LS
2359	$2s2p(^1P^o)3p$	$2s2p(^1P^o)3d$	^2P	$^2D^o$		1/2	3/2	987720	1030100	7.5+07	1.9+00	C	FH99LS
2365	$2s2p(^1P^o)3p$	$2s2p(^1P^o)3d$	^2P	$^2D^o$		3/2	5/2	987989.0	1030250.6	8.9+07	3.5+00	C	FH99LS
2374	$2s2p(^1P^o)3p$	$2s2p(^1P^o)3d$	^2P	$^2D^o$		3/2	3/2	987989.0	1030100	1.5+07	3.9−01	C	FH99LS
2481	$2s^23d$	$2s2p(^3P^o)3s$	^2D	$^2P^o$		3/2	3/2	816218	856510	9.4+05	2.8−02	C	FH99LS
2486	$2s^23d$	$2s2p(^3P^o)3s$	^2D	$^2P^o$		5/2	3/2	816294	856510	8.4+06	2.5−01	C	FH99LS

λ_{Ritz} [a] (Å)	Configurations		Terms		Type	J_i [b]	J_k [b]	E_i (cm^{-1})	E_k (cm^{-1})	A_{ki} [c] (s^{-1})	S_{ik} [c]	Acc. [d]	Ref. [e]
2487	2s2p(^{3}P°)3d	2s2p(^{1}P°)3p	^{2}F°	^{2}D		5/2	5/2	945900	986100	7.0+04	3.2−03	C	FH99LS
2531	2s^{2}3d	2s2p(^{3}P°)3s	^{2}D	^{2}P°		3/2	1/2	816218	855720	8.8+06	1.4−01	C	FH99LS
2533	2s2p(^{3}P°)3d	2s2p(^{1}P°)3p	^{2}F°	^{2}D		7/2	5/2	946633	986100	1.33+06	6.4−02	B	FH99LS
2556	2p^2(^{3}P)3s	2p^2(^{3}P)3p	^{4}P	^{4}D°		1/2	3/2	1021770	1060880	5.8+07	1.91+00	B	FH99LS
2561	2p^2(^{3}P)3s	2p^2(^{3}P)3p	^{4}P	^{4}D°		3/2	5/2	1022320	1061350	9.7+07	4.8+00	B	FH99LS
2575	2p^2(^{3}P)3s	2p^2(^{3}P)3p	^{4}P	^{4}D°		1/2	1/2	1021770	1060590	1.13+08	1.91+00	B	FH99LS
2583	2p^2(^{3}P)3s	2p^2(^{3}P)3p	^{4}P	^{4}D°		5/2	7/2	1023270	1061980	1.35+08	9.2+00	B	FH99LS
2593	2p^2(^{3}P)3s	2p^2(^{3}P)3p	^{4}P	^{4}D°		3/2	3/2	1022320	1060880	7.1+07	2.45+00	B	FH99LS
2612	2p^2(^{3}P)3s	2p^2(^{3}P)3p	^{4}P	^{4}D°		3/2	1/2	1022320	1060590	2.17+07	3.8−01	B	FH99LS
2625	2p^2(^{3}P)3s	2p^2(^{3}P)3p	^{4}P	^{4}D°		5/2	5/2	1023270	1061350	3.8+07	2.06+00	B	FH99LS
2658	2p^2(^{3}P)3s	2p^2(^{3}P)3p	^{4}P	^{4}D°		5/2	3/2	1023270	1060880	6.2+06	2.29−01	B	FH99LS
2945	2s2p(^{3}P°)3d	2s2p(^{1}P°)3p	^{2}P°	^{2}D		3/2	5/2	952150	986100	3.8+06	2.9−01	C	FH99LS
3565	2s2p(^{3}P°)3p	2s2p(^{3}P°)3d	^{2}D	^{2}D°		3/2	5/2	900100	928140	6.3+05	8.4−02	B	FH99LS
3579	2s2p(^{3}P°)3p	2s2p(^{3}P°)3d	^{2}D	^{2}D°		3/2	3/2	900100	928030	8.3+06	7.6−01	B	FH99LS
3691	2s2p(^{3}P°)3p	2s2p(^{3}P°)3d	^{2}D	^{2}D°		5/2	5/2	901054	928140	7.9+06	1.18+00	B	FH99LS
3706	2s2p(^{3}P°)3p	2s2p(^{3}P°)3d	^{2}D	^{2}D°		5/2	3/2	901054	928030	8.4+05	8.4−02	B	FH99LS
4340	2s2p(^{3}P°)4d	2s^{2}6d	^{2}F°	^{2}D		5/2	3/2	1138200	1161238	6.3+05	1.02−01	B	FH99LS
4340	2s2p(^{3}P°)4d	2s^{2}6d	^{2}F°	^{2}D		5/2	5/2	1138200	1161252	3.0+04	7.3−03	C	FH99LS
4490	2s2p(^{3}P°)4d	2s^{2}6d	^{2}F°	^{2}D		7/2	5/2	1139000	1161252	5.4+05	1.46−01	B	FH99LS
6440	2p^2(^{1}D)3p	2s^{2}5d	^{2}F°	^{2}D		5/2	3/2	1095860	1111379	3.9+05	2.1−01	C	FH99LS
6440	2p^2(^{1}D)3p	2s^{2}5d	^{2}F°	^{2}D		5/2	5/2	1095860	1111379	1.9+04	1.5−02	C	FH99LS
6570	2p^2(^{1}D)3p	2s^{2}5d	^{2}F°	^{2}D		7/2	5/2	1096170	1111379	3.5+05	3.0−01	C	FH99LS
76522	2s^{2}2p	2s^{2}2p	^{2}P°	^{2}P°	M1	1/2	3/2	0.00	1306.81	2.01−02	1.33+00	A	R12
Ne VII (Be-like)													
65.850	2s^2	2s6p	^{1}S	^{1}P°		0	1	0.0	1518600	1.5+10	6.−03	D	TS90
67.755	2s2p	2s8d	^{3}P°	^{3}D		0	1	111252.8	1587160	3.4+09	1.56−03	B	TS90LS
67.776	2s2p	2s8d	^{3}P°	^{3}D		1	1,2	111708.0	1587160	3.8+09	4.7−03	B	TS90LS
67.822	2s2p	2s8d	^{3}P°	^{3}D		2	1,2,3	112702.4	1587160	3.4+09	7.8−03	B	TS90LS

λ_{Ritz}[a] (Å)	Configurations		Terms		Type	J_i[b]	J_k[b]	E_i (cm^{-1})	E_k (cm^{-1})	A_{ki}[c] (s^{-1})	S_{ik}[c]	Acc.[d]	Ref.[e]
68.024	$2s^2$	2p4d	^1S	$^1P^\circ$		0	1	0.0	1470080	2.4+09	1.1−03	D	TS90
68.311	2s2p	2p5p	$^3P^\circ$	^3D		2	3	112702.4	1576600	1.06+10	1.17−02	B	TS90LS
68.979	2s2p	2s7d	$^3P^\circ$	^3D		0	1	111252.8	1560975	7.4+09	3.6−03	B	TS90LS
69.000	2s2p	2s7d	$^3P^\circ$	^3D		1	1,2	111708.0	1560975	8.3+09	1.08−02	B	TS90LS
69.048	2s2p	2s7d	$^3P^\circ$	^3D		2	1,2,3	112702.4	1560975	7.4+09	1.79−02	B	TS90LS
70.980	2s2p	2s6d	$^3P^\circ$	^3D		0	1	111252.8	1520100	1.1+10	5.9−03	C	TS90LS
70.990	2s2p	2s6d	$^3P^\circ$	^3D		1	2	111708.0	1520360	1.51+10	1.34−02	B	TS90LS
71.003	2s2p	2s6d	$^3P^\circ$	^3D		1	1	111708.0	1520100	8.4+09	4.5−03	C	TS90LS
71.035	2s2p	2s6d	$^3P^\circ$	^3D		2	3	112702.4	1520460	2.01+10	2.49−02	B	TS90LS
71.040	2s2p	2s6d	$^3P^\circ$	^3D		2	2	112702.4	1520360	5.0+09	4.5−03	C	TS90LS
71.053	2s2p	2s6d	$^3P^\circ$	^3D		2	1	112702.4	1520100	6.+08	3.0−04	D	TS90LS
74.231	2s2p	2s7d	$^1P^\circ$	^1D		1	2	214951.6	1562100	8.9+09	9.0−03	C	TS90
74.54	2s2p	2s5d	$^3P^\circ$	^3D		0	1	111252.8	1452900	1.5+10	9.2−03	C	TS90LS
74.549	2s2p	2s5d	$^3P^\circ$	^3D		1	2	111708.0	1453100	2.02+10	2.07−02	B	TS90LS
74.56	2s2p	2s5d	$^3P^\circ$	^3D		1	1	111708.0	1452900	1.1+10	6.9−03	C	TS90LS
74.599	2s2p	2s5d	$^3P^\circ$	^3D		2	3	112702.4	1453200	2.7+10	3.9−02	B	TS90LS
74.605	2s2p	2s5d	$^3P^\circ$	^3D		2	2	112702.4	1453100	6.7+09	6.9−03	C	TS90LS
74.62	2s2p	2s5d	$^3P^\circ$	^3D		2	1	112702.4	1452900	7.+08	4.6−04	D	TS90LS
74.959	2s2p	2p4p	$^3P^\circ$	^3D		2	3	112702.4	1446760	2.9+10	4.2−02	B	TS90LS
74.963	2s2p	2p4p	$^3P^\circ$	^3D		1	2	111708.0	1445700	2.17+10	2.26−02	B	TS90LS
75.019	2s2p	2p4p	$^3P^\circ$	^3D		2	2	112702.4	1445700	7.2+09	7.5−03	C	TS90LS
75.764	$2s^2$	2s4p	^1S	$^1P^\circ$		0	1	0.0	1319890	4.9+10	3.2−02	C	TS90
76.516	2s2p	2s6d	$^1P^\circ$	^1D		1	2	214951.6	1521860	1.55+10	1.71−02	B	TS90
82.073	$2s^2$	2p3d	^1S	$^1P^\circ$		0	1	0.0	1218430	1.2+10	1.0−02	D	TS90
82.160	2s2p	2s4d	$^3P^\circ$	^3D		0	1	111252.8	1328390	4.4+10	3.6−02	B	TS90LS
82.191	2s2p	2s4d	$^3P^\circ$	^3D		1	1	111708.0	1328390	3.3+10	2.7−02	B	TS90LS
82.199	2s2p	2s4d	$^3P^\circ$	^3D		1	2	111708.0	1328265	5.9+10	8.1−02	B	TS90LS
82.258	2s2p	2s4d	$^3P^\circ$	^3D		2	1	112702.4	1328390	2.2+09	1.8−03	C	TS90LS
82.263	2s2p	2s4d	$^3P^\circ$	^3D		2	3	112702.4	1328318	7.8+10	1.51−01	B	TS90LS

λ_{Ritz} [a] (Å)	Configurations		Terms		Type	J_i [b]	J_k [b]	E_i (cm^{-1})	E_k (cm^{-1})	A_{ki} [c] (s^{-1})	S_{ik} [c]	Acc. [d]	Ref. [e]
82.266	2s2p	2s4d	$^3P^o$	^3D		2	2	112702.4	1328265	1.96+10	2.7−02	B	TS90LS
84.181	2s2p	2s4s	$^3P^o$	^3S		0	1	111252.8	1299170	1.6+09	1.4−03	C	TS90LS
84.213	2s2p	2s4s	$^3P^o$	^3S		1	1	111708.0	1299170	4.9+09	4.3−03	C	TS90LS
84.284	2s2p	2s4s	$^3P^o$	^3S		2	1	112702.4	1299170	8.1+09	7.2−03	C	TS90LS
85.38	$2p^2$	2p4d	^3P	$^3D^o$		0	1	289330.4	1460600	5.8+10	5.4−02	B	TS90LS
85.41	$2p^2$	2p4d	^3P	$^3D^o$		1	1	289841.1	1460600	4.4+10	4.0−02	B	TS90LS
85.434	$2p^2$	2p4d	^3P	$^3D^o$		2	3	290724	1461220	1.04+11	2.25−01	B	TS90LS
85.48	$2p^2$	2p4d	^3P	$^3D^o$		2	1	290724	1460600	2.9+09	2.7−03	C	TS90LS
86.777	$2p^2$	2p4d	^1D	$^1P^o$		2	1	317694.6	1470080	3.7+09	3.6−03	C	TS90
87.807	$2s^2$	2p3s	^1S	$^1P^o$		0	1	0.0	1138860	3.9+09	3.9−03	D	F09
88.85	$2p^2$	2s6p	^1S	$^1P^o$		0	1	393125	1518600	1.4+09	1.5−03	D	TS90
89.228	$2s^2$	2p3s	^1S	$^3P^o$		0	1	0.0	1120720	7.3+06	7.7−06	C	F09
89.370	2s2p	2s4d	$^1P^o$	^1D		1	2	214951.6	1333900	5.9+10	1.03−01	B	TS90
91.564	2s2p	2s4s	$^1P^o$	^1S		1	0	214951.6	1307080	1.4+10	5.2−03	C	TS90
92.854	$2p^2$	2p4d	^1S	$^1P^o$		0	1	393125	1470080	6.6+10	7.9−02	C	TS90
94.271	2s2p	2p3p	$^3P^o$	^3P		0	1	111252.8	1172020	1.76+10	2.19−02	B	TS90LS
94.283	2s2p	2p3p	$^3P^o$	^3P		1	2	111708.0	1172350	1.32+10	2.7−02	B	TS90LS
94.312	2s2p	2p3p	$^3P^o$	^3P		1	1	111708.0	1172020	1.32+10	1.64−02	B	TS90LS
94.36	2s2p	2p3p	$^3P^o$	^3P		1	0	111708.0	1171500	5.3+10	2.19−02	B	TS90LS
94.371	2s2p	2p3p	$^3P^o$	^3P		2	2	112702.4	1172350	4.0+10	8.2−02	B	TS90LS
94.400	2s2p	2p3p	$^3P^o$	^3P		2	1	112702.4	1172020	2.20+10	2.7−02	B	TS90LS
94.853	2s2p	2p3p	$^3P^o$	^3S		0	1	111252.8	1165520	6.1+09	7.7−03	C	TS90LS
94.894	2s2p	2p3p	$^3P^o$	^3S		1	1	111708.0	1165520	1.82+10	2.30−02	B	TS90LS
94.983	2s2p	2p3p	$^3P^o$	^3S		2	1	112702.4	1165520	3.0+10	3.8−02	B	TS90LS
95.836	2s2p	2p3p	$^3P^o$	^3D		2	3	112702.4	1156150	3.0+10	9.2−02	B	TS90LS
95.839	2s2p	2p3p	$^3P^o$	^3D		1	2	111708.0	1155120	2.27+10	4.9−02	B	TS90LS
95.856	2s2p	2p3p	$^3P^o$	^3D		0	1	111252.8	1154488	1.68+10	2.19−02	B	TS90LS
95.898	2s2p	2p3p	$^3P^o$	^3D		1	1	111708.0	1154488	1.26+10	1.64−02	B	TS90LS
95.931	2s2p	2p3p	$^3P^o$	^3D		2	2	112702.4	1155120	7.5+09	1.64−02	B	TS90LS

λ_{Ritz} [a] (Å)	Configurations		Terms		Type	J_i [b]	J_k [b]	E_i (cm^{-1})	E_k (cm^{-1})	A_{ki} [c] (s^{-1})	S_{ik} [c]	Acc. [d]	Ref. [e]
95.989	2s2p	2p3p	$^3P°$	^3D		2	1	112702.4	1154488	8.4+08	1.1−03	C	TS90LS
97.229	$2s^2$	2s3p	^1S	$^3P°$		0	1	0.0	1028499.59	6.21+08	8.4−04	A	F09
97.502	$2s^2$	2s3p	^1S	$^1P°$		0	1	0.0	1025620.6	1.14+11	1.56−01	A	F09
103.074	2s2p	2p3p	$^1P°$	^1D		1	2	214951.6	1185130	8.0+10	2.16−01	B	TS90
104.144	2s2p	2s3d	$^3P°$	^1D		1	2	111708.0	1071914	1.99+06	5.5−06	B	F09
104.252	2s2p	2s3d	$^3P°$	^1D		2	2	112702.4	1071914	3.2+05	9.−07	D	F09
106.038	2s2p	2s3d	$^3P°$	^3D		0	1	111252.8	1054310.0	1.36+11	2.40−01	A	F09
106.084	2s2p	2s3d	$^3P°$	^3D		1	2	111708.0	1054354.4	1.83+11	5.40−01	A	F09
106.089	2s2p	2s3d	$^3P°$	^3D		1	1	111708.0	1054310.0	1.02+11	1.80−01	A	F09
106.188	2s2p	2s3d	$^3P°$	^3D		2	3	112702.4	1054428.3	2.44+11	1.01+00	A	F09
106.196	2s2p	2s3d	$^3P°$	^3D		2	2	112702.4	1054354.4	6.10+10	1.80−01	A	F09
106.201	2s2p	2s3d	$^3P°$	^3D		2	1	112702.4	1054310.0	6.78+09	1.20−02	A	F09
107.101	2s2p	2p3p	$^1P°$	^1P		1	1	214951.6	1148647	6.4+10	1.17−01	B	TS90
107.902	$2p^2$	2s4p	^1S	$^1P°$		0	1	393125	1319890	5.+09	9.−03	D	TS90
109.753	$2p^2$	2p3d	^3P	$^3P°$		0	1	289330.4	1200470	5.1+10	1.00−01	B	TS90LS
109.780	$2p^2$	2p3d	^3P	$^3P°$		1	0	289841.1	1200750	1.54+11	1.00−01	B	TS90LS
109.814	$2p^2$	2p3d	^3P	$^3P°$		1	1	289841.1	1200470	3.8+10	7.5−02	B	TS90LS
109.881	$2p^2$	2p3d	^3P	$^3P°$		1	2	289841.1	1199920	3.8+10	1.26−01	B	TS90LS
109.921	$2p^2$	2p3d	^3P	$^3P°$		2	1	290724	1200470	6.4+10	1.26−01	B	TS90LS
109.987	$2p^2$	2p3d	^3P	$^3P°$		2	2	290724	1199920	1.15+11	3.8−01	B	TS90LS
110.530	$2p^2$	2p3d	^3P	$^3D°$		0	1	289330.4	1194060	1.62+11	3.2−01	B	TS90LS
110.561	$2p^2$	2p3d	^3P	$^3D°$		1	2	289841.1	1194320	2.19+11	7.3−01	B	TS90LS
110.593	$2p^2$	2p3d	^3P	$^3D°$		1	1	289841.1	1194060	1.22+11	2.44−01	B	TS90LS
110.642	$2p^2$	2p3d	^3P	$^3D°$		2	3	290724	1194538	2.9+11	1.36+00	B	TS90LS
110.669	$2p^2$	2p3d	^3P	$^3D°$		2	2	290724	1194320	7.3+10	2.44−01	B	TS90LS
110.701	$2p^2$	2p3d	^3P	$^3D°$		2	1	290724	1194060	8.1+09	1.62−02	B	TS90LS
111.020	$2p^2$	2p3d	^1D	$^1P°$		2	1	317694.6	1218430	9.4+09	1.90−02	B	TS90
111.811	$2p^2$	2p3d	^1D	$^1F°$		2	3	317694.6	1212060	3.4+11	1.62+00	B	TS90
115.334	2s2p	2s3s	$^3P°$	^3S		0	1	111252.8	978300.29	6.34+09	1.44−02	A	F09

λ_{Ritz} [a] (Å)	Configurations		Terms		Type	J_i [b]	J_k [b]	E_i (cm^{-1})	E_k (cm^{-1})	A_{ki} [c] (s^{-1})	S_{ik} [c]	Acc. [d]	Ref. [e]
115.395	2s2p	2s3s	$^3P^\circ$	^3S		1	1	111708.0	978300.29	1.90+10	4.33−02	A	F09
115.527	2s2p	2s3s	$^3P^\circ$	^3S		2	1	112702.4	978300.29	3.18+10	7.26−02	A	F09
116.691	2s2p	2s3d	$^1P^\circ$	^1D		1	2	214951.6	1071914	1.50+11	5.88−01	A	F09
117.712	$2p^2$	2p3s	^3P	$^1P^\circ$		0	1	289330.4	1138860	7.3+06	1.76−05	B	F09
117.783	$2p^2$	2p3s	^3P	$^1P^\circ$		1	1	289841.1	1138860	1.41+07	3.41−05	A	F09
117.906	$2p^2$	2p3s	^3P	$^1P^\circ$		2	1	290724	1138860	6.+04	1.5−07	E	F09
119.132	2s2p	2s3d	$^1P^\circ$	^3D		1	2	214951.6	1054354.4	2.1+06	8.8−06	C	F09
119.139	2s2p	2s3d	$^1P^\circ$	^3D		1	1	214951.6	1054310.0	3.66+06	9.2−06	A	F09
120.204	$2p^2$	2p3s	^3P	$^3P^\circ$		1	2	289841.1	1121757	1.03+10	4.4−02	B	F09
120.281	$2p^2$	2p3s	^3P	$^3P^\circ$		0	1	289330.4	1120720	1.38+10	3.5−02	B	F09
120.332	$2p^2$	2p3s	^3P	$^3P^\circ$		2	2	290724	1121757	3.1+10	1.33−01	B	F09
120.354	$2p^2$	2p3s	^3P	$^3P^\circ$		1	1	289841.1	1120720	1.03+10	2.7−02	B	F09
120.429	$2p^2$	2p3s	^3P	$^3P^\circ$		1	0	289841.1	1120207	4.1+10	3.6−02	B	F09
120.483	$2p^2$	2p3s	^3P	$^3P^\circ$		2	1	290724	1120720	1.73+10	4.5−02	B	F09
121.167	$2p^2$	2p3d	^1S	$^1P^\circ$		0	1	393125	1218430	1.7+11	4.5−01	C	TS90
121.778	$2p^2$	2p3s	^1D	$^1P^\circ$		2	1	317694.6	1138860	3.30+10	8.8−02	A	F09
124.368	$2p^2$	2p3s	^1D	$^3P^\circ$		2	2	317694.6	1121757	1.66+07	7.9−05	B	F09
124.529	$2p^2$	2p3s	^1D	$^3P^\circ$		2	1	317694.6	1120720	1.02+07	2.92−05	A	F09
127.676	2s2p	2s3s	$^1P^\circ$	^1S		1	0	214951.6	998183.1	1.4+10	1.5−02	C	F09
131.002	2s2p	2s3s	$^1P^\circ$	^3S		1	1	214951.6	978300.29	8.4+05	2.8−06	B	F09
134.096	$2p^2$	2p3s	^1S	$^1P^\circ$		0	1	393125	1138860	8.4+09	3.0−02	C	F09
135.287	$2p^2$	2s3p	^3P	$^3P^\circ$		0	1	289330.4	1028499.59	1.2+08	4.4−04	C	F09
135.334	$2p^2$	2s3p	^3P	$^3P^\circ$		1	2	289841.1	1028755.1	8.8+07	5.4−04	C	F09
135.381	$2p^2$	2s3p	^3P	$^3P^\circ$		1	1	289841.1	1028499.59	8.9+07	3.3−04	C	F09
135.405	$2p^2$	2s3p	^3P	$^3P^\circ$		1	0	289841.1	1028366.7	3.6+08	4.5−04	C	F09
135.496	$2p^2$	2s3p	^3P	$^3P^\circ$		2	2	290724	1028755.1	2.6+08	1.6−03	C	F09
135.543	$2p^2$	2s3p	^3P	$^3P^\circ$		2	1	290724	1028499.59	1.4+08	5.3−04	C	F09
135.816	$2p^2$	2s3p	^3P	$^1P^\circ$		0	1	289330.4	1025620.6	9.+05	3.4−06	D	F09
135.910	$2p^2$	2s3p	^3P	$^1P^\circ$		1	1	289841.1	1025620.6	8.5+05	3.1−06	C	F09

λ_{Ritz} [a] (Å)	Configurations		Terms		Type	J_i [b]	J_k [b]	E_i (cm^{-1})	E_k (cm^{-1})	A_{ki} [c] (s^{-1})	S_{ik} [c]	Acc. [d]	Ref. [e]
136.074	$2p^2$	2s3p	^3P	$^1P°$		2	1	290724	1025620.6	9.+06	3.5−05	D	F09
137.439	$2p^2$	2p3s	^1S	$^3P°$		0	1	393125	1120720	5.+06	2.1−05	D	F09
140.635	$2p^2$	2s3p	^1D	$^3P°$		2	2	317694.6	1028755.1	4.9+04	3.4−07	C	F09
140.686	$2p^2$	2s3p	^1D	$^3P°$		2	1	317694.6	1028499.59	3.0+07	1.2−04	D	F09
141.258	$2p^2$	2s3p	^1D	$^1P°$		2	1	317694.6	1025620.6	5.+09	2.0−02	D	F09
178.96	2s3p	2s8d	$^3P°$	^3D		0	1	1028366.7	1587160	8.0+08	6.8−03	B	TS90LS
179.00	2s3p	2s8d	$^3P°$	^3D		1	1,2	1028499.59	1587160	9.0+08	2.05−02	B	TS90LS
179.08	2s3p	2s8d	$^3P°$	^3D		2	1,2,3	1028755.1	1587160	8.0+08	3.4−02	B	TS90LS
186.40	2s3p	2s7d	$^1P°$	^1D		1	2	1025620.6	1562100	3.1+09	4.9−02	C	TS90
187.76	2s3p	2s7d	$^3P°$	^3D		0	1	1028366.7	1560975	1.78+09	1.74−02	B	TS90LS
187.80	2s3p	2s7d	$^3P°$	^3D		1	1,2	1028499.59	1560975	2.00+09	5.2−02	B	TS90LS
187.89	2s3p	2s7d	$^3P°$	^3D		2	1,2,3	1028755.1	1560975	1.77+09	8.7−02	B	TS90LS
192.15	2s3s	2s6p	^1S	$^1P°$		0	1	998183.1	1518600	2.01+09	2.11−02	B	TS90
201.52	2s3p	2s6d	$^1P°$	^1D		1	2	1025620.6	1521860	4.8+09	9.7−02	C	TS90
203.31	2s3p	2s6d	$^3P°$	^3D		1	2	1028499.59	1520360	3.8+09	7.9−02	B	TS90LS
203.36	2s3p	2s6d	$^3P°$	^3D		0	1	1028366.7	1520100	2.8+09	3.5−02	B	TS90LS
203.37	2s3p	2s6d	$^3P°$	^3D		2	3	1028755.1	1520460	5.1+09	1.48−01	B	TS90LS
203.42	2s3p	2s6d	$^3P°$	^3D		2	2	1028755.1	1520360	1.28+09	2.6−02	B	TS90LS
203.42	2s3p	2s6d	$^3P°$	^3D		1	1	1028499.59	1520100	2.12+09	2.6−02	B	TS90LS
203.52	2s3p	2s6d	$^3P°$	^3D		2	1	1028755.1	1520100	1.4+08	1.8−03	C	TS90LS
219.86	2p3s	2p5p	$^3P°$	^3D		2	3	1121757	1576600	3.7+09	1.35−01	B	TS90LS
223.9	2s3d	2s6p	^1D	$^1P°$		2	1	1071914	1518600	3.9+08	6.5−03	C	TS90
235.52	2s3p	2s5d	$^3P°$	^3D		1	2	1028499.59	1453100	6.7+09	2.16−01	B	TS90LS
235.6	2s3p	2s5d	$^3P°$	^3D		0	1	1028366.7	1452900	4.9+09	9.6−02	B	TS90LS
235.60	2s3p	2s5d	$^3P°$	^3D		2	3	1028755.1	1453200	8.9+09	4.0−01	B	TS90LS
235.6	2s3p	2s5d	$^3P°$	^3D		1	1	1028499.59	1452900	3.7+09	7.2−02	B	TS90LS
235.66	2s3p	2s5d	$^3P°$	^3D		2	2	1028755.1	1453100	2.22+09	7.2−02	B	TS90LS
235.8	2s3p	2s5d	$^3P°$	^3D		2	1	1028755.1	1452900	2.5+08	4.8−03	C	TS90LS
236.27	2p3s	2s7d	$^1P°$	^1D		1	2	1138860	1562100	7.+07	2.3−03	D	TS90

λ_{Ritz} [a] (Å)	Configurations		Terms		Type	J_i [b]	J_k [b]	E_i (cm^{-1})	E_k (cm^{-1})	A_{ki} [c] (s^{-1})	S_{ik} [c]	Acc. [d]	Ref. [e]
239.23	2s3p	2p4p	$^3P°$	^3D		2	3	1028755.1	1446760	2.7+08	1.30−02	B	TS90LS
239.69	2s3p	2p4p	$^3P°$	^3D		1	2	1028499.59	1445700	2.0+08	7.0−03	C	TS90LS
239.84	2s3p	2p4p	$^3P°$	^3D		2	2	1028755.1	1445700	6.8+07	2.3−03	C	TS90LS
258.24	2p3d	2s8d	$^3P°$	^3D		2	1,2,3	1199920	1587160	7.5+08	9.6−02	B	TS90LS
258.61	2p3d	2s8d	$^3P°$	^3D		1	1,2	1200470	1587160	8.4+08	5.8−02	B	TS90LS
258.79	2p3d	2s8d	$^3P°$	^3D		0	1	1200750	1587160	7.5+08	1.92−02	B	TS90LS
261.10	2p3s	2s6d	$^1P°$	^1D		1	2	1138860	1521860	1.8+08	8.−03	D	TS90
261.6	2p3d	2p5p	$^3D°$	^3D		2	3	1194320	1576600	1.1+07	7.−04	D	TS90LS
261.7	2p3d	2p5p	$^3D°$	^3D		3	3	1194538	1576600	8.5+07	5.3−03	C	TS90LS
265.5	2p3d	2p5p	$^3P°$	^3D		2	3	1199920	1576600	2.40+08	1.55−02	B	TS90LS
300.6	2p3s	2s5d	$^3P°$	^3D		0	1	1120207	1452900	1.7+08	6.9−03	C	TS90LS
300.9	2p3s	2s5d	$^3P°$	^3D		1	2	1120720	1453100	2.32+08	1.56−02	B	TS90LS
301.0	2p3s	2s5d	$^3P°$	^3D		1	1	1120720	1452900	1.3+08	5.2−03	C	TS90LS
301.7	2p3s	2s5d	$^3P°$	^3D		2	3	1121757	1453200	3.1+08	2.9−02	B	TS90LS
301.8	2p3s	2s5d	$^3P°$	^3D		2	2	1121757	1453100	7.7+07	5.2−03	C	TS90LS
302.0	2p3s	2s5d	$^3P°$	^3D		2	1	1121757	1452900	9.+06	3.5−04	D	TS90LS
307.69	2p3s	2p4p	$^3P°$	^3D		2	3	1121757	1446760	6.3+09	6.3−01	B	TS90LS
307.7	2p3s	2p4p	$^3P°$	^3D		1	2	1120720	1445700	4.7+09	3.4−01	B	TS90LS
308.7	2p3s	2p4p	$^3P°$	^3D		2	2	1121757	1445700	1.56+09	1.13−01	B	TS90LS
310.84	2s3s	2s4p	^1S	$^1P°$		0	1	998183.1	1319890	5.8+09	2.6−01	B	TS90
311.11	2p3p	2p4d	^1P	$^1P°$		1	1	1148647	1470080	4.9+09	2.18−01	B	TS90
320.1	2p3p	2s6p	^1S	$^1P°$		0	1	1206150	1518600	4.1+07	2.0−03	D	TS90
324.38	2s3p	2s4d	$^1P°$	^1D		1	2	1025620.6	1333900	1.4+10	1.2+00	C	TS90
326.7	2p3p	2p4d	^3D	$^3D°$		1	1	1154488	1460600	3.0+09	1.57−01	B	TS90LS
326.69	2p3p	2p4d	^3D	$^3D°$		2	3	1155120	1461220	4.5+08	5.4−02	B	TS90LS
327.4	2p3p	2p4d	^3D	$^3D°$		2	1	1155120	1460600	1.01+09	5.2−02	B	TS90LS
327.79	2p3p	2p4d	^3D	$^3D°$		3	3	1156150	1461220	3.6+09	4.3−01	B	TS90LS
333.3	2s3p	2s4d	$^3P°$	^3D		0	1	1028366.7	1328390	9.7+09	5.3−01	B	TS90LS
333.5	2s3p	2s4d	$^3P°$	^3D		1	1	1028499.59	1328390	7.3+09	4.0−01	B	TS90LS

λ_{Ritz} [a] (Å)	Configurations		Terms		Type	J_i [b]	J_k [b]	E_i (cm^{-1})	E_k (cm^{-1})	A_{ki} [c] (s^{-1})	S_{ik} [c]	Acc. [d]	Ref. [e]
333.59	2s3p	2s4d	$^3P^o$	^3D		1	2	1028499.59	1328265	1.31+10	1.20+00	B	TS90LS
333.7	2s3p	2s4d	$^3P^o$	^3D		2	1	1028755.1	1328390	4.8+08	2.7−02	B	TS90LS
333.82	2s3p	2s4d	$^3P^o$	^3D		2	3	1028755.1	1328318	1.74+10	2.24+00	B	TS90LS
333.88	2s3p	2s4d	$^3P^o$	^3D		2	2	1028755.1	1328265	4.3+09	4.0−01	B	TS90LS
345.9	2p3p	2p4d	^3P	$^3D^o$		0	1	1171500	1460600	7.1+09	4.3−01	B	TS90LS
346.18	2p3p	2p4d	^3P	$^3D^o$		2	3	1172350	1461220	1.27+10	1.82+00	B	TS90LS
346.5	2p3p	2p4d	^3P	$^3D^o$		1	1	1172020	1460600	5.3+09	3.3−01	B	TS90LS
346.9	2p3p	2p4d	^3P	$^3D^o$		2	1	1172350	1460600	3.5+08	2.17−02	B	TS90LS
350.94	2p3p	2p4d	^1D	$^1P^o$		2	1	1185130	1470080	3.6+08	2.29−02	B	TS90
355.29	2s3p	2s4s	$^1P^o$	^1S		1	0	1025620.6	1307080	7.4+09	1.6−01	C	TS90
355.345	2s2p	$2p^2$	$^3P^o$	^1S		1	0	111708.0	393125	1.77+05	3.92−06	A	F09
369.27	2s3p	2s4s	$^3P^o$	^3S		0	1	1028366.7	1299170	1.03+09	7.7−02	B	TS90LS
369.45	2s3p	2s4s	$^3P^o$	^3S		1	1	1028499.59	1299170	3.1+09	2.30−01	B	TS90LS
369.80	2s3p	2s4s	$^3P^o$	^3S		2	1	1028755.1	1299170	5.1+09	3.8−01	B	TS90LS
378.89	2p3p	2p4d	^1S	$^1P^o$		0	1	1206150	1470080	8.2+09	6.6−01	C	TS90
386.0	2p3d	2s5d	$^3D^o$	^3D		1	2	1194060	1453100	7.+06	1.0−03	D	TS90LS
386.3	2p3d	2s5d	$^3D^o$	^3D		2	3	1194320	1453200	5.+06	1.0−03	D	TS90LS
386.3	2p3d	2s5d	$^3D^o$	^3D		1	1	1194060	1452900	3.4+07	2.9−03	C	TS90LS
386.4	2p3d	2s5d	$^3D^o$	^3D		2	2	1194320	1453100	3.1+07	4.5−03	C	TS90LS
386.6	2p3d	2s5d	$^3D^o$	^3D		3	3	1194538	1453200	4.0+07	8.0−03	C	TS90LS
386.7	2p3d	2s5d	$^3D^o$	^3D		2	1	1194320	1452900	1.1+07	1.0−03	D	TS90LS
386.8	2p3d	2s5d	$^3D^o$	^3D		3	2	1194538	1453100	7.+06	1.0−03	D	TS90LS
396.13	2p3d	2p4p	$^3D^o$	^3D		2	3	1194320	1446760	2.6+07	5.7−03	C	TS90LS
396.48	2p3d	2p4p	$^3D^o$	^3D		3	3	1194538	1446760	2.09+08	4.5−02	B	TS90LS
397.4	2p3d	2p4p	$^3D^o$	^3D		1	2	1194060	1445700	3.5+07	5.4−03	C	TS90LS
397.8	2p3d	2p4p	$^3D^o$	^3D		2	2	1194320	1445700	1.62+08	2.5−02	B	TS90LS
398.1	2p3d	2p4p	$^3D^o$	^3D		3	2	1194538	1445700	3.6+07	5.7−03	C	TS90LS
403.26	2s3d	2s4p	^1D	$^1P^o$		2	1	1071914	1319890	1.69+09	1.64−01	B	TS90
405.12	2p3d	2p4p	$^3P^o$	^3D		2	3	1199920	1446760	5.1+08	1.18−01	B	TS90LS

λ_{Ritz} [a] (Å)	Configurations		Terms		Type	J_i [b]	J_k [b]	E_i (cm^{-1})	E_k (cm^{-1})	A_{ki} [c] (s^{-1})	S_{ik} [c]	Acc. [d]	Ref. [e]
406.9	2p3d	2p4p	$^3P°$	^3D		2	2	1199920	1445700	1.26+08	2.10−02	B	TS90LS
407.8	2p3d	2p4p	$^3P°$	^3D		1	2	1200470	1445700	3.8+08	6.3−02	B	TS90LS
412.9	2s4p	2s7d	$^1P°$	^1D		1	2	1319890	1562100	1.24+09	2.15−01	B	TS90
449.5	2s3s	2p3d	^3S	$^3P°$		1	0	978300.29	1200750	6.8+07	3.0−03	C	TS90LS
450.11	2s3s	2p3d	^3S	$^3P°$		1	1	978300.29	1200470	6.8+07	9.1−03	C	TS90LS
451.22	2s3s	2p3d	^3S	$^3P°$		1	2	978300.29	1199920	6.7+07	1.52−02	B	TS90LS
454.04	2s3s	2p3d	^1S	$^1P°$		0	1	998183.1	1218430	3.5+07	4.8−03	C	TS90
465.221	$2s^2$	2s2p	^1S	$^1P°$		0	1	0.0	214951.6	3.93+09	5.86−01	A	F09
472.8	2s4s	2s6p	^1S	$^1P°$		0	1	1307080	1518600	8.6+08	1.35−01	B	TS90
485.468	2s2p	$2p^2$	$^3P°$	^1D		1	2	111708.0	317694.6	1.59+05	4.49−05	A	F09
487.823	2s2p	$2p^2$	$^3P°$	^1D		2	2	112702.4	317694.6	2.37+06	6.80−04	A	F09
495.1	2s4p	2s6d	$^1P°$	^1D		1	2	1319890	1521860	1.91+09	5.7−01	B	TS90
512.72	2p3s	2s4d	$^1P°$	^1D		1	2	1138860	1333900	1.1+09	3.7−01	C	TS90
541.4	2s4d	2s6p	^1D	$^1P°$		2	1	1333900	1518600	3.7+08	8.6−02	B	TS90
558.609	2s2p	$2p^2$	$^3P°$	^3P		1	2	111708.0	290724	7.93+08	3.41−01	A	F09
558.8	2p3s	2s4s	$^3P°$	^3S		0	1	1120207	1299170	4.8+06	1.2−03	C	TS90LS
559.947	2s2p	$2p^2$	$^3P°$	^3P		0	1	111252.8	289841.1	1.05+09	2.73−01	A	F09
560.4	2p3s	2s4s	$^3P°$	^3S		1	1	1120720	1299170	1.4+07	3.7−03	C	TS90LS
561.251	2s2p	$2p^2$	$^1P°$	^1S		1	0	214951.6	393125	5.58+09	4.87−01	A	F09
561.378	2s2p	$2p^2$	$^3P°$	^3P		1	1	111708.0	289841.1	7.80+08	2.04−01	A	F09
561.730	2s2p	$2p^2$	$^3P°$	^3P		2	2	112702.4	290724	2.33+09	1.02+00	A	F09
562.992	2s2p	$2p^2$	$^3P°$	^3P		1	0	111708.0	289330.4	3.09+09	2.72−01	A	F09
563.66	2p3s	2s4s	$^3P°$	^3S		2	1	1121757	1299170	2.3+07	6.2−03	C	TS90LS
564.529	2s2p	$2p^2$	$^3P°$	^3P		2	1	112702.4	289841.1	1.28+09	3.40−01	A	F09
583.97	2p3p	2s4p	^1P	$^1P°$		1	1	1148647	1319890	6.1+06	1.8−03	C	TS90
594.5	2p3s	2s4s	$^1P°$	^1S		1	0	1138860	1307080	9.9+07	1.0−02	C	TS90
613.5	2s4s	2p4d	^1S	$^1P°$		0	1	1307080	1470080	3.3+08	1.14−01	B	TS90
622.8	2s3s	2p3s	^3S	$^1P°$		1	1	978300.29	1138860	1.17+06	4.19−04	A	F09
626.9	2s3p	2p3p	$^1P°$	^1D		1	2	1025620.6	1185130	1.3+08	8.2−02	C	TS90

λ_{Ritz} [a] (Å)	Configurations		Terms		Type	J_i [b]	J_k [b]	E_i (cm^{-1})	E_k (cm^{-1})	A_{ki} [c] (s^{-1})	S_{ik} [c]	Acc. [d]	Ref. [e]
682.5	2s3d	2p3d	^{1}D	$^{1}P^{\circ}$		2	1	1071914	1218430	1.11+09	5.2−01	B	TS90
682.9	2s3d	2p3d	^{3}D	$^{3}P^{\circ}$		1	0	1054310.0	1200750	1.06+09	1.67−01	B	TS90LS
684.2	2s3d	2p3d	^{3}D	$^{3}P^{\circ}$		1	1	1054310.0	1200470	2.6+08	1.25−01	B	TS90LS
684.4	2s3d	2p3d	^{3}D	$^{3}P^{\circ}$		2	1	1054354.4	1200470	7.9+08	3.7−01	B	TS90LS
686.8	2s3d	2p3d	^{3}D	$^{3}P^{\circ}$		1	2	1054310.0	1199920	1.0+07	8.3−03	C	TS90LS
687.0	2s3d	2p3d	^{3}D	$^{3}P^{\circ}$		2	2	1054354.4	1199920	1.56+08	1.25−01	B	TS90LS
687.3	2s3d	2p3d	^{3}D	$^{3}P^{\circ}$		3	2	1054428.3	1199920	8.7+08	7.0−01	B	TS90LS
695.2	2s3p	2p3p	$^{3}P^{\circ}$	^{3}P		1	2	1028499.59	1172350	2.02+08	1.68−01	B	TS90LS
696.1	2s3p	2p3p	$^{3}P^{\circ}$	^{3}P		0	1	1028366.7	1172020	2.7+08	1.34−01	B	TS90LS
696.4	2s3p	2p3p	$^{3}P^{\circ}$	^{3}P		2	2	1028755.1	1172350	6.0+08	5.0−01	B	TS90LS
696.8	2s3p	2p3p	$^{3}P^{\circ}$	^{3}P		1	1	1028499.59	1172020	2.01+08	1.01−01	B	TS90LS
697.1	2s3s	2p3s	^{3}S	$^{3}P^{\circ}$		1	2	978300.29	1121757	6.9+08	5.8−01	B	F09
698.0	2s3p	2p3p	$^{3}P^{\circ}$	^{3}P		2	1	1028755.1	1172020	3.3+08	1.68−01	B	TS90LS
699.30	2s3p	2p3p	$^{3}P^{\circ}$	^{3}P		1	0	1028499.59	1171500	7.9+08	1.34−01	B	TS90LS
702.2	2s3s	2p3s	^{3}S	$^{3}P^{\circ}$		1	1	978300.29	1120720	6.7+08	3.5−01	B	F09
704.7	2s3s	2p3s	^{3}S	$^{3}P^{\circ}$		1	0	978300.29	1120207	6.7+08	1.15−01	B	F09
710.8	2s3s	2p3s	^{1}S	$^{1}P^{\circ}$		0	1	998183.1	1138860	1.4+09	8.−01	D	F09
713.4	2s3d	2p3d	^{3}D	$^{3}D^{\circ}$		2	3	1054354.4	1194538	1.01+08	1.27−01	B	TS90LS
713.5	2s3d	2p3d	^{1}D	$^{1}F^{\circ}$		2	3	1071914	1212060	1.02+08	1.28−01	B	TS90
713.7	2s3d	2p3d	^{3}D	$^{3}D^{\circ}$		3	3	1054428.3	1194538	8.0+08	1.01+00	B	TS90LS
714.2	2s3d	2p3d	^{3}D	$^{3}D^{\circ}$		1	2	1054310.0	1194320	1.35+08	1.22−01	B	TS90LS
714.5	2s3d	2p3d	^{3}D	$^{3}D^{\circ}$		2	2	1054354.4	1194320	6.3+08	5.6−01	B	TS90LS
714.8	2s3d	2p3d	^{3}D	$^{3}D^{\circ}$		3	2	1054428.3	1194320	1.40+08	1.27−01	B	TS90LS
715.6	2s3d	2p3d	^{3}D	$^{3}D^{\circ}$		1	1	1054310.0	1194060	6.7+08	3.7−01	B	TS90LS
715.8	2s3d	2p3d	^{3}D	$^{3}D^{\circ}$		2	1	1054354.4	1194060	2.24+08	1.22−01	B	TS90LS
729.1	2s3p	2p3p	$^{3}P^{\circ}$	^{3}S		0	1	1028366.7	1165520	1.18+08	6.7−02	B	TS90LS
729.8	2s3p	2p3p	$^{3}P^{\circ}$	^{3}S		1	1	1028499.59	1165520	3.5+08	2.02−01	B	TS90LS
731.2	2s3p	2p3p	$^{3}P^{\circ}$	^{3}S		2	1	1028755.1	1165520	5.8+08	3.4−01	B	TS90LS
734.3	2s4d	2p4d	^{1}D	$^{1}P^{\circ}$		2	1	1333900	1470080	5.7+08	3.3−01	B	TS90

λ_{Ritz} [a] (Å)	Configurations		Terms		Type	J_i [b]	J_k [b]	E_i (cm^{-1})	E_k (cm^{-1})	A_{ki} [c] (s^{-1})	S_{ik} [c]	Acc. [d]	Ref. [e]
742.1	2p3p	2s4p	^1D	$^1P°$		2	1	1185130	1319890	8.3+07	5.0−02	B	TS90
752.1	2s4d	2p4d	^3D	$^3D°$		2	3	1328265	1461220	7.8+07	1.14−01	B	TS90LS
752.4	2s4d	2p4d	^3D	$^3D°$		3	3	1328318	1461220	6.2+08	9.1−01	B	TS90LS
755.66	2s4d	2p4d	^3D	$^3D°$		2	1	1328265	1460600	1.72+08	1.10−01	B	TS90LS
756	2s4d	2p4d	^3D	$^3D°$		1	1	1328390	1460600	5.1+08	3.3−01	B	TS90LS
778.4	2p3d	2s4d	$^3P°$	^3D		2	1	1199920	1328390	2.6+06	1.8−03	C	TS90LS
778.8	2p3d	2s4d	$^3P°$	^3D		2	3	1199920	1328318	9.4+07	1.53−01	B	TS90LS
779.1	2p3d	2s4d	$^3P°$	^3D		2	2	1199920	1328265	2.34+07	2.7−02	B	TS90LS
781.7	2p3d	2s4d	$^3P°$	^3D		1	1	1200470	1328390	3.9+07	2.7−02	B	TS90LS
782.5	2p3d	2s4d	$^3P°$	^3D		1	2	1200470	1328265	6.9+07	8.2−02	B	TS90LS
783.45	2p3d	2s4d	$^3P°$	^3D		0	1	1200750	1328390	5.1+07	3.6−02	B	TS90LS
785.0	2s3p	2p3p	$^3P°$	^3D		2	3	1028755.1	1156150	6.4+08	1.07+00	B	TS90LS
789.8	2s3p	2p3p	$^3P°$	^3D		1	2	1028499.59	1155120	4.7+08	5.7−01	B	TS90LS
790	2p4d	2s8d	$^3D°$	^3D		1	1,2	1460600	1587160	4.2+07	8.2−02	B	TS90LS
791.4	2s3p	2p3p	$^3P°$	^3D		2	2	1028755.1	1155120	1.57+08	1.92−01	B	TS90LS
792.9	2s3p	2p3p	$^3P°$	^3D		0	1	1028366.7	1154488	3.5+08	2.6−01	B	TS90LS
793.3	2p4d	2s8d	$^3D°$	^3D		2	1,2,3	1461110	1587160	3.7+07	1.4−01	C	TS90LS
793.7	2s3p	2p3p	$^3P°$	^3D		1	1	1028499.59	1154488	2.6+08	1.92−01	B	TS90LS
794.0	2p4d	2s8d	$^3D°$	^3D		3	2,3	1461220	1587160	6.4+07	1.90−01	B	TS90LS
795.3	2s3p	2p3p	$^3P°$	^3D		2	1	1028755.1	1154488	1.71+07	1.28−02	B	TS90LS
808.0	2p4d	2s8d	$^3P°$	^3D		1	1,2	1463390	1587160	5.4+08	1.1+00	C	TS90LS
812.8	2s3p	2p3p	$^1P°$	^1P		1	1	1025620.6	1148647	1.2+09	9.6−01	C	TS90
813.1	2p4d	2s8d	$^3P°$	^3D		0	1	1464180	1587160	4.7+08	3.7−01	C	TS90LS
816.1	2s3s	2p3s	^1S	$^3P°$		0	1	998183.1	1120720	1.4+06	1.1−03	C	F09
820.7	2p3d	2s4d	$^1F°$	^1D		3	2	1212060	1333900	4.3+07	5.9−02	B	TS90
866	2p4d	2p5p	$^3D°$	^3D		2	3	1461110	1576600	1.2+07	2.8−02	C	TS90LS
866.0	2p3d	2s4d	$^1P°$	^1D		1	2	1218430	1333900	9.3+07	1.49−01	B	TS90
867	2p4d	2p5p	$^3D°$	^3D		3	3	1461220	1576600	9.8+07	2.19−01	B	TS90LS
879.2	2p3p	2s4p	^1S	$^1P°$		0	1	1206150	1319890	9.9+07	1.0−01	C	TS90

λ_{Ritz} [a] (Å)	Configurations		Terms		Type	J_i [b]	J_k [b]	E_i (cm^{-1})	E_k (cm^{-1})	A_{ki} [c] (s^{-1})	S_{ik} [c]	Acc.[d]	Ref.[e]
887.29	$2s^2$	2s2p	^1S	$^3P°$	M2	0	2	0.0	112702.4	5.75−02	1.06+01	A	F09
895.191	$2s^2$	2s2p	^1S	$^3P°$		0	1	0.0	111708.0	1.8+04	1.9−05	C	F09
973.302	2s2p	$2p^2$	$^1P°$	^1D		1	2	214951.6	317694.6	5.6+08	1.27+00	B	F09
1007.6	2p3d	2s4s	$^3P°$	^3S		2	1	1199920	1299170	9.8+06	1.49−02	B	TS90LS
1013.2	2p3d	2s4s	$^3P°$	^3S		1	1	1200470	1299170	5.8+06	8.9−03	C	TS90LS
1016.1	2p3d	2s4s	$^3P°$	^3S		0	1	1200750	1299170	1.9+06	3.0−03	C	TS90LS
1025	2p4d	2s7d	$^3P°$	^3D		1	1,2	1463390	1560975	1.7+07	7.4−02	C	TS90LS
1033	2p4d	2s7d	$^3P°$	^3D		0	1	1464180	1560975	1.5+07	2.5−02	C	TS90LS
1087	2p4d	2s7d	$^1P°$	^1D		1	2	1470080	1562100	2.6+07	8.3−02	B	TS90
1319.74	2s2p	$2p^2$	$^1P°$	^3P		1	2	214951.6	290724	1.1+05	6.0−04	C	F09
1335.30	2s2p	$2p^2$	$^1P°$	^3P		1	1	214951.6	289841.1	2.4+03	8.5−06	C	F09
1344.47	2s2p	$2p^2$	$^1P°$	^3P		1	0	214951.6	289330.4	3.2+04	3.8−05	D	F09
1433.0	2p3p	2p3d	^1P	$^1P°$		1	1	1148647	1218430	2.9+08	1.27+00	B	TS90
1482.6	2s3d	2p3s	^3D	$^3P°$		1	2	1054310.0	1121757	9.0+04	7.26−04	A	F09
1483.6	2s3d	2p3s	^3D	$^3P°$		2	2	1054354.4	1121757	1.33+06	1.07−02	B	F09
1485.3	2s3d	2p3s	^3D	$^3P°$		3	2	1054428.3	1121757	7.3+06	5.9−02	B	F09
1505.8	2s3d	2p3s	^3D	$^3P°$		1	1	1054310.0	1120720	2.15+06	1.09−02	B	F09
1506.8	2s3d	2p3s	^3D	$^3P°$		2	1	1054354.4	1120720	6.3+06	3.2−02	B	F09
1518	2s3d	2p3s	^3D	$^3P°$		1	0	1054310.0	1120207	8.4+06	1.45−02	B	F09
1755	2p4d	2s6d	$^3P°$	^3D		1	2	1463390	1520360	5.7+07	7.7−01	C	TS90LS
1763	2p4d	2s6d	$^3P°$	^3D		1	1	1463390	1520100	3.1+07	2.6−01	C	TS90LS
1788	2p4d	2s6d	$^3P°$	^3D		0	1	1464180	1520100	4.0+07	3.4−01	C	TS90LS
1930	2p3s	2p3p	$^3P°$	^3P		0	1	1120207	1172020	9.6+07	1.02+00	B	TS90LS
1931	2p4d	2s6d	$^1P°$	^1D		1	2	1470080	1521860	5.1+07	9.1−01	B	TS90
1937	2p3s	2p3p	$^3P°$	^3P		1	2	1120720	1172350	7.1+07	1.27+00	B	TS90LS
1949	2p3s	2p3p	$^3P°$	^3P		1	1	1120720	1172020	7.0+07	7.6−01	B	TS90LS
1969	2p3s	2p3p	$^3P°$	^3P		1	0	1120720	1171500	2.7+08	1.02+00	B	TS90LS
1977	2p3s	2p3p	$^3P°$	^3P		2	2	1121757	1172350	2.01+08	3.8+00	B	TS90LS
1981.972	2s3s	2s3p	^3S	$^3P°$		1	2	978300.29	1028755.1	2.10+08	4.0+00	B	F09

λ_{Ritz} [a] (Å)	Configurations		Terms		Type	J_i [b]	J_k [b]	E_i (cm^{-1})	E_k (cm^{-1})	A_{ki} [c] (s^{-1})	S_{ik} [c]	Acc. [d]	Ref. [e]
1990	2p3s	2p3p	$^3P°$	^3P		2	1	1121757	1172020	1.09+08	1.27+00	B	TS90LS
1992.060	2s3s	2s3p	^3S	$^3P°$		1	1	978300.29	1028499.59	2.06+08	2.41+00	A	F09
1997.347	2s3s	2s3p	^3S	$^3P°$		1	0	978300.29	1028366.7	2.05+08	8.07−01	A	F09
2113	2s3s	2s3p	^3S	$^1P°$		1	1	978300.29	1025620.6	9.9+05	1.39−02	A	F09
2159	2s3p	2s3d	$^1P°$	^1D		1	2	1025620.6	1071914	1.39+08	3.45+00	A	F09
2161	2p3s	2p3p	$^1P°$	^1D		1	2	1138860	1185130	2.2+08	5.6+00	C	TS90
2161	2p3p	2p3d	^3D	$^3P°$		1	0	1154488	1200750	3.0+07	1.48−01	B	TS90LS
2174	2p3p	2p3d	^3D	$^3P°$		1	1	1154488	1200470	7.3+06	1.11−01	B	TS90LS
2200	2p3p	2p3d	^3D	$^3P°$		1	2	1154488	1199920	2.8+05	7.4−03	C	TS90LS
2204	2p3p	2p3d	^3D	$^3P°$		2	1	1155120	1200470	2.09+07	3.3−01	B	TS90LS
2206	2p3s	2p3p	$^3P°$	^3S		0	1	1120207	1165520	1.85+07	2.9−01	B	TS90LS
2231	2p3p	2p3d	^3D	$^3P°$		2	2	1155120	1199920	4.0+06	1.11−01	B	TS90LS
2231	2p3s	2p3p	$^3P°$	^3S		1	1	1120720	1165520	5.4+07	8.8−01	B	TS90LS
2284	2p3p	2p3d	^3D	$^3P°$		3	2	1156150	1199920	2.11+07	6.2−01	B	TS90LS
2284	2p3s	2p3p	$^3P°$	^3S		2	1	1121757	1165520	8.3+07	1.47+00	B	TS90LS
2300	2s6p	2s7d	$^1P°$	^1D		1	2	1518600	1562100	3.0+08	9.0+00	B	TS90
2303	2s3p	2s3d	$^3P°$	^1D		1	2	1028499.59	1071914	6.9+05	2.08−02	B	F09
2316	2s3p	2s3d	$^3P°$	^1D		2	2	1028755.1	1071914	1.5+02	4.−06	E	F09
2510	2p3p	2p3d	^3D	$^3D°$		1	2	1154488	1194320	3.2+06	1.26−01	B	TS90LS
2526	2p3p	2p3d	^3D	$^3D°$		1	1	1154488	1194060	1.59+07	3.8−01	B	TS90LS
2536	2p3p	2p3d	^3D	$^3D°$		2	3	1155120	1194538	2.33+06	1.31−01	B	TS90LS
2550	2p3p	2p3d	^3D	$^3D°$		2	2	1155120	1194320	1.43+07	5.9−01	B	TS90LS
2567	2p3p	2p3d	^3D	$^3D°$		2	1	1155120	1194060	5.0+06	1.26−01	B	TS90LS
2604	2p3p	2p3d	^3D	$^3D°$		3	3	1156150	1194538	1.72+07	1.05+00	B	TS90LS
2619	2p3p	2p3d	^3D	$^3D°$		3	2	1156150	1194320	3.0+06	1.31−01	B	TS90LS
2838	2p3p	2p3d	^3S	$^3P°$		1	0	1165520	1200750	5.4+07	6.1−01	B	TS90LS
2860	2p3p	2p3d	^3S	$^3P°$		1	1	1165520	1200470	5.2+07	1.82+00	B	TS90LS
2906	2p3s	2p3p	$^3P°$	^3D		1	2	1120720	1155120	5.6+07	3.4+00	B	TS90LS
2906	2p3p	2p3d	^3S	$^3P°$		1	2	1165520	1199920	5.0+07	3.0+00	B	TS90LS

λ_{Ritz} [a] (Å)	Configurations		Terms		Type	J_i [b]	J_k [b]	E_i (cm^{-1})	E_k (cm^{-1})	A_{ki} [c] (s^{-1})	S_{ik} [c]	Acc. [d]	Ref. [e]
2907	2p3s	2p3p	$^3P^o$	^3D		2	3	1121757	1156150	7.5+07	6.3+00	B	TS90LS
2916	2p3s	2p3p	$^3P^o$	^3D		0	1	1120207	1154488	4.1+07	1.51+00	B	TS90LS
2961	2p3s	2p3p	$^3P^o$	^3D		1	1	1120720	1154488	2.9+07	1.13+00	B	TS90LS
2996	2p3s	2p3p	$^3P^o$	^3D		2	2	1121757	1155120	1.70+07	1.13+00	B	TS90LS
3002	2p3p	2p3d	^1D	$^1P^o$		2	1	1185130	1218430	2.7+06	1.07−01	B	TS90
3054	2p3s	2p3p	$^3P^o$	^3D		2	1	1121757	1154488	1.79+06	7.5−02	B	TS90LS
3298	2s3s	2s3p	^1S	$^3P^o$		0	1	998183.1	1028499.59	2.02+05	1.07−02	A	F09
3450	2p3p	2p3d	^3P	$^3P^o$		0	1	1171500	1200470	4.0+06	2.44−01	B	TS90LS
3479	2s3p	2s3d	$^1P^o$	^3D		1	2	1025620.6	1054354.4	1.4+05	1.5−02	C	F09
3480	2p3p	2p3d	^3P	$^3P^o$		1	0	1172020	1200750	1.17+07	2.44−01	B	TS90LS
3485	2s3p	2s3d	$^1P^o$	^3D		1	1	1025620.6	1054310.0	7.16+04	4.49−03	A	F09
3514	2p3p	2p3d	^3P	$^3P^o$		1	1	1172020	1200470	2.8+06	1.83−01	B	TS90LS
3555	2p3p	2p3d	^3P	$^3P^o$		2	1	1172350	1200470	4.6+06	3.1−01	B	TS90LS
3583	2p3p	2p3d	^3P	$^3P^o$		1	2	1172020	1199920	2.7+06	3.1−01	B	TS90LS
3626	2p3p	2p3d	^3P	$^3P^o$		2	2	1172350	1199920	7.8+06	9.2−01	B	TS90LS
3644	2s3s	2s3p	^1S	$^1P^o$		0	1	998183.1	1025620.6	2.6+07	1.88+00	B	F09
3712	2p3p	2p3d	^1D	$^1F^o$		2	3	1185130	1212060	3.5+07	6.2+00	B	TS90
3853	2s3p	2s3d	$^3P^o$	^3D		0	1	1028366.7	1054310.0	1.29+07	1.09+00	A	F09
3867	2s3p	2s3d	$^3P^o$	^3D		1	2	1028499.59	1054354.4	1.71+07	2.44+00	A	F09
3873	2s3p	2s3d	$^3P^o$	^3D		1	1	1028499.59	1054310.0	9.4+06	8.13−01	A	F09
3894	2s3p	2s3d	$^3P^o$	^3D		2	3	1028755.1	1054428.3	2.25+07	4.59+00	A	F09
3905	2s3p	2s3d	$^3P^o$	^3D		2	2	1028755.1	1054354.4	5.57+06	8.19−01	A	F09
3912	2s3p	2s3d	$^3P^o$	^3D		2	1	1028755.1	1054310.0	6.15+05	5.46−02	A	F09
4430	2p3p	2p3d	^3P	$^3D^o$		0	1	1171500	1194060	7.8+06	1.01+00	B	TS90LS
4483	2p3p	2p3d	^3P	$^3D^o$		1	2	1172020	1194320	1.02+07	2.26+00	B	TS90LS
4506	2p3p	2p3d	^3P	$^3D^o$		2	3	1172350	1194538	1.34+07	4.2+00	B	TS90LS
4540	2p3p	2p3d	^3P	$^3D^o$		1	1	1172020	1194060	5.5+06	7.5−01	B	TS90LS
4550	2p3p	2p3d	^3P	$^3D^o$		2	2	1172350	1194320	3.2+06	7.5−01	B	TS90LS
4600	2p3p	2p3d	^3P	$^3D^o$		2	1	1172350	1194060	3.5+05	5.0−02	B	TS90LS

λ_{Ritz} [a] (Å)	Configurations		Terms		Type	J_i [b]	J_k [b]	E_i (cm^{-1})	E_k (cm^{-1})	A_{ki} [c] (s^{-1})	S_{ik} [c]	Acc. [d]	Ref. [e]
6440	2p4p	2p4d	^3D	$^3D°$		2	3	1445700	1461220	6.7+05	6.2−01	B	TS90LS
6710	2p4p	2p4d	^3D	$^3D°$		2	1	1445700	1460600	1.34+06	6.0−01	B	TS90LS
6910	2p4p	2p4d	^3D	$^3D°$		3	3	1446760	1461220	4.3+06	5.0+00	B	TS90LS
7140	2s4p	2s4d	$^1P°$	^1D		1	2	1319890	1333900	1.86+07	1.67+01	B	TS90
7800	2s4s	2s4p	^1S	$^1P°$		0	1	1307080	1319890	1.27+07	9.0+00	B	TS90
8140	2p3p	2p3d	^1S	$^1P°$		0	1	1206150	1218430	1.9+06	1.5+00	C	TS90
10210	2p3s	2p3p	$^1P°$	^1P		1	1	1138860	1148647	1.3+06	2.0+00	C	TS90
12300	2s5d	2p4d	^3D	$^3D°$		2	3	1453100	1461220	4.9+03	3.2−02	B	TS90LS
12500	2s5d	2p4d	^3D	$^3D°$		3	3	1453200	1461220	3.8+04	2.5−01	B	TS90LS
13000	2s5d	2p4d	^3D	$^3D°$		1	1	1452900	1460600	2.8+04	9.2−02	B	TS90LS
13300	2s5d	2p4d	^3D	$^3D°$		2	1	1453100	1460600	8.7+03	3.1−02	B	TS90LS
19800	2p3d	2p3p	$^1D°$	^1D		2	2	1180070	1185130	4.3+04	8.2−01	C	TS90
31000	2s6p	2s6d	$^1P°$	^1D		1	2	1518600	1521860	1.46+06	1.04+02	B	TS90
36000	2p5d	2s8d	$^3D°$	^3D		3	2,3	1584400	1587160	2.8+03	7.9−01	C	TS90LS
100560	2s2p	2s2p	$^3P°$	$^3P°$	M1	1	2	111708.0	112702.4	1.33−02	2.50+00	B	F09
Ne VIII (Li-like)													
53.8251	$1s^22s$	$1s^210p$	^2S	$^2P°$		1/2	1/2,3/2	0.0	1857869	2.56+09	1.18−03	A	N02
54.3105	$1s^22s$	$1s^29p$	^2S	$^2P°$		1/2	1/2,3/2	0.0	1841265	3.52+09	1.67−03	A	N02
55.0043	$1s^22s$	$1s^28p$	^2S	$^2P°$		1/2	3/2	0.0	1818040.8	4.99+09	1.64−03	A	N02
55.0050	$1s^22s$	$1s^28p$	^2S	$^2P°$		1/2	1/2	0.0	1818014.9	5.01+09	8.23−04	A	N02
56.0502	$1s^22s$	$1s^27p$	^2S	$^2P°$		1/2	3/2	0.0	1784116	7.44+09	2.59−03	A	N02
56.0514	$1s^22s$	$1s^27p$	^2S	$^2P°$		1/2	1/2	0.0	1784078	7.47+09	1.30−03	A	N02
57.7445	$1s^22s$	$1s^26p$	^2S	$^2P°$		1/2	3/2	0.0	1731768	1.18+10	4.48−03	A	N02
57.7465	$1s^22s$	$1s^26p$	^2S	$^2P°$		1/2	1/2	0.0	1731708	1.18+10	2.25−03	A	N02
57.8023	$1s^22p$	$1s^210d$	$^2P°$	^2D		1/2	3/2	128151.9	1858187	3.84+09	1.47−03	A	N02LS
57.8426	$1s^22p$	$1s^210s$	$^2P°$	^2S		1/2	1/2	128151.9	1856983	3.51+08	6.70−05	A	N02
57.8575	$1s^22p$	$1s^210d$	$^2P°$	^2D		3/2	3/2,5/2	129801.2	1858187	3.07+09	2.93−03	A	N02LS
57.8978	$1s^22p$	$1s^210s$	$^2P°$	^2S		3/2	1/2	129801.2	1856983	7.06+08	1.35−04	A	N02

λ_{Ritz} [a] (Å)	Configurations		Terms		Type	J_i [b]	J_k [b]	E_i (cm^{-1})	E_k (cm^{-1})	A_{ki} [c] (s^{-1})	S_{ik} [c]	Acc. [d]	Ref. [e]
58.3584	$1s^22p$	$1s^29d$	$^2P^o$	^2D		1/2	3/2	128151.9	1841702	5.32+09	2.09−03	A	N02LS
58.4146	$1s^22p$	$1s^29d$	$^2P^o$	^2D		3/2	3/2,5/2	129801.2	1841702	4.25+09	4.18−03	A	N02LS
58.4148	$1s^22p$	$1s^29s$	$^2P^o$	^2S		1/2	1/2	128151.9	1840047.0	4.85+08	9.5−05	A	N02
58.4711	$1s^22p$	$1s^29s$	$^2P^o$	^2S		3/2	1/2	129801.2	1840047.0	9.8+08	1.93−04	A	N02
59.1542	$1s^22p$	$1s^28d$	$^2P^o$	^2D		1/2	3/2	128151.9	1818650	7.68+09	3.14−03	A	N02
59.2117	$1s^22p$	$1s^28d$	$^2P^o$	^2D		3/2	5/2	129801.2	1818658	9.2+09	5.65−03	A	N02
59.2119	$1s^22p$	$1s^28d$	$^2P^o$	^2D		3/2	3/2	129801.2	1818650	1.53+09	6.27−04	A	N02
59.2368	$1s^22p$	$1s^28s$	$^2P^o$	^2S		1/2	1/2	128151.9	1816292	6.96+08	1.43−04	A	N02
59.2947	$1s^22p$	$1s^28s$	$^2P^o$	^2S		3/2	1/2	129801.2	1816292	1.40+09	2.89−04	A	N02
60.3545	$1s^22p$	$1s^27d$	$^2P^o$	^2D		1/2	3/2	128151.9	1785029	1.17+10	5.08−03	A	N02
60.4142	$1s^22p$	$1s^27d$	$^2P^o$	^2D		3/2	5/2	129801.2	1785041	1.40+10	9.1−03	A	N02
60.4146	$1s^22p$	$1s^27d$	$^2P^o$	^2D		3/2	3/2	129801.2	1785029	2.33+09	1.01−03	A	N02
60.4835	$1s^22p$	$1s^27s$	$^2P^o$	^2S		1/2	1/2	128151.9	1781495	1.05+09	2.30−04	A	N02
60.5439	$1s^22p$	$1s^27s$	$^2P^o$	^2S		3/2	1/2	129801.2	1781495	2.12+09	4.64−04	A	N02
60.7043	$1s^22s$	$1s^25d$	^2S	^2D	E2	1/2	5/2	0.0	1647330	1.16+06	5.1−03	B	A10
60.7055	$1s^22s$	$1s^25d$	^2S	^2D	E2	1/2	3/2	0.0	1647298	1.16+06	3.4−03	B	A10
60.7988	$1s^22s$	$1s^25p$	^2S	$^2P^o$		1/2	3/2	0.0	1644768	2.02+10	9.0−03	A	N02
60.8027	$1s^22s$	$1s^25p$	^2S	$^2P^o$		1/2	1/2	0.0	1644664	2.03+10	4.51−03	A	N02
62.3025	$1s^22p$	$1s^26d$	$^2P^o$	^2D		1/2	3/2	128151.9	1733224	1.91+10	9.1−03	A	N02
62.3659	$1s^22p$	$1s^26d$	$^2P^o$	^2D		3/2	5/2	129801.2	1733242	2.29+10	1.65−02	A	N02
62.3666	$1s^22p$	$1s^26d$	$^2P^o$	^2D		3/2	3/2	129801.2	1733224	3.81+09	1.83−03	A	N02
62.5224	$1s^22p$	$1s^26s$	$^2P^o$	^2S		1/2	1/2	128151.9	1727579	1.70+09	4.10−04	A	N02
62.5869	$1s^22p$	$1s^26s$	$^2P^o$	^2S		3/2	1/2	129801.2	1727579	3.42+09	8.28−04	A	N02
65.8185	$1s^22p$	$1s^25f$	$^2P^o$	$^2F^o$	E2	1/2	5/2	128151.9	1647480.7	8.5+06	5.6−02	B	A10
65.8265	$1s^22p$	$1s^25d$	$^2P^o$	^2D		1/2	3/2	128151.9	1647298	3.49+10	1.96−02	A	N02
65.8894	$1s^22p$	$1s^25f$	$^2P^o$	$^2F^o$	E2	3/2	7/2	129801.2	1647496.7	1.09+07	9.7−02	B	A10
65.8901	$1s^22p$	$1s^25f$	$^2P^o$	$^2F^o$	E2	3/2	5/2	129801.2	1647480.7	2.43+06	1.62−02	B	A10
65.8966	$1s^22p$	$1s^25d$	$^2P^o$	^2D		3/2	5/2	129801.2	1647330	4.17+10	3.54−02	A	N02
65.8980	$1s^22p$	$1s^25d$	$^2P^o$	^2D		3/2	3/2	129801.2	1647298	6.95+09	3.93−03	A	N02

λ_{Ritz} [a] (Å)	Configurations		Terms		Type	J_i [b]	J_k [b]	E_i (cm^{-1})	E_k (cm^{-1})	A_{ki} [c] (s^{-1})	S_{ik} [c]	Acc.[d]	Ref.[e]
66.0126	$1s^22p$	$1s^25p$	$^2P^o$	$^2P^o$	E2	3/2	1/2	129801.2	1644664	1.46+06	3.3−03	B	A10
66.2555	$1s^22p$	$1s^25s$	$^2P^o$	^2S		1/2	1/2	128151.9	1637460	3.02+09	8.7−04	A	N02
66.3280	$1s^22p$	$1s^25s$	$^2P^o$	^2S		3/2	1/2	129801.2	1637460	6.08+09	1.75−03	A	N02
67.1517	$1s^22s$	$1s^24d$	^2S	^2D	E2	1/2	5/2	0.0	1489166	3.7+06	2.7−02	B	A10
67.1545	$1s^22s$	$1s^24d$	^2S	^2D	E2	1/2	3/2	0.0	1489104	3.7+06	1.79−02	B	A10
67.3801	$1s^22s$	$1s^24p$	^2S	$^2P^o$		1/2	3/2	0.0	1484117	3.91+10	2.36−02	A	N02
67.3894	$1s^22s$	$1s^24p$	^2S	$^2P^o$		1/2	1/2	0.0	1483912	3.93+10	1.19−02	A	N02
73.4601	$1s^22p$	$1s^24f$	$^2P^o$	$^2F^o$	E2	1/2	5/2	128151.9	1489436	1.26+07	1.45−01	B	A10
73.4780	$1s^22p$	$1s^24d$	$^2P^o$	^2D		1/2	3/2	128151.9	1489104	7.54+10	5.91−02	A	N02
73.5475	$1s^22p$	$1s^24f$	$^2P^o$	$^2F^o$	E2	3/2	7/2	129801.2	1489467	1.62+07	2.49−01	B	A10
73.5638	$1s^22p$	$1s^24d$	$^2P^o$	^2D		3/2	5/2	129801.2	1489166	9.0+10	1.07−01	A	N02
73.5671	$1s^22p$	$1s^24d$	$^2P^o$	^2D		3/2	3/2	129801.2	1489104	1.50+10	1.18−02	A	N02
73.7482	$1s^22p$	$1s^24p$	$^2P^o$	$^2P^o$	E2	1/2	3/2	128151.9	1484117	1.43+06	1.11−02	B	A10
73.8380	$1s^22p$	$1s^24p$	$^2P^o$	$^2P^o$	E2	3/2	3/2	129801.2	1484117	1.42+06	1.11−02	B	A10
73.8492	$1s^22p$	$1s^24p$	$^2P^o$	$^2P^o$	E2	3/2	1/2	129801.2	1483912	2.8+06	1.11−02	B	A10
74.5454	$1s^22p$	$1s^24s$	$^2P^o$	^2S		1/2	1/2	128151.9	1469616	6.14+09	2.51−03	A	F04
74.6372	$1s^22p$	$1s^24s$	$^2P^o$	^2S		3/2	1/2	129801.2	1469616	1.24+10	5.09−03	A	F04
87.1464	$1s^22s$	$1s^23d$	^2S	^2D	E2	1/2	5/2	0.0	1147494	2.07+07	5.6−01	B	A10
87.1576	$1s^22s$	$1s^23d$	^2S	^2D	E2	1/2	3/2	0.0	1147347	2.07+07	3.7−01	B	A10
88.0803	$1s^22s$	$1s^23p$	^2S	$^2P^o$		1/2	3/2	0.0	1135327.7	8.64+10	1.165−01	AA	F04
88.1182	$1s^22s$	$1s^23p$	^2S	$^2P^o$		1/2	1/2	0.0	1134839.5	8.67+10	5.86−02	AA	F04
98.1166	$1s^22p$	$1s^23d$	$^2P^o$	^2D		1/2	3/2	128151.9	1147347	2.290+11	4.270−01	AAA	F04
98.2615	$1s^22p$	$1s^23d$	$^2P^o$	^2D		3/2	5/2	129801.2	1147494	2.742+11	7.705−01	AAA	F04
98.2757	$1s^22p$	$1s^23d$	$^2P^o$	^2D		3/2	3/2	129801.2	1147347	4.571+10	8.57−02	AAA	F04
99.2875	$1s^22p$	$1s^23p$	$^2P^o$	$^2P^o$	E2	1/2	3/2	128151.9	1135327.7	3.3+06	1.15−01	B	A10
99.4504	$1s^22p$	$1s^23p$	$^2P^o$	$^2P^o$	E2	3/2	3/2	129801.2	1135327.7	3.3+06	1.15−01	B	A10
99.4987	$1s^22p$	$1s^23p$	$^2P^o$	$^2P^o$	E2	3/2	1/2	129801.2	1134839.5	6.7+06	1.16−01	B	A10
102.9088	$1s^22p$	$1s^23s$	$^2P^o$	^2S		1/2	1/2	128151.9	1099886.0	1.55+10	1.67−02	A	F04
103.0838	$1s^22p$	$1s^23s$	$^2P^o$	^2S		3/2	1/2	129801.2	1099886.0	3.10+10	3.36−02	A	F04

λ_{Ritz} [a] (Å)	Configurations		Terms		Type	J_i [b]	J_k [b]	E_i (cm^{-1})	E_k (cm^{-1})	A_{ki} [c] (s^{-1})	S_{ik} [c]	Acc.[d]	Ref.[e]
131.929	$1s^23s$	$1s^210p$	^2S	$^2P^o$		1/2	1/2,3/2	1099886.0	1857869	7.91+08	5.38−03	A	N02
134.884	$1s^23s$	$1s^29p$	^2S	$^2P^o$		1/2	1/2,3/2	1099886.0	1841265	1.08+09	7.87−03	A	N02
138.246	$1s^23p$	$1s^210d$	$^2P^o$	^2D		1/2	3/2	1134839.5	1858187	1.35+09	7.06−03	A	N02LS
138.340	$1s^23p$	$1s^210d$	$^2P^o$	^2D		3/2	3/2,5/2	1135327.7	1858187	1.08+09	1.41−02	A	N02LS
138.477	$1s^23p$	$1s^210s$	$^2P^o$	^2S		1/2	1/2	1134839.5	1856983	2.14+08	5.60−04	A	N02
138.570	$1s^23p$	$1s^210s$	$^2P^o$	^2S		3/2	1/2	1135327.7	1856983	4.30+08	1.13−03	A	N02
139.246	$1s^23s$	$1s^28p$	^2S	$^2P^o$		1/2	3/2	1099886.0	1818040.8	1.53+09	8.17−03	A	N02
139.251	$1s^23s$	$1s^28p$	^2S	$^2P^o$		1/2	1/2	1099886.0	1818014.9	1.54+09	4.10−03	A	N02
140.674	$1s^23d$	$1s^210f$	^2D	$^2F^o$		3/2	5/2	1147347	1858210.3	1.37+09	1.13−02	A	N02LS
140.703	$1s^23d$	$1s^210f$	^2D	$^2F^o$		5/2	5/2,7/2	1147494	1858210.3	8.8+08	1.69−02	A	N02LS
140.742	$1s^23d$	$1s^210p$	^2D	$^2P^o$		3/2	1/2,3/2	1147347	1857869	3.04+07	2.51−04	A	N02LS
140.771	$1s^23d$	$1s^210p$	^2D	$^2P^o$		5/2	3/2	1147494	1857869	6.84+07	3.77−04	A	N02LS
141.470	$1s^23p$	$1s^29d$	$^2P^o$	^2D		1/2	3/2	1134839.5	1841702	1.87+09	1.05−02	A	N02LS
141.568	$1s^23p$	$1s^29d$	$^2P^o$	^2D		3/2	3/2,5/2	1135327.7	1841702	1.49+09	2.09−02	A	N02LS
141.802	$1s^23p$	$1s^29s$	$^2P^o$	^2S		1/2	1/2	1134839.5	1840047.0	2.97+08	8.4−04	A	N02
141.900	$1s^23p$	$1s^29s$	$^2P^o$	^2S		3/2	1/2	1135327.7	1840047.0	5.97+08	1.68−03	A	N02
144.012	$1s^23d$	$1s^29f$	^2D	$^2F^o$		3/2	5/2	1147347	1841734	1.94+09	1.72−02	A	N02LS
144.042	$1s^23d$	$1s^29f$	^2D	$^2F^o$		5/2	5/2,7/2	1147494	1841734	1.25+09	2.58−02	A	N02LS
144.109	$1s^23d$	$1s^29p$	^2D	$^2P^o$		3/2	1/2,3/2	1147347	1841265	4.25+07	3.77−04	A	N02LS
144.140	$1s^23d$	$1s^29p$	^2D	$^2P^o$		5/2	3/2	1147494	1841265	9.6+07	5.65−04	A	N02LS
146.150	$1s^23s$	$1s^27p$	^2S	$^2P^o$		1/2	3/2	1099886.0	1784116	2.27+09	1.40−02	A	N02
146.158	$1s^23s$	$1s^27p$	^2S	$^2P^o$		1/2	1/2	1099886.0	1784078	2.28+09	7.03−03	A	N02
146.239	$1s^23p$	$1s^28d$	$^2P^o$	^2D		1/2	3/2	1134839.5	1818650	2.69+09	1.66−02	A	N02
146.342	$1s^23p$	$1s^28d$	$^2P^o$	^2D		3/2	5/2	1135327.7	1818658	3.22+09	2.99−02	A	N02
146.344	$1s^23p$	$1s^28d$	$^2P^o$	^2D		3/2	3/2	1135327.7	1818650	5.37+08	3.33−03	A	N02
146.745	$1s^23p$	$1s^28s$	$^2P^o$	^2S		1/2	1/2	1134839.5	1816292	4.29+08	1.34−03	A	N02
146.851	$1s^23p$	$1s^28s$	$^2P^o$	^2S		3/2	1/2	1135327.7	1816292	8.6+08	2.70−03	A	N02
148.955	$1s^23d$	$1s^28f$	^2D	$^2F^o$		3/2	5/2	1147347	1818691.1	2.90+09	2.84−02	A	N02LS
148.988	$1s^23d$	$1s^28f$	^2D	$^2F^o$		5/2	5/2,7/2	1147494	1818691.1	1.86+09	4.25−02	A	N02LS

λ_{Ritz} [a] (Å)	Configurations		Terms		Type	J_i [b]	J_k [b]	E_i (cm^{-1})	E_k (cm^{-1})	A_{ki} [c] (s^{-1})	S_{ik} [c]	Acc. [d]	Ref. [e]
149.099	1s^{2}3d	1s^{2}8p	^{2}D	^{2}P°		3/2	3/2	1147347	1818040.8	1.55+07	1.01−04	A	N02
149.105	1s^{2}3d	1s^{2}8p	^{2}D	^{2}P°		3/2	1/2	1147347	1818014.9	1.56+08	5.12−04	A	N02
149.132	1s^{2}3d	1s^{2}8p	^{2}D	^{2}P°		5/2	3/2	1147494	1818040.8	1.40+08	9.1−04	A	N02
153.801	1s^{2}3p	1s^{2}7d	^{2}P°	^{2}D		1/2	3/2	1134839.5	1785029	4.07+09	2.93−02	A	N02
153.914	1s^{2}3p	1s^{2}7d	^{2}P°	^{2}D		3/2	5/2	1135327.7	1785041	4.88+09	5.27−02	A	N02
153.917	1s^{2}3p	1s^{2}7d	^{2}P°	^{2}D		3/2	3/2	1135327.7	1785029	8.14+08	5.86−03	A	N02
154.642	1s^{2}3p	1s^{2}7s	^{2}P°	^{2}S		1/2	1/2	1134839.5	1781495	6.54+08	2.39−03	A	N02
154.759	1s^{2}3p	1s^{2}7s	^{2}P°	^{2}S		3/2	1/2	1135327.7	1781495	1.32+09	4.82−03	A	N02
156.800	1s^{2}3d	1s^{2}7f	^{2}D	^{2}F°		3/2	5/2	1147347	1785103	4.64+09	5.30−02	A	N02LS
156.836	1s^{2}3d	1s^{2}7f	^{2}D	^{2}F°		5/2	5/2,7/2	1147494	1785103	2.98+09	7.94−02	A	N02LS
157.043	1s^{2}3d	1s^{2}7p	^{2}D	^{2}P°		3/2	3/2	1147347	1784116	2.40+07	1.84−04	A	N02
157.052	1s^{2}3d	1s^{2}7p	^{2}D	^{2}P°		3/2	1/2	1147347	1784078	2.43+08	9.3−04	A	N02
157.079	1s^{2}3d	1s^{2}7p	^{2}D	^{2}P°		5/2	3/2	1147494	1784116	2.17+08	1.66−03	A	N02
158.257	1s^{2}3s	1s^{2}6p	^{2}S	^{2}P°		1/2	3/2	1099886.0	1731768	3.56+09	2.78−02	A	N02
158.272	1s^{2}3s	1s^{2}6p	^{2}S	^{2}P°		1/2	1/2	1099886.0	1731708	3.57+09	1.40−02	A	N02
167.117	1s^{2}3p	1s^{2}6d	^{2}P°	^{2}D		1/2	3/2	1134839.5	1733224	6.61+09	6.09−02	A	N02
167.248	1s^{2}3p	1s^{2}6d	^{2}P°	^{2}D		3/2	5/2	1135327.7	1733242	7.93+09	1.10−01	A	N02
167.253	1s^{2}3p	1s^{2}6d	^{2}P°	^{2}D		3/2	3/2	1135327.7	1733224	1.32+09	1.22−02	A	N02
168.708	1s^{2}3p	1s^{2}6s	^{2}P°	^{2}S		1/2	1/2	1134839.5	1727579	1.07+09	5.09−03	A	N02
168.847	1s^{2}3p	1s^{2}6s	^{2}P°	^{2}S		3/2	1/2	1135327.7	1727579	2.16+09	1.03−02	A	N02
170.651	1s^{2}3d	1s^{2}6f	^{2}D	^{2}F°		3/2	5/2	1147347	1733339	8.24+09	1.21−01	A	N02LS
170.694	1s^{2}3d	1s^{2}6f	^{2}D	^{2}F°		5/2	5/2,7/2	1147494	1733339	5.29+09	1.82−01	A	N02LS
171.110	1s^{2}3d	1s^{2}6p	^{2}D	^{2}P°		3/2	3/2	1147347	1731768	4.05+07	4.01−04	A	N02
171.127	1s^{2}3d	1s^{2}6p	^{2}D	^{2}P°		3/2	1/2	1147347	1731708	4.10+08	2.03−03	A	N02
171.153	1s^{2}3d	1s^{2}6p	^{2}D	^{2}P°		5/2	3/2	1147494	1731768	3.66+08	3.62−03	A	N02
183.526	1s^{2}3s	1s^{2}5p	^{2}S	^{2}P°		1/2	3/2	1099886.0	1644768	5.95+09	7.27−02	A	N02
183.561	1s^{2}3s	1s^{2}5p	^{2}S	^{2}P°		1/2	1/2	1099886.0	1644664	5.99+09	3.66−02	A	N02
195.138	1s^{2}3p	1s^{2}5d	^{2}P°	^{2}D		1/2	3/2	1134839.5	1647298	1.18+10	1.73−01	A	N02
195.312	1s^{2}3p	1s^{2}5d	^{2}P°	^{2}D		3/2	5/2	1135327.7	1647330	1.41+10	3.12−01	A	N02

λ_{Ritz} [a] (Å)	Configurations		Terms		Type	J_i [b]	J_k [b]	E_i (cm^{-1})	E_k (cm^{-1})	A_{ki} [c] (s^{-1})	S_{ik} [c]	Acc. [d]	Ref. [e]
195.324	$1s^23p$	$1s^25d$	$^2P^o$	^2D		3/2	3/2	1135327.7	1647298	2.36+09	3.47−02	A	N02
198.957	$1s^23p$	$1s^25s$	$^2P^o$	^2S		1/2	1/2	1134839.5	1637460	1.96+09	1.52−02	A	N02
199.151	$1s^23p$	$1s^25s$	$^2P^o$	^2S		3/2	1/2	1135327.7	1637460	3.94+09	3.08−02	A	N02
199.947	$1s^23d$	$1s^25f$	^2D	$^2F^o$		3/2	5/2	1147347	1647480.7	1.74+10	4.13−01	A	N02
199.999	$1s^23d$	$1s^25f$	^2D	$^2F^o$		5/2	7/2	1147494	1647496.7	1.87+10	5.90−01	A	N02
200.005	$1s^23d$	$1s^25f$	^2D	$^2F^o$		5/2	5/2	1147494	1647480.7	1.24+09	2.95−02	A	N02
201.037	$1s^23d$	$1s^25p$	^2D	$^2P^o$		3/2	3/2	1147347	1644768	7.76+07	1.24−03	A	N02
201.079	$1s^23d$	$1s^25p$	^2D	$^2P^o$		3/2	1/2	1147347	1644664	7.85+08	6.30−03	A	N02
201.096	$1s^23d$	$1s^25p$	^2D	$^2P^o$		5/2	3/2	1147494	1644768	7.00+08	1.12−02	A	N02
256.885	$1s^23s$	$1s^24d$	^2S	^2D	E2	1/2	5/2	1099886.0	1489166	1.29+06	7.7+00	B	A10
256.925	$1s^23s$	$1s^24d$	^2S	^2D	E2	1/2	3/2	1099886.0	1489104	1.29+06	5.2+00	B	A10
257.564	$1s^24s$	$1s^210p$	^2S	$^2P^o$		1/2	1/2,3/2	1469616	1857869	3.50+08	1.77−02	A	N02
260.260	$1s^23s$	$1s^24p$	^2S	$^2P^o$		1/2	3/2	1099886.0	1484117	1.04+10	3.63−01	A	N02
260.399	$1s^23s$	$1s^24p$	^2S	$^2P^o$		1/2	1/2	1099886.0	1483912	1.05+10	1.83−01	A	N02
267.183	$1s^24p$	$1s^210d$	$^2P^o$	^2D		1/2	3/2	1483912	1858187	6.14+08	2.31−02	A	N02LS
267.330	$1s^24p$	$1s^210d$	$^2P^o$	^2D		3/2	3/2,5/2	1484117	1858187	4.90+08	4.62−02	A	N02LS
268.045	$1s^24p$	$1s^210s$	$^2P^o$	^2S		1/2	1/2	1483912	1856983	1.43+08	2.72−03	A	N02
268.19	$1s^24p$	$1s^210s$	$^2P^o$	^2S		3/2	1/2	1484117	1856983	2.87+08	5.47−03	A	N02
269.071	$1s^24s$	$1s^29p$	^2S	$^2P^o$		1/2	1/2,3/2	1469616	1841265	4.78+08	2.76−02	A	N02
270.925	$1s^24d$	$1s^210f$	^2D	$^2F^o$		3/2	5/2	1489104	1858210.3	8.4+08	4.95−02	A	N02LS
270.970	$1s^24d$	$1s^210f$	^2D	$^2F^o$		5/2	5/2,7/2	1489166	1858210.3	5.41+08	7.43−02	A	N02LS
271.167	$1s^24f$	$1s^210g$	$^2F^o$	^2G		5/2	7/2	1489436	1858213.1	6.11+08	4.81−02	A	N02LS
271.175	$1s^24d$	$1s^210p$	^2D	$^2P^o$		3/2	1/2,3/2	1489104	1857869	3.29+07	1.95−03	A	N02LS
271.186	$1s^24f$	$1s^210d$	$^2F^o$	^2D		5/2	3/2,5/2	1489436	1858187	5.38+06	5.30−04	A	N02LS
271.189	$1s^24f$	$1s^210g$	$^2F^o$	^2G		7/2	7/2,9/2	1489467	1858213.1	3.62+08	6.41−02	A	N02LS
271.209	$1s^24f$	$1s^210d$	$^2F^o$	^2D		7/2	5/2	1489467	1858187	1.20+07	7.06−04	A	N02LS
271.221	$1s^24d$	$1s^210p$	^2D	$^2P^o$		5/2	3/2	1489166	1857869	7.41+07	2.92−03	A	N02LS
279.494	$1s^24p$	$1s^29d$	$^2P^o$	^2D		1/2	3/2	1483912	1841702	9.8+08	4.24−02	A	N02LS
279.65	$1s^24p$	$1s^29d$	$^2P^o$	^2D		3/2	3/2,5/2	1484117	1841702	7.85+08	8.5−02	A	N02LS

λ_{Ritz} [a] (Å)	Configurations		Terms		Type	J_i [b]	J_k [b]	E_i (cm^{-1})	E_k (cm^{-1})	A_{ki} [c] (s^{-1})	S_{ik} [c]	Acc.[d]	Ref.[e]
280.792	$1s^24p$	$1s^29s$	$^2P°$	^2S		1/2	1/2	1483912	1840047.0	1.99+08	4.36−03	A	N02
280.95	$1s^24p$	$1s^29s$	$^2P°$	^2S		3/2	1/2	1484117	1840047.0	4.01+08	8.8−03	A	N02
282.011	$1s^23p$	$1s^24f$	$^2P°$	$^2F°$	E2	1/2	5/2	1134839.5	1489436	1.33+06	1.27+01	B	A10
282.275	$1s^23p$	$1s^24d$	$^2P°$	^2D		1/2	3/2	1134839.5	1489104	2.36+10	1.05+00	A	N02
282.375	$1s^23p$	$1s^24f$	$^2P°$	$^2F°$	E2	3/2	7/2	1135327.7	1489467	1.70+06	2.18+01	B	A10
282.615	$1s^23p$	$1s^24d$	$^2P°$	^2D		3/2	5/2	1135327.7	1489166	2.84+10	1.90+00	A	N02
282.664	$1s^23p$	$1s^24d$	$^2P°$	^2D		3/2	3/2	1135327.7	1489104	4.74+09	2.12−01	A	N02
283.583	$1s^24d$	$1s^29f$	^2D	$^2F°$		3/2	5/2	1489104	1841734	1.19+09	8.04−02	A	N02LS
283.633	$1s^24d$	$1s^29f$	^2D	$^2F°$		5/2	5/2,7/2	1489166	1841734	7.65+08	1.21−01	A	N02LS
283.847	$1s^24f$	$1s^29g$	$^2F°$	^2G		5/2	7/2	1489436	1841738.2	9.0+08	8.17−02	A	N02LS
283.872	$1s^24f$	$1s^29g$	$^2F°$	^2G		7/2	7/2,9/2	1489467	1841738.2	5.36+08	1.09−01	A	N02LS
283.876	$1s^24f$	$1s^29d$	$^2F°$	^2D		5/2	3/2,5/2	1489436	1841702	7.80+06	8.8−04	A	N02LS
283.901	$1s^24f$	$1s^29d$	$^2F°$	^2D		7/2	5/2	1489467	1841702	1.73+07	1.17−03	A	N02LS
283.961	$1s^24d$	$1s^29p$	^2D	$^2P°$		3/2	1/2,3/2	1489104	1841265	4.65+07	3.16−03	A	N02LS
284.011	$1s^24d$	$1s^29p$	^2D	$^2P°$		5/2	3/2	1489166	1841265	1.05+08	4.73−03	A	N02LS
287.006	$1s^24s$	$1s^28p$	^2S	$^2P°$		1/2	3/2	1469616	1818040.8	6.73+08	3.14−02	A	N02
287.027	$1s^24s$	$1s^28p$	^2S	$^2P°$		1/2	1/2	1469616	1818014.9	6.76+08	1.58−02	A	N02
292.32	$1s^23d$	$1s^24f$	^2D	$^2F°$		3/2	5/2	1147347	1489436	5.28+10	3.91+00	A	N02
292.42	$1s^23d$	$1s^24f$	^2D	$^2F°$		5/2	7/2	1147494	1489467	5.66+10	5.58+00	A	N02
292.45	$1s^23d$	$1s^24f$	^2D	$^2F°$		5/2	5/2	1147494	1489436	3.77+09	2.79−01	A	N02
296.94	$1s^23d$	$1s^24p$	^2D	$^2P°$		3/2	3/2	1147347	1484117	1.81+08	9.4−03	A	N02
297.07	$1s^23d$	$1s^24p$	^2D	$^2P°$		5/2	3/2	1147494	1484117	1.64+09	8.5−02	A	N02
297.12	$1s^23d$	$1s^24p$	^2D	$^2P°$		3/2	1/2	1147347	1483912	1.84+09	4.75−02	A	N02
298.707	$1s^23p$	$1s^24s$	$^2P°$	^2S		1/2	1/2	1134839.5	1469616	3.82+09	1.00−01	A	F04
298.74	$1s^24p$	$1s^28d$	$^2P°$	^2D		1/2	3/2	1483912	1818650	1.21+09	6.38−02	A	N02
298.92	$1s^24p$	$1s^28d$	$^2P°$	^2D		3/2	5/2	1484117	1818658	1.45+09	1.15−01	A	N02
298.92	$1s^24p$	$1s^28d$	$^2P°$	^2D		3/2	3/2	1484117	1818650	2.42+08	1.28−02	A	N02
299.143	$1s^23p$	$1s^24s$	$^2P°$	^2S		3/2	1/2	1135327.7	1469616	7.64+09	2.02−01	A	F04
300.86	$1s^24p$	$1s^28s$	$^2P°$	^2S		1/2	1/2	1483912	1816292	2.91+08	7.82−03	A	N02

λ_{Ritz} [a] (Å)	Configurations		Terms		Type	J_i [b]	J_k [b]	E_i (cm^{-1})	E_k (cm^{-1})	A_{ki} [c] (s^{-1})	S_{ik} [c]	Acc. [d]	Ref. [e]
301.05	$1s^24p$	$1s^28s$	$^2P°$	^2S		3/2	1/2	1484117	1816292	5.85+08	1.58−02	A	N02
303.410	$1s^24d$	$1s^28f$	^2D	$^2F°$		3/2	5/2	1489104	1818691.1	1.77+09	1.46−01	A	N02LS
303.467	$1s^24d$	$1s^28f$	^2D	$^2F°$		5/2	5/2,7/2	1489166	1818691.1	1.14+09	2.20−01	A	N02LS
303.703	$1s^24f$	$1s^28g$	$^2F°$	^2G		5/2	7/2	1489436	1818704.9	1.44+09	1.59−01	A	N02LS
303.732	$1s^24f$	$1s^28g$	$^2F°$	^2G		7/2	7/2,9/2	1489467	1818704.9	8.5+08	2.12−01	A	N02LS
303.746	$1s^24f$	$1s^28d$	$^2F°$	^2D		5/2	5/2	1489436	1818658	1.33+06	1.10−04	A	N02
303.754	$1s^24f$	$1s^28d$	$^2F°$	^2D		5/2	3/2	1489436	1818650	2.80+07	1.55−03	A	N02
303.78	$1s^24f$	$1s^28d$	$^2F°$	^2D		7/2	5/2	1489467	1818658	2.66+07	2.21−03	A	N02
304.01	$1s^24d$	$1s^28p$	^2D	$^2P°$		3/2	3/2	1489104	1818040.8	1.72+07	9.5−04	A	N02
304.03	$1s^24d$	$1s^28p$	^2D	$^2P°$		3/2	1/2	1489104	1818014.9	1.73+08	4.81−03	A	N02
304.07	$1s^24d$	$1s^28p$	^2D	$^2P°$		5/2	3/2	1489166	1818040.8	1.55+08	8.6−03	A	N02
317.97	$1s^24s$	$1s^27p$	^2S	$^2P°$		1/2	3/2	1469616	1784116	9.9+08	6.27−02	A	N02
318.00	$1s^24s$	$1s^27p$	^2S	$^2P°$		1/2	1/2	1469616	1784078	9.9+08	3.15−02	A	N02
332.10	$1s^24p$	$1s^27d$	$^2P°$	^2D		1/2	3/2	1483912	1785029	1.82+09	1.32−01	A	N02
332.31	$1s^24p$	$1s^27d$	$^2P°$	^2D		3/2	5/2	1484117	1785041	2.19+09	2.38−01	A	N02
332.32	$1s^24p$	$1s^27d$	$^2P°$	^2D		3/2	3/2	1484117	1785029	3.65+08	2.64−02	A	N02
336.04	$1s^24p$	$1s^27s$	$^2P°$	^2S		1/2	1/2	1483912	1781495	4.50+08	1.69−02	A	N02
336.27	$1s^24p$	$1s^27s$	$^2P°$	^2S		3/2	1/2	1484117	1781495	9.0+08	3.40−02	A	N02
337.84	$1s^24d$	$1s^27f$	^2D	$^2F°$		3/2	5/2	1489104	1785103	2.82+09	3.22−01	A	N02LS
337.91	$1s^24d$	$1s^27f$	^2D	$^2F°$		5/2	5/2,7/2	1489166	1785103	1.81+09	4.83−01	A	N02LS
338.21	$1s^24f$	$1s^27g$	$^2F°$	^2G		5/2	7/2	1489436	1785111.4	2.55+09	3.90−01	A	N02LS
338.24	$1s^24f$	$1s^27g$	$^2F°$	^2G		7/2	7/2,9/2	1489467	1785111.4	1.51+09	5.20−01	A	N02LS
338.29	$1s^24f$	$1s^27d$	$^2F°$	^2D		5/2	5/2	1489436	1785041	2.23+06	2.55−04	A	N02
338.30	$1s^24f$	$1s^27d$	$^2F°$	^2D		5/2	3/2	1489436	1785029	4.70+07	3.59−03	A	N02
338.32	$1s^24f$	$1s^27d$	$^2F°$	^2D		7/2	5/2	1489467	1785041	4.46+07	5.12−03	A	N02
338.97	$1s^24d$	$1s^27p$	^2D	$^2P°$		3/2	3/2	1489104	1784116	2.73+07	2.10−03	A	N02
339.01	$1s^24d$	$1s^27p$	^2D	$^2P°$		3/2	1/2	1489104	1784078	2.75+08	1.06−02	A	N02
339.04	$1s^24d$	$1s^27p$	^2D	$^2P°$		5/2	3/2	1489166	1784116	2.46+08	1.89−02	A	N02
381.46	$1s^24s$	$1s^26p$	^2S	$^2P°$		1/2	3/2	1469616	1731768	1.51+09	1.66−01	A	N02

λ_{Ritz} [a] (Å)	Configurations		Terms		Type	J_i [b]	J_k [b]	E_i (cm^{-1})	E_k (cm^{-1})	A_{ki} [c] (s^{-1})	S_{ik} [c]	Acc.[d]	Ref.[e]
381.55	$1s^24s$	$1s^26p$	^2S	$^2P°$		1/2	1/2	1469616	1731708	1.52+09	8.3−02	A	N02
401.10	$1s^24p$	$1s^26d$	$^2P°$	^2D		1/2	3/2	1483912	1733224	2.90+09	3.70−01	A	N02
401.40	$1s^24p$	$1s^26d$	$^2P°$	^2D		3/2	5/2	1484117	1733242	3.49+09	6.68−01	A	N02
401.43	$1s^24p$	$1s^26d$	$^2P°$	^2D		3/2	3/2	1484117	1733224	5.81+08	7.42−02	A	N02
409.44	$1s^24d$	$1s^26f$	^2D	$^2F°$		3/2	5/2	1489104	1733339	4.93+09	1.00+00	A	N02LS
409.55	$1s^24d$	$1s^26f$	^2D	$^2F°$		5/2	5/2,7/2	1489166	1733339	3.17+09	1.50+00	A	N02LS
409.98	$1s^24f$	$1s^26g$	$^2F°$	^2G		5/2	7/2	1489436	1733351.4	5.43+09	1.48+00	A	N02LS
410.03	$1s^24f$	$1s^26g$	$^2F°$	^2G		7/2	7/2,9/2	1489467	1733351.4	3.22+09	1.97+00	A	N02LS
410.16	$1s^24f$	$1s^26d$	$^2F°$	^2D		5/2	5/2	1489436	1733242	4.25+06	8.7−04	A	N02
410.19	$1s^24f$	$1s^26d$	$^2F°$	^2D		5/2	3/2	1489436	1733224	9.0+07	1.22−02	A	N02
410.21	$1s^24f$	$1s^26d$	$^2F°$	^2D		7/2	5/2	1489467	1733242	8.5+07	1.74−02	A	N02
410.40	$1s^24p$	$1s^26s$	$^2P°$	^2S		1/2	1/2	1483912	1727579	7.55+08	5.15−02	A	N02
410.74	$1s^24p$	$1s^26s$	$^2P°$	^2S		3/2	1/2	1484117	1727579	1.52+09	1.04−01	A	N02
412.09	$1s^24d$	$1s^26p$	^2D	$^2P°$		3/2	3/2	1489104	1731768	4.76+07	6.58−03	A	N02
412.19	$1s^24d$	$1s^26p$	^2D	$^2P°$		3/2	1/2	1489104	1731708	4.82+08	3.33−02	A	N02
412.20	$1s^24d$	$1s^26p$	^2D	$^2P°$		5/2	3/2	1489166	1731768	4.30+08	5.95−02	A	N02
453.70	$1s^25s$	$1s^210p$	^2S	$^2P°$		1/2	1/2,3/2	1637460	1857869	1.89+08	5.22−02	A	N02
468.33	$1s^25p$	$1s^210d$	$^2P°$	^2D		1/2	3/2	1644664	1858187	3.30+08	6.70−02	A	N02LS
468.56	$1s^25p$	$1s^210d$	$^2P°$	^2D		3/2	3/2,5/2	1644768	1858187	2.64+08	1.34−01	A	N02LS
470.99	$1s^25p$	$1s^210s$	$^2P°$	^2S		1/2	1/2	1644664	1856983	1.01+08	1.04−02	A	N02
471.22	$1s^25p$	$1s^210s$	$^2P°$	^2S		3/2	1/2	1644768	1856983	2.03+08	2.10−02	A	N02
474.13	$1s^25d$	$1s^210f$	^2D	$^2F°$		3/2	5/2	1647298	1858210.3	5.05+08	1.59−01	A	N02LS
474.20	$1s^25d$	$1s^210f$	^2D	$^2F°$		5/2	5/2,7/2	1647330	1858210.3	3.25+08	2.39−01	A	N02LS
474.54	$1s^25f$	$1s^210g$	$^2F°$	^2G		5/2	7/2	1647480.7	1858213.1	5.30+08	2.24−01	A	N02LS
474.57	$1s^25f$	$1s^210g$	$^2F°$	^2G		7/2	7/2,9/2	1647496.7	1858213.1	3.14+08	2.98−01	A	N02LS
474.59	$1s^25f$	$1s^210d$	$^2F°$	^2D		5/2	3/2,5/2	1647480.7	1858187	8.30+06	4.38−03	A	N02LS
474.60	$1s^25g$	$1s^210h$	^2G	$^2H°$		7/2,9/2	9/2,11/2	1647511	1858214.3	3.33+08	3.87−01	A	N02
474.61	$1s^25g$	$1s^210f$	^2G	$^2F°$		7/2,9/2	5/2,7/2	1647511	1858210.3	3.75+06	2.77−03	A	N02
474.63	$1s^25f$	$1s^210d$	$^2F°$	^2D		7/2	5/2	1647496.7	1858187	1.84+07	5.84−03	A	N02LS

λ_{Ritz} [a] (Å)	Configurations		Terms		Type	J_i [b]	J_k [b]	E_i (cm^{-1})	E_k (cm^{-1})	A_{ki} [c] (s^{-1})	S_{ik} [c]	Acc.[d]	Ref.[e]
474.90	$1s^25d$	$1s^210p$	^2D	$^2P°$		3/2	1/2,3/2	1647298	1857869	2.98+07	9.5−03	A	N02LS
474.97	$1s^25d$	$1s^210p$	^2D	$^2P°$		5/2	3/2	1647330	1857869	6.71+07	1.42−02	A	N02LS
490.67	$1s^25s$	$1s^29p$	^2S	$^2P°$		1/2	1/2,3/2	1637460	1841265	2.57+08	9.0−02	A	N02
507.52	$1s^25p$	$1s^29d$	$^2P°$	^2D		1/2	3/2	1644664	1841702	4.55+08	1.17−01	A	N02LS
507.78	$1s^25p$	$1s^29d$	$^2P°$	^2D		3/2	3/2,5/2	1644768	1841702	3.63+08	2.35−01	A	N02LS
511.82	$1s^25p$	$1s^29s$	$^2P°$	^2S		1/2	1/2	1644664	1840047.0	1.42+08	1.88−02	A	N02
512.09	$1s^25p$	$1s^29s$	$^2P°$	^2S		3/2	1/2	1644768	1840047.0	2.86+08	3.79−02	A	N02
514.31	$1s^25d$	$1s^29f$	^2D	$^2F°$		3/2	5/2	1647298	1841734	7.13+08	2.87−01	A	N02LS
514.39	$1s^25d$	$1s^29f$	^2D	$^2F°$		5/2	5/2,7/2	1647330	1841734	4.58+08	4.31−01	A	N02LS
514.78	$1s^25f$	$1s^29g$	$^2F°$	^2G		5/2	7/2	1647480.7	1841738.2	7.83+08	4.22−01	A	N02LS
514.82	$1s^25f$	$1s^29g$	$^2F°$	^2G		7/2	7/2,9/2	1647496.7	1841738.2	4.64+08	5.62−01	A	N02LS
514.86	$1s^25g$	$1s^29h$	^2G	$^2H°$		7/2,9/2	9/2,11/2	1647511	1841739.9	5.34+08	7.91−01	A	N02
514.87	$1s^25g$	$1s^29f$	^2G	$^2F°$		7/2,9/2	5/2,7/2	1647511	1841734	5.79+06	5.46−03	A	N02
514.88	$1s^25f$	$1s^29d$	$^2F°$	^2D		5/2	3/2,5/2	1647480.7	1841702	1.23+07	8.26−03	A	N02LS
514.92	$1s^25f$	$1s^29d$	$^2F°$	^2D		7/2	5/2	1647496.7	1841702	2.72+07	1.10−02	A	N02LS
515.55	$1s^25d$	$1s^29p$	^2D	$^2P°$		3/2	1/2,3/2	1647298	1841265	4.27+07	1.73−02	A	N02LS
515.64	$1s^25d$	$1s^29p$	^2D	$^2P°$		5/2	3/2	1647330	1841265	9.6+07	2.60−02	A	N02LS
553.77	$1s^25s$	$1s^28p$	^2S	$^2P°$		1/2	3/2	1637460	1818040.8	3.59+08	1.20−01	A	N02
553.85	$1s^25s$	$1s^28p$	^2S	$^2P°$		1/2	1/2	1637460	1818014.9	3.61+08	6.06−02	A	N02
570.93	$1s^24s$	$1s^25p$	^2S	$^2P°$		1/2	3/2	1469616	1644768	2.31+09	8.5−01	A	N02
571.27	$1s^24s$	$1s^25p$	^2S	$^2P°$		1/2	1/2	1469616	1644664	2.33+09	4.29−01	A	N02
574.76	$1s^25p$	$1s^28d$	$^2P°$	^2D		1/2	3/2	1644664	1818650	6.46+08	2.42−01	A	N02
575.08	$1s^25p$	$1s^28d$	$^2P°$	^2D		3/2	5/2	1644768	1818658	7.77+08	4.37−01	A	N02
575.10	$1s^25p$	$1s^28d$	$^2P°$	^2D		3/2	3/2	1644768	1818650	1.30+08	4.87−02	A	N02
582.66	$1s^25p$	$1s^28s$	$^2P°$	^2S		1/2	1/2	1644664	1816292	2.10+08	4.11−02	A	N02
583.01	$1s^25p$	$1s^28s$	$^2P°$	^2S		3/2	1/2	1644768	1816292	4.23+08	8.28−02	A	N02
583.45	$1s^25d$	$1s^28f$	^2D	$^2F°$		3/2	5/2	1647298	1818691.1	1.06+09	6.21−01	A	N02LS
583.56	$1s^25d$	$1s^28f$	^2D	$^2F°$		5/2	5/2,7/2	1647330	1818691.1	6.78+08	9.3−01	A	N02LS
584.03	$1s^25f$	$1s^28g$	$^2F°$	^2G		5/2	7/2	1647480.7	1818704.9	1.24+09	9.8−01	A	N02LS

λ_{Ritz} [a] (Å)	Configurations		Terms		Type	J_i [b]	J_k [b]	E_i (cm^{-1})	E_k (cm^{-1})	A_{ki} [c] (s^{-1})	S_{ik} [c]	Acc.[d]	Ref.[e]
584.08	$1s^25f$	$1s^28g$	$^2F°$	^2G		7/2	7/2,9/2	1647496.7	1818704.9	7.34+08	1.30+00	A	N02LS
584.12	$1s^25g$	$1s^28h$	^2G	$^2H°$		7/2,9/2	9/2,11/2	1647511	1818709	9.6+08	2.07+00	A	N02
584.18	$1s^25g$	$1s^28f$	^2G	$^2F°$		7/2,9/2	5/2,7/2	1647511	1818691.1	9.8+06	1.35−02	A	N02
584.19	$1s^25f$	$1s^28d$	$^2F°$	^2D		5/2	5/2	1647480.7	1818658	2.15+06	1.27−03	A	N02
584.22	$1s^25f$	$1s^28d$	$^2F°$	^2D		5/2	3/2	1647480.7	1818650	4.54+07	1.79−02	A	N02
584.24	$1s^25f$	$1s^28d$	$^2F°$	^2D		7/2	5/2	1647496.7	1818658	4.32+07	2.55−02	A	N02
585.68	$1s^25d$	$1s^28p$	^2D	$^2P°$		3/2	3/2	1647298	1818040.8	1.61+07	6.39−03	A	N02
585.77	$1s^25d$	$1s^28p$	^2D	$^2P°$		3/2	1/2	1647298	1818014.9	1.62+08	3.22−02	A	N02
585.79	$1s^25d$	$1s^28p$	^2D	$^2P°$		5/2	3/2	1647330	1818040.8	1.45+08	5.76−02	A	N02
612.05	$1s^24p$	$1s^25d$	$^2P°$	^2D		1/2	3/2	1483912	1647298	4.84+09	2.19+00	A	N02
612.70	$1s^24p$	$1s^25d$	$^2P°$	^2D		3/2	5/2	1484117	1647330	5.82+09	3.96+00	A	N02
612.82	$1s^24p$	$1s^25d$	$^2P°$	^2D		3/2	3/2	1484117	1647298	9.7+08	4.41−01	A	N02
631.41	$1s^24d$	$1s^25f$	^2D	$^2F°$		3/2	5/2	1489104	1647480.7	9.9+09	7.37+00	A	N02
631.59	$1s^24d$	$1s^25f$	^2D	$^2F°$		5/2	7/2	1489166	1647496.7	1.06+10	1.05+01	A	N02
631.65	$1s^24d$	$1s^25f$	^2D	$^2F°$		5/2	5/2	1489166	1647480.7	7.06+08	5.27−01	A	N02
632.61	$1s^24f$	$1s^25g$	$^2F°$	^2G		5/2	7/2	1489436	1647511	1.68+10	1.68+01	A	N02LS
632.74	$1s^24f$	$1s^25g$	$^2F°$	^2G		7/2	7/2,9/2	1489467	1647511	1.0+10	2.24+01	A	N02LS
633.34	$1s^24f$	$1s^25d$	$^2F°$	^2D		5/2	5/2	1489436	1647330	1.0+07	7.51−03	A	N02
633.46	$1s^24f$	$1s^25d$	$^2F°$	^2D		7/2	5/2	1489467	1647330	2.00+08	1.51−01	A	N02
633.46	$1s^24f$	$1s^25d$	$^2F°$	^2D		5/2	3/2	1489436	1647298	2.11+08	1.06−01	A	N02
642.41	$1s^24d$	$1s^25p$	^2D	$^2P°$		3/2	3/2	1489104	1644768	9.6+07	5.03−02	A	N02
642.67	$1s^24d$	$1s^25p$	^2D	$^2P°$		5/2	3/2	1489166	1644768	8.7+08	4.55−01	A	N02
642.84	$1s^24d$	$1s^25p$	^2D	$^2P°$		3/2	1/2	1489104	1644664	9.7+08	2.55−01	A	N02
651.26	$1s^24p$	$1s^25s$	$^2P°$	^2S		1/2	1/2	1483912	1637460	1.43+09	3.90−01	A	N02
652.13	$1s^24p$	$1s^25s$	$^2P°$	^2S		3/2	1/2	1484117	1637460	2.88+09	7.88−01	A	N02
681.87	$1s^25s$	$1s^27p$	^2S	$^2P°$		1/2	3/2	1637460	1784116	5.16+08	3.23−01	A	N02
682.04	$1s^25s$	$1s^27p$	^2S	$^2P°$		1/2	1/2	1637460	1784078	5.20+08	1.63−01	A	N02
712.43	$1s^25p$	$1s^27d$	$^2P°$	^2D		1/2	3/2	1644664	1785029	9.6+08	6.83−01	A	N02
712.90	$1s^25p$	$1s^27d$	$^2P°$	^2D		3/2	5/2	1644768	1785041	1.15+09	1.24+00	A	N02

λ_{Ritz} [a] (Å)	Configurations		Terms		Type	J_i [b]	J_k [b]	E_i (cm^{-1})	E_k (cm^{-1})	A_{ki} [c] (s^{-1})	S_{ik} [c]	Acc. [d]	Ref. [e]
712.96	$1s^25p$	$1s^27d$	$^2P^o$	^2D		3/2	3/2	1644768	1785029	1.92+08	1.37−01	A	N02
725.66	$1s^25d$	$1s^27f$	^2D	$^2F^o$		3/2	5/2	1647298	1785103	1.66+09	1.88+00	A	N02LS
725.83	$1s^25d$	$1s^27f$	^2D	$^2F^o$		5/2	5/2,7/2	1647330	1785103	1.07+09	2.82+00	A	N02LS
726.58	$1s^25f$	$1s^27g$	$^2F^o$	^2G		5/2	7/2	1647480.7	1785111.4	2.17+09	3.28+00	A	N02LS
726.67	$1s^25f$	$1s^27g$	$^2F^o$	^2G		7/2	7/2,9/2	1647496.7	1785111.4	1.28+09	4.38+00	A	N02LS
726.72	$1s^25g$	$1s^27h$	^2G	$^2H^o$		7/2,9/2	9/2,11/2	1647511	1785115.0	2.09+09	8.7+00	A	N02
726.79	$1s^25g$	$1s^27f$	^2G	$^2F^o$		7/2,9/2	5/2,7/2	1647511	1785103	1.91+07	5.07−02	A	N02
726.95	$1s^25f$	$1s^27d$	$^2F^o$	^2D		5/2	5/2	1647480.7	1785041	3.78+06	4.30−03	A	N02
727.02	$1s^25f$	$1s^27d$	$^2F^o$	^2D		5/2	3/2	1647480.7	1785029	7.98+07	6.05−02	A	N02
727.04	$1s^25f$	$1s^27d$	$^2F^o$	^2D		7/2	5/2	1647496.7	1785041	7.58+07	8.6−02	A	N02
730.83	$1s^25p$	$1s^27s$	$^2P^o$	^2S		1/2	1/2	1644664	1781495	3.33+08	1.28−01	A	N02
730.90	$1s^25d$	$1s^27p$	^2D	$^2P^o$		3/2	3/2	1647298	1784116	2.64+07	2.03−02	A	N02
731.07	$1s^25d$	$1s^27p$	^2D	$^2P^o$		5/2	3/2	1647330	1784116	2.38+08	1.84−01	A	N02
731.10	$1s^25d$	$1s^27p$	^2D	$^2P^o$		3/2	1/2	1647298	1784078	2.66+08	1.03−01	A	N02
731.38	$1s^25p$	$1s^27s$	$^2P^o$	^2S		3/2	1/2	1644768	1781495	6.70+08	2.59−01	A	N02
767.52	$1s^26s$	$1s^210p$	^2S	$^2P^o$		1/2	1/2,3/2	1727579	1857869	1.15+08	1.54−01	A	N02
770.409	$1s^22s$	$1s^22p$	^2S	$^2P^o$		1/2	3/2	0.0	129801.2	5.74+08	5.19−01	AA	F04
780.324	$1s^22s$	$1s^22p$	^2S	$^2P^o$		1/2	1/2	0.0	128151.9	5.52+08	2.59−01	AA	F04
790.65	$1s^26p$	$1s^210d$	$^2P^o$	^2D		1/2	3/2	1731708	1858187	2.00+08	1.95−01	A	N02LS
791.02	$1s^26p$	$1s^210d$	$^2P^o$	^2D		3/2	3/2,5/2	1731768	1858187	1.59+08	3.90−01	A	N02LS
798.24	$1s^26p$	$1s^210s$	$^2P^o$	^2S		1/2	1/2	1731708	1856983	7.56+07	3.80−02	A	N02
798.63	$1s^26p$	$1s^210s$	$^2P^o$	^2S		3/2	1/2	1731768	1856983	1.52+08	7.64−02	A	N02
800.09	$1s^26d$	$1s^210f$	^2D	$^2F^o$		3/2	5/2	1733224	1858210.3	3.21+08	4.86−01	A	N02LS
800.20	$1s^26d$	$1s^210f$	^2D	$^2F^o$		5/2	5/2,7/2	1733242	1858210.3	2.06+08	7.29−01	A	N02LS
800.81	$1s^26f$	$1s^210g$	$^2F^o$	^2G		5/2,7/2	7/2,9/2	1733339	1858213.1	4.04+08	1.84+00	A	N02
800.88	$1s^26g$	$1s^210h$	^2G	$^2H^o$		7/2,9/2	9/2,11/2	1733351.4	1858214.3	3.74+08	2.09+00	A	N02
800.90	$1s^26g$	$1s^210f$	^2G	$^2F^o$		7/2,9/2	5/2,7/2	1733351.4	1858210.3	7.49+06	2.66−02	A	N02
800.91	$1s^26h$	$1s^210i$	$^2H^o$	^2I		9/2,11/2	11/2,13/2	1733357.0	1858215.1	2.16+08	1.42+00	A	N02
800.92	$1s^26h$	$1s^210g$	$^2H^o$	^2G		9/2,11/2	7/2,9/2	1733357.0	1858213.1	1.54+06	7.03−03	A	N02

λ_{Ritz} [a] (Å)	Configurations		Terms		Type	J_i [b]	J_k [b]	E_i (cm^{-1})	E_k (cm^{-1})	A_{ki} [c] (s^{-1})	S_{ik} [c]	Acc. [d]	Ref. [e]
800.97	$1s^26f$	$1s^210d$	$^2F^\circ$	^2D		5/2,7/2	3/2,5/2	1733339	1858187	2.26+07	5.73−02	A	N02
802.28	$1s^26d$	$1s^210p$	^2D	$^2P^\circ$		3/2	1/2,3/2	1733224	1857869	2.63+07	4.02−02	A	N02LS
802.39	$1s^26d$	$1s^210p$	^2D	$^2P^\circ$		5/2	3/2	1733242	1857869	5.92+07	6.04−02	A	N02LS
879.62	$1s^26s$	$1s^29p$	^2S	$^2P^\circ$		1/2	1/2,3/2	1727579	1841265	1.56+08	3.14−01	A	N02
909.14	$1s^26p$	$1s^29d$	$^2P^\circ$	^2D		1/2	3/2	1731708	1841702	2.73+08	4.05−01	A	N02LS
909.64	$1s^26p$	$1s^29d$	$^2P^\circ$	^2D		3/2	3/2,5/2	1731768	1841702	2.18+08	8.11−01	A	N02LS
921.57	$1s^26d$	$1s^29f$	^2D	$^2F^\circ$		3/2	5/2	1733224	1841734	4.51+08	1.05+00	A	N02LS
921.73	$1s^26d$	$1s^29f$	^2D	$^2F^\circ$		5/2	5/2,7/2	1733242	1841734	2.90+08	1.57+00	A	N02LS
922.52	$1s^26f$	$1s^29g$	$^2F^\circ$	^2G		5/2,7/2	7/2,9/2	1733339	1841738.2	5.93+08	4.14+00	A	N02
922.61	$1s^26g$	$1s^29h$	^2G	$^2H^\circ$		7/2,9/2	9/2,11/2	1733351.4	1841739.9	5.96+08	5.09+00	A	N02
922.65	$1s^26h$	$1s^29i$	$^2H^\circ$	^2I		9/2,11/2	11/2,13/2	1733357.0	1841740.9	3.96+08	4.00+00	A	N02
922.66	$1s^26g$	$1s^29f$	^2G	$^2F^\circ$		7/2,9/2	5/2,7/2	1733351.4	1841734	1.20+07	6.51−02	A	N02
922.67	$1s^26h$	$1s^29g$	$^2H^\circ$	^2G		9/2,11/2	7/2,9/2	1733357.0	1841738.2	2.66+06	1.86−02	A	N02
922.82	$1s^26f$	$1s^29d$	$^2F^\circ$	^2D		5/2,7/2	3/2,5/2	1733339	1841702	3.43+07	1.33−01	A	N02
923.03	$1s^26p$	$1s^29s$	$^2P^\circ$	^2S		1/2	1/2	1731708	1840047.0	1.08+08	8.4−02	A	N02
923.5	$1s^26p$	$1s^29s$	$^2P^\circ$	^2S		3/2	1/2	1731768	1840047.0	2.17+08	1.69−01	A	N02
925.57	$1s^26d$	$1s^29p$	^2D	$^2P^\circ$		3/2	1/2,3/2	1733224	1841265	3.84+07	9.0−02	A	N02LS
925.73	$1s^26d$	$1s^29p$	^2D	$^2P^\circ$		5/2	3/2	1733242	1841265	8.6+07	1.35−01	A	N02LS
1060.4	$1s^25s$	$1s^26p$	^2S	$^2P^\circ$		1/2	3/2	1637460	1731768	7.25+08	1.71+00	A	N02
1061.0	$1s^25s$	$1s^26p$	^2S	$^2P^\circ$		1/2	1/2	1637460	1731708	7.31+08	8.6−01	A	N02
1105.4	$1s^26s$	$1s^28p$	^2S	$^2P^\circ$		1/2	3/2	1727579	1818040.8	2.14+08	5.70−01	A	N02
1105.8	$1s^26s$	$1s^28p$	^2S	$^2P^\circ$		1/2	1/2	1727579	1818014.9	2.15+08	2.87−01	A	N02
1129.2	$1s^25p$	$1s^26d$	$^2P^\circ$	^2D		1/2	3/2	1644664	1733224	1.43+09	4.08+00	A	N02
1130.3	$1s^25p$	$1s^26d$	$^2P^\circ$	^2D		3/2	5/2	1644768	1733242	1.73+09	7.38+00	A	N02
1130.5	$1s^25p$	$1s^26d$	$^2P^\circ$	^2D		3/2	3/2	1644768	1733224	2.88+08	8.22−01	A	N02
1150.2	$1s^26p$	$1s^28d$	$^2P^\circ$	^2D		1/2	3/2	1731708	1818650	3.83+08	1.15+00	A	N02
1150.9	$1s^26p$	$1s^28d$	$^2P^\circ$	^2D		3/2	5/2	1731768	1818658	4.61+08	2.08+00	A	N02
1151.0	$1s^26p$	$1s^28d$	$^2P^\circ$	^2D		3/2	3/2	1731768	1818650	7.70+07	2.32−01	A	N02
1162.2	$1s^25d$	$1s^26f$	^2D	$^2F^\circ$		3/2	5/2	1647298	1733339	2.77+09	1.29+01	A	N02LS

λ_{Ritz} [a] (Å)	Configurations		Terms		Type	J_i [b]	J_k [b]	E_i (cm^{-1})	E_k (cm^{-1})	A_{ki} [c] (s^{-1})	S_{ik} [c]	Acc. [d]	Ref. [e]
1162.7	$1s^2 5d$	$1s^2 6f$	^2D	$^2F^o$		5/2	5/2,7/2	1647330	1733339	1.78+09	1.93+01	A	N02LS
1164.5	$1s^2 5f$	$1s^2 6g$	$^2F^o$	^2G		5/2	7/2	1647480.7	1733351.4	4.37+09	2.73+01	A	N02LS
1164.8	$1s^2 5f$	$1s^2 6g$	$^2F^o$	^2G		7/2	7/2,9/2	1647496.7	1733351.4	2.59+09	3.63+01	A	N02LS
1164.9	$1s^2 5g$	$1s^2 6h$	^2G	$^2H^o$		7/2,9/2	9/2,11/2	1647511	1733357.0	6.74+09	1.16+02	A	N02
1165.1	$1s^2 5g$	$1s^2 6f$	^2G	$^2F^o$		7/2,9/2	5/2,7/2	1647511	1733339	4.67+07	5.11−01	A	N02
1166.0	$1s^2 5f$	$1s^2 6d$	$^2F^o$	^2D		5/2	5/2	1647480.7	1733242	7.73+06	3.63−02	A	N02
1166.2	$1s^2 5f$	$1s^2 6d$	$^2F^o$	^2D		7/2	5/2	1647496.7	1733242	1.55+08	7.28−01	A	N02
1166.3	$1s^2 5f$	$1s^2 6d$	$^2F^o$	^2D		5/2	3/2	1647480.7	1733224	1.63+08	5.10−01	A	N02
1170.0	$1s^2 6d$	$1s^2 8f$	^2D	$^2F^o$		3/2	5/2	1733224	1818691.1	6.60+08	3.13+00	A	N02LS
1170.3	$1s^2 6d$	$1s^2 8f$	^2D	$^2F^o$		5/2	5/2,7/2	1733242	1818691.1	4.24+08	4.70+00	A	N02LS
1171.4	$1s^2 6f$	$1s^2 8g$	$^2F^o$	^2G		5/2,7/2	7/2,9/2	1733339	1818704.9	9.3+08	1.32+01	A	N02
1171.5	$1s^2 6g$	$1s^2 8h$	^2G	$^2H^o$		7/2,9/2	9/2,11/2	1733351.4	1818709	1.06+09	1.84+01	A	N02
1171.6	$1s^2 6h$	$1s^2 8i$	$^2H^o$	^2I		9/2,11/2	11/2,13/2	1733357.0	1818710	8.9+08	1.84+01	A	N02
1171.7	$1s^2 6h$	$1s^2 8g$	$^2H^o$	^2G		9/2,11/2	7/2,9/2	1733357.0	1818704.9	5.36+06	7.65−02	A	N02
1171.8	$1s^2 6g$	$1s^2 8f$	^2G	$^2F^o$		7/2,9/2	5/2,7/2	1733351.4	1818691.1	2.15+07	2.39−01	A	N02
1172.1	$1s^2 6f$	$1s^2 8d$	$^2F^o$	^2D		5/2,7/2	5/2	1733339	1818658	5.67+07	2.70−01	A	N02LS
1172.2	$1s^2 6f$	$1s^2 8d$	$^2F^o$	^2D		5/2	3/2	1733339	1818650	5.66+07	1.80−01	A	N02LS
1179.0	$1s^2 6d$	$1s^2 8p$	^2D	$^2P^o$		3/2	3/2	1733224	1818040.8	1.49+07	4.82−02	A	N02
1179.3	$1s^2 6d$	$1s^2 8p$	^2D	$^2P^o$		5/2	3/2	1733242	1818040.8	1.34+08	4.35−01	A	N02
1179.4	$1s^2 6d$	$1s^2 8p$	^2D	$^2P^o$		3/2	1/2	1733224	1818014.9	1.50+08	2.44−01	A	N02
1182.3	$1s^2 6p$	$1s^2 8s$	$^2P^o$	^2S		1/2	1/2	1731708	1816292	1.64+08	2.67−01	A	N02
1183.1	$1s^2 6p$	$1s^2 8s$	$^2P^o$	^2S		3/2	1/2	1731768	1816292	3.29+08	5.38−01	A	N02
1183.9	$1s^2 5d$	$1s^2 6p$	^2D	$^2P^o$		3/2	3/2	1647298	1731768	4.82+07	1.58−01	A	N02
1184.3	$1s^2 5d$	$1s^2 6p$	^2D	$^2P^o$		5/2	3/2	1647330	1731768	4.35+08	1.43+00	A	N02
1184.7	$1s^2 5d$	$1s^2 6p$	^2D	$^2P^o$		3/2	1/2	1647298	1731708	4.87+08	7.99−01	A	N02
1206.1	$1s^2 5p$	$1s^2 6s$	$^2P^o$	^2S		1/2	1/2	1644664	1727579	5.82+08	1.01+00	A	N02
1207.6	$1s^2 5p$	$1s^2 6s$	$^2P^o$	^2S		3/2	1/2	1644768	1727579	1.17+09	2.04+00	A	N02
1309.3	$1s^2 7s$	$1s^2 10p$	^2S	$^2P^o$		1/2	1/2,3/2	1781495	1857869	7.62+07	5.06−01	A	N02
1349.4	$1s^2 7p$	$1s^2 10d$	$^2P^o$	^2D		1/2	3/2	1784078	1858187	1.31+08	6.33−01	A	N02LS

λ_{Ritz} [a] (Å)	Configurations		Terms		Type	J_i [b]	J_k [b]	E_i (cm^{-1})	E_k (cm^{-1})	A_{ki} [c] (s^{-1})	S_{ik} [c]	Acc. [d]	Ref. [e]
1350.1	$1s^27p$	$1s^210d$	$^2P^o$	^2D		3/2	3/2,5/2	1784116	1858187	1.04+08	1.27+00	A	N02LS
1366.5	$1s^27d$	$1s^210f$	^2D	$^2F^o$		3/2	5/2	1785029	1858210.3	2.15+08	1.63+00	A	N02LS
1366.7	$1s^27d$	$1s^210f$	^2D	$^2F^o$		5/2	5/2,7/2	1785041	1858210.3	1.38+08	2.44+00	A	N02LS
1367.8	$1s^27f$	$1s^210g$	$^2F^o$	^2G		5/2,7/2	7/2,9/2	1785103	1858213.1	2.91+08	6.61+00	A	N02
1367.9	$1s^27g$	$1s^210h$	^2G	$^2H^o$		7/2,9/2	9/2,11/2	1785111.4	1858214.3	3.24+08	9.0+00	A	N02
1368.0	$1s^27h$	$1s^210i$	$^2H^o$	^2I		9/2,11/2	11/2,13/2	1785115.0	1858215.1	2.97+08	9.8+00	A	N02
1368.0	$1s^27g$	$1s^210f$	^2G	$^2F^o$		7/2,9/2	5/2,7/2	1785111.4	1858210.3	1.08+07	1.90−01	A	N02
1368.0	$1s^27h$	$1s^210g$	$^2H^o$	^2G		9/2,11/2	7/2,9/2	1785115.0	1858213.1	3.88+06	8.8−02	A	N02
1368.0	$1s^27i$	$1s^210h$	^2I	$^2H^o$		11/2,13/2	9/2,11/2	1785117.1	1858214.3	8.6+05	2.40−02	A	N02
1368.3	$1s^27f$	$1s^210d$	$^2F^o$	^2D		5/2,7/2	3/2,5/2	1785103	1858187	2.44+07	3.09−01	A	N02
1371.6	$1s^27p$	$1s^210s$	$^2P^o$	^2S		1/2	1/2	1784078	1856983	5.99+07	1.53−01	A	N02
1372.4	$1s^27p$	$1s^210s$	$^2P^o$	^2S		3/2	1/2	1784116	1856983	1.20+08	3.07−01	A	N02
1372.9	$1s^27d$	$1s^210p$	^2D	$^2P^o$		3/2	1/2,3/2	1785029	1857869	2.36+07	1.81−01	A	N02LS
1373.1	$1s^27d$	$1s^210p$	^2D	$^2P^o$		5/2	3/2	1785041	1857869	5.30+07	2.71−01	A	N02LS
1673.1	$1s^27s$	$1s^29p$	^2S	$^2P^o$		1/2	1/2,3/2	1781495	1841265	1.01+08	1.40+00	A	N02
1735.4	$1s^27p$	$1s^29d$	$^2P^o$	^2D		1/2	3/2	1784078	1841702	1.77+08	1.83+00	A	N02LS
1736.5	$1s^27p$	$1s^29d$	$^2P^o$	^2D		3/2	3/2,5/2	1784116	1841702	1.41+08	3.65+00	A	N02LS
1763.5	$1s^27d$	$1s^29f$	^2D	$^2F^o$		3/2	5/2	1785029	1841734	3.00+08	4.87+00	A	N02LS
1763.9	$1s^27d$	$1s^29f$	^2D	$^2F^o$		5/2	5/2,7/2	1785041	1841734	1.92+08	7.30+00	A	N02LS
1765.7	$1s^27f$	$1s^29g$	$^2F^o$	^2G		5/2,7/2	7/2,9/2	1785103	1841738.2	4.22+08	2.07+01	A	N02
1765.9	$1s^27g$	$1s^29h$	^2G	$^2H^o$		7/2,9/2	9/2,11/2	1785111.4	1841739.9	5.09+08	3.04+01	A	N02
1766.0	$1s^27h$	$1s^29i$	$^2H^o$	^2I		9/2,11/2	11/2,13/2	1785115.0	1841740.9	5.39+08	3.81+01	A	N02
1766.0	$1s^27i$	$1s^29k$	^2I	$^2K^o$		11/2,13/2	13/2,15/2	1785117.1	1841741.7	4.24+08	3.46+01	A	N02
1766.1	$1s^27h$	$1s^29g$	$^2H^o$	^2G		9/2,11/2	7/2,9/2	1785115.0	1841738.2	7.15+06	3.50−01	A	N02
1766.1	$1s^27i$	$1s^29h$	^2I	$^2H^o$		11/2,13/2	9/2,11/2	1785117.1	1841739.9	1.80+06	1.08−01	A	N02
1766.1	$1s^27g$	$1s^29f$	^2G	$^2F^o$		7/2,9/2	5/2,7/2	1785111.4	1841734	1.81+07	6.90−01	A	N02
1766.8	$1s^27f$	$1s^29d$	$^2F^o$	^2D		5/2,7/2	3/2,5/2	1785103	1841702	3.84+07	1.05+00	A	N02
1768.8	$1s^26s$	$1s^27p$	^2S	$^2P^o$		1/2	3/2	1727579	1784116	2.83+08	3.09+00	A	N02
1769.9	$1s^26s$	$1s^27p$	^2S	$^2P^o$		1/2	1/2	1727579	1784078	2.86+08	1.57+00	A	N02

λ_{Ritz} [a] (Å)	Configurations		Terms		Type	J_i [b]	J_k [b]	E_i (cm^{-1})	E_k (cm^{-1})	A_{ki} [c] (s^{-1})	S_{ik} [c]	Acc.[d]	Ref.[e]
1778.2	$1s^27d$	$1s^29p$	^2D	$^2P°$		3/2	1/2,3/2	1785029	1841265	3.53+07	5.88−01	A	N02LS
1778.6	$1s^27d$	$1s^29p$	^2D	$^2P°$		5/2	3/2	1785041	1841265	7.95+07	8.8−01	A	N02LS
1786.7	$1s^27p$	$1s^29s$	$^2P°$	^2S		1/2	1/2	1784078	1840047.0	8.8+07	4.94−01	A	N02
1787.9	$1s^27p$	$1s^29s$	$^2P°$	^2S		3/2	1/2	1784116	1840047.0	1.76+08	9.9−01	A	N02
1875.4	$1s^26p$	$1s^27d$	$^2P°$	^2D		1/2	3/2	1731708	1785029	5.37+08	6.99+00	A	N02
1877.1	$1s^26p$	$1s^27d$	$^2P°$	^2D		3/2	5/2	1731768	1785041	6.46+08	1.27+01	A	N02
1877.5	$1s^26p$	$1s^27d$	$^2P°$	^2D		3/2	3/2	1731768	1785029	1.08+08	1.41+00	A	N02
1927.6	$1s^26d$	$1s^27f$	^2D	$^2F°$		3/2	5/2	1733224	1785103	9.9+08	2.10+01	A	N02LS
1928.2	$1s^26d$	$1s^27f$	^2D	$^2F°$		5/2	5/2,7/2	1733242	1785103	6.36+08	3.15+01	A	N02LS
1931.5	$1s^26f$	$1s^27g$	$^2F°$	^2G		5/2,7/2	7/2,9/2	1733339	1785111.4	1.54+09	9.9+01	A	N02
1931.9	$1s^26g$	$1s^27h$	^2G	$^2H°$		7/2,9/2	9/2,11/2	1733351.4	1785115.0	2.18+09	1.71+02	A	N02
1932.0	$1s^26h$	$1s^27i$	$^2H°$	^2I		9/2,11/2	11/2,13/2	1733357.0	1785117.1	3.04+09	2.81+02	A	N02
1932.2	$1s^26h$	$1s^27g$	$^2H°$	^2G		9/2,11/2	7/2,9/2	1733357.0	1785111.4	1.38+07	8.8−01	A	N02
1932.3	$1s^26g$	$1s^27f$	^2G	$^2F°$		7/2,9/2	5/2,7/2	1733351.4	1785103	4.56+07	2.27+00	A	N02
1934.2	$1s^26f$	$1s^27d$	$^2F°$	^2D		5/2,7/2	5/2	1733339	1785041	1.05+08	2.25+00	A	N02LS
1934.6	$1s^26f$	$1s^27d$	$^2F°$	^2D		5/2	3/2	1733339	1785029	1.05+08	1.50+00	A	N02LS
1964.9	$1s^26d$	$1s^27p$	^2D	$^2P°$		3/2	3/2	1733224	1784116	2.53+07	3.80−01	A	N02
1965.6	$1s^26d$	$1s^27p$	^2D	$^2P°$		5/2	3/2	1733242	1784116	2.29+08	3.43+00	A	N02
1966.4	$1s^26d$	$1s^27p$	^2D	$^2P°$		3/2	1/2	1733224	1784078	2.56+08	1.92+00	A	N02
2007.9	$1s^26p$	$1s^27s$	$^2P°$	^2S		1/2	1/2	1731708	1781495	2.7+08	2.16+00	B	N02
2010.3	$1s^26p$	$1s^27s$	$^2P°$	^2S		3/2	1/2	1731768	1781495	5.4+08	4.4+00	B	N02
2404.4	$1s^28s$	$1s^210p$	^2S	$^2P°$		1/2	1/2,3/2	1816292	1857869	5.28+07	2.18+00	A	N02
2488.5	$1s^28p$	$1s^210d$	$^2P°$	^2D		1/2	3/2	1818014.9	1858187	9.0+07	2.74+00	A	N02LS
2490.1	$1s^28p$	$1s^210d$	$^2P°$	^2D		3/2	3/2,5/2	1818040.8	1858187	7.20+07	5.49+00	A	N02LS
2527.0	$1s^28d$	$1s^210f$	^2D	$^2F°$		3/2	5/2	1818650	1858210.3	1.50+08	7.19+00	A	N02LS
2527.5	$1s^28d$	$1s^210f$	^2D	$^2F°$		5/2	5/2,7/2	1818658	1858210.3	9.7+07	1.08+01	A	N02LS
2529.5	$1s^28f$	$1s^210g$	$^2F°$	^2G		5/2,7/2	7/2,9/2	1818691.1	1858213.1	2.11+08	3.03+01	A	N02
2530.3	$1s^28g$	$1s^210h$	^2G	$^2H°$		7/2,9/2	9/2,11/2	1818704.9	1858214.3	2.58+08	4.55+01	A	N02
2530.5	$1s^28h$	$1s^210i$	$^2H°$	^2I		9/2,11/2	11/2,13/2	1818709	1858215.1	2.93+08	6.10+01	A	N02

λ_{Ritz} [a] (Å)	Configurations		Terms		Type	J_i [b]	J_k [b]	E_i (cm^{-1})	E_k (cm^{-1})	A_{ki} [c] (s^{-1})	S_{ik} [c]	Acc.[d]	Ref.[e]
2530.5	$1s^28g$	$1s^210f$	^2G	$^2F°$		7/2,9/2	5/2,7/2	1818704.9	1858210.3	1.40+07	1.57+00	A	N02
2530.6	$1s^28i$	$1s^210h$	^2I	$^2H°$		11/2,13/2	9/2,11/2	1818710	1858214.3	2.73+06	4.81−01	A	N02
2530.6	$1s^28k$	$1s^210i$	$^2K°$	^2I		13/2,15/2	11/2,13/2	1818711	1858215.1	6.93+05	1.44−01	A	N02
2530.6	$1s^28h$	$1s^210g$	$^2H°$	^2G		9/2,11/2	7/2,9/2	1818709	1858213.1	6.84+06	9.9−01	A	N02
2531.1	$1s^28f$	$1s^210d$	$^2F°$	^2D		5/2,7/2	3/2,5/2	1818691.1	1858187	2.60+07	2.08+00	A	N02
2549.0	$1s^28d$	$1s^210p$	^2D	$^2P°$		3/2	1/2,3/2	1818650	1857869	2.17+07	1.07+00	A	N02LS
2549.5	$1s^28d$	$1s^210p$	^2D	$^2P°$		5/2	3/2	1818658	1857869	4.89+07	1.60+00	A	N02LS
2565.4	$1s^28p$	$1s^210s$	$^2P°$	^2S		1/2	1/2	1818014.9	1856983	5.03+07	8.4−01	A	N02
2567.1	$1s^28p$	$1s^210s$	$^2P°$	^2S		3/2	1/2	1818040.8	1856983	1.01+08	1.69+00	A	N02
2735.5	$1s^27s$	$1s^28p$	^2S	$^2P°$		1/2	3/2	1781495	1818040.8	1.28+08	5.20+00	A	N02
2737.4	$1s^27s$	$1s^28p$	^2S	$^2P°$		1/2	1/2	1781495	1818014.9	1.30+08	2.63+00	A	N02
2820.7	$1s^23s$	$1s^23p$	^2S	$^2P°$		1/2	3/2	1099886.0	1135327.7	7.2+07	3.2+00	B	F04
2860.1	$1s^23s$	$1s^23p$	^2S	$^2P°$		1/2	1/2	1099886.0	1134839.5	6.9+07	1.59+00	B	F04
2891.7	$1s^27p$	$1s^28d$	$^2P°$	^2D		1/2	3/2	1784078	1818650	2.35+08	1.12+01	A	N02
2894.2	$1s^27p$	$1s^28d$	$^2P°$	^2D		3/2	5/2	1784116	1818658	2.83+08	2.04+01	A	N02
2894.8	$1s^27p$	$1s^28d$	$^2P°$	^2D		3/2	3/2	1784116	1818650	4.73+07	2.27+00	A	N02
2969.8	$1s^27d$	$1s^28f$	^2D	$^2F°$		3/2	5/2	1785029	1818691.1	4.20+08	3.26+01	A	N02LS
2970.9	$1s^27d$	$1s^28f$	^2D	$^2F°$		5/2	5/2,7/2	1785041	1818691.1	2.70+08	4.89+01	A	N02LS
2975.2	$1s^27f$	$1s^28g$	$^2F°$	^2G		5/2,7/2	7/2,9/2	1785103	1818704.9	6.29+08	1.47+02	A	N02
2975.5	$1s^27g$	$1s^28h$	^2G	$^2H°$		7/2,9/2	9/2,11/2	1785111.4	1818709	8.6+08	2.45+02	A	N02
2975.8	$1s^27h$	$1s^28i$	$^2H°$	^2I		9/2,11/2	11/2,13/2	1785115.0	1818710	1.15+09	3.89+02	A	N02
2975.9	$1s^27i$	$1s^28k$	^2I	$^2K°$		11/2,13/2	13/2,15/2	1785117.1	1818711	1.53+09	5.96+02	A	N02
2976.0	$1s^27i$	$1s^28h$	^2I	$^2H°$		11/2,13/2	9/2,11/2	1785117.1	1818709	4.88+06	1.40+00	A	N02
2976.2	$1s^27h$	$1s^28g$	$^2H°$	^2G		9/2,11/2	7/2,9/2	1785115.0	1818704.9	1.58+07	3.71+00	A	N02
2977.1	$1s^27g$	$1s^28f$	^2G	$^2F°$		7/2,9/2	5/2,7/2	1785111.4	1818691.1	3.48+07	6.34+00	A	N02
2979.3	$1s^27f$	$1s^28d$	$^2F°$	^2D		5/2,7/2	5/2	1785103	1818658	6.62+07	5.19+00	A	N02LS
2980.0	$1s^27f$	$1s^28d$	$^2F°$	^2D		5/2	3/2	1785103	1818650	6.62+07	3.46+00	A	N02LS
3028.3	$1s^27d$	$1s^28p$	^2D	$^2P°$		3/2	3/2	1785029	1818040.8	1.41+07	7.76−01	A	N02
3029.4	$1s^27d$	$1s^28p$	^2D	$^2P°$		5/2	3/2	1785041	1818040.8	1.28+08	7.01+00	A	N02

λ_{Ritz} [a] (Å)	Configurations		Terms		Type	J_i [b]	J_k [b]	E_i (cm^{-1})	E_k (cm^{-1})	A_{ki} [c] (s^{-1})	S_{ik} [c]	Acc. [d]	Ref. [e]
3030.7	$1s^27d$	$1s^28p$	^2D	$^2P°$		3/2	1/2	1785029	1818014.9	1.43+08	3.93+00	A	N02
3103	$1s^27p$	$1s^28s$	$^2P°$	^2S		1/2	1/2	1784078	1816292	1.38+08	4.08+00	A	N02
3107	$1s^27p$	$1s^28s$	$^2P°$	^2S		3/2	1/2	1784116	1816292	2.78+08	8.23+00	A	N02
4003	$1s^28s$	$1s^29p$	^2S	$^2P°$		1/2	1/2,3/2	1816292	1841265	6.51+07	1.24+01	A	N02
4221	$1s^28p$	$1s^29d$	$^2P°$	^2D		1/2	3/2	1818014.9	1841702	1.16+08	1.73+01	A	N02LS
4225	$1s^28p$	$1s^29d$	$^2P°$	^2D		3/2	3/2,5/2	1818040.8	1841702	9.3+07	3.45+01	A	N02LS
4331	$1s^28d$	$1s^29f$	^2D	$^2F°$		3/2	5/2	1818650	1841734	2.01+08	4.85+01	A	N02LS
4332	$1s^28d$	$1s^29f$	^2D	$^2F°$		5/2	5/2,7/2	1818658	1841734	1.29+08	7.27+01	A	N02LS
4338	$1s^28f$	$1s^29g$	$^2F°$	^2G		5/2,7/2	7/2,9/2	1818691.1	1841738.2	2.93+08	2.13+02	A	N02
4340	$1s^28g$	$1s^29h$	^2G	$^2H°$		7/2,9/2	9/2,11/2	1818704.9	1841739.9	3.87+08	3.44+02	A	N02
4341	$1s^28h$	$1s^29i$	$^2H°$	^2I		9/2,11/2	11/2,13/2	1818709	1841740.9	5.04+08	5.30+02	A	N02
4341	$1s^28i$	$1s^29k$	^2I	$^2K°$		11/2,13/2	13/2,15/2	1818710	1841741.7	6.51+08	7.89+02	A	N02
4341	$1s^28k$	$1s^29l$	$^2K°$	^2L		13/2,15/2	15/2,17/2	1818711	1841742.2	8.3+08	1.15+03	A	N02
4341	$1s^28k$	$1s^29i$	$^2K°$	^2I		13/2,15/2	11/2,13/2	1818711	1841740.9	1.98+06	2.08+00	A	N02
4341	$1s^28i$	$1s^29h$	^2I	$^2H°$		11/2,13/2	9/2,11/2	1818710	1841739.9	6.33+06	5.63+00	A	N02
4341	$1s^28h$	$1s^29g$	$^2H°$	^2G		9/2,11/2	7/2,9/2	1818709	1841738.2	1.37+07	9.9+00	A	N02
4341	$1s^28g$	$1s^29f$	^2G	$^2F°$		7/2,9/2	5/2,7/2	1818704.9	1841734	2.49+07	1.41+01	A	N02
4345	$1s^28f$	$1s^29d$	$^2F°$	^2D		5/2,7/2	3/2,5/2	1818691.1	1841702	4.24+07	1.72+01	A	N02
4421	$1s^28d$	$1s^29p$	^2D	$^2P°$		3/2	1/2,3/2	1818650	1841265	3.35+07	8.6+00	A	N02LS
4422	$1s^28d$	$1s^29p$	^2D	$^2P°$		5/2	3/2	1818658	1841265	7.53+07	1.29+01	A	N02LS
4538	$1s^28p$	$1s^29s$	$^2P°$	^2S		1/2	1/2	1818014.9	1840047.0	7.64+07	7.05+00	A	N02
4543	$1s^28p$	$1s^29s$	$^2P°$	^2S		3/2	1/2	1818040.8	1840047.0	1.54+08	1.42+01	A	N02
5609	$1s^29s$	$1s^210p$	^2S	$^2P°$		1/2	1/2,3/2	1840047.0	1857869	3.57+07	1.87+01	A	N02
5908	$1s^29p$	$1s^210d$	$^2P°$	^2D		1/2,3/2	3/2,5/2	1841265	1858187	7.46+07	7.60+01	A	N02
6056	$1s^29d$	$1s^210f$	^2D	$^2F°$		3/2,5/2	5/2,7/2	1841702	1858210.3	1.13+08	1.74+02	A	N02
6067	$1s^29f$	$1s^210g$	$^2F°$	^2G		5/2,7/2	7/2,9/2	1841734	1858213.1	1.50+08	2.98+02	A	N02
6068	$1s^29g$	$1s^210h$	^2G	$^2H°$		7/2,9/2	9/2,11/2	1841738.2	1858214.3	1.94+08	4.70+02	A	N02
6068	$1s^29h$	$1s^210i$	$^2H°$	^2I		9/2,11/2	11/2,13/2	1841739.9	1858215.1	2.47+08	7.08+02	A	N02
6069	$1s^29k$	$1s^210i$	$^2K°$	^2I		13/2,15/2	11/2,13/2	1841741.7	1858215.1	2.83+06	8.12+00	A	N02

λ_{Ritz} [a] (Å)	Configurations		Terms		Type	J_i [b]	J_k [b]	E_i (cm^{-1})	E_k (cm^{-1})	A_{ki} [c] (s^{-1})	S_{ik} [c]	Acc.[d]	Ref.[e]
6069	$1s^2 9i$	$1s^2 10h$	^2I	$^2H^o$		11/2,13/2	9/2,11/2	1841740.9	1858214.3	6.01+06	1.46+01	A	N02
6069	$1s^2 9h$	$1s^2 10g$	$^2H^o$	^2G		9/2,11/2	7/2,9/2	1841739.9	1858213.1	1.08+07	2.14+01	A	N02
6069	$1s^2 9g$	$1s^2 10f$	^2G	$^2F^o$		7/2,9/2	5/2,7/2	1841738.2	1858210.3	1.76+07	2.72+01	A	N02
6076	$1s^2 9f$	$1s^2 10d$	$^2F^o$	^2D		5/2,7/2	3/2,5/2	1841734	1858187	2.78+07	3.08+01	A	N02
6184	$1s^2 9d$	$1s^2 10p$	^2D	$^2P^o$		3/2,5/2	1/2,3/2	1841702	1857869	5.17+07	3.62+01	A	N02
6360	$1s^2 9p$	$1s^2 10s$	$^2P^o$	^2S		1/2,3/2	1/2	1841265	1856983	1.36+08	3.45+01	A	N02
6894	$1s^2 4s$	$1s^2 4p$	^2S	$^2P^o$		1/2	3/2	1469616	1484117	1.67+07	1.08+01	B	N02
6993	$1s^2 4s$	$1s^2 4p$	^2S	$^2P^o$		1/2	1/2	1469616	1483912	1.59+07	5.4+00	B	N02
7993	$1s^2 3p$	$1s^2 3d$	$^2P^o$	^2D		1/2	3/2	1134839.5	1147347	2.10+06	2.12+00	B	F04
8217	$1s^2 3p$	$1s^2 3d$	$^2P^o$	^2D		3/2	5/2	1135327.7	1147494	2.32+06	3.8+00	B	F04
8318	$1s^2 3p$	$1s^2 3d$	$^2P^o$	^2D		3/2	3/2	1135327.7	1147347	3.7+05	4.2−01	B	F04
13680	$1s^2 5s$	$1s^2 5p$	^2S	$^2P^o$		1/2	3/2	1637460	1644768	5.4+06	2.7+01	C	N02
13880	$1s^2 5s$	$1s^2 5p$	^2S	$^2P^o$		1/2	1/2	1637460	1644664	5.1+06	1.4+01	C	N02
19260	$1s^2 4p$	$1s^2 4d$	$^2P^o$	^2D		1/2	3/2	1483912	1489104	6.4+05	9.0+00	C	N02
19800	$1s^2 4p$	$1s^2 4d$	$^2P^o$	^2D		3/2	5/2	1484117	1489166	7.1+05	1.6+01	C	N02
20050	$1s^2 4p$	$1s^2 4d$	$^2P^o$	^2D		3/2	3/2	1484117	1489104	1.1+05	1.8+00	C	N02
23870	$1s^2 6s$	$1s^2 6p$	^2S	$^2P^o$		1/2	3/2	1727579	1731768	2.1+06	5.7+01	C	N02
24220	$1s^2 6s$	$1s^2 6p$	^2S	$^2P^o$		1/2	1/2	1727579	1731708	2.0+06	2.9+01	C	N02
38200	$1s^2 7s$	$1s^2 7p$	^2S	$^2P^o$		1/2	3/2	1781495	1784116	9.8+05	1.07+02	A	N02
38700	$1s^2 7s$	$1s^2 7p$	^2S	$^2P^o$		1/2	1/2	1781495	1784078	9.4+05	5.36+01	A	N02
57200	$1s^2 8s$	$1s^2 8p$	^2S	$^2P^o$		1/2	3/2	1816292	1818040.8	4.99+05	1.84+02	A	N02
58000	$1s^2 8s$	$1s^2 8p$	^2S	$^2P^o$		1/2	1/2	1816292	1818014.9	4.77+05	9.2+01	A	N02
66000	$1s^2 6p$	$1s^2 6d$	$^2P^o$	^2D		1/2	3/2	1731708	1733224	9.6+04	5.4+01	C	N02
67800	$1s^2 6p$	$1s^2 6d$	$^2P^o$	^2D		3/2	5/2	1731768	1733242	1.1+05	9.7+01	C	N02
68700	$1s^2 6p$	$1s^2 6d$	$^2P^o$	^2D		3/2	3/2	1731768	1733224	1.7+04	1.1+01	C	N02
82100	$1s^2 9s$	$1s^2 9p$	^2S	$^2P^o$		1/2	1/2,3/2	1840047.0	1841265	2.71+05	4.45+02	A	N02
105000	$1s^2 7p$	$1s^2 7d$	$^2P^o$	^2D		1/2	3/2	1784078	1785029	4.51+04	1.04+02	A	N02
108000	$1s^2 7p$	$1s^2 7d$	$^2P^o$	^2D		3/2	5/2	1784116	1785041	4.98+04	1.86+02	A	N02
110000	$1s^2 7p$	$1s^2 7d$	$^2P^o$	^2D		3/2	3/2	1784116	1785029	7.98+03	2.07+01	A	N02

λ_{Ritz} [a] (Å)	Configurations		Terms		Type	J_i [b]	J_k [b]	E_i (cm^{-1})	E_k (cm^{-1})	A_{ki} [c] (s^{-1})	S_{ik} [c]	Acc. [d]	Ref. [e]
113000	$1s^2 10s$	$1s^2 10p$	^2S	$^2P^o$		1/2	1/2,3/2	1856983	1857869	1.60+05	6.80+02	A	N02
157000	$1s^2 8p$	$1s^2 8d$	$^2P^o$	^2D		1/2	3/2	1818014.9	1818650	2.3+04	1.8+02	C	N02
162000	$1s^2 8p$	$1s^2 8d$	$^2P^o$	^2D		3/2	5/2	1818040.8	1818658	2.6+04	3.2+02	C	N02
164000	$1s^2 8p$	$1s^2 8d$	$^2P^o$	^2D		3/2	3/2	1818040.8	1818650	4.1+03	3.6+01	C	N02
229000	$1s^2 9p$	$1s^2 9d$	$^2P^o$	^2D		1/2,3/2	3/2,5/2	1841265	1841702	1.48+04	8.8+02	A	N02
314000	$1s^2 10p$	$1s^2 10d$	$^2P^o$	^2D		1/2,3/2	3/2,5/2	1857869	1858187	8.8+03	1.35+03	A	N02
Ne IX (He-like)													
10.643	$1s^2$	1s6p	^1S	$^1P^o$		0	1	0	9396200.	3.01+11	5.37−04	AA	J02
10.644	$1s^2$	1s6p	^1S	$^3P^o$		0	1	0	9395000.	2.067+08	3.69−07	AA	J02
10.764	$1s^2$	1s5p	^1S	$^1P^o$		0	1	0	9290000.	5.20+11	9.60−04	AA	J02
10.768	$1s^2$	1s5p	^1S	$^3P^o$		0	1	0	9286660.	3.52+08	6.51−07	AA	J02
11.000	$1s^2$	1s4p	^1S	$^1P^o$		0	1	0	9090630.	1.024+12	2.018−03	AA	J02
11.008	$1s^2$	1s4p	^1S	$^3P^o$		0	1	0	9084040.	6.84+08	1.351−06	AA	J02
11.547	$1s^2$	1s3p	^1S	$^1P^o$		0	1	0	8660530.	2.478+12	5.65−03	AA	J02
11.568	$1s^2$	1s3p	^1S	$^3P^o$		0	1	0	8644500.	1.604+09	3.68−06	AA	J02
13.447	$1s^2$	1s2p	^1S	$^1P^o$		0	1	0	7436560.	8.85+12	3.19−02	AA	J02
13.553	$1s^2$	1s2p	^1S	$^3P^o$		0	1	0	7378500.	5.25+09	1.936−05	AA	J02
47.731	1s2s	1s6p	^3S	$^3P^o$		1	0	7299940.	9395000.	1.957+10	1.050−03	AA	J02LS
47.731	1s2s	1s6p	^3S	$^3P^o$		1	1	7299940.	9395000.	1.957+10	3.15−03	AA	J02LS
47.731	1s2s	1s6p	^3S	$^3P^o$		1	2	7299940.	9395000.	1.957+10	5.25−03	AA	J02LS
49.46	1s2p	1s6d	$^3P^o$	^3D		0	1	7378200.	9400000.	2.016+10	3.61−03	AA	J02LS
49.47	1s2p	1s6d	$^3P^o$	^3D		1	1	7378500.	9400000.	1.511+10	2.71−03	AA	J02LS
49.47	1s2p	1s6d	$^3P^o$	^3D		1	2	7378500.	9400000.	2.72+10	8.13−03	AA	J02LS
49.51	1s2p	1s6d	$^3P^o$	^3D		2	1	7380050.	9400000.	1.005+09	1.806−04	AA	J02LS
49.51	1s2p	1s6d	$^3P^o$	^3D		2	2	7380050.	9400000.	9.05+09	2.71−03	AA	J02LS
49.51	1s2p	1s6d	$^3P^o$	^3D		2	3	7380050.	9400000.	3.62+10	1.517−02	AA	J02LS
49.664	1s2s	1s6p	^1S	$^1P^o$		0	1	7382680.	9396200.	1.864+10	3.38−03	AA	J02
49.694	1s2s	1s6p	^1S	$^3P^o$		0	1	7382680.	9395000.	1.357+07	2.466−06	AA	J02

λ_{Ritz} [a] (Å)	Configurations		Terms		Type	J_i [b]	J_k [b]	E_i (cm^{-1})	E_k (cm^{-1})	A_{ki} [c] (s^{-1})	S_{ik} [c]	Acc. [d]	Ref. [e]
50.332	1s2s	1s5p	^3S	$^3P^o$		1	2	7299940.	9286760.	3.36+10	1.056−02	AA	J02LS
50.334	1s2s	1s5p	^3S	$^3P^o$		1	1	7299940.	9286660.	3.36+10	6.34−03	AA	J02LS
50.335	1s2s	1s5p	^3S	$^3P^o$		1	0	7299940.	9286640.	3.36+10	2.112−03	AA	J02LS
52.348	1s2p	1s5d	$^3P^o$	^3D		0	1	7378200.	9288500.	3.66+10	7.78−03	AA	J02LS
52.356	1s2p	1s5d	$^3P^o$	^3D		1	1	7378500.	9288500.	2.75+10	5.83−03	AA	J02LS
52.356	1s2p	1s5d	$^3P^o$	^3D		1	2	7378500.	9288500.	4.94+10	1.750−02	AA	J02LS
52.399	1s2p	1s5d	$^3P^o$	^3D		2	1	7380050.	9288500.	1.826+09	3.89−04	AA	J02LS
52.399	1s2p	1s5d	$^3P^o$	^3D		2	2	7380050.	9288500.	1.643+10	5.83−03	AA	J02LS
52.399	1s2p	1s5d	$^3P^o$	^3D		2	3	7380050.	9288500.	6.57+10	3.27−02	AA	J02LS
52.407	1s2p	1s5s	$^3P^o$	^1S		1	0	7378500.	9286650.	5.02+06	3.57−07	AA	J02
52.430	1s2s	1s5p	^1S	$^1P^o$		0	1	7382680.	9290000.	3.19+10	6.82−03	AA	J02
52.521	1s2p	1s5s	$^3P^o$	^3S		0	1	7378200.	9282200.	1.251+09	2.68−04	AA	J02LS
52.522	1s2s	1s5p	^1S	$^3P^o$		0	1	7382680.	9286660.	2.303+07	4.94−06	AA	J02
52.529	1s2p	1s5s	$^3P^o$	^3S		1	1	7378500.	9282200.	3.75+09	8.05−04	AA	J02LS
52.572	1s2p	1s5s	$^3P^o$	^3S		2	1	7380050.	9282200.	6.24+09	1.342−03	AA	J02LS
54.051	1s2p	1s5s	$^1P^o$	^1S		1	0	7436560.	9286650.	1.057+10	8.24−04	AA	J02
56.045	1s2s	1s4p	^3S	$^3P^o$		1	2	7299940.	9084230.	6.49+10	2.82−02	AA	J02LS
56.051	1s2s	1s4p	^3S	$^3P^o$		1	1	7299940.	9084040.	6.49+10	1.693−02	AA	J02LS
56.052	1s2s	1s4p	^3S	$^3P^o$		1	0	7299940.	9084000.	6.49+10	5.64−03	AA	J02LS
58.408	1s2p	1s4d	$^3P^o$	^3D		0	1	7378200.	9090300.	7.93+10	2.338−02	AA	J02LS
58.417	1s2p	1s4d	$^3P^o$	^3D		1	2	7378500.	9090340.	1.070+11	5.26−02	AA	J02LS
58.418	1s2p	1s4d	$^3P^o$	^3D		1	1	7378500.	9090300.	5.94+10	1.754−02	AA	J02LS
58.468	1s2p	1s4d	$^3P^o$	^3D		2	3	7380050.	9090390.	1.422+11	9.82−02	AA	J02LS
58.470	1s2p	1s4d	$^3P^o$	^3D		2	2	7380050.	9090340.	3.56+10	1.754−02	AA	J02LS
58.471	1s2p	1s4d	$^3P^o$	^3D		2	1	7380050.	9090300.	3.95+09	1.169−03	AA	J02LS
58.550	1s2s	1s4p	^1S	$^1P^o$		0	1	7382680.	9090630.	6.18+10	1.837−02	AA	J02
58.632	1s2p	1s4s	$^3P^o$	^1S		1	0	7378500.	9084060.	1.003+07	9.98−07	AA	J02
58.777	1s2s	1s4p	^1S	$^3P^o$		0	1	7382680.	9084040.	4.42+07	1.329−05	AA	J02
58.928	1s2p	1s4s	$^3P^o$	^3S		0	1	7378200.	9075200.	2.53+09	7.67−04	AA	J02LS

λ_{Ritz} [a] (Å)	Configurations		Terms		Type	J_i [b]	J_k [b]	E_i (cm^{-1})	E_k (cm^{-1})	A_{ki} [c] (s^{-1})	S_{ik} [c]	Acc. [d]	Ref. [e]
58.938	1s2p	1s4s	$^3P^o$	^3S		1	1	7378500.	9075200.	7.59+09	2.302−03	AA	J02LS
58.992	1s2p	1s4s	$^3P^o$	^3S		2	1	7380050.	9075200.	1.262+10	3.84−03	AA	J02LS
60.698	1s2p	1s4s	$^1P^o$	^1S		1	0	7436560.	9084060.	2.117+10	2.337−03	AA	J02
74.348	1s2s	1s3p	^3S	$^3P^o$		1	2	7299940.	8644960.	1.463+11	1.483−01	AA	J02LS
74.374	1s2s	1s3p	^3S	$^3P^o$		1	1	7299940.	8644500.	1.461+11	8.90−02	AA	J02LS
74.379	1s2s	1s3p	^3S	$^3P^o$		1	0	7299940.	8644400.	1.461+11	2.97−02	AA	J02LS
78.198	1s2p	1s3d	$^3P^o$	^3D		0	1	7378200.	8657000.	2.410+11	1.706−01	AA	J02LS
78.209	1s2p	1s3d	$^3P^o$	^3D		1	2	7378500.	8657120.	3.25+11	3.84−01	AA	J02LS
78.217	1s2p	1s3d	$^3P^o$	^3D		1	1	7378500.	8657000.	1.806+11	1.280−01	AA	J02LS
78.256	1s2s	1s3p	^1S	$^1P^o$		0	1	7382680.	8660530.	1.397+11	9.91−02	AA	J02
78.300	1s2p	1s3d	$^3P^o$	^3D		2	3	7380050.	8657190.	4.32+11	7.17−01	AA	J02LS
78.304	1s2p	1s3d	$^3P^o$	^3D		2	2	7380050.	8657120.	1.080+11	1.280−01	AA	J02LS
78.312	1s2p	1s3d	$^3P^o$	^3D		2	1	7380050.	8657000.	1.200+10	8.53−03	AA	J02LS
78.965	1s2p	1s3s	$^3P^o$	^1S		1	0	7378500.	8644880.	2.442+07	5.93−06	AA	J02
79.251	1s2s	1s3p	^1S	$^3P^o$		0	1	7382680.	8644500.	9.90+07	7.29−05	AA	J02
80.334	1s2p	1s3s	$^3P^o$	^3S		0	1	7378200.	8623000.	6.40+09	4.91−03	AA	J02LS
80.354	1s2p	1s3s	$^3P^o$	^3S		1	1	7378500.	8623000.	1.918+10	1.473−02	AA	J02LS
80.454	1s2p	1s3s	$^3P^o$	^3S		2	1	7380050.	8623000.	3.18+10	2.455−02	AA	J02LS
81.577	1s2p	1s3d	$^1P^o$	^1D		1	2	7436560.	8662400.	4.02+11	5.39−01	AA	J02
82.760	1s2p	1s3s	$^1P^o$	^1S		1	0	7436560.	8644880.	5.23+10	1.462−02	AA	J02
129.53	1s3s	1s6p	^3S	$^3P^o$		1	0	8623000.	9395000.	6.16+09	6.60−03	AA	J02LS
129.53	1s3s	1s6p	^3S	$^3P^o$		1	1	8623000.	9395000.	6.16+09	1.981−02	AA	J02LS
129.53	1s3s	1s6p	^3S	$^3P^o$		1	2	8623000.	9395000.	6.16+09	3.30−02	AA	J02LS
132.3	1s3p	1s6d	$^3P^o$	^3D		0	1	8644400.	9400000.	7.00+09	2.404−02	AA	J02LS
132.4	1s3p	1s6d	$^3P^o$	^3D		1	1	8644500.	9400000.	5.25+09	1.803−02	AA	J02LS
132.4	1s3p	1s6d	$^3P^o$	^3D		1	2	8644500.	9400000.	9.45+09	5.41−02	AA	J02LS
132.4	1s3p	1s6d	$^3P^o$	^3D		2	1	8644960.	9400000.	3.49+08	1.202−03	AA	J02LS
132.4	1s3p	1s6d	$^3P^o$	^3D		2	2	8644960.	9400000.	3.14+09	1.803−02	AA	J02LS
132.4	1s3p	1s6d	$^3P^o$	^3D		2	3	8644960.	9400000.	1.258+10	1.010−01	AA	J02LS

λ_{Ritz} [a] (Å)	Configurations		Terms		Type	J_i [b]	J_k [b]	E_i (cm^{-1})	E_k (cm^{-1})	A_{ki} [c] (s^{-1})	S_{ik} [c]	Acc. [d]	Ref. [e]
133.10	1s3s	1s6p	^1S	$^1P°$		0	1	8644880.	9396200.	6.05+09	2.113−02	AA	J02
133.31	1s3s	1s6p	^1S	$^3P°$		0	1	8644880.	9395000.	4.53+06	1.588−05	AA	J02
135.0	1s3d	1s6f	^1D	$^1F°$		2	3	8662400.	9403000.	1.115+10	9.48−02	AA	J02
135.50	1s3d	1s6p	^3D	$^3P°$		1	0	8657000.	9395000.	6.16+08	7.56−04	AA	J02LS
135.50	1s3d	1s6p	^3D	$^3P°$		1	1	8657000.	9395000.	1.540+08	5.67−04	AA	J02LS
135.50	1s3d	1s6p	^3D	$^3P°$		1	2	8657000.	9395000.	6.16+06	3.78−05	AA	J02LS
135.52	1s3d	1s6p	^3D	$^3P°$		2	1	8657120.	9395000.	4.62+08	1.701−03	AA	J02LS
135.52	1s3d	1s6p	^3D	$^3P°$		2	2	8657120.	9395000.	9.23+07	5.67−04	AA	J02LS
135.54	1s3d	1s6p	^3D	$^3P°$		3	2	8657190.	9395000.	5.17+08	3.18−03	AA	J02LS
136.28	1s3d	1s6p	^1D	$^1P°$		2	1	8662400.	9396200.	4.76+08	1.784−03	AA	J02
150.66	1s3s	1s5p	^3S	$^3P°$		1	2	8623000.	9286760.	1.038+10	8.76−02	AA	J02LS
150.68	1s3s	1s5p	^3S	$^3P°$		1	1	8623000.	9286660.	1.038+10	5.26−02	AA	J02LS
150.68	1s3s	1s5p	^3S	$^3P°$		1	0	8623000.	9286640.	1.038+10	1.752−02	AA	J02LS
155.01	1s3s	1s5p	^1S	$^1P°$		0	1	8644880.	9290000.	1.021+10	5.63−02	AA	J02
155.26	1s3p	1s5d	$^3P°$	^3D		0	1	8644400.	9288500.	1.249+10	6.92−02	AA	J02LS
155.28	1s3p	1s5d	$^3P°$	^3D		1	1	8644500.	9288500.	9.36+09	5.19−02	AA	J02LS
155.28	1s3p	1s5d	$^3P°$	^3D		1	2	8644500.	9288500.	1.685+10	1.557−01	AA	J02LS
155.39	1s3p	1s5d	$^3P°$	^3D		2	1	8644960.	9288500.	6.23+08	3.46−03	AA	J02LS
155.39	1s3p	1s5d	$^3P°$	^3D		2	2	8644960.	9288500.	5.60+09	5.19−02	AA	J02LS
155.39	1s3p	1s5d	$^3P°$	^3D		2	3	8644960.	9288500.	2.242+10	2.91−01	AA	J02LS
155.73	1s3p	1s5s	$^3P°$	^1S		1	0	8644500.	9286650.	3.92+06	7.31−06	AA	J02
155.82	1s3s	1s5p	^1S	$^3P°$		0	1	8644880.	9286660.	7.61+06	4.26−05	AA	J02
156.79	1s3p	1s5s	$^3P°$	^3S		0	1	8644400.	9282200.	8.51+08	4.86−03	AA	J02LS
156.81	1s3p	1s5s	$^3P°$	^3S		1	1	8644500.	9282200.	2.55+09	1.457−02	AA	J02LS
156.93	1s3p	1s5s	$^3P°$	^3S		2	1	8644960.	9282200.	4.24+09	2.428−02	AA	J02LS
158.23	1s3d	1s5f	^1D	$^1F°$		2	3	8662400.	9294400.	2.333+10	3.19−01	AA	J02
158.79	1s3d	1s5p	^3D	$^3P°$		1	2	8657000.	9286760.	1.171+07	1.157−04	AA	J02LS
158.82	1s3d	1s5p	^3D	$^3P°$		1	1	8657000.	9286660.	2.93+08	1.735−03	AA	J02LS
158.82	1s3d	1s5p	^3D	$^3P°$		2	2	8657120.	9286760.	1.755+08	1.735−03	AA	J02LS

λ_{Ritz} [a] (Å)	Configurations		Terms		Type	J_i [b]	J_k [b]	E_i (cm^{-1})	E_k (cm^{-1})	A_{ki} [c] (s^{-1})	S_{ik} [c]	Acc. [d]	Ref. [e]
158.82	1s3d	1s5p	^{3}D	^{3}P°		1	0	8657000.	9286640.	1.170+09	2.314−03	AA	J02LS
158.84	1s3d	1s5p	^{3}D	^{3}P°		3	2	8657190.	9286760.	9.83+08	9.72−03	AA	J02LS
158.85	1s3d	1s5p	^{3}D	^{3}P°		2	1	8657120.	9286660.	8.77+08	5.21−03	AA	J02LS
159.34	1s3d	1s5p	^{1}D	^{1}P°		2	1	8662400.	9290000.	9.01+08	5.39−03	AA	J02
159.71	1s3p	1s5s	^{1}P°	^{1}S		1	0	8660530.	9286650.	7.10+09	1.429−02	AA	J02
216.81	1s3s	1s4p	^{3}S	^{3}P°		1	2	8623000.	9084230.	1.881+10	4.73−01	AA	J02LS
216.90	1s3s	1s4p	^{3}S	^{3}P°		1	1	8623000.	9084040.	1.879+10	2.84−01	AA	J02LS
216.92	1s3s	1s4p	^{3}S	^{3}P°		1	0	8623000.	9084000.	1.878+10	9.46−02	AA	J02LS
224.27	1s3p	1s4d	^{3}P°	^{3}D		0	1	8644400.	9090300.	2.53+10	4.22−01	AA	J02LS
224.30	1s3p	1s4d	^{3}P°	^{3}D		1	2	8644500.	9090340.	3.41+10	9.49−01	AA	J02LS
224.32	1s3p	1s4d	^{3}P°	^{3}D		1	1	8644500.	9090300.	1.893+10	3.16−01	AA	J02LS
224.34	1s3s	1s4p	^{1}S	^{1}P°		0	1	8644880.	9090630.	1.859+10	3.11−01	AA	J02
224.50	1s3p	1s4d	^{3}P°	^{3}D		2	3	8644960.	9090390.	4.53+10	1.772+00	AA	J02LS
224.53	1s3p	1s4d	^{3}P°	^{3}D		2	2	8644960.	9090340.	1.133+10	3.16−01	AA	J02LS
224.55	1s3p	1s4d	^{3}P°	^{3}D		2	1	8644960.	9090300.	1.258+09	2.109−02	AA	J02LS
227.50	1s3p	1s4s	^{3}P°	^{1}S		1	0	8644500.	9084060.	7.95+06	4.62−05	AA	J02
227.71	1s3s	1s4p	^{1}S	^{3}P°		0	1	8644880.	9084040.	1.399+07	2.446−04	AA	J02
231.48	1s3d	1s4f	^{1}D	^{1}F°		2	3	8662400.	9094400.	6.95+10	2.98+00	AA	J02
232.13	1s3p	1s4s	^{3}P°	^{3}S		0	1	8644400.	9075200.	1.771+09	3.28−02	AA	J02LS
232.18	1s3p	1s4s	^{3}P°	^{3}S		1	1	8644500.	9075200.	5.31+09	9.84−02	AA	J02LS
232.43	1s3p	1s4s	^{3}P°	^{3}S		2	1	8644960.	9075200.	8.82+09	1.640−01	AA	J02LS
233.52	1s3d	1s4p	^{1}D	^{1}P°		2	1	8662400.	9090630.	2.085+09	3.93−02	AA	J02
234.07	1s3d	1s4p	^{3}D	^{3}P°		1	2	8657000.	9084230.	2.74+07	8.67−04	AA	J02LS
234.13	1s3d	1s4p	^{3}D	^{3}P°		2	2	8657120.	9084230.	4.10+08	1.300−02	AA	J02LS
234.17	1s3d	1s4p	^{3}D	^{3}P°		3	2	8657190.	9084230.	2.297+09	7.28−02	AA	J02LS
234.17	1s3d	1s4p	^{3}D	^{3}P°		1	1	8657000.	9084040.	6.84+08	1.300−02	AA	J02LS
234.19	1s3d	1s4p	^{3}D	^{3}P°		1	0	8657000.	9084000.	2.73+09	1.733−02	AA	J02LS
234.24	1s3d	1s4p	^{3}D	^{3}P°		2	1	8657120.	9084040.	2.049+09	3.90−02	AA	J02LS
236.11	1s3p	1s4s	^{1}P°	^{1}S		1	0	8660530.	9084060.	1.460+10	9.48−02	AA	J02

λ_{Ritz} [a] (Å)	Configurations		Terms		Type	J_i [b]	J_k [b]	E_i (cm^{-1})	E_k (cm^{-1})	A_{ki} [c] (s^{-1})	S_{ik} [c]	Acc. [d]	Ref. [e]
312.7	1s4s	1s6p	^3S	$^3P^o$		1	0	9075200.	9395000.	2.76+09	4.16−02	AA	J02LS
312.7	1s4s	1s6p	^3S	$^3P^o$		1	1	9075200.	9395000.	2.76+09	1.249−01	AA	J02LS
312.7	1s4s	1s6p	^3S	$^3P^o$		1	2	9075200.	9395000.	2.76+09	2.082−01	AA	J02LS
316.5	1s4p	1s6d	$^3P^o$	^3D		0	1	9084000.	9400000.	3.11+09	1.462−01	AA	J02LS
316.5	1s4p	1s6d	$^3P^o$	^3D		1	1	9084040.	9400000.	2.335+09	1.096−01	AA	J02LS
316.5	1s4p	1s6d	$^3P^o$	^3D		1	2	9084040.	9400000.	4.20+09	3.29−01	AA	J02LS
316.7	1s4p	1s6d	$^3P^o$	^3D		2	1	9084230.	9400000.	1.554+08	7.31−03	AA	J02LS
316.7	1s4p	1s6d	$^3P^o$	^3D		2	2	9084230.	9400000.	1.399+09	1.096−01	AA	J02LS
316.7	1s4p	1s6d	$^3P^o$	^3D		2	3	9084230.	9400000.	5.59+09	6.14−01	AA	J02LS
320.4	1s4s	1s6p	^1S	$^1P^o$		0	1	9084060.	9396200.	2.75+09	1.340−01	AA	J02
321.6	1s4s	1s6p	^1S	$^3P^o$		0	1	9084060.	9395000.	2.105+06	1.037−04	AA	J02
328.2	1s4d	1s6p	^3D	$^3P^o$		1	0	9090300.	9395000.	7.26+08	1.267−02	AA	J02LS
328.2	1s4d	1s6p	^3D	$^3P^o$		1	1	9090300.	9395000.	1.816+08	9.50−03	AA	J02LS
328.2	1s4d	1s6p	^3D	$^3P^o$		1	2	9090300.	9395000.	7.26+06	6.34−04	AA	J02LS
328.2	1s4d	1s6p	^3D	$^3P^o$		2	1	9090340.	9395000.	5.45+08	2.85−02	AA	J02LS
328.2	1s4d	1s6p	^3D	$^3P^o$		2	2	9090340.	9395000.	1.089+08	9.50−03	AA	J02LS
328.3	1s4d	1s6p	^3D	$^3P^o$		3	2	9090390.	9395000.	6.10+08	5.32−02	AA	J02LS
472.7	1s4s	1s5p	^3S	$^3P^o$		1	2	9075200.	9286760.	4.39+09	1.145+00	AA	J02LS
472.9	1s4s	1s5p	^3S	$^3P^o$		1	1	9075200.	9286660.	4.39+09	6.87−01	AA	J02LS
472.9	1s4s	1s5p	^3S	$^3P^o$		1	0	9075200.	9286640.	4.38+09	2.289−01	AA	J02LS
485.6	1s4s	1s5p	^1S	$^1P^o$		0	1	9084060.	9290000.	4.40+09	7.47−01	AA	J02
489.0	1s4p	1s5d	$^3P^o$	^3D		0	1	9084000.	9288500.	5.23+09	9.05−01	AA	J02LS
489.1	1s4p	1s5d	$^3P^o$	^3D		1	1	9084040.	9288500.	3.92+09	6.79−01	AA	J02LS
489.1	1s4p	1s5d	$^3P^o$	^3D		1	2	9084040.	9288500.	7.05+09	2.037+00	AA	J02LS
489.5	1s4p	1s5d	$^3P^o$	^3D		2	1	9084230.	9288500.	2.61+08	4.53−02	AA	J02LS
489.5	1s4p	1s5d	$^3P^o$	^3D		2	2	9084230.	9288500.	2.345+09	6.79−01	AA	J02LS
489.5	1s4p	1s5d	$^3P^o$	^3D		2	3	9084230.	9288500.	9.38+09	3.80+00	AA	J02LS
493.6	1s4p	1s5s	$^3P^o$	^1S		1	0	9084040.	9286650.	2.91+06	1.727−04	AA	J02
493.6	1s4s	1s5p	^1S	$^3P^o$		0	1	9084060.	9286660.	3.41+06	6.08−04	AA	J02

λ_{Ritz} [a] (Å)	Configurations		Terms		Type	J_i [b]	J_k [b]	E_i (cm^{-1})	E_k (cm^{-1})	A_{ki} [c] (s^{-1})	S_{ik} [c]	Acc. [d]	Ref. [e]
504.5	1s4p	1s5s	$^3P^o$	^3S		0	1	9084000.	9282200.	6.07+08	1.154−01	AA	J02LS
504.6	1s4p	1s5s	$^3P^o$	^3S		1	1	9084040.	9282200.	1.819+09	3.46−01	AA	J02LS
505.1	1s4p	1s5s	$^3P^o$	^3S		2	1	9084230.	9282200.	3.02+09	5.77−01	AA	J02LS
509.0	1s4d	1s5p	^3D	$^3P^o$		1	2	9090300.	9286760.	1.456+07	4.74−03	AA	J02LS
509.1	1s4d	1s5p	^3D	$^3P^o$		2	2	9090340.	9286760.	2.183+08	7.11−02	AA	J02LS
509.2	1s4d	1s5p	^3D	$^3P^o$		3	2	9090390.	9286760.	1.222+09	3.98−01	AA	J02LS
509.3	1s4d	1s5p	^3D	$^3P^o$		1	1	9090300.	9286660.	3.64+08	7.11−02	AA	J02LS
509.3	1s4d	1s5p	^3D	$^3P^o$		1	0	9090300.	9286640.	1.454+09	9.48−02	AA	J02LS
509.4	1s4d	1s5p	^3D	$^3P^o$		2	1	9090340.	9286660.	1.090+09	2.133−01	AA	J02LS
510.2	1s4p	1s5s	$^1P^o$	^1S		1	0	9090630.	9286650.	5.05+09	3.31−01	AA	J02
882	1s5p	1s6d	$^3P^o$	^3D		0	1	9286640.	9400000.	1.567+09	1.593+00	AA	J02LS
882	1s5p	1s6d	$^3P^o$	^3D		1	1	9286660.	9400000.	1.175+09	1.194+00	AA	J02LS
882	1s5p	1s6d	$^3P^o$	^3D		1	2	9286660.	9400000.	2.114+09	3.58+00	AA	J02LS
883	1s5p	1s6d	$^3P^o$	^3D		2	1	9286760.	9400000.	7.81+07	7.96−02	AA	J02LS
883	1s5p	1s6d	$^3P^o$	^3D		2	2	9286760.	9400000.	7.03+08	1.194+00	AA	J02LS
883	1s5p	1s6d	$^3P^o$	^3D		2	3	9286760.	9400000.	2.81+09	6.69+00	AA	J02LS
887	1s5s	1s6p	^3S	$^3P^o$		1	0	9282200.	9395000.	1.428+09	4.91−01	AA	J02LS
887	1s5s	1s6p	^3S	$^3P^o$		1	1	9282200.	9395000.	1.428+09	1.473+00	AA	J02LS
887	1s5s	1s6p	^3S	$^3P^o$		1	2	9282200.	9395000.	1.428+09	2.455+00	AA	J02LS
913	1s5s	1s6p	^1S	$^1P^o$		0	1	9286650.	9396200.	1.446+09	1.628+00	AA	J02
923	1s5s	1s6p	^1S	$^3P^o$		0	1	9286650.	9395000.	1.145+06	1.333−03	AA	J02
939	1s5d	1s6p	^3D	$^3P^o$		3	2	9288500.	9395000.	6.17+08	1.261+00	AA	J02LS
939	1s5d	1s6p	^3D	$^3P^o$		2	1	9288500.	9395000.	5.51+08	6.76−01	AA	J02LS
939	1s5d	1s6p	^3D	$^3P^o$		2	2	9288500.	9395000.	1.102+08	2.252−01	AA	J02LS
939	1s5d	1s6p	^3D	$^3P^o$		1	0	9288500.	9395000.	7.35+08	3.00−01	AA	J02LS
939	1s5d	1s6p	^3D	$^3P^o$		1	1	9288500.	9395000.	1.837+08	2.252−01	AA	J02LS
939	1s5d	1s6p	^3D	$^3P^o$		1	2	9288500.	9395000.	7.35+06	1.502−02	AA	J02LS
1248.3	1s2s	1s2p	^3S	$^3P^o$		1	2	7299940.	7380050.	1.107+08	5.31−01	AA	J02LS
1272.9	1s2s	1s2p	^3S	$^3P^o$		1	1	7299940.	7378500.	1.044+08	3.19−01	AA	J02LS

λ_{Ritz} [a] (Å)	Configurations		Terms		Type	J_i [b]	J_k [b]	E_i (cm^{-1})	E_k (cm^{-1})	A_{ki} [c] (s^{-1})	S_{ik} [c]	Acc. [d]	Ref. [e]
1277.8	1s2s	1s2p	^3S	$^3P^o$		1	0	7299940.	7378200.	1.032+08	1.062−01	AA	J02LS
1856	1s2s	1s2p	^1S	$^1P^o$		0	1	7382680.	7436560.	3.54+07	3.35−01	AA	J02
4550	1s3s	1s3p	^3S	$^3P^o$		1	2	8623000.	8644960.	1.395+07	3.25+00	AA	J02LS
4650	1s3s	1s3p	^3S	$^3P^o$		1	1	8623000.	8644500.	1.309+07	1.950+00	AA	J02LS
4670	1s3s	1s3p	^3S	$^3P^o$		1	0	8623000.	8644400.	1.291+07	6.50−01	AA	J02LS
6390	1s3s	1s3p	^1S	$^1P^o$		0	1	8644880.	8660530.	5.21+06	2.014+00	AA	J02
7920	1s3p	1s3d	$^3P^o$	^3D		1	2	8644500.	8657120.	1.427+06	1.752+00	AA	CT92LS
7930	1s3p	1s3d	$^3P^o$	^3D		0	1	8644400.	8657000.	1.052+06	7.79−01	AA	CT92LS
8000	1s3p	1s3d	$^3P^o$	^3D		1	1	8644500.	8657000.	7.70+05	5.84−01	AA	CT92LS
8170	1s3p	1s3d	$^3P^o$	^3D		2	3	8644960.	8657190.	1.732+06	3.27+00	AA	CT92LS
8220	1s3p	1s3d	$^3P^o$	^3D		2	2	8644960.	8657120.	4.25+05	5.84−01	AA	CT92LS
8300	1s3p	1s3d	$^3P^o$	^3D		2	1	8644960.	8657000.	4.59+04	3.89−02	AA	CT92LS
11100	1s4s	1s4p	^3S	$^3P^o$		1	2	9075200.	9084230.	3.24+06	1.086+01	AA	J02LS
11300	1s4s	1s4p	^3S	$^3P^o$		1	1	9075200.	9084040.	3.04+06	6.52+00	AA	J02LS
11400	1s4s	1s4p	^3S	$^3P^o$		1	0	9075200.	9084000.	3.00+06	2.173+00	AA	J02LS
15200	1s4s	1s4p	^1S	$^1P^o$		0	1	9084060.	9090630.	1.274+06	6.65+00	AA	J02
15900	1s4p	1s4d	$^3P^o$	^3D		1	2	9084040.	9090340.	6.19+05	6.11+00	AA	CT92LS
15900	1s4p	1s4d	$^3P^o$	^3D		0	1	9084000.	9090300.	4.58+05	2.71+00	AA	CT92LS
16000	1s4p	1s4d	$^3P^o$	^3D		1	1	9084040.	9090300.	3.37+05	2.035+00	AA	CT92LS
16200	1s4p	1s4d	$^3P^o$	^3D		2	3	9084230.	9090390.	7.71+05	1.140+01	AA	CT92LS
16400	1s4p	1s4d	$^3P^o$	^3D		2	2	9084230.	9090340.	1.881+05	2.035+00	AA	CT92LS
16500	1s4p	1s4d	$^3P^o$	^3D		2	1	9084230.	9090300.	2.049+04	1.357−01	AA	CT92LS
21900	1s5s	1s5p	^3S	$^3P^o$		1	2	9282200.	9286760.	1.052+06	2.74+01	AA	J02LS
22400	1s5s	1s5p	^3S	$^3P^o$		1	1	9282200.	9286660.	9.84+05	1.643+01	AA	J02LS
22500	1s5s	1s5p	^3S	$^3P^o$		1	0	9282200.	9286640.	9.71+05	5.48+00	AA	J02LS
30000	1s5s	1s5p	^1S	$^1P^o$		0	1	9286650.	9290000.	4.23+05	1.665+01	AA	J02
53000	1s3p	1s3d	$^1P^o$	^1D		1	2	8660530.	8662400.	1.323+04	4.99+00	AA	CT92
54000	1s5p	1s5d	$^3P^o$	^3D		1	1	9286660.	9288500.	4.11+04	9.77+00	AA	CT92LS
54000	1s5p	1s5d	$^3P^o$	^3D		1	2	9286660.	9288500.	7.40+04	2.93+01	AA	CT92LS

λ_{Ritz} [a] (Å)	Configurations		Terms		Type	J_i [b]	J_k [b]	E_i (cm^{-1})	E_k (cm^{-1})	A_{ki} [c] (s^{-1})	S_{ik} [c]	Acc. [d]	Ref. [e]
54000	1s5p	1s5d	$^3P^o$	^3D		0	1	9286640.	9288500.	5.66+04	1.303+01	AA	CT92LS
57000	1s5p	1s5d	$^3P^o$	^3D		2	1	9286760.	9288500.	2.318+03	6.52−01	AA	CT92LS
57000	1s5p	1s5d	$^3P^o$	^3D		2	2	9286760.	9288500.	2.087+04	9.77+00	AA	CT92LS
57000	1s5p	1s5d	$^3P^o$	^3D		2	3	9286760.	9288500.	8.35+04	5.47+01	AA	CT92LS

[a] Ritz wavelengths are calculated from the energies of the lower and upper levels (E_i and E_k, respectively) as given in the NIST ASD database [18]. They are rounded according to their accuracy using the "rule of 25", meaning that the uncertainty in the least significant digit does not exceed 25 units of that digit. Wavelengths between 2000 Å and 20000 Å are given in standard air; outside this range they are given in vacuum. Conversion from vacuum to air and vice versa was made with the five-parameter formula for the refraction index of air from Peck and Reeder [19].

[b] J_i and J_k are the values of the total angular momentum of the lower and upper level, respectively. For transitions involving unresolved levels, the J_i and J_k values of all fine-structure transitions contributing to the given A_{ki} and S_{ik} values are listed, separated by comma.

[c] A_{ki} is the spontaneous radiative decay rate; S_{ik} is the line strength (dimensionless). The digits following the '+' or '−' sign are powers of 10, *e.g.*, 5.21+06 means 5.21×10^6.

[d] Transition probability accuracy code is explained in Table 1.

[e] Key to the references: A10 – Aggarwal *et al.* 2010 [50]; CT92 – Cann and Thakkar 1992 [52]; F04 – Froese Fischer 2004 [30]; F09 – Froese Fischer 2009 [31]; FH99 – Fernley *et al.* 1999 [22]; J02 – Johnson *et al.* 2002 [53]; N02 – Nahar 2002 [40]; R12 – Rynkun *et al.* 2012 [20]; TF00 – Tachiev and Froese Fischer 2000 [21]; TS90 – Tully *et al.* 1990 [32]. References followed by 'LS' mean that the given values of A_{ki} and S_{ik} were derived from the multiplet values listed in that reference using the formulas for relative strengths of fine-structure transitions within *LS* multiplets (see, for example, Cowan [23]) assuming pure *LS* coupling.

Journal of Physics: Conference Series **576** (2015) 012008 doi:10.1088/1742-6596/576/1/012008

Excitation of the n=2 levels of hydrogen by proton impact for hydrogen beam diagnostics

D.R. Schultz[1] and S.Y. Ovchinnikov[1,2]

[1]Department of Physics, University of North Texas, Denton, TX

[2]Ioffe Physical Technical Institute, St. Peterburg, Russia

Abstract. The atomic orbital close coupling approach has been used to compute a comprehensive data set for proton-impact excitation of atomic hydrogen owing to the need for complete data for motional Stark effect spectroscopy of fusion plasmas enabled by neutral beam injection.

1. Introduction

Fusion plasma diagnostics such as charge exchange recombination spectroscopy, beam emission spectroscopy, and motional Stark effect spectroscopy, can determine parameters such as field strength, temperature, density, and magnetic field orientation and are therefore of great importance. These diagnostics, in turn, depend critically on the existence and accuracy of atomic data for charge transfer and excitation. In particular, collisional-radiative models need not only cross sections for atomic collisions but also the full set of density matrix elements [1,2]. The data needed for such modeling underpinning motional Stark effect (MSE) diagnostics, the subject of the present work, includes the density matrix elements for excitation of atomic hydrogen (from neutral beam injection) by plasma protons, fusion alpha particles, and ions of the most significant impurities, beryllium and carbon.

As a contribution to the IAEA Coordinated Research Program on data for light ions, comprehensive atomic orbital close coupling (AOCC) calculations for H^+ + H were made. These calculations were needed not only to be able to cover the entire relevant collision energy range (400 eV to 2 MeV) with a fine energy mesh spacing that was not available from data in the literature, but because the collisional-radiative modeling of the MSE required calculation of the density matrix elements, which were essentially completely lacking. The AOCC data, plus results from the Glauber approximation for higher levels (n>4), were used to analyze population distributions for excited magnetic sublevels under typical conditions of existing and future fusion devices [2].

Here, description of the AOCC results for excitation to the n=2 states is made to document the computational procedure adopted and to convey new insight into the behavior of these fundamental cross sections. Toward that end, comparison is made with some of the most reliable, recent, independent data. The entire AOCC+Glauber density matrix data set will be described in a future publication as will analogous data produced for He^{2+}, Be^{4+}, and C^{6+} + H excitation.

2. The AOCC method and procedure used

The AOCC method is the most reliable of the commonly used methods to calculate transition probabilities, cross sections, or density matrix elements for inelastic ion-atom collisions in the intermediate- to high-energy regime (see, e.g., the review by Fritsch and Lin [3]). In this method, the time-dependent electronic wavefunction is expanded in terms of functions such as those of the atomic orbitals of the isolated target atom and projectile ion, or other functions such as Gaussians or Sturmians, that may span not only the bound states but also the continuum. We adopt the AOCC with pseudostates method of Kaung and Lin [4,5], an approach we have used recently to study excitation and charge transfer in H^+ + H(2s) collisions [6] and for charge transfer in He^{2+}, Be^{4+} + H [7,8], and H^+ + He^+ and He^{2+} + Li^{2+} [9] collisions, in comparison with results of other methods, the classical trajectory Monte Carlo and lattice, time-dependent Schrödinger equation methods.

There are two principal computational or mathematical issues that limit the accuracy of the AOCC approach adopted. The first is basis set size. Generally, as many basis states as possible provide a more physically realistic space within which to solve the time-dependent Schrödinger equation, involving the coupling by the ion-atom interaction of many open channels of excitation, charge transfer, and ionization. Practically, numerical integration becomes extremely difficult in certain regimes, in particular, for large numbers of states at low collision energy. Thus, we use the largest basis set practical at the highest energy, and then decrease its size incrementally as lower impact energies are considered. In the present work, the largest basis set (denoted "basis 6") has all target H atom s states from n=1 to 13, all p states from n=2 to 13, all d states from n=3 to 13, all f states from n=4 to 13, and all g states from n=5 to 9. The higher lying states (n>7 for s states and n>6 for p, d, f, and g states) are of positive energy and therefore lie in the two-center pseudocontinuum.

To avoid the second, mathematical, difficulty, the basis set is chosen to be somewhat different on the other center, the H^+ projectile, to avoid unphysical behavior arising from the symmetry of the collision system. Therefore for basis 6, the projectile center has states with a lower upper limit in n (i.e., n=1,11 for s, n=2-11 for p, n=3-11 for d, n=4-9 for f, and n=5-7 for g states, with the same states as for the target center in the pseudocontinuum). By the lowest impact energy considered, five additional, progressively smaller basis sets were employed, the smallest having only bound states on the target (1s-5f) and projectile (1s-4f).

To further reduce the basis-set dependent unphysical oscillations of the resulting cross sections, we repeat the calculations with the basis states reversed on the two centers, for example, the states that were on the target in basis 6 are moved to the projectile and vice versa to produce what we denote basis 6', and the two results are averaged. The effect of this procedure is illustrated in Figure 1, showing a portion of the impact energy dependence of the 2s excitation cross section (i.e., one element of the density matrix along the diagonal). The figure shows the use of basis sets 6 and 6' at high energies, giving way to basis set pairs 5 and 5', 4 and 4', and 3 and 3', successively at lower energies. For example, basis sets 6 and 6' are used from the highest impact energy down to 40 keV, basis sets 5 and 5' are used for energies beginning just above this point, etc., demonstrating a reasonable concordance of results as the basis sets are decreased in size in small overlapping ranges of energies. The solid lines in this figure show the average of the results from pairs of basis sets, illustrating the reduction in oscillations where the two results do not agree, and connection of the averages as transition from one pair of basis sets to the next is accomplished.

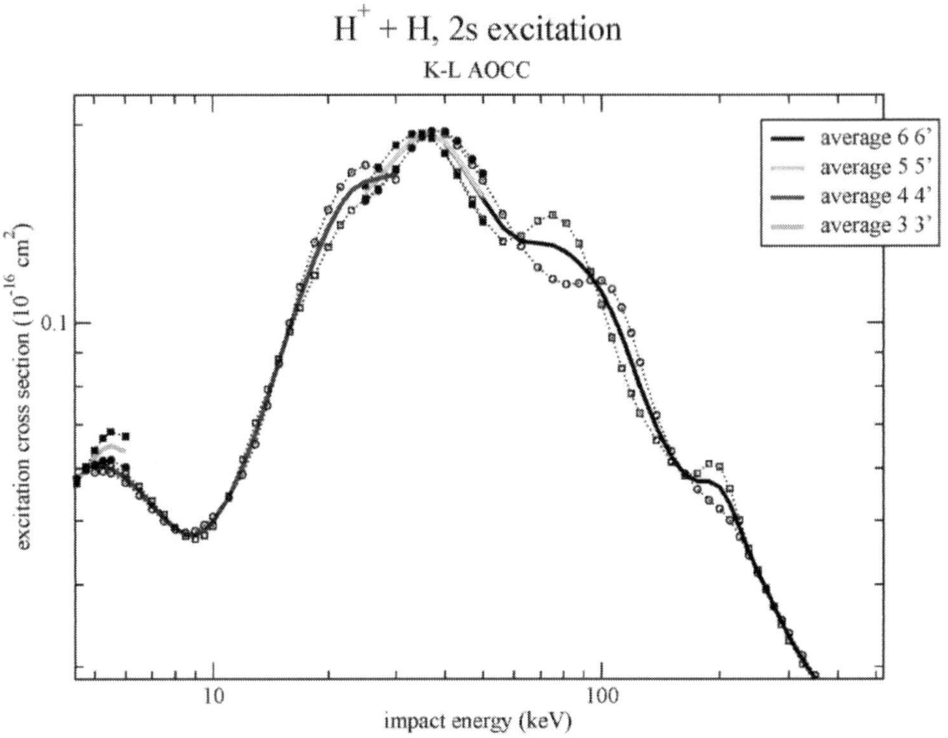

Figure 1. The results for excitation to H(2s) by proton impact over part of the impact energy range considered, illustrating the overlapping agreement as basis set size is reduced as a function of impact energy and the level of agreement between the results for a given basis set and its interchanged counterpart.

Corroboration of this method of averaging the results from the pairs of basis sets comes from supporting calculations made using the lattice, time-dependent Schrödinger equation (LTDSE) method [10]. This method is in principal more accurate and more controllably convergent than the present AOCC approach, but requires a much greater computational effort and therefore has not to this point been practical for such comprehensive calculations. The H(2p) excitation probability as a function of impact parameter computed via the LTDSE method using a modest grid size of 200^3 points utilizing Fourier collocation is shown in Figure 2 for 25 and 100 keV proton impact in comparison with results from the pairs of AOCC basis sets used here. The figure shows that one or the other of the AOCC basis sets typically overestimates or underestimates the excitation probability in the small to intermediate impact parameter range. Thus, averaging the results of the two basis sets accounts for this in an approximate but effective way.

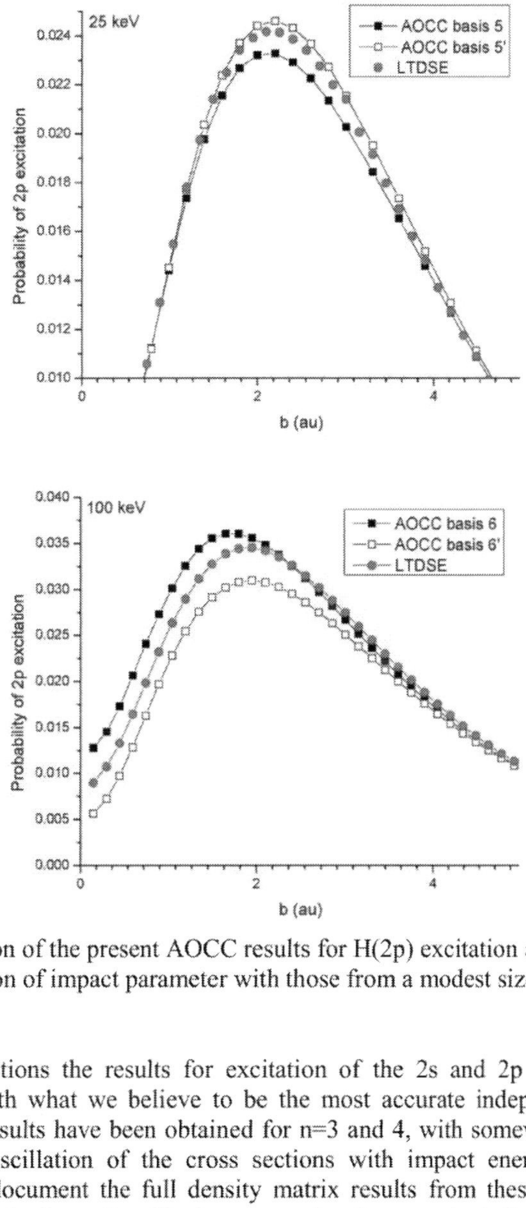

Figure 2. Comparison of the present AOCC results for H(2p) excitation at 25 and 100 keV impact energies as a function of impact parameter with those from a modest size LTDSE calculation.

In the following sections the results for excitation of the 2s and 2p states are described and comparison is given with what we believe to be the most accurate independent calculation in the literature. Analogous results have been obtained for n=3 and 4, with somewhat greater possibility of remaining unphysical oscillation of the cross sections with impact energy. As noted above, a subsequent paper will document the full density matrix results from these calculations along with those for higher n levels from the Glauber approximation required for the collisional-radiative modeling of neutral beam effects in fusion plasmas.

3. Excitation to H(2s)

Figure 3 displays the full range of data produced for energies between 400 eV and 2 MeV from the pairs of basis sets along with their average. It is not possible from the present result alone to determine if the small oscillations above about 50 keV that remain after the averaging are physical or not, but the large oscillation near 5 keV is confirmed by many previous calculations, with the one near

500 eV possibly being newly noted. The origin of the physical oscillations has been explained as coming from the oscillation of electronic probability density that is 'cut off' on one center or the other depending on the impact velocity [11].

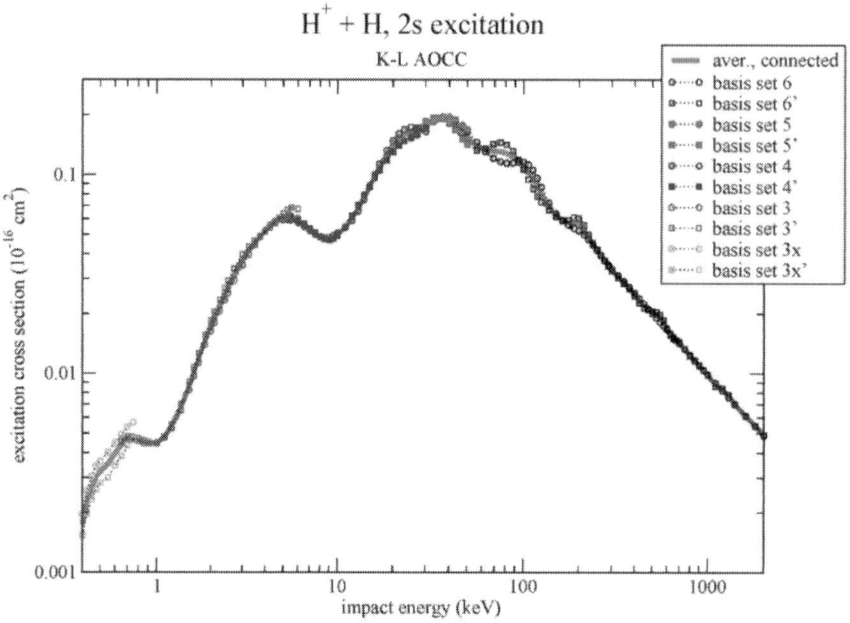

Figure 3. The present results for excitation of H(2s) showing the results from the pairs of basis sets and the connected set of average results.

To corroborate the accuracy of the results, comparison is made in Figure 4 with the best available results in the literature, those from the Sturmian AOCC approach of Winter [12]. Winter has also employed several successively larger basis sets to demonstrate the degree of convergence with basis size, ranging from 24 to 280 states, the largest of which is roughly comparable to the number of states in the present largest basis sets. As the figure shows, there is very good overall agreement between Winter's results with the larger basis sets and the present connected average data. For large impact energies, Winter's results do not show the oscillations that the present results show but it would be interesting to have his results at a large number of energies to see if any evidence of them appears. For lower energies, his 220 state calculation coincides almost entirely with the present connected average results. Also shown in the figure is the recommended data from Janev and Smith [13] for this cross section. The recommended curve falls below the present results and that of Winter at energies below 500 keV and underestimates the peak of the cross section and the 5 keV peak. Thus, in light of the present calculations and those of Winter, the recommended curve should be updated.

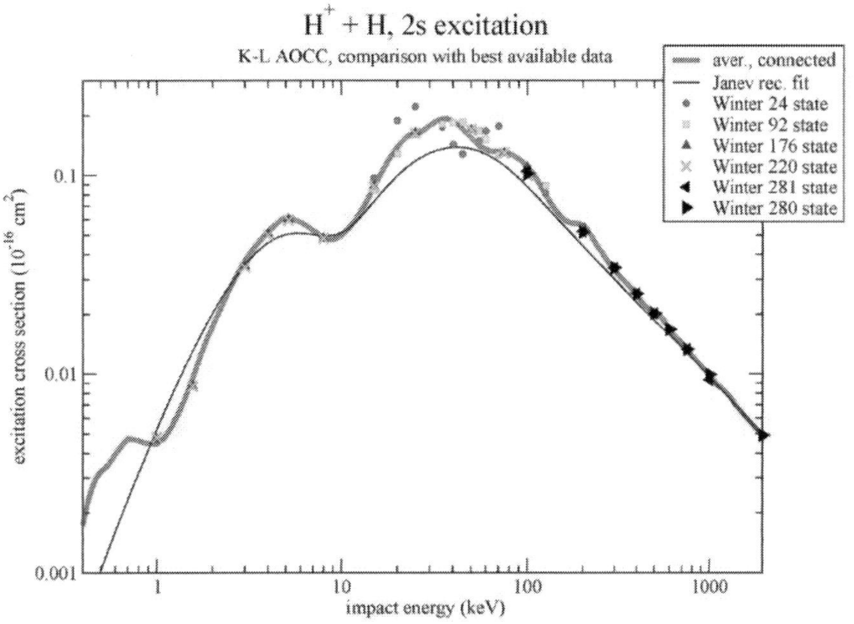

Figure 4. Comparison of the best available data from the literature with the present results for H(2s) excitation, along with a previous recommended cross section.

4. Excitation to H(2p)

Results for excitation to H(2p) are given in Figure 5, showing the raw results for the pairs of basis sets and the connected average curve. Because of the significant deviation of one of the basis set results near 5 keV, which would have made it very difficult to make a smooth connection between results at the nearby transition from one pair to the next, one basis set was adopted in favor of taking the average in that energy range. Additional results from other methods might help shed light on the best representation of the behavior of the cross section in this regime once available. Results from Winter's [12] calculations are available for the $2p_0$ and $2p_1$ states at higher energies as shown in Figure 6, again showing substantially very good agreement with the present results. As before, Winter's results show less oscillation at a fine scale, but are not free of oscillations either. Figure 7 shows the sum of the $2p_0$ and $2p_1$ results for comparison with more of Winter's data and the recommended curve of Janev and Smith [13]. This figure shows excellent agreement between the two AOCC methods and how the recommended curve again requires updating in light of the newly available results.

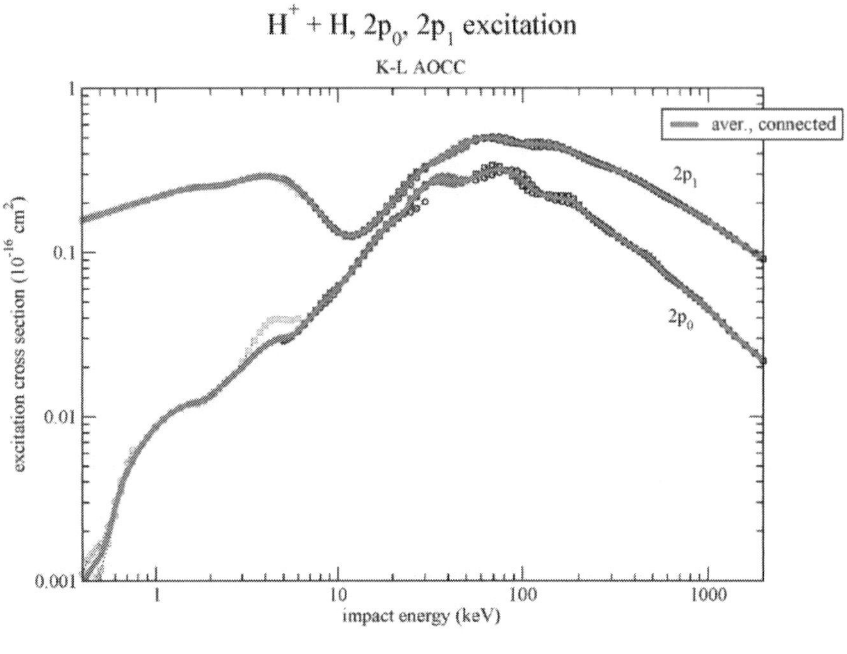

Figure 5. The raw and connected average results for excitation to $H(2p_0, 2p_1)$.

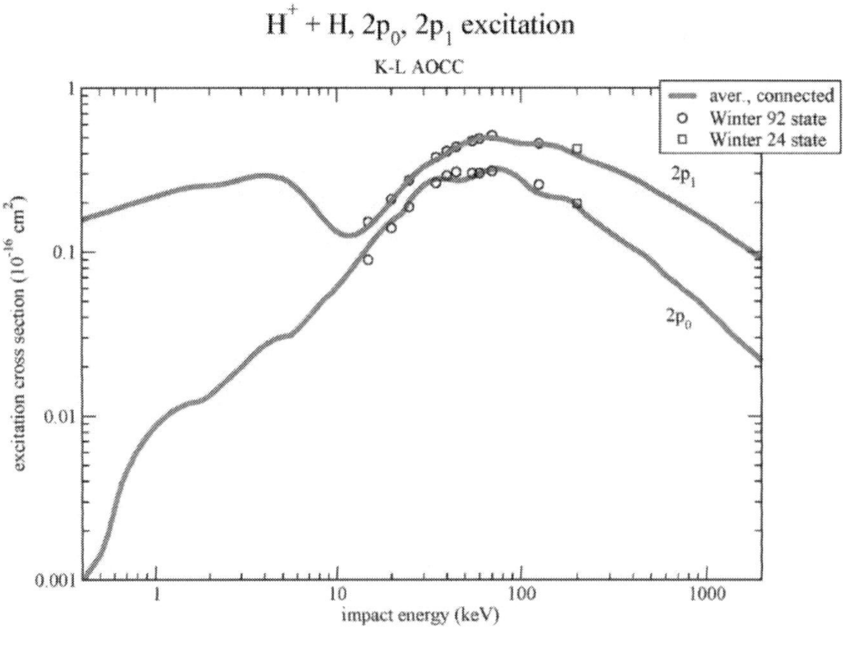

Figure 6. The connected average results for $H(2p_0, 2p_1)$ excitation compared to results from Winter [12].

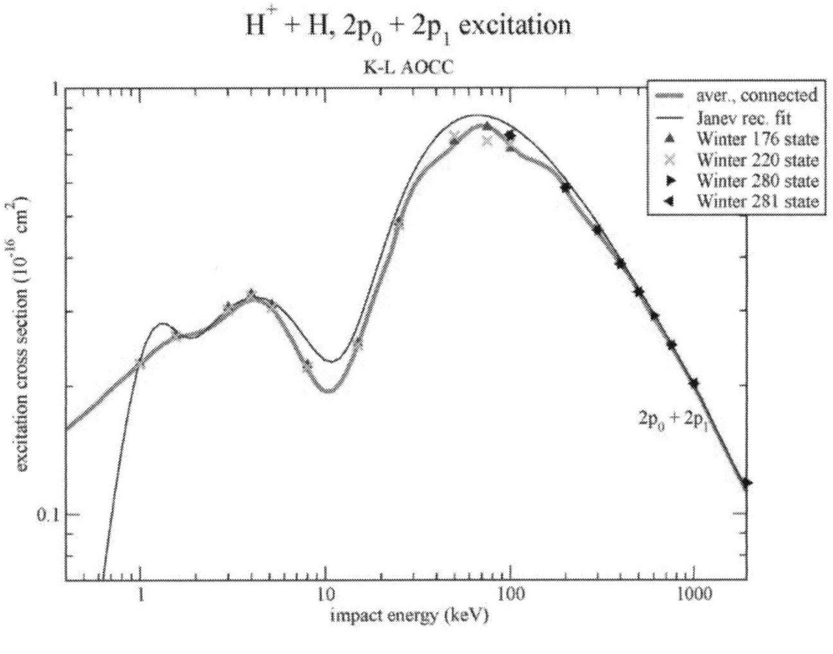

Figure 7. Results summed over magnetic quantum number for n=2 excitation compared with results from Winter [12] and the recommended curve of Janev and Smith [13].

5. Matrix elements

As described above, the density matrix elements are required for the collisional-radiative modeling of neutral hydrogen beams in fusion plasmas. Because we find better overall agreement with other calculations, including the LTDSE results, by using the connected successive unprimed basis sets (basis sets 6, 5, 4, 3) we utilize these to provide the density matrix elements. This has been done because the density matrix elements vary more rapidly than the cross sections as a function of energy and can have both positive and negative real and imaginary parts, so averaging the primed and unprimed basis set results could result in less smooth variation than selecting the more reliable sequence of basis sets.

In addition, we have checked that the density matrix elements do not change dramatically at the changeover points between these basis sets. To illustrate this, for several density matrix elements (for a sample of density matrix elements within and amongst the n=2, 3, and 4 shells), we display in Table 1 their real and imaginary parts for several basis sets, for an energy of 50 keV. In this energy region, there is a changeover between basis 6 and basis 5 and we see from the table that there is generally good agreement between results of these two basis sets. Given this sampling of matrix elements, the aggregate percent difference between results from basis 6 and 5 is about 5.8%. There are, however, cases (such as for Re(211-320)) that have larger differences of 25-35%.

While the results from the smaller basis sets are not used in the energy range where basis sets 6 and 5 are applicable, to illustrate how great the variation would be if we considered smaller basis sets for this energy, the table also includes results from basis set 4. In this case, the average percent difference between the results of basis 6 and 4 is on the order of 16% in this sample.

Element	basis 6	basis 5	basis 4	% diff 6-5	% diff 6-4
Re 210-200	-9.21e-18	-9.14e-18	-8.23e-18	0.8	10
Im 210-200	1.38e-17	1.41e-17	1.59e-17	2.2	15
Re 200-300	7.94e-18	8.06e-18	7.49e-18	1.5	5.7
Im 200-300	-1.61e-18	-1.66e-18	-1.46e-18	3.1	9.3
Re 211-320	1.54e-18	1.02e-18	2.61e-19	34	41
Im 211-320	8.15e-18	7.91e-18	6.66e-18	2.9	18
Re 300-310	-2.31e-18	-2.23e-18	-1.56e-18	3.5	32
Im 300-310	-2.94e-18	-2.91e-18	-2.88e-18	1.0	2.0
Re 320-322	-2.80e-19	-3.07e-18	-2.92e-19	9.6	4.3
Im 320-322	-6.61e-19	-6.22e-19	-4.74e-19	5.9	28
Re 300-411	2.51e-18	2.53e-18	2.33e-18	0.8	7.2
Im 300-411	-6.17e-19	-6.28e-19	-6.01e-19	1.8	2.6
Re 431-322	6.29e-20	6.14e-20	4.26e-20	2.4	32
Im 431-322	-1.41e-19	-1.31e-19	-1.08e-19	7.1	23
Re 400-211	1.81e-19	1.94e-19	1.50e-19	8.7	17
Im 400-211	2.65e-19	2.50e-19	2.17e-19	5.7	18
Re 410-321	9.61e-19	8.99e-19	8.36e-19	6.5	13
Im 410-321	-9.77e-19	-1.04e-18	-1.03e-18	6.4	5.4

Table 1. Examples of density matrix elements for 50 keV proton impact of atomic hydrogen computed with successively smaller basis sets. In the energy range near 50 keV, the transition from the largest (basis 6) to second largest (basis 5) basis set is taking place (i.e., they have overlapping applicability). Results from the next smaller basis set (basis 4) are not used in the data set but are shown here for comparison. Just as for the cross sections, the real and imaginary parts of the density matrix elements are given in units of cm^2.

To illustrate the form and scope of the density matrix elements, in Table 2 we tabulate them connecting the n=1 to 4 states for 60 keV. Note that the diagonal elements are the cross sections for excitation.

Acknowledgements

One of us (DRS) gratefully acknowledges collaboration with O. Marchuk and Yu. Ralchenko who have carried out the collisional-radiative modeling of fusion plasmas motivating the undertaking of this work. This work was carried out in part as a contribution to an IAEA Coordinated Research Program on "Light Atom, Molecule, and Radical Behavior in the Divertor and Edge Plasma Regions."

References

[1] O. Marchuk, Yu. Ralchenko, R.K. Janev, W. Biel, E. Delabie, and A.M. Urnov, J. Phys. B 43, 011002 (2010)
[2] O. Marchuk, Yu. Ralchenko, and D.R. Schultz, Plasma Phys. Control. Fusion 54, 095010 (2012).
[3] W. Fritsch and C.D. Lin, Phys. Rep. 201, 1 (1999).
[4] J. Kuang and C.D. Lin, J. Phys. B 29, 1207 (1996).
[5] J. Kaung and C.D. Lin, J. Phys. B 29, 5443 (1996).
[6] M.S. Pindzola, T-G. Lee, T. Minami, and D.R. Schultz, Phys. Rev. A 72, 062703 (2005).
[7] T. Minami, T-G. Lee, M.S. Pindzola, and D.R. Schultz, J. Phys. B 41, 135201 (2008).
[8] T. Minami, M.S. Pindzola, T-G. Lee, and D.R. Schultz, J. Phys. B 39, 2877 (2006).
[9] T. Minami, M.S. Pindzola, T-G. Lee, and D.R. Schultz, J. Phys. B 40, 3629 (2007).
[10] D.R. Schultz, M.R. Strayer, and J.C. Wells, Phys. Rev. Lett. 82, 3976 (1999).
[11] D.R. Schultz, C.O. Reinhold, and P.S. Krstić, Phys. Rev. Lett. 78, 2720 (1997); P.S. Krstić, C.O. Reinhold, and D.R. Schultz, J. Phys. B 31, L155 (1998).
[12] T.G. Winter, Phys. Rev. A 80, 032701 (2009).
[13] R.K. Janev and J.J. Smith, Atomic and Plasma Material Interaction Data for Fusion 4, 1 (1993).

Journal of Physics: Conference Series **576** (2015) 012008 doi:10.1088/1742-6596/576/1/012008

Table 2. Density matrix elements (in cm^2) connecting the n=1 and 4 states for excitation of H by proton impact at 60 keV from the present AOCC calculation

n	l	m	n	l	m	matrix elem.		n	l	m	n	l	m	matrix elem.		n	l	m	n	l	m	matrix elem.	
2	0	0	2	0	0	Re	1.36E-17	3	2	0	3	2	0	Re	1.30E-18	4	2	0	4	2	0	Re	6.13E-19
2	0	0	2	0	0	Im	0.00E+00	3	2	0	3	2	0	Im	0.00E+00	4	2	0	4	2	0	Im	0.00E+00
2	0	0	2	1	0	Re	-7.48E-18	3	2	0	3	2	1	Re	-5.30E-19	4	2	0	4	2	1	Re	-2.27E-19
2	0	0	2	1	0	Im	-1.24E-17	3	2	0	3	2	1	Im	7.86E-19	4	2	0	4	2	1	Im	3.79E-19
2	0	0	2	1	1	Re	1.74E-17	3	2	0	3	2	2	Re	-2.11E-19	4	2	0	4	2	2	Re	-8.32E-20
2	0	0	2	1	1	Im	-6.90E-18	3	2	0	3	2	2	Im	-5.32E-19	4	2	0	4	2	2	Im	-2.44E-19
2	0	0	3	0	0	Re	6.41E-18	3	2	0	4	0	0	Re	-3.58E-20	4	2	0	4	3	0	Re	7.13E-20
2	0	0	3	0	0	Im	-1.37E-18	3	2	0	4	0	0	Im	-2.37E-20	4	2	0	4	3	0	Im	1.08E-19
2	0	0	3	1	0	Re	-4.67E-18	3	2	0	4	1	0	Re	-1.16E-18	4	2	0	4	3	1	Re	-1.01E-19
2	0	0	3	1	0	Im	-4.06E-18	3	2	0	4	1	0	Im	3.40E-19	4	2	0	4	3	1	Im	9.73E-20
2	0	0	3	1	1	Re	7.42E-18	3	2	0	4	1	1	Re	3.30E-19	4	2	0	4	3	2	Re	2.11E-21
2	0	0	3	1	1	Im	-3.38E-18	3	2	0	4	1	1	Im	-1.61E-18	4	2	0	4	3	2	Im	-4.85E-20
2	0	0	3	2	0	Re	2.57E-18	3	2	0	4	2	0	Re	-8.93E-19	4	2	0	4	3	3	Re	2.00E-20
2	0	0	3	2	0	Im	3.07E-18	3	2	0	4	2	0	Im	1.84E-20	4	2	0	4	3	3	Im	1.94E-20
2	0	0	3	2	1	Re	-2.59E-18	3	2	0	4	2	1	Re	3.21E-19	4	2	1	2	0	0	Re	1.78E-18
2	0	0	3	2	1	Im	2.11E-19	3	2	0	4	2	1	Im	-5.72E-19	4	2	1	2	0	0	Im	3.09E-19
2	0	0	3	2	2	Re	8.09E-19	3	2	0	4	2	2	Re	1.37E-19	4	2	1	2	1	0	Re	-1.45E-18
2	0	0	3	2	2	Im	-1.33E-18	3	2	0	4	2	2	Im	3.57E-19	4	2	1	2	1	0	Im	-3.11E-18
2	0	0	4	0	0	Re	-8.61E-22	3	2	0	4	3	0	Re	-1.08E-19	4	2	1	2	1	1	Re	4.22E-18
2	0	0	4	0	0	Im	-1.07E-19	3	2	0	4	3	0	Im	-1.56E-19	4	2	1	2	1	1	Im	-1.51E-18
2	0	0	4	1	0	Re	-2.95E-18	3	2	0	4	3	1	Re	1.46E-19	4	2	1	3	0	0	Re	8.31E-19
2	0	0	4	1	0	Im	-1.94E-18	3	2	0	4	3	1	Im	-1.47E-19	4	2	1	3	0	0	Im	-7.99E-20
2	0	0	4	1	1	Re	4.30E-18	3	2	0	4	3	2	Re	-8.21E-22	4	2	1	3	1	0	Re	-8.20E-19
2	0	0	4	1	1	Im	-2.12E-18	3	2	0	4	3	2	Im	7.31E-20	4	2	1	3	1	0	Im	-1.04E-18
2	0	0	4	2	0	Re	-1.83E-18	3	2	0	4	3	3	Re	-3.14E-20	4	2	1	3	1	1	Re	1.65E-18
2	0	0	4	2	0	Im	-2.10E-18	3	2	0	4	3	3	Im	-2.80E-20	4	2	1	3	1	1	Im	-7.03E-19
2	0	0	4	2	1	Re	1.78E-18	3	2	1	2	0	0	Re	-2.59E-18	4	2	1	3	2	0	Re	3.21E-19
2	0	0	4	2	1	Im	-3.09E-19	3	2	1	2	0	0	Im	-2.11E-19	4	2	1	3	2	0	Im	5.72E-19
2	0	0	4	2	2	Re	-5.50E-19	3	2	1	2	1	0	Re	2.60E-18	4	2	1	3	2	1	Re	-6.35E-19
2	0	0	4	2	2	Im	8.91E-19	3	2	1	2	1	0	Im	4.39E-18	4	2	1	3	2	1	Im	-5.82E-20
2	0	0	4	3	0	Re	1.31E-19	3	2	1	2	1	1	Re	-5.99E-18	4	2	1	3	2	2	Re	2.62E-19
2	0	0	4	3	0	Im	-5.38E-19	3	2	1	2	1	1	Im	2.88E-18	4	2	1	3	2	2	Im	-3.55E-19
2	0	0	4	3	1	Re	5.77E-19	3	2	1	3	0	0	Re	-1.17E-18	4	2	1	4	0	0	Re	6.85E-21
2	0	0	4	3	1	Im	4.13E-20	3	2	1	3	0	0	Im	2.24E-19	4	2	1	4	0	0	Im	-3.64E-20
2	0	0	4	3	2	Re	-1.46E-19	3	2	1	3	1	0	Re	1.34E-18	4	2	1	4	1	0	Re	-5.07E-19
2	0	0	4	3	2	Im	1.11E-19	3	2	1	3	1	0	Im	1.41E-18	4	2	1	4	1	0	Im	-5.08E-19
2	0	0	4	3	3	Re	1.30E-20	3	2	1	3	1	1	Re	-2.31E-18	4	2	1	4	1	1	Re	9.34E-19
2	0	0	4	3	3	Im	-1.07E-19	3	2	1	3	1	1	Im	1.28E-18	4	2	1	4	1	1	Im	-4.37E-19
2	1	0	2	0	0	Re	-7.48E-18	3	2	1	3	2	0	Re	-5.30E-19	4	2	1	4	2	0	Re	-2.27E-19
2	1	0	2	0	0	Im	1.24E-17	3	2	1	3	2	0	Im	-7.86E-19	4	2	1	4	2	0	Im	-3.79E-19
2	1	0	2	1	0	Re	2.80E-17	3	2	1	3	2	1	Re	9.46E-19	4	2	1	4	2	1	Re	4.31E-19
2	1	0	2	1	0	Im	0.00E+00	3	2	1	3	2	1	Im	0.00E+00	4	2	1	4	2	1	Im	0.00E+00
2	1	0	2	1	1	Re	-3.06E-18	3	2	1	3	2	2	Re	-3.36E-19	4	2	1	4	2	2	Re	-1.72E-19
2	1	0	2	1	1	Im	3.65E-17	3	2	1	3	2	2	Im	5.71E-19	4	2	1	4	2	2	Im	2.36E-19
2	1	0	3	0	0	Re	-1.79E-18	3	2	1	4	0	0	Re	-5.29E-21	4	2	1	4	3	0	Re	4.22E-20
2	1	0	3	0	0	Im	6.45E-18	3	2	1	4	0	0	Im	5.63E-20	4	2	1	4	3	0	Im	-1.08E-19
2	1	0	3	1	0	Re	1.03E-17	3	2	1	4	1	0	Re	8.09E-19	4	2	1	4	3	1	Re	1.23E-19
2	1	0	3	1	0	Im	-2.55E-18	3	2	1	4	1	0	Im	6.77E-19	4	2	1	4	3	1	Im	2.39E-20
2	1	0	3	1	1	Re	-1.75E-19	3	2	1	4	1	1	Re	-1.30E-18	4	2	1	4	3	2	Re	-4.97E-20
2	1	0	3	1	1	Im	1.46E-17	3	2	1	4	1	1	Im	7.83E-19	4	2	1	4	3	2	Im	2.36E-20
2	1	0	3	2	0	Re	-5.09E-18	3	2	1	4	2	0	Re	3.72E-19	4	2	1	4	3	3	Re	1.01E-20
2	1	0	3	2	0	Im	5.44E-19	3	2	1	4	2	0	Im	5.17E-19	4	2	1	4	3	3	Im	-3.43E-20
2	1	0	3	2	1	Re	2.60E-18	3	2	1	4	2	1	Re	-6.35E-19	4	2	2	2	0	0	Re	-5.50E-19
2	1	0	3	2	1	Im	-4.39E-18	3	2	1	4	2	1	Im	5.82E-20	4	2	2	2	0	0	Im	-8.91E-19
2	1	0	3	2	2	Re	1.71E-18	3	2	1	4	2	2	Re	2.20E-19	4	2	2	2	1	0	Re	-1.15E-18
2	1	0	3	2	2	Im	3.20E-18	3	2	1	4	2	2	Im	-3.79E-19	4	2	2	2	1	0	Im	2.11E-18
2	1	0	4	0	0	Re	2.40E-19	3	2	1	4	3	0	Re	-4.51E-20	4	2	2	2	1	1	Re	-2.71E-18
2	1	0	4	0	0	Im	1.90E-19	3	2	1	4	3	0	Im	1.63E-19	4	2	2	2	1	1	Im	-1.70E-18

2 1 0 4 1 0 Re 5.33E-18	3 2 1 4 3 1 Re -1.81E-19	4 2 2 3 0 0 Re -3.61E-19
2 1 0 4 1 0 Im -2.03E-18	3 2 1 4 3 1 Im -1.76E-20	4 2 2 3 0 0 Im -3.12E-19
2 1 0 4 1 1 Re 2.27E-19	3 2 1 4 3 2 Re 7.12E-20	4 2 2 3 1 0 Re -2.25E-19
2 1 0 4 1 1 Im 8.37E-18	3 2 1 4 3 2 Im -4.19E-20	4 2 2 3 1 0 Im 8.53E-19
2 1 0 4 2 0 Re 3.42E-18	3 2 1 4 3 3 Re -1.03E-20	4 2 2 3 1 1 Re -1.12E-18
2 1 0 4 2 0 Im -4.86E-19	3 2 1 4 3 3 Im 5.34E-20	4 2 2 3 1 1 Im -5.67E-19
2 1 0 4 2 1 Re -1.45E-18	3 2 2 2 0 0 Re 8.09E-19	4 2 2 3 2 0 Re 1.37E-19
2 1 0 4 2 1 Im 3.11E-18	3 2 2 2 0 0 Im 1.33E-18	4 2 2 3 2 0 Im -3.57E-19
2 1 0 4 2 2 Re -1.15E-18	3 2 2 2 1 0 Re 1.71E-18	4 2 2 3 2 1 Re 2.20E-19
2 1 0 4 2 2 Im -2.11E-18	3 2 2 2 1 0 Im -3.20E-18	4 2 2 3 2 1 Im 3.79E-19
2 1 0 4 3 0 Re 6.68E-19	3 2 2 2 1 1 Re 4.11E-18	4 2 2 3 2 2 Re -3.33E-19
2 1 0 4 3 0 Im 6.47E-19	3 2 2 2 1 1 Im 2.51E-18	4 2 2 3 2 2 Im -5.45E-21
2 1 0 4 3 1 Re -5.59E-19	3 2 2 3 0 0 Re 5.32E-19	4 2 2 4 0 0 Re -2.69E-20
2 1 0 4 3 1 Im 8.12E-19	3 2 2 3 0 0 Im 4.63E-19	4 2 2 4 0 0 Im 9.03E-21
2 1 0 4 3 2 Re 9.02E-21	3 2 2 3 1 0 Re 3.34E-19	4 2 2 4 1 0 Re -6.61E-20
2 1 0 4 3 2 Im -4.43E-19	3 2 2 3 1 0 Im -1.29E-18	4 2 2 4 1 0 Im 4.64E-19
2 1 0 4 3 3 Re 2.14E-19	3 2 2 3 1 1 Re 1.69E-18	4 2 2 4 1 1 Re -6.56E-19
2 1 0 4 3 3 Im 2.03E-19	3 2 2 3 1 1 Im 8.32E-19	4 2 2 4 1 1 Im -2.95E-19
2 1 1 2 0 0 Re 1.74E-17	3 2 2 3 2 0 Re -2.11E-19	4 2 2 4 2 0 Re -8.32E-20
2 1 1 2 0 0 Im 6.90E-18	3 2 2 3 2 0 Im 5.32E-19	4 2 2 4 2 0 Im 2.44E-19
2 1 1 2 1 0 Re -3.06E-18	3 2 2 3 2 1 Re -3.36E-19	4 2 2 4 2 1 Re -1.72E-19
2 1 1 2 1 0 Im -3.65E-17	3 2 2 3 2 1 Im -5.71E-19	4 2 2 4 2 1 Im -2.36E-19
2 1 1 2 1 1 Re 4.89E-17	3 2 2 3 2 2 Re 5.03E-19	4 2 2 4 2 2 Re 2.21E-19
2 1 1 2 1 1 Im 0.00E+00	3 2 2 3 2 2 Im 0.00E+00	4 2 2 4 2 2 Im 0.00E+00
2 1 1 3 0 0 Re 8.74E-18	3 2 2 4 0 0 Re 4.08E-20	4 2 2 4 3 0 Re -7.46E-20
2 1 1 3 0 0 Im 8.15E-19	3 2 2 4 0 0 Im -1.47E-20	4 2 2 4 3 0 Im 2.84E-20
2 1 1 3 1 0 Re -4.67E-18	3 2 2 4 1 0 Re 9.83E-20	4 2 2 4 3 1 Re -4.25E-20
2 1 1 3 1 0 Im -1.29E-17	3 2 2 4 1 0 Im -6.98E-19	4 2 2 4 3 1 Im -7.11E-20
2 1 1 3 1 1 Re 1.95E-17	3 2 2 4 1 1 Re 9.86E-19	4 2 2 4 3 2 Re 3.34E-20
2 1 1 3 1 1 Im -1.68E-18	3 2 2 4 1 1 Im 4.31E-19	4 2 2 4 3 2 Im 2.04E-20
2 1 1 3 2 0 Re 1.71E-18	3 2 2 4 2 0 Re 1.28E-19	4 2 2 4 3 3 Re -2.66E-20
2 1 1 3 2 0 Im 6.37E-18	3 2 2 4 2 0 Im -3.62E-19	4 2 2 4 3 3 Im 6.85E-21
2 1 1 3 2 1 Re -5.99E-18	3 2 2 4 2 1 Re 2.62E-19	4 3 0 2 0 0 Re 1.31E-19
2 1 1 3 2 1 Im -2.88E-18	3 2 2 4 2 1 Im 3.55E-19	4 3 0 2 0 0 Im 5.38E-19
2 1 1 3 2 2 Re 4.11E-18	3 2 2 4 2 2 Re -3.33E-19	4 3 0 2 1 0 Re 6.68E-19
2 1 1 3 2 2 Im -2.51E-18	3 2 2 4 2 2 Im 5.45E-21	4 3 0 2 1 0 Im -6.47E-19
2 1 1 4 0 0 Re 2.37E-19	3 2 2 4 3 0 Re 1.11E-19	4 3 0 2 1 1 Re 7.68E-19
2 1 1 4 0 0 Im -3.16E-19	3 2 2 4 3 0 Im -4.29E-20	4 3 0 2 1 1 Im 9.84E-19
2 1 1 4 1 0 Re -3.38E-18	3 2 2 4 3 1 Re 6.33E-20	4 3 0 3 0 0 Re 1.25E-19
2 1 1 4 1 0 Im -6.59E-18	3 2 2 4 3 1 Im 1.06E-19	4 3 0 3 0 0 Im 2.27E-19
2 1 1 4 1 1 Re 1.12E-17	3 2 2 4 3 2 Re -5.11E-20	4 3 0 3 1 0 Re 1.84E-19
2 1 1 4 1 1 Im -1.45E-18	3 2 2 4 3 2 Im -3.06E-20	4 3 0 3 1 0 Im -3.16E-19
2 1 1 4 2 0 Re -1.32E-18	3 2 2 4 3 3 Re 4.03E-20	4 3 0 3 1 1 Re 3.48E-19
2 1 1 4 2 0 Im -4.27E-18	3 2 2 4 3 3 Im -1.12E-20	4 3 0 3 1 1 Im 3.83E-19
2 1 1 4 2 1 Re 4.22E-18	4 0 0 2 0 0 Re -8.61E-22	4 3 0 3 2 0 Re -1.08E-19
2 1 1 4 2 1 Im 1.51E-18	4 0 0 2 0 0 Im 1.07E-19	4 3 0 3 2 0 Im 1.56E-19
2 1 1 4 2 2 Re -2.71E-18	4 0 0 2 1 0 Re 2.40E-19	4 3 0 3 2 1 Re -4.51E-20
2 1 1 4 2 2 Im 1.70E-18	4 0 0 2 1 0 Im -1.90E-19	4 3 0 3 2 1 Im -1.63E-19
2 1 1 4 3 0 Re 7.68E-19	4 0 0 2 1 1 Re 2.37E-19	4 3 0 3 2 2 Re 1.11E-19
2 1 1 4 3 0 Im -9.84E-19	4 0 0 2 1 1 Im 3.16E-19	4 3 0 3 2 2 Im 4.29E-20
2 1 1 4 3 1 Re 1.17E-18	4 0 0 3 0 0 Re 1.38E-20	4 3 0 4 0 0 Re 8.60E-21
2 1 1 4 3 1 Im 6.10E-19	4 0 0 3 0 0 Im 4.39E-20	4 3 0 4 0 0 Im -4.73E-22
2 1 1 4 3 2 Re -5.72E-19	4 0 0 3 1 0 Re 7.08E-20	4 3 0 4 1 0 Re 7.68E-20
2 1 1 4 3 2 Im 2.39E-20	4 0 0 3 1 0 Im -8.19E-20	4 3 0 4 1 0 Im -1.84E-19
2 1 1 4 3 3 Re 2.55E-19	4 0 0 3 1 1 Re 9.79E-20	4 3 0 4 1 1 Re 2.10E-19
2 1 1 4 3 3 Im -2.87E-19	4 0 0 3 1 1 Im 1.10E-19	4 3 0 4 1 1 Im 2.14E-19
3 0 0 2 0 0 Re 6.41E-18	4 0 0 3 2 0 Re -3.58E-20	4 3 0 4 2 0 Re 7.13E-20
3 0 0 2 0 0 Im 1.37E-18	4 0 0 3 2 0 Im 2.37E-20	4 3 0 4 2 0 Im -1.08E-19
3 0 0 2 1 0 Re -1.79E-18	4 0 0 3 2 1 Re -5.29E-21	4 3 0 4 2 1 Re 4.22E-20
3 0 0 2 1 0 Im -6.45E-18	4 0 0 3 2 1 Im -5.63E-20	4 3 0 4 2 1 Im 1.08E-19
3 0 0 2 1 1 Re 8.74E-18	4 0 0 3 2 2 Re 4.08E-20	4 3 0 4 2 2 Re -7.46E-20
3 0 0 2 1 1 Im -8.15E-19	4 0 0 3 2 2 Im 1.47E-20	4 3 0 4 2 2 Im -2.84E-20
3 0 0 3 0 0 Re 3.19E-18	4 0 0 4 0 0 Re 4.18E-21	4 3 0 4 3 0 Re 3.59E-20

3	0	0	3	0	0	Im	0.00E+00	4	0	0	4	0	0	Im	0.00E+00	4	3	0	4	3	0	Im	0.00E+00
3	0	0	3	1	0	Re	-1.63E-18	4	0	0	4	1	0	Re	3.31E-20	4	3	0	4	3	1	Re	7.94E-21
3	0	0	3	1	0	Im	-2.37E-18	4	0	0	4	1	0	Im	-4.63E-20	4	3	0	4	3	1	Im	3.66E-20
3	0	0	3	1	1	Re	3.81E-18	4	0	0	4	1	1	Re	5.78E-20	4	3	0	4	3	2	Re	-9.98E-21
3	0	0	3	1	1	Im	-5.85E-19	4	0	0	4	1	1	Im	5.93E-20	4	3	0	4	3	2	Im	-9.82E-21
3	0	0	3	2	0	Re	8.68E-19	4	0	0	4	2	0	Re	2.28E-20	4	3	0	4	3	3	Re	8.66E-21
3	0	0	3	2	0	Im	1.70E-18	4	0	0	4	2	0	Im	-1.66E-20	4	3	0	4	3	3	Im	4.34E-22
3	0	0	3	2	1	Re	-1.17E-18	4	0	0	4	2	1	Re	6.85E-21	4	3	1	2	0	0	Re	5.77E-19
3	0	0	3	2	1	Im	-2.24E-19	4	0	0	4	2	1	Im	3.64E-20	4	3	1	2	0	0	Im	-4.13E-20
3	0	0	3	2	2	Re	5.32E-19	4	0	0	4	2	2	Re	-2.69E-20	4	3	1	2	1	0	Re	-5.59E-19
3	0	0	3	2	2	Im	-4.63E-19	4	0	0	4	2	2	Im	-9.03E-21	4	3	1	2	1	0	Im	-8.12E-19
3	0	0	4	0	0	Re	1.38E-20	4	0	0	4	3	0	Re	8.60E-21	4	3	1	2	1	1	Re	1.17E-18
3	0	0	4	0	0	Im	-4.39E-20	4	0	0	4	3	0	Im	4.73E-22	4	3	1	2	1	1	Im	-6.10E-19
3	0	0	4	1	0	Re	-1.12E-18	4	0	0	4	3	1	Re	4.68E-22	4	3	1	3	0	0	Re	2.57E-19
3	0	0	4	1	0	Im	-1.21E-18	4	0	0	4	3	1	Im	8.96E-21	4	3	1	3	0	0	Im	-9.05E-20
3	0	0	4	1	1	Re	2.24E-18	4	0	0	4	3	2	Re	-3.06E-21	4	3	1	3	1	0	Re	-2.97E-19
3	0	0	4	1	1	Im	-4.18E-19	4	0	0	4	3	2	Im	-4.45E-21	4	3	1	3	1	0	Im	-2.54E-19
3	0	0	4	2	0	Re	-6.30E-19	4	0	0	4	3	3	Re	3.94E-21	4	3	1	3	1	1	Re	4.68E-19
3	0	0	4	2	0	Im	-1.17E-18	4	0	0	4	3	3	Im	2.65E-22	4	3	1	3	1	1	Im	-2.82E-19
3	0	0	4	2	1	Re	8.31E-19	4	1	0	2	0	0	Re	-2.95E-18	4	3	1	3	2	0	Re	1.46E-19
3	0	0	4	2	1	Im	7.99E-20	4	1	0	2	0	0	Im	1.94E-18	4	3	1	3	2	0	Im	1.47E-19
3	0	0	4	2	2	Re	-3.61E-19	4	1	0	2	1	0	Re	5.33E-18	4	3	1	3	2	1	Re	-1.81E-19
3	0	0	4	2	2	Im	3.12E-19	4	1	0	2	1	0	Im	2.03E-18	4	3	1	3	2	1	Im	1.76E-20
3	0	0	4	3	0	Re	1.25E-19	4	1	0	2	1	1	Re	-3.38E-18	4	3	1	3	2	2	Re	6.33E-20
3	0	0	4	3	0	Im	-2.27E-19	4	1	0	2	1	1	Im	6.59E-18	4	3	1	3	2	2	Im	-1.06E-19
3	0	0	4	3	1	Re	2.57E-19	4	1	0	3	0	0	Re	-1.12E-18	4	3	1	4	0	0	Re	4.68E-22
3	0	0	4	3	1	Im	9.05E-20	4	1	0	3	0	0	Im	1.21E-18	4	3	1	4	0	0	Im	-8.96E-21
3	0	0	4	3	2	Re	-7.68E-20	4	1	0	3	1	0	Re	2.25E-18	4	3	1	4	1	0	Re	-1.79E-19
3	0	0	4	3	2	Im	2.84E-20	4	1	0	3	1	0	Im	2.67E-19	4	3	1	4	1	0	Im	-1.17E-19
3	0	0	4	3	3	Re	1.98E-20	4	1	0	3	1	1	Re	-1.24E-18	4	3	1	4	1	1	Re	2.65E-19
3	0	0	4	3	3	Im	-4.26E-20	4	1	0	3	1	1	Im	2.76E-18	4	3	1	4	1	1	Im	-1.74E-19
3	1	0	2	0	0	Re	-4.67E-18	4	1	0	3	2	0	Re	-1.16E-18	4	3	1	4	2	0	Re	-1.01E-19
3	1	0	2	0	0	Im	4.06E-18	4	1	0	3	2	0	Im	-3.40E-19	4	3	1	4	2	0	Im	-9.73E-20
3	1	0	2	1	0	Re	1.03E-17	4	1	0	3	2	1	Re	8.09E-19	4	3	1	4	2	1	Re	1.23E-19
3	1	0	2	1	0	Im	2.55E-18	4	1	0	3	2	1	Im	-6.77E-19	4	3	1	4	2	1	Im	-2.39E-20
3	1	0	2	1	1	Re	-4.67E-18	4	1	0	3	2	2	Re	9.83E-20	4	3	1	4	2	2	Re	-4.25E-20
3	1	0	2	1	1	Im	1.29E-17	4	1	0	3	2	2	Im	6.98E-19	4	3	1	4	2	2	Im	7.11E-20
3	1	0	3	0	0	Re	-1.63E-18	4	1	0	4	0	0	Re	3.31E-20	4	3	1	4	3	0	Re	7.94E-21
3	1	0	3	0	0	Im	2.37E-18	4	1	0	4	0	0	Im	4.63E-20	4	3	1	4	3	0	Im	-3.66E-20
3	1	0	3	1	0	Re	4.14E-18	4	1	0	4	1	0	Re	1.25E-18	4	3	1	4	3	1	Re	4.06E-20
3	1	0	3	1	0	Im	0.00E+00	4	1	0	4	1	0	Im	0.00E+00	4	3	1	4	3	1	Im	0.00E+00
3	1	0	3	1	1	Re	-1.59E-18	4	1	0	4	1	1	Re	-6.66E-19	4	3	1	4	3	2	Re	-1.21E-20
3	1	0	3	1	1	Im	5.35E-18	4	1	0	4	1	1	Im	1.62E-18	4	3	1	4	3	2	Im	8.99E-21
3	1	0	3	2	0	Re	-2.11E-18	4	1	0	4	2	0	Re	7.91E-19	4	3	1	4	3	3	Re	1.57E-21
3	1	0	3	2	0	Im	-3.38E-19	4	1	0	4	2	0	Im	2.12E-19	4	3	1	4	3	3	Im	-8.86E-21
3	1	0	3	2	1	Re	1.34E-18	4	1	0	4	2	1	Re	-5.07E-19	4	3	2	2	0	0	Re	-1.46E-19
3	1	0	3	2	1	Im	-1.41E-18	4	1	0	4	2	1	Im	5.08E-19	4	3	2	2	0	0	Im	-1.11E-19
3	1	0	3	2	2	Re	3.34E-19	4	1	0	4	2	2	Re	-6.61E-20	4	3	2	2	1	0	Re	9.02E-21
3	1	0	3	2	2	Im	1.29E-18	4	1	0	4	2	2	Im	-4.64E-19	4	3	2	2	1	0	Im	4.43E-19
3	1	0	4	0	0	Re	7.08E-20	4	1	0	4	3	0	Re	7.68E-20	4	3	2	2	1	1	Re	-5.72E-19
3	1	0	4	0	0	Im	8.19E-20	4	1	0	4	3	0	Im	1.84E-19	4	3	2	2	1	1	Im	-2.39E-20
3	1	0	4	1	0	Re	2.25E-18	4	1	0	4	3	1	Re	-1.79E-19	4	3	2	3	0	0	Re	-7.68E-20
3	1	0	4	1	0	Im	-2.67E-19	4	1	0	4	3	1	Im	1.17E-19	4	3	2	3	0	0	Im	-2.84E-20
3	1	0	4	1	1	Re	-8.13E-19	4	1	0	4	3	2	Re	2.75E-20	4	3	2	3	1	0	Re	3.55E-20
3	1	0	4	1	1	Im	3.11E-18	4	1	0	4	3	2	Im	-8.29E-20	4	3	2	3	1	0	Im	1.60E-19
3	1	0	4	2	0	Re	1.44E-18	4	1	0	4	3	3	Re	2.72E-20	4	3	2	3	1	1	Re	-2.21E-19
3	1	0	4	2	0	Im	1.92E-19	4	1	0	4	3	3	Im	4.82E-20	4	3	2	3	1	1	Im	4.22E-21
3	1	0	4	2	1	Re	-8.20E-19	4	1	1	2	0	0	Re	4.30E-18	4	3	2	3	2	0	Re	-8.21E-22
3	1	0	4	2	1	Im	1.04E-18	4	1	1	2	0	0	Im	2.12E-18	4	3	2	3	2	0	Im	-7.31E-20
3	1	0	4	2	2	Re	-2.25E-19	4	1	1	2	1	0	Re	2.27E-19	4	3	2	3	2	1	Re	7.12E-20
3	1	0	4	2	2	Im	-8.53E-19	4	1	1	2	1	0	Im	-8.37E-18	4	3	2	3	2	1	Im	4.19E-20
3	1	0	4	3	0	Re	1.84E-19	4	1	1	2	1	1	Re	1.12E-17	4	3	2	3	2	2	Re	-5.11E-20
3	1	0	4	3	0	Im	3.16E-19	4	1	1	2	1	1	Im	1.45E-18	4	3	2	3	2	2	Im	3.06E-20

3	1	0	4	3	1	Re	-2.97E-19
3	1	0	4	3	1	Im	2.54E-19
3	1	0	4	3	2	Re	3.55E-20
3	1	0	4	3	2	Im	-1.60E-19
3	1	0	4	3	3	Re	6.03E-20
3	1	0	4	3	3	Im	8.61E-20
3	1	1	2	0	0	Re	7.42E-18
3	1	1	2	0	0	Im	3.38E-18
3	1	1	2	1	0	Re	-1.75E-19
3	1	1	2	1	0	Im	-1.46E-17
3	1	1	2	1	1	Re	1.95E-17
3	1	1	2	1	1	Im	1.68E-18
3	1	1	3	0	0	Re	3.81E-18
3	1	1	3	0	0	Im	5.85E-19
3	1	1	3	1	0	Re	-1.59E-18
3	1	1	3	1	0	Im	-5.35E-18
3	1	1	3	1	1	Re	7.99E-18
3	1	1	3	1	1	Im	0.00E+00
3	1	1	3	2	0	Re	6.40E-19
3	1	1	3	2	0	Im	2.73E-18
3	1	1	3	2	1	Re	-2.31E-18
3	1	1	3	2	1	Im	-1.28E-18
3	1	1	3	2	2	Re	1.69E-18
3	1	1	3	2	2	Im	-8.32E-19
3	1	1	4	0	0	Re	9.79E-20
3	1	1	4	0	0	Im	-1.10E-19
3	1	1	4	1	0	Re	-1.24E-18
3	1	1	4	1	0	Im	-2.76E-18
3	1	1	4	1	1	Re	4.61E-18
3	1	1	4	1	1	Im	-1.97E-19
3	1	1	4	2	0	Re	-4.97E-19
3	1	1	4	2	0	Im	-1.84E-18
3	1	1	4	2	1	Re	1.65E-18
3	1	1	4	2	1	Im	7.03E-19
3	1	1	4	2	2	Re	-1.12E-18
3	1	1	4	2	2	Im	5.67E-19
3	1	1	4	3	0	Re	3.48E-19
3	1	1	4	3	0	Im	-3.83E-19
3	1	1	4	3	1	Re	4.68E-19
3	1	1	4	3	1	Im	2.82E-19
3	1	1	4	3	2	Re	-2.21E-19
3	1	1	4	3	2	Im	-4.22E-21
3	1	1	4	3	3	Re	1.05E-19
3	1	1	4	3	3	Im	-9.95E-20
3	2	0	2	0	0	Re	2.57E-18
3	2	0	2	0	0	Im	-3.07E-18
3	2	0	2	1	0	Re	-5.09E-18
3	2	0	2	1	0	Im	-5.44E-19
3	2	0	2	1	1	Re	1.71E-18
3	2	0	2	1	1	Im	-6.37E-18
3	2	0	3	0	0	Re	8.68E-19
3	2	0	3	0	0	Im	-1.70E-18
3	2	0	3	1	0	Re	-2.11E-18
3	2	0	3	1	0	Im	3.38E-19
3	2	0	3	1	1	Re	6.40E-19
3	2	0	3	1	1	Im	-2.73E-18
4	1	1	3	0	0	Re	2.24E-18
4	1	1	3	0	0	Im	4.18E-19
4	1	1	3	1	0	Re	-8.13E-19
4	1	1	3	1	0	Im	-3.11E-18
4	1	1	3	1	1	Re	4.61E-18
4	1	1	3	1	1	Im	1.97E-19
4	1	1	3	2	0	Re	3.30E-19
4	1	1	3	2	0	Im	1.61E-18
4	1	1	3	2	1	Re	-1.30E-18
4	1	1	3	2	1	Im	-7.83E-19
4	1	1	3	2	2	Re	9.86E-19
4	1	1	3	2	2	Im	-4.31E-19
4	1	1	4	0	0	Re	5.78E-20
4	1	1	4	0	0	Im	-5.93E-20
4	1	1	4	1	0	Re	-6.66E-19
4	1	1	4	1	0	Im	-1.62E-18
4	1	1	4	1	1	Re	2.67E-18
4	1	1	4	1	1	Im	0.00E+00
4	1	1	4	2	0	Re	-2.61E-19
4	1	1	4	2	0	Im	-1.09E-18
4	1	1	4	2	1	Re	9.34E-19
4	1	1	4	2	1	Im	4.37E-19
4	1	1	4	2	2	Re	-6.56E-19
4	1	1	4	2	2	Im	2.95E-19
4	1	1	4	3	0	Re	2.10E-19
4	1	1	4	3	0	Im	-2.14E-19
4	1	1	4	3	1	Re	2.65E-19
4	1	1	4	3	1	Im	1.74E-19
4	1	1	4	3	2	Re	-1.25E-19
4	1	1	4	3	2	Im	-7.11E-21
4	1	1	4	3	3	Re	6.20E-20
4	1	1	4	3	3	Im	-5.35E-20
4	2	0	2	0	0	Re	-1.83E-18
4	2	0	2	0	0	Im	2.10E-18
4	2	0	2	1	0	Re	3.42E-18
4	2	0	2	1	0	Im	4.86E-19
4	2	0	2	1	1	Re	-1.32E-18
4	2	0	2	1	1	Im	4.27E-18
4	2	0	3	0	0	Re	-6.30E-19
4	2	0	3	0	0	Im	1.17E-18
4	2	0	3	1	0	Re	1.44E-18
4	2	0	3	1	0	Im	-1.92E-19
4	2	0	3	1	1	Re	-4.97E-19
4	2	0	3	1	1	Im	1.84E-18
4	2	0	3	2	0	Re	-8.93E-19
4	2	0	3	2	0	Im	-1.84E-20
4	2	0	3	2	1	Re	3.72E-19
4	2	0	3	2	1	Im	-5.17E-19
4	2	0	3	2	2	Re	1.28E-19
4	2	0	3	2	2	Im	3.62E-19
4	2	0	4	0	0	Re	2.28E-20
4	2	0	4	0	0	Im	1.66E-20
4	2	0	4	1	0	Re	7.91E-19
4	2	0	4	1	0	Im	-2.12E-19
4	2	0	4	1	1	Re	-2.61E-19
4	2	0	4	1	1	Im	1.09E-18
4	3	2	4	0	0	Re	-3.06E-21
4	3	2	4	0	0	Im	4.45E-21
4	3	2	4	1	0	Re	2.75E-20
4	3	2	4	1	0	Im	8.29E-20
4	3	2	4	1	1	Re	-1.25E-19
4	3	2	4	1	1	Im	7.11E-21
4	3	2	4	2	0	Re	2.11E-21
4	3	2	4	2	0	Im	4.85E-20
4	3	2	4	2	1	Re	-4.97E-20
4	3	2	4	2	1	Im	-2.36E-20
4	3	2	4	2	2	Re	3.34E-20
4	3	2	4	2	2	Im	-2.04E-20
4	3	2	4	3	0	Re	-9.98E-21
4	3	2	4	3	0	Im	9.82E-21
4	3	2	4	3	1	Re	-1.21E-20
4	3	2	4	3	1	Im	-8.99E-21
4	3	2	4	3	2	Re	7.64E-21
4	3	2	4	3	2	Im	0.00E+00
4	3	2	4	3	3	Re	-3.28E-21
4	3	2	4	3	3	Im	3.92E-21
4	3	3	2	0	0	Re	1.30E-20
4	3	3	2	0	0	Im	1.07E-19
4	3	3	2	1	0	Re	2.14E-19
4	3	3	2	1	0	Im	-2.03E-19
4	3	3	2	1	1	Re	2.55E-19
4	3	3	2	1	1	Im	2.87E-19
4	3	3	3	0	0	Re	1.98E-20
4	3	3	3	0	0	Im	4.26E-20
4	3	3	3	1	0	Re	6.03E-20
4	3	3	3	1	0	Im	-8.61E-20
4	3	3	3	1	1	Re	1.05E-19
4	3	3	3	1	1	Im	9.95E-20
4	3	3	3	2	0	Re	-3.14E-20
4	3	3	3	2	0	Im	2.80E-20
4	3	3	3	2	1	Re	-1.03E-20
4	3	3	3	2	1	Im	-5.34E-20
4	3	3	3	2	2	Re	4.03E-20
4	3	3	3	2	2	Im	1.12E-20
4	3	3	4	0	0	Re	3.94E-21
4	3	3	4	0	0	Im	-2.65E-22
4	3	3	4	1	0	Re	2.72E-20
4	3	3	4	1	0	Im	-4.82E-20
4	3	3	4	1	1	Re	6.20E-20
4	3	3	4	1	1	Im	5.35E-20
4	3	3	4	2	0	Re	2.00E-20
4	3	3	4	2	0	Im	-1.94E-20
4	3	3	4	2	1	Re	1.01E-20
4	3	3	4	2	1	Im	3.43E-20
4	3	3	4	2	2	Re	-2.66E-20
4	3	3	4	2	2	Im	-6.85E-21
4	3	3	4	3	0	Re	8.66E-21
4	3	3	4	3	0	Im	-4.34E-22
4	3	3	4	3	1	Re	1.57E-21
4	3	3	4	3	1	Im	8.86E-21
4	3	3	4	3	2	Re	-3.28E-21
4	3	3	4	3	2	Im	-3.92E-21
4	3	3	4	3	3	Re	3.73E-21
4	3	3	4	3	3	Im	0.00E+00

Journal of Physics: Conference Series **576** (2015) 012009 doi:10.1088/1742-6596/576/1/012009

State-selective charge transfer cross sections for light ion impact of atomic hydrogen

D.R. Schultz[1], P.C. Stancil[2], and C.C. Havener[3]

[1]Department of Physics, University of North Texas, Denton, TX
[2]Department of Physics and Astronomy, University of Georgia, Athens, GA
[3]Physics Division, Oak Ridge National Laboratory, Oak Ridge, TN

Abstract. Owing to the utility of diagnosing plasma properties such as impurity concentration and spatial distribution, and plasma temperature and rotation, by detection of photon emission following capture of electrons from atomic hydrogen to excited states of multiply charged ions, new calculations of state-selective charge transfer involving light ions have been carried out using the atomic orbital close-coupling and the classical trajectory Monte Carlo methods. By comparing these with results of other approaches applicable in a lower impact energy regime, and by benchmarking them using key experimental data, knowledge of the cross sections can be made available across the range parameters needed by fusion plasma diagnostics.

1. Introduction

Magnetic fusion experiments rely on a wide range of diagnostic techniques to understand important properties of the plasma such as the concentration of impurities, their transport within the plasma, and local plasma parameters such as density and rotation. One means of such diagnostics is the observation of photon emission when the electron from a hydrogen atom in a neutral heating beam is captured by a multiply charged impurity ion. Such collisions proceed with high probability to excited states of the ion and subsequent radiative relaxation yields photons of characteristic wavelengths (and other properties such as Doppler shift and line width and intensity) allowing identification of impurity species, motion, and density along the line of sight of the detector. This technique, commonly called charge exchange recombination spectroscopy (CHERS or CXRS) [1-4] is an important diagnostic tool implemented at many fusion research facilities.

As a contribution to the IAEA Coordinated Research Program on data for light ions, calculations were made for several common impurity ions (C^{5+}, N^{6+}, O^{6+}, and O^{7+}) colliding with atomic hydrogen in order to help create the database needed to derive cross sections over a broad range of impact energies and final captured states for CXRS diagnostics. This was carried out within a broad collaboration, consisting of work to provide benchmark measurements of the total charge transfer cross section at Oak Ridge National Laboratory [5] and theoretical work at several institutions including the University of Georgia, Bergische Universität Wuppertal, and the University of North Texas, using complementary methods for intercomparison and broad coverage of impact energies and final states. The complete results for C^{5+} + H [6], N^{6+} + H [7], and O^{6+} + H [8] have been published and those for O^{7+} + H are in final preparation.

Here we describe some of the results and tabulate in detail a subset of data resulting from the calculations carried out by the first author who was a participant in the IAEA Coordinated Research Program. These calculations provide results for intermediate energy collisions, ~1 to 200 keV/u C^{5+}, N^{6+}, O^{6+}, and O^{7+} + H, using a method that is applicable particularly for capture to the low-lying excited states populated dominantly in these collisions (the atomic orbital close-coupling (AOCC) method) and another that is useful for prediction of results for the higher-lying excited states (the classical trajectory Monte Carlo (CTMC) method). When benchmarked by the total cross section measurements and combined with results applicable at lower collision energies (the molecular orbital close-coupling (MOCC) method) from our collaboration, recommended state-selective results can be derived for use in fusion plasma diagnostics and other applications.

2. Results and discussion

2.1. Total cross sections

At intermediate collisions energies (defined by collision velocities roughly comparable to those of the initially bound electron) the AOCC method should provide the most reliable results. As described in Refs. [6-8], we have employed the technique of Kuang and Lin [9,10] using very large basis sets to obtain state-selective charge transfer cross sections for all bound states in the calculation, and summed them to obtain the total cross section. Figures 1-2 show comparisons of the AOCC total cross sections with the benchmark measurements made at Oak Ridge. As these figures show, there is very good overall agreement between AOCC and the measurements regarding both the magnitude of the cross section and its behavior with energy throughout the overlapping energy ranges. Similar good agreement is also found with the to-be-published O^{7+} measurements. The limits of applicability of the AOCC method are reached below several hundred eV and so in that regime the results are the least reliable and another method, in particular, MOCC, is needed. The AOCC results are applicable to energies higher than those available in the experiment and are in generally good agreement with the results from CTMC [6-8] in this higher energy regime.

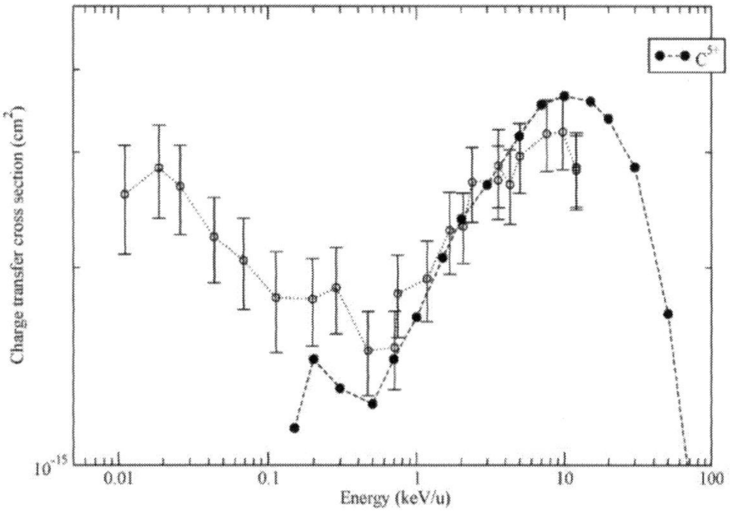

Figure 1. Comparison of the AOCC total cross section [6] (filled circles) for charge transfer in C^{5+} + H collisions with the benchmark measurements [5] (open circles).

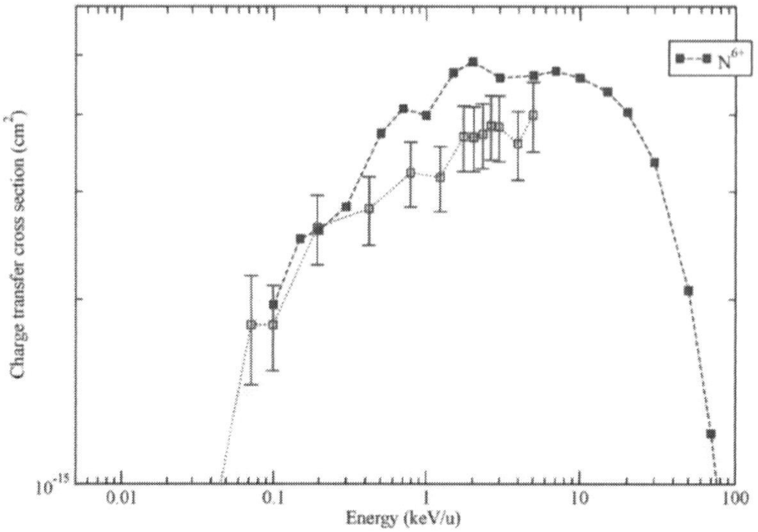

Figure 2. Comparison of the AOCC total cross section [7] (filled squares) for charge transfer in N^{6+} + H collisions with the benchmark measurements [7] (open squares).

2.2 State-selective cross sections

State-selective cross sections were computed using the AOCC method for the four collision systems at 19 impact energies between 0.15 and 200 keV/u and provide results for charge transfer up to principal quantum numbers of n=8. Since the AOCC formulation used is a one-electron model, resolution of the results by singlet and triplet spins is not possible and thus its results represent the sum of these channels that would be resolved by a two-electron treatment. The AOCC results are in generally good agreement with those obtained via the CTMC method and CTMC has been used to provide results for n levels higher than those contained in the AOCC basis sets. Together with analogous results of the MOCC method for lower energies, these data are being used to deduce recommended sets of data for use in plasma applications. To illustrate their behavior at intermediate energies, Tables 1-3 contain the AOCC state-selective charge transfer cross sections for $n< 6$ and those from CTMC for $n\geq 6$ at six energies between 0.5 and 100 keV/u for C^{5+}, N^{6+}, and O^{6+} + H.

Acknowledgement

This work was carried out in part as a contribution to the IAEA Coordinated Research Project on "Light Element Atom, Molecule, and Radical Behavior in the Divertor and Edge Regions."

References

[1] R.J. Groebner, N.H. Brooks, K.H. Burrell, and L. Rottler, Appl. Phys. Lett. **43**, 920 (1983).
[2] R.C. Isler, Phys. Scr. **35**, 650 (1987).
[3] A. Boileau, M.G. von Hellerman, L.D. Horton, J. Spence, and H.P. Summers, J. Phys. B **31**, 779 (1989).
[4] O. Marchuk, Yu. Ralchenko, R.K. Janev, G. Berschinger, and W. Biel, J. Phys. B **42**, 165701 (2009).

[5] I.N. Draganić, D.G. Seely, and C.C. Havener, Phys. Rev. A 83, 054701 (2011) and references therein.
[6] J.L. Nolte, P.C. Stancil, H.P. Liebermann, R.J. Buenker, Y. Hui, and D.R. Schultz, J. Phys. B 45, 245202 (2012)
[7] Y. Wu, P.C. Stancil, H.P. Liebermann, P. Funke, S.N. Rai, R.J. Buenker, D.R. Schultz, Y. Hui, I.N. Draganić, and C.C. Havener, Phys. Rev. A 84, 022711 (2011).
[8] Y. Wu, P.C. Stancil, D.R. Schultz, Y. Hui, H.P. Liebermann, and R.J. Buenker, J. Phys. B 45, 235201 (2012).
[9] J. Kuang and C.D. Lin, J. Phys. B 29, 1207 (1996).
[10] J. Kuang and C.D. Lin, J. Phys. B 29, 5442 (1996).

Table 1. State-selective charge transfer cross sections (in units of 10^{-16} cm^2) for C^{5+} + H calculated by the present AOCC (n<6) and CTMC (n≥6) methods at selected intermediate collision energies (*E*).

C^{4+} state	*E*=0.5 keV/u	1 keV/u	5 keV/u	10 keV/u	50 keV/u	100 keV/u
2s	1.92E-05	1.21E-06	3.51E-04	7.11E-03	7.59E-02	3.72E-02
$2p_0$	1.29E-05	1.01E-06	3.65E-04	5.90E-03	5.87E-02	7.28E-02
$2p_1$	1.48E-05	3.27E-07	2.38E-04	2.81E-03	1.13E-01	8.16E-02
3s	9.05E-02	4.12E-01	1.87E+00	1.18E+00	1.56E-01	4.17E-02
$3p_0$	6.84E-02	2.96E-01	1.12E+00	1.79E+00	5.02E-01	1.17E-01
$3p_1$	8.68E-02	4.57E-01	3.14E+00	2.50E+00	3.44E-01	5.94E-02
$3d_0$	2.12E-02	7.02E-02	9.77E-01	1.46E+00	1.20E+00	3.41E-01
$3d_1$	2.37E-02	1.39E-01	9.28E-01	2.23E+00	1.65E+00	3.67E-01
$3d_2$	2.30E-02	1.10E-01	1.10E+00	1.50E+00	4.34E-01	6.19E-02
4s	1.02E+00	1.27E+00	9.55E-01	5.99E-01	1.11E-01	2.53E-02
$4p_0$	1.47E+00	1.64E+00	1.93E+00	1.29E+00	3.43E-01	9.82E-02
$4p_1$	1.13E+00	2.13E+00	1.14E+00	7.02E-01	1.41E-01	3.48E-02
$4d_0$	1.92E+00	1.83E+00	2.41E+00	2.73E+00	6.99E-01	1.96E-01
$4d_1$	2.24E+00	2.40E+00	3.34E+00	3.00E+00	7.26E-01	1.90E-01
$4d_2$	8.55E-01	8.00E-01	7.02E-01	6.19E-01	1.07E-01	2.40E-02
$4f_0$	1.02E+00	1.36E+00	3.01E+00	4.19E+00	1.21E+00	1.90E-01
$4f_1$	1.27E+00	2.08E+00	4.66E+00	6.46E+00	1.68E+00	2.31E-01
$4f_2$	6.01E-01	9.01E-01	2.08E+00	2.87E+00	5.22E-01	5.88E-02
$4f_3$	2.70E-01	3.17E-01	4.59E-01	5.61E-01	6.82E-02	5.74E-03
5s	3.50E-02	8.30E-02	6.72E-02	8.07E-02	7.30E-02	1.67E-02
$5p_0$	1.07E-02	2.72E-02	8.87E-02	1.51E-01	2.12E-01	7.42E-02
$5p_1$	4.56E-02	8.83E-02	3.62E-02	3.78E-02	7.93E-02	2.17E-02
$5d_0$	2.32E-02	6.63E-02	1.83E-01	1.53E-01	4.22E-01	1.30E-01
$5d_1$	3.56E-02	8.34E-02	1.33E-01	1.04E-01	3.65E-01	1.12E-01
$5d_2$	3.10E-02	2.49E-02	3.02E-02	2.03E-02	4.75E-02	1.25E-02
$5f_0$	3.40E-02	5.21E-02	2.64E-01	3.28E-01	5.39E-01	1.22E-01
$5f_1$	4.37E-02	4.20E-02	3.01E-01	3.55E-01	6.98E-01	1.48E-01
$5f_2$	1.50E-02	2.68E-02	1.22E-01	8.02E-02	1.82E-01	3.52E-02
$5f_3$	1.30E-02	1.94E-02	2.56E-02	1.38E-02	1.48E-02	3.22E-03
$5g_0$	1.80E-02	1.03E-02	1.53E-01	3.61E-01	3.34E-01	4.03E-02
$5g_1$	1.43E-02	1.49E-02	2.01E-01	5.21E-01	4.85E-01	5.71E-02
$5g_2$	2.96E-03	1.58E-02	8.28E-02	2.31E-01	1.90E-01	2.02E-02
$5g_3$	4.77E-03	1.33E-02	2.50E-02	5.16E-02	3.87E-02	3.60E-03
$5g_4$	7.81E-03	9.77E-03	3.58E-03	7.75E-03	4.20E-03	2.44E-04
n=6	8.8E-04	3.37E-03	3.89E-02	2.06E-01	1.74E+00	6.73E-01
n=7	-	6.4E-04	9.54E-03	7.28E-02	9.18E-01	4.69E-01
n=8	-	-	3.29E-03	3.78E-02	5.25E-01	3.36E-01
n=9	-	-	2.0E-03	2.03E-02	3.29E-01	2.46E-01
n=10	-	-	6.8E-04	1.29E-02	2.23E-01	1.84E-01
n=20	-	-	-	8.4E-04	2.19e-02	2.52E-02

Journal of Physics: Conference Series **576** (2015) 012009 doi:10.1088/1742-6596/576/1/012009

Table 2. State-selective charge transfer cross sections (in units of 10^{-16} cm^2) for N^{6+} + H calculated by the present AOCC (n<6) and CTMC (n≥6) methods at selected intermediate collision energies (*E*).

N^{5+} state	*E*=0.5 keV/u	1 keV/u	5 keV/u	10 keV/u	50 keV/u	100 keV/u
2s	9.49E-05	1.16E-04	1.29E-05	2.20E-04	1.62E-02	1.82E-02
$2p_0$	6.71E-05	7.38E-05	6.72E-06	2.66E-04	8.18E-03	2.01E-02
$2p_1$	9.45E-05	1.70E-04	1.25E-05	5.41E-05	1.88E-02	2.73E-02
3s	1.11E-04	1.67E-03	2.72E-01	3.97E-01	1.54E-01	4.01E-02
$3p_0$	9.91E-05	1.73E-03	1.27E-01	3.09E-01	3.63E-01	1.02E-01
$3p_1$	1.52E-04	3.13E-03	3.16E-01	6.63E-01	3.58E-01	5.09E-02
$3d_0$	4.86E-05	1.04E-03	5.04E-02	2.41E-01	5.09E-01	2.64E-01
$3d_1$	9.23E-05	1.02E-03	1.01E-01	2.35E-01	8.19E-01	3.51E-01
$3d_2$	3.31E-05	6.13E-04	8.50E-02	2.81E-01	3.01E-01	7.59E-02
4s	2.71E+00	3.28E+00	1.73E+00	9.35E-01	1.17E-01	3.37E-02
$4p_0$	3.61E+00	3.61E+00	3.61E+00	1.97E+00	3.55E-01	8.23E-02
$4p_1$	2.62E+00	4.31E+00	3.22E+00	1.72E+00	1.65E-01	3.30E-02
$4d_0$	2.44E+00	2.65E+00	4.43E+00	3.68E+00	6.14E-01	1.67E-01
$4d_1$	3.55E+00	3.57E+00	6.55E+00	5.17E+00	6.86E-01	1.71E-01
$4d_2$	1.41E+00	2.21E+00	2.82E+00	1.67E+00	1.25E-01	3.14E-02
$4f_0$	4.30E+00	3.98E+00	3.64E+00	4.33E+00	1.52E+00	3.25E-01
$4f_1$	6.67E+00	6.17E+00	6.06E+00	6.74E+00	2.26E+00	4.47E-01
$4f_2$	2.62E+00	2.44E+00	3.25E+00	3.75E+00	8.76E-01	1.24E-01
$4f_3$	6.03E-01	9.47E-01	1.28E+00	1.10E+00	1.41E-01	1.30E-02
5s	5.99E-01	6.11E-01	1.15E-01	1.60E-01	7.64E-02	2.86E-02
$5p_0$	1.21E-01	6.07E-01	4.01E-01	4.36E-01	2.42E-01	5.01E-02
$5p_1$	8.14E-01	1.14E+00	1.84E-01	1.81E-01	9.30E-02	2.02E-02
$5d_0$	8.77E-01	8.03E-01	4.03E-01	5.19E-01	3.83E-01	1.46E-01
$5d_1$	6.83E-01	8.26E-01	5.73E-01	5.54E-01	3.59E-01	1.12E-01
$5d_2$	4.73E-01	3.05E-01	1.50E-01	1.08E-01	5.00E-02	1.54E-02
$5f_0$	1.33E+00	7.88E-01	7.38E-01	1.18E+00	6.60E-01	1.94E-01
$5f_1$	5.54E-01	4.79E-01	1.43E+00	1.55E+00	8.91E-01	2.38E-01
$5f_2$	2.27E-01	2.07E-01	5.13E-01	5.07E-01	2.61E-01	6.08E-02
$5f_3$	2.17E-01	1.83E-01	8.08E-02	7.76E-02	2.31E-02	5.32E-03
$5g_0$	3.84E-01	1.76E-01	1.12E+00	1.86E+00	8.62E-01	1.19E-01
$5g_1$	1.68E-01	1.86E-01	1.66E+00	2.87E+00	1.28E+00	1.80E-01
$5g_2$	7.39E-02	1.46E-01	6.72E-01	1.41E+00	5.52E-01	7.04E-02
$5g_3$	1.15E-01	1.48E-01	1.74E-01	4.08E-01	1.23E-01	1.33E-02
$5g_4$	1.26E-01	8.65E-02	3.63E-02	5.78E-02	1.29E-02	1.33E-03
n=6	6.89E-03	1.69E-02	1.47E-01	4.34E-01	3.30E+00	1.17E+00
n=7	2.8E-04	1.17E-03	1.95E-02	1.26E-01	1.23E+00	8.70E-01
n=8	-	2.4E-04	5.49E-03	5.20E-02	8.75E-01	6.31E-01
n=9	-	-	2.05E-03	2.79E-02	5.08E-01	4.71E-01
n=10	-	-	1.04E-03	1.50E-02	3.31E-01	3.52E-01
n=20	-	-	-	9.77E-04	2.86E-02	4.87E-02

Table 3. State-selective charge transfer cross sections (in units of 10^{-16} cm^2) for O^{6+} + H calculated by the present AOCC (n<6) and CTMC (n≥6) methods at selected intermediate collision energies (*E*).

O^{5+} state	*E*=0.5 keV/u	1 keV/u	5 keV/u	10 keV/u	50 keV/u	100 keV/u
2s	-	-	2.77E-05	8.10E-05	4.34E-03	1.88E-02
$2p_0$	-	-	1.40E-05	2.07E-04	4.97E-03	1.32E-02
$2p_1$	-	-	7.42E-06	9.34E-05	8.34E-03	3.34E-02
3s	1.19E-03	1.59E-03	6.61E-02	2.53E-01	1.55E-01	4.36E-02
$3p_0$	7.74E-04	4.87E-03	2.81E-01	1.97E-01	3.51E-01	9.31E-02
$3p_1$	5.44E-04	4.06E-03	4.77E-01	4.43E-01	2.72E-01	4.41E-02
$3d_0$	8.77E-04	4.44E-03	6.77E-02	2.51E-01	5.48E-01	3.33E-01
$3d_1$	4.20E-04	2.57E-03	7.43E-02	1.41E-01	7.88E-01	3.81E-01
$3d_2$	4.75E-04	7.80E-03	2.36E-01	2.33E-01	2.79E-01	6.32E-02
4s	6.85E+00	9.34E+00	7.22E+00	2.38E+00	9.68E-02	4.20E-02
$4p_0$	4.43E+00	6.72E+00	3.49E+00	2.47E+00	4.08E-01	7.00E-02
$4p_1$	1.18E+01	7.42E+00	2.75E+00	1.69E+00	1.21E-01	1.97E-02
$4d_0$	8.63E-01	9.22E-01	3.17E+00	3.45E+00	6.58E-01	2.55E-01
$4d_1$	6.19E-01	9.30E-01	3.01E+00	3.73E+00	7.12E-01	1.74E-01
$4d_2$	1.95E+00	1.83E+00	1.35E+00	1.06E+00	1.20E-01	2.22E-02
$4f_0$	2.10E+00	2.37E+00	3.08E+00	4.02E+00	1.52E+00	3.12E-01
$4f_1$	1.09E+00	2.81E+00	5.05E+00	6.32E+00	2.23E+00	4.06E-01
$4f_2$	5.47E+00	1.92E+00	2.43E+00	3.38E+00	8.77E-01	1.32E-01
$4f_3$	5.50E+00	4.95E+00	1.22E+00	8.80E-01	1.32E-01	1.91E-02
5s	1.62E-01	7.28E-01	4.41E-01	3.08E-01	6.17E-02	3.55E-02
$5p_0$	1.49E+00	4.04E-01	4.09E-01	5.27E-01	3.04E-01	4.96E-02
$5p_1$	7.93E-01	9.45E-01	2.43E-01	1.68E-01	5.90E-02	1.23E-02
$5d_0$	3.83E-01	8.82E-01	4.94E-01	6.03E-01	4.15E-01	1.90E-01
$5d_1$	7.44E-01	8.53E-01	5.53E-01	5.18E-01	3.94E-01	1.14E-01
$5d_2$	2.51E-01	1.85E-01	1.08E-01	7.11E-02	4.55E-02	1.26E-02
$5f_0$	2.42E-01	4.08E-01	9.94E-01	1.32E+00	7.39E-01	1.89E-01
$5f_1$	2.70E-01	3.47E-01	1.64E+00	1.80E+00	8.52E-01	2.14E-01
$5f_2$	1.29E-01	1.44E-01	5.62E-01	5.86E-01	2.56E-01	6.22E-02
$5f_3$	1.47E-01	1.45E-01	6.98E-02	7.35E-02	3.29E-02	7.47E-03
$5g_0$	8.73E-02	9.52E-02	1.24E+00	2.20E+00	9.03E-01	1.32E-01
$5g_1$	7.53E-02	1.75E-01	1.70E+00	3.25E+00	1.35E+00	1.92E-01
$5g_2$	4.46E-02	1.30E-01	7.34E-01	1.67E+00	6.10E-01	8.21E-02
$5g_3$	4.90E-02	7.38E-02	2.25E-01	4.64E-01	1.37E-01	1.60E-02
$5g_4$	1.26E-01	8.87E-02	3.36E-02	5.93E-02	1.19E-02	1.60E-03
n=6	4.36E-03	1.38E-02	1.48E-01	4.61E-01	3.34E+00	1.16E+00
n=7	-	1.63E-03	2.18E-02	1.33E-01	1.67E+00	8.66E-01
n=8	-	2.2E-04	6.66E-03	5.47E-02	9.02E-01	6.31E-01
n=9	-	-	3.02E-03	2.82E-02	5.30E-01	4.68E-01
n=10	-	-	1.41E-03	1.63E-02	3.44E-01	3.55E-01
n=20	-	-	1.0E-04	9.58E-04	2.91E-02	4.97E-02

Contribution to the theoretical investigation of electron and photon interaction with carbon atom and its ions

V Stancalie

National Institute for Laser, Plasma and Radiation Physics, Department of Lasers, Atomistilor 409, P.O.Box MG-36, Magurele –Ilfov, 077125 Romania. Association EURATOM/MEdC

E-mail: viorica.stancalie@inflpr.ro

Abstract. This work presents results from the theoretical investigation of electron and photon interaction with carbon atoms and its Be-like ion. After an introductory survey on the existing reported theoretical works, we present results from the calculation of energy levels and oscillator strengths for selected transitions in neutral carbon. We also report results from the detailed state selective photo-recombination study in Be-like C^{2+} ions.

1. Introduction

Detailed knowledge of carbon transport properties, and consequently of the accurate cross-section data, is required for the fusion plasma diagnostics. The difficulties in obtaining sufficient quantities of atomic carbon in a well-defined initial state led to a complete absence of experimental cross-section measurements. This absence of experimental data made theoretical investigations a computational grand challenge.

In general, the electron-impact excitation cross sections calculated for neutral carbon with different methods are not in good agreement with each other especially for optically forbidden transitions. Detailed structure calculations have been reported for the astrophysically important transitions [1-4]. More extensive theoretical and experimental reported works exist for the carbon ions. Theoretical works on C^{2+} have been reported by Glass [5], Berrington [6], Berrington et al. [7], and Mitnik et al. [8]. Experimental data have also been published [9]. Results from the electron and photon interactions with C^{1+} have been reported by Wilson and Bell [10], Wilson et al. [11, 12], and Nahar [13]. There are a number of publications dealing with the ionization of the CIII to CV ionization stages. Detailed analysis of the importance of this process in populating the excited levels of ions in plasmas typical for those found in the edge of tokamak for the CIV and CV ionization stages has been recently made by Lawson et al. [14, and references herein].

We will discuss the existing theoretical models for electron-scattering by neutral carbon atom and will give complementary results for its atomic structure. We report a new set of *R*-matrix calculation intended specifically to investigate the role of including configuration interaction (CI) wave functions both in the target state expansion and in the (N+1)-electron quadratically integrable function expansion. Target-atom states were computed using the Hartree-Fock ground configuration $2s^22p^2$ plus the near-degenerate configurations $2s2p^3$ and $2p^4$. In addition, for consistency with the virtual excitation pattern used for scattering wave functions, all virtual excitations of $2s^22p^2$ of the form 2p-np and 2p-nf were included. Results for the energy levels and oscillator strengths for important transitions have been obtained

The second issue addressed here is the effect of quantum interference between dielectronic (DR) and radiative recombination (RR) on photo-recombination (PR) cross section in Be-like C ion. The calculation was carried out for the excited Rydberg states of type $1s^22sns$ ($^1S^e$) which interact with the odd-parity continua up to the C^{3+} 2p threshold limit. The numerical evaluation has been performed at a fine energy mesh across all the autoionizing Rydberg series of resonances $1s^22pns(^1P^0)$ converging to Li-like ion 2p threshold. The method of calculation is based on the on the essential ingredients of the Feshbach projection operator approximation. The photo-ionization cross sections have been evaluated with and without relativistic effects included into the *R*-matrix numerical procedures, while the allowance for both quantum interference between DR and RR and overlapping resonances has been done using results from the earlier *R*-matrix Floquet calculation [15]. We present all these results with respect to the effect of quantum interference term on the energy dependence profile of PR cross section.

The paper is structured as follows. Section 2 presents an introductory survey on the existing target model calculation in electron scattering on atomic carbon. The present calculated energy levels are compared with existing experimental data in Atomic Structure Database of the National Institute for Standards and Technology wherever available. Section 3 is focused on photon-interaction with the C^{2+} ion. The study of resonant single-photon ionization process from the $1s^22sns$ ($^1S^e$) states of Be-like Al^{9+} and C^{2+} ion, ($9 \leq n \leq 12$, and $5 \leq n \leq 12$, respectively) has been done [16, 17] in response to the need in the best way of including radiation damping in a non-perturbative treatment of DR of Li-like into Be-like ions. Furthermore, the dominant states in DR of these ions have been investigated [18-21] along with detailed close-coupling calculation of the electron-ion scattering [22-24]. Currently, we employ all these results and carry out a new set of calculation in order to explore the effect of quantum interference between DR and RR on the energy dependence profile of PR cross section in C^{2+}. Section 4 presents our concluding remarks and further direction of research.

2. Structure calculation

In neutral carbon the electron scattering calculation is made difficult by the fact that low-energy electron scattering is dominated by a resonance due to the $1s^22s^22p^3$ $^3P^0$ state of the negative ion, C^-. The theoretical prediction of the location of this resonance, and of the $1s^22s^22p^3$ $^4S^0$ and $^2D^0$ bound states of C^-, depends on a balance between short-range correlation and long range polarization effects.

The ground-state configuration of carbon in the independent-particle model is $(1s)^2(2s)^2(2p)^2$, which leads to the ground-state terms ^3P, ^1D, ^1S. In order to get reliable data, the wave functions of the N-electron target state and of the negative ionic states that have N+1electrons should be calculated with the same degree of accuracy. In the calculation of low-energy electron scattering by neutral carbon atoms, one must take into account long-range polarization effects and short-range correlation between the incident electron and the electron attachment effects that lead to scattering resonances. The early theoretical models and methods applied to low-electron energy scattering from atomic carbon could be summarized as follows: *a*) the polarized-orbital method [25-27]. This is single configuration-based method which treats the target-atom dipole polarizability and the resulting long-range polarization potential accurately, but involves approximations to the short-range interaction that are difficult to evaluate. The polarized-orbital method cannot describe resonances; *b*) the close-coupling method [28, 29] which is capable, in principle, of representing the full structure of the electronic continuum wave functions through a series of expansion in the stationary sates of the target atom. Elaborate calculation including some target correlation effects and a limited pseudostate representation of target polarizability has been reported by Saraph [30], Rountree, Smith and Henry [31], Le Dourneuf [32], Le Dourneuf et al. [33]; *c*) the matrix variational method including configuration interaction approximation (CI) that takes into account the effect of near-degenerate configuration interaction [34-36]. The major innovation in this method is that LS eigenfunctions are used through the calculations, but does not describe resonances. All these calculations were limited in elastic scattering from ground states and transitions amongst the n=2 states at low collision energies.

In the last two decades, the most detailed works have been reported by Dunseath et al. [37], Zatsarinny et al. [38] and Liu, Wang and Zhou [39]. In the work by Dunseath et al. [37], two independent standard *R*-matrix calculations were performed. In the first, all levels of carbon with $n \leq 3$ have been included in the expansion of the total wave function. In the second calculation, all levels with $n \leq 4$ have been considered in the target model calculation. Extensive calculation of cross sections and collision strengths for important transitions to the $n = 3$ and $n = 4$ levels of carbon has been performed using the asymptotic distorted wave method available in the FARM package [40] after which the top-up procedure was used. These authors reported some resonance structures in the low-energy region, and pointed out the importance of the target-model in scattering calculation. The correlation effects were only included in the first model-calculation.

In their work, Zatsarinny et al. [38] presented theoretical results based on the extended version of the R-matrix method, where a *B*-spline basis is employed for the representation of the continuum functions. These authors revealed a prominent resonance structure for the most electron-induced transitions in atomic carbon at low energies. The close-coupling expansion has included the bound and autoionizing states derived from the $1s^22s^22p^2$, $1s^22s^22p3l$ (l=0, 1, 2), $1s^22s^22p4s$, $2s^22s2p^3$, and $1s^22p^4$ configurations, plus eight pseudo-states to account for the target continuum. The calculated cross sections and effective collision strengths for important transitions from the ground state $2p^2\ {}^3P$, and the metastable $2p^2\ {}^1D$ and ${}^1S$ states show significant discrepancies with those reported by Dunseath et al. [37] in a similar scattering model. This work also provides a detailed analysis of the resonance structure in the low-energy near-threshold regions.

In the third and most recent theoretical calculation by Liu et al. [39], the authors use the momentum space coupled-channels optical (CCO) method to solve coupled-integral equations for a finite set of discrete physical states of target to convergence. The continuum is treated via an 'ab initio' complex equivalent-local potential. The model-calculation includes 12 states in the coupled-channels. They were described by single configuration Hartree-Fock approximation.

We conclude this introductory survey by mentioning that in these works the cross sections agree well in shape, but large discrepancies in magnitude exist, namely the Dunseath's excitation cross sections are 32% higher than those obtained from the calculation of *B*-spline R-matrix method, and the results of CCO method are 18% lower than the calculation of *B*-spline R-matrix method. A database has been constructed [41] consisting of the recommended cross sections for electron-impact excitation and ionization of carbon atoms and ions, as well as for charge exchange processes between carbon ions and hydrogen atoms.

In the present work, we report results concerning the non-relativistic and relativistic atomic structure calculation in neutral carbon. An initial LS coupling- standard R-matrix atomic structure calculation has been performed in order to investigate the role of including configuration interaction wave functions both in the target state expansion and in the (N+1)-electron quadratically integrable function expansion. The electron correlation effects have been explored by carrying out separate calculations with and without the configurations: $2s2p^3$, $2s2p^2nl$ (n =3, 4), $2p^4$ in the target state expansion, and $2s^22p^3$, $2s2p^4$, $2s2p^3nl$ (n =3, 4) in the (N+1)-electron wave function. In table 1, results for selected state energies, in eV units, are compared with existing reported experimental data in Atomic Structure Database of the National Institute for Standards and Technology [42] wherever available. Comparison with the earlier published theoretical works is also provided in this table.

In our target model calculation, the (N+1)-electron configurations data have been obtained by adding one electron to the N-electron configurations in all possible ways. This wave function was augmented by including all configurations arising from virtual excitation of a 2p orbital. Exploratory calculations indicated that further 2s virtual excitations could be neglected.

As a further check on reported structure calculation, table 2 shows the calculated oscillator strengths and comparison with previously reported theoretical and experimental values.

Journal of Physics: Conference Series **576** (2015) 012010 doi:10.1088/1742-6596/576/1/012010

Table 1. Excitation energies (in eV) for the spectroscopic target states.

	State	Term	Present	Ref. [37]	Ref. [38]	NIST [42]
1	$2s^22p^2$	^3P	0.00	0.00	0.00	0.00
2	$2s^22p^2$	^1D	1.557	1.545	1.353	1.260
3	$2s^22p^2$	^1S	2.602	2.545	2.833	2.680
4	$2s2p^3$	$^5S^0$	3.092	3.133	4.069	4.179
5	$2s^22p3s$	$^3P^0$	7.401	8.488	7.488	7.481
6	$2s^22p3s$	$^1P^0$	7.740	8.936	7.727	7.680
7	$2s2p^3$	$^3D^0$	8.340	8.412	8.082	7.942
8	$2s^22p3p$	^1P	8.451	9.456	8.528	8.534
9	$2s^22p3p$	^3D	8.600	9.589	8.647	8.642
10	$2s^22p3p$	^3S	8.772	9.785	8.737	8.767
11	$2s^22p3p$	^3P	9.309	10.390	8.822	8.845
12	$2s^22p3p$	^1D	9.443	10.757	9.012	8.998
13	$2s^22p3p$	^1S	10.424	11.370	9.256	9.168
14	$2s2p^3$	$^3P^0$	9.517	9.981	9.504	9.326
15	$2s^22p3d$	$^1D^0$	9.772	10.719	9.647	9.627
16	$2s^22p4s$	$^3P^0$	10.142	10.810	9.708	9.683
17	$2s^22p4s$	$^1P^0$	9.549	10.834	9.708	9.709
18	$2s^22p3d$	$^3F^0$	9.517	10.809	9.729	9.695
19	$2s^22p3d$	$^3D^0$	9.607	10.888	9.731	9.705
20	$2s^22p3d$	$^1F^0$	9.607	10.947	9.759	9.732
21	$2s^22p3d$	$^1P^0$	9.653	10.970	9.782	9.758
22	$2s^22p3d$	$^3P^0$	13.407	11.018	9.983	9.830
23	$2s2p^3$	$^1D^0$	14.470	14.645	12.984	12.132
24	$2s2p^3$	$^3S^0$	13.407	15.366	13.273	13.114
25	$2s2p^3$	$^1P^0$	15.927	16.182	14.949	14.860
26	$2p^4$	^3P	19.361	21.149	19.986	
27	$2p^4$	^1D	19.621	22.677	20.877	
28	$2p^4$	^1S	20.479	26.227	24.389	

Table 2. Oscillator strengths in C.

Lower level	Upper level	Present	Ref. [37]	Ref. [38]	NIST [42]
$2s^22p^2\ ^3P$	$2s^22p3s\ ^3P^0$	0.124	0.154	0.133	0.140
	$2s2p^3\ ^3D^0$	0.098	0.152	0.107	0.072
	$2s2p^3\ ^3P^0$	0.028	0.117	0.055	0.063
	$2s^22p4s\ ^3P^0$	0.023	0.010	0.009	0.021
	$2s^22p3d\ ^3D^0$	0.112	0.132	0.107	0.094
	$2s^22p3d\ ^3P^0$	0.340	0.069	0.098	0.040
	$2s2p^3\ ^3S^0$	0.171	0.269	0.134	0.152
$2s^22p^2\ ^1D$	$2s^22p3s\ ^1P^0$	0.128	0.103	0.118	0.118
	$2s^22p3d\ ^1D^0$	0.009	0.007	0.013	0.013
	$2s^22p4s\ ^1P^0$	0.004	0.010	0.004	0.011
	$2s^22p3d\ ^1F^0$	0.061	0.099	0.118	0.085
	$2s^22p3d\ ^1P^0$	0.018	0.014	0.011	0.009
	$2s2p^3\ ^1D^0$	0.344	0.529	0.396	
	$2s2p^3\ ^1P^0$	0.351	0.333	0.257	
$2s^22p^2\ ^1S$	$2s^22p3s\ ^1P^0$	0.021	0.076	0.098	0.094
	$2s^22p4s\ ^1P^0$	0.007	0.001	0.004	0.005
	$2s^22p3d\ ^1P^0$	0.050	0.142	0.196	0.125
	$2s2p^3\ ^1P^0$	0.122	0.633	0.458	

Next, we initiated fine-structure splitting calculation employing the full relativistic GRASP (General-purpose Relativistic Atomic Structure Package) code [43], subsequently revised by several workers under the names GRASP, GRASP92 and GRASP2K. In our work we used GRASP0 code [44]. The code is fully relativistic and it is based on the *jj* coupling scheme. Further relativistic corrections arising from the Breit interaction and QED effects (vacuum polarization and Lamb shift) have also been included. Additionally, we have used the option of extended average level, in which a weighted (proportional to $2j+1$) trace Hamiltonian matrix is minimized. This produces a compromise set of orbitals describing closely lying states with moderate accuracy. For comparison purposes, we have performed parallel calculations with the Flexible Atomic Code (FAC) [45]. This is also a fully relativistic code, which provides a variety of atomic parameters, and (generally) yields results for energy levels and radiative rates comparable to GRASP. Results from FAC are helpful in assessing the accuracy of our energy levels and radiative rates, and in estimating the contribution of resonances to effective collision strengths.

The $2s^22p^2$, $2s^22p3l$, $2s^22p4l$, $2s^22p5s$, $2p^3$ and $2p^4$ configurations of carbon give rise to 88 fine structure levels , J = 0 – 5, odd and even parity. In table 3 we compare level energies obtained with GRASP (*without* and *with* the inclusion of Breit and QED effects) with the critically evaluated data compiled by NIST for the lowest 51 levels. Also included in this table are results obtained from the FAC code (FAC) including the same CI (configuration interaction) as in GRASP. The level energies obtained without the Breit and QED effects (GRASP1) are consistently higher than the NIST values up to ~ 0.03Ryd. However, the orderings are nearly the same as those of NIST. The inclusion of Breit and QED effects (GRASP2) lowers the energies by a maximum of ~ 0.0001 Ryd. The FAC level energies are consistently lower by up to 0.003 Ryd than GRASP results, and hence they are comparatively in better agreement with the NIST listings. The level ordering from FAC are also in general agreement with the calculation from GRASP, but differ in some instances, particularly for the $2s2p^3$ levels. A further inclusion of the 2p5*l*, $l = 2 - 4$, configurations, labeled FAC2 calculation in table 3 makes no appreciable difference either in the magnitude or ordering of the levels.

Table 3. Fine-structure splitting calculation in C. NIST: http://www.nist.gov; GRASP1: energies from the GRASP code with 88 level calculations without Breit and QED effects included; GRASP2: energies from the GRASP code with 88 level calculations with Breit and QED effects; FAC1: energies from the FAC code with 88 level calculations; FAC2: energies from the FAC code with 134 level calculations.

Index	*Configuration/Term*	NIST	GRASP1 (Ryd)	GRASP2 (Ryd)	FAC1 (Ryd)	FAC2 (Ryd)
1	$2s^22p^2\ ^3P_0$	0.000	0.000	0.0000	0.00000	0.0000
2	$2s^22p^2\ ^3P_1$	0.0001494	0.0001873	0.0001570	0.000064	0.000064
3	$2s^22p^2\ ^3P_2$	0.0003955	0.0005584	0.0004093	0.00015105	0.000150
4	$2s^22p^2\ ^1D_2$	0.09288208	0.1164040	0.1162848	0.11214485	0.112139
5	$2s^22p^2\ ^1S_0$	0.1972712	0.1862982	0.1862062	0.18620175	0.185830
6	$2s2p^3\ ^5S_2^0$	0.3074178	0.2209301	0.2207590	0.24012327	0.240146
7	$2s^22p3s\ ^3P_0^0$	0.5497987	0.5672529	0.5670994	0.5637540	0.563779
8	$2s^22p3s\ ^3P_1^0$	0.5499736	0.56743662	0.56723954	0.5638635	0.563888
9	$2s^22p3s\ ^3P_2^0$	0.5503428	0.56781234	0.5673131	0.5640918	0.564117
10	$2s^22p3s\ ^1P_1^0$	0.5648199	0.5915391	0.5912947	0.5898824	0.589799
11	$2s2p^3\ ^3D_3^0$	0.5840030	0.6287575	0.62851954	0.6215855	0.620708
12	$2s2p^3\ ^3D_1^0$	0.5840297	0.62877758	0.6286506	0.6216502	0.620772
13	$2s2p^3\ ^3D_2^0$	0.5840397	0.6287918	0.6286165	0.6216703	0.620793
14	$2s^22p3p\ ^1P_1$	0.6274650	0.6419229	0.6416769	0.6335048	0.632521
15	$2s^22p3p\ ^3D_1$	0.6350573	0.6500307	0.6498481	0.6404025	0.639685
16	$2s^22p3p\ ^3D_2$	0.6352503	0.65024712	0.6500201	0.6408055	0.640223
17	$2s^22p3p\ ^3D_3$	0.6355544	0.6505814	0.65028541	0.6409477	0.640081
18	$2s^22p3p\ ^3S_1$	0.6446663	0.66232293	0.6620727	0.6515510	0.651184
19	$2s^22p3p\ ^3P_0$	0.6502119	0.68021956	0.6800215	0.6806422	0.680232
20	$2s^22p3p\ ^3P_1$	0.6503248	0.68032504	0.6801031	0.6811181	0.680783
21	$2s^22p3p\ ^3P_2$	0.6505114	0.68052371	0.6802520	0.6811958	0.680232
22	$2s^22p3p\ ^1D_2$	0.6616776	0.6933946	0.6931423	0.6938854	0.680783
23	$2s^22p3p\ ^1S_0$	0.6741181	0.70517591	0.7049219	0.6933448	0.692573
24	$2s2p^3\ ^3P_1^0$	0.6857646	0.714868994	0.7147168	-	-
25	$2s2p^3\ ^3P_2^0$	0.6857765	0.71493401	0.71481283	-	-
26	$2s2p^3\ ^3P_0^0$	0.6857842	0.71506221	0.714676964	-	-
27	$2s^22p3d\ ^1D_2^0$	0.7078706	0.72124488	0.72098426	0.6971971	0.696376
28	$2s^22p4s\ ^3P_0^0$	0.7117449	0.727291515	0.72739911	0.6963345	0.696337

Table 3 - continued

Index	*Configuration/Term*	NIST	GRASP1 (Ryd)	GRASP2 (Ryd)	FAC1 (Ryd)	FAC2 (Ryd)
29	$2s^22p4s\ ^3P_1^0$	0.7118521	0.727461988	0.72741456	0.6964099	0.696260
30	$2s^22p4s\ ^3P_2^0$	0.7121378	0.727641373	0.72708323	0.6964513	
31	$2s^22p3d\ ^3F_2^0$	0.7126024	0.72773354	0.7274672	0.70124003	0.697085
32	$2s^22p3d\ ^3F_3^0$	0.7127522	0.727461988	0.72721555	0.70194265	0.701930
33	$2s^22p3d\ ^3F_4^0$	0.7130659	0.727715995	0.72740876	0.7016843	0.701672
34	$2s^22p3d\ ^3D_1^0$	0.7134628	0.73556129	0.73533694	0.7062923	0.706266
35	$2s^22p3d\ ^3D_2^0$	0.7135916	0.73563534	0.73538238	0.7065979	0.701226
36	$2s^22p3d\ ^3D_3^0$	0.7136884	0.735720808	0.734715436	0.70641018	0.706377
37	$2s^22p4s\ ^1P_1^0$	0.7138892	0.73662305	0.7363463	0.70456633	0.704580
38	$2s^22p3d\ ^1F_3^0$	0.7156146	0.734973379	0.73471543	0.7078843	0.707857
39	$2s^22p3d\ ^1P_1^0$	0.7174521	0.736623055	0.73634636	0.71068036	0.710583
40	$2s^22p3d\ ^3P_2^0$	0.7227336	0.72771599	0.72740876	-	-
41	$2s^22p3d\ ^3P_1^0$	0.7228059	0.72766176	0.727414567	-	-
42	$2s^22p3d\ ^3P_0^0$	0.7228458	0.72764137	0.72739912	0.70470595	0.704203
43	$2s^22p4p\ ^1P_1$	0.7341427	0.7470183265	0.74676676	0.7327676	0.732111
44	$2s^22p4p\ ^3D_1$	0.7361444	0.75000403	0.74980834	0.7338335	0.733358
45	$2s^22p4p\ ^3D_2$	0.7362153	0.75021145	0.74997625	0.73421335	0.733736
46	$2s^22p4p\ ^3D_3$	0.7366192	0.75054157	0.75024022	0.73423192	0.733761
47	$2s^22p4p\ ^3S_1$	0.7390834	0.75599413	0.75573827	0.7362784	0.736073
48	$2s^22p4p\ ^3P_0$	0.7409604	0.77657497	0.7763905	0.74538969	0.744327
49	$2s^22p4p\ ^3P_1$	0.7410949	0.77672703	0.77651882	0.7458102	0.744682
50	$2s^22p4p\ ^3P_2$	0.7412610	0.777024551	0.77675744	0.74578021	0.744703
51	$2s^22p4p\ ^1D_2$	0.7451412	0.776778183	0.77651438	0.75620790	0.745683

3. State-selective photo-recombination cross sections in C^{2+}

In competition with the electron-ion scattering occurs the photo-recombination process (PR). PR is an important cooling mechanism in the hot plasmas and is also used as a plasma diagnostics [46]. It can proceed non-resonantly, through the electron-ion continuum (radiative recombination (RR)). It can also proceed resonantly, through intermediate doubly excited states (dielectronic recombination (DR)). In addition, the emitted photon can be recaptured by the ion, giving rise to a modification of the electron-ion collision cross section, called radiative damping.

Following the pioneering work by Davies and Seaton [47], a general theory of DR, based on the quantum defect and including radiative damping in the close-coupling equations, was developed by Bell and Seaton [48]. The method has been applied by Pradhan and Seaton [49] and more recently by Zabaydullin and Dubau [50]. These authors have reported results on the effects of radiative decay on collision strengths, and from the photo-absorption and photo-ionization (PI) cross sections calculation, respectively. Detailed PR and radiative damping calculations have been carried out using the *R*-matrix code which includes relativistic Breit-Pauli effects. The method of procedure has been suggested by Robicheaux *et al.* [51]. They proposed treating the electron scattering process using a coupled channel method (e.g. an R-matrix method) but adding an imaginary part, to the energy of the closed channels to take account of the radiative decay. Then the execution of the substantive calculations has been promoted as international collaboration [52]. The allowance for both interference and overlapping resonances in a practical calculation usually has been done using the Feshbach projection operator methods by LaGattuta [53], Haan and Jacobs [54] and Jacobs and Behar [55]. The theoretical works by Badnell and Pindzola [56], and more recently by Behar *et al.* [57] have reported results for highly charged uranium ions, and for the He-like Ar and Fe ions, respectively.

Employing projection-operator techniques, the energy-averaged unified PR cross section is a sum over three terms:

$$\sigma_{PR} = \sigma_{RR} + \sigma_{DR} + \sigma_{\text{int}} \tag{2}$$

where σ_{RR}, σ_{DR} and σ_{int} represent the RR cross section, the DR cross section and the interference (denoted by int) between the RR and DR transition amplitude, respectively.

To begin with, one class of configurations which creates channels for DR are analyzed in this work, namely L shell channel $1s^2 2pnl - 1s^2 2s\, n'l'$, with $\Delta n = 0$, $\Delta l = 0$. The first $2s + e^- \rightarrow 2pnl$ levels to lie in the continuum for C^{3+} ions belong to $n = 5$. Doubly excited levels in the $\Delta n = 0$ channel are sometimes below the continuum for low enough values of n; in this case the level cannot be formed by capture plus excitation as serves in the present work as a final level for the recombined ion. A similar situation occurs in C^{3+} for the $1s^2 2p4s$ level of Be-like C^{2+} which is below the continuum, whereas the $1s^2 2p5s$ level is in the continuum and contributes to DR through:

$$1s^2 2s + e^- \Leftrightarrow 1s^2 2p5s \rightarrow 1s^2 2s5s + h\nu \tag{3}$$

In order to understand the method of calculation, we recall here the principal idea of the working DR model. An initial calculation based on the standard R-matrix program [58] has been done to obtain the field-free atomic states of type $1s^2 2sns$, $1s^2 2pns$ and $1s^2 2pnd$ in Be-like ions, relative to ($^2S^e$) and ($^2P^0$) ionization threshold respectively, including a full description of the electron-electron correlation.

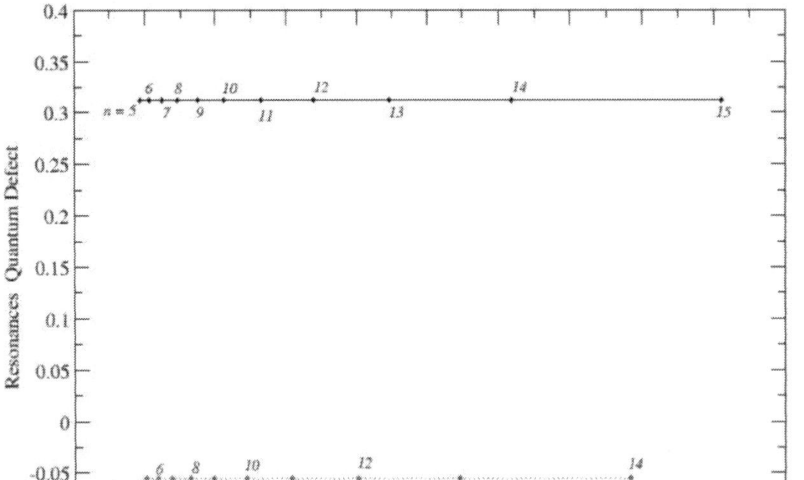

Figure 1. R-Matrix Edlén plot for $1s^2 2pns$ ($^1P^0$) (solid) and $1s^2 2pnd$($^1P^0$)(dotted) Rydberg serie of resonances in C III, as function of the resonance position (in atomic units, au) below the $^2P^0$ threshold. The n values are the principle quantum numbers for the associated data point.

The series of resonances, of type $^1P^0$, are allowed in *LS* coupling. They have been identified by the quantum defect theory (QDT) and method. The effective quantum number ν of the autoionizing state has been obtained from the quantum-defect formula, $\nu = z/\sqrt{(E-E_T)}$ where *z* is the ion charge, *E* is the resonance energy, and E_T is the threshold energy in Rydbergs.

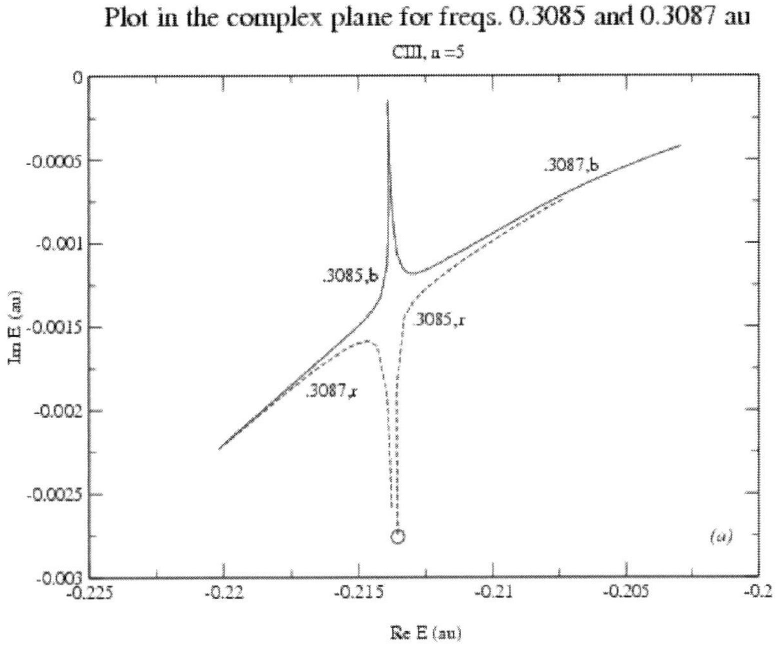

Figure 2. Trajectories of the complex energies of $1s^22s5s(^1S^e)$ and $1s^22p7s(^1P^0)$ states as function of field intensity and frequency. For each frequency, there are two curves, one connected adiabatically to the 'bound' and the other to the autoionizing state. The intensity is varied from 0 to 5×10^{13} Wcm^{-2}. The small dots give intensity in steps of 5×10^{11} W cm^{-2}. The atomic units are used and the energy scale is chosen such that ground states energy is set to zero.

The DR cross section has been calculated utilizing the Feshbach projection operator approximation [57] where the physical quantities correspond to our model as follows:

$$\sigma_{idf}^{DR}(\varepsilon) = \frac{(2\pi\hbar)^3}{8\pi m\varepsilon}\frac{(2J_d+1)}{2(2J_i+1)}\Gamma_{di}^{a}L_d(\varepsilon)B_{df}^{D} \tag{4}$$

with $$L_d(\varepsilon) = \frac{\Gamma(d)/2\pi}{(\varepsilon-\omega_{tune})^2+\left(\frac{\Gamma(d)}{2}\right)^2}$$ and $$B_{df}^{D} = \frac{\hbar\Gamma_{df}}{\Gamma(d)}$$

In the above equations, ε and ω_{tune} are the electron collision and the photon frequency, respectively, and $B^D{}_{df}$ is the branching ratio.

We denoted with *i, d, f* the initial, the doubly excited (continuum), and the excited (bound) state, respectively. The transition probabilities for autoionization process are denoted by Γ_{di}^{a} and the radiative process from the same d state is represented by Γ_{df}. The total (natural) width of the doubly-excited levels (d) in Eq. 4 is defined as follows:

$$\Gamma(d) = \hbar\left(\sum_f \Gamma_{df} + \sum_i \Gamma^a_{di}\right) \tag{5}$$

where Γ_{df} is the spontaneous radiative decay from the doubly excited d state to final f state and Γ^a_{di} is the rate for autoionization from the resonance d state to the non-resonant electron-continuum i state. According to our model calculation [16, 17], $\Gamma(d)$ corresponds to the LIDS width. If the autoionization is affected by the radiative rates, the interaction with the radiative field has to be comparable in strength with the electron-electron interaction. In other words, for each photon energy, i.e. one-photon channel, the imaginary part of the quasi-energy corresponding to autoionizing states, Γ^a_{di}, includes Γ^{field} signifying the correction due to the interaction with the ion field. Finally, the quantity $\Gamma^a_{di}/(\Gamma_{df}+\Gamma^a_{di}+\Gamma^{field})$ arises directly from the calculation. Moreover, the problem of overlapping resonances is solved exactly. Table 4 presents the results for the position and width of the autoionization states of type $1s^22pns$, and the total natural width of doubly excited states of type $1s^22pns$, with $n = 5 - 11$.

Table 4. The total natural width, $\Gamma(d)$, of doubly excited states, the autoionization probability, Γ^a, and the position, (E_a), of states of type $1s^22pns$ belonging to the C III ions, n = 5 – 12. Atomic units (*au*) are used.

Ion Li-like	Initial states	Intermediate states L shell channel $1s^22pns$	$\Gamma(d)$ *au*	Γ^a *au*	E_a *au*
C^{3+}	$1s^22s$	$n = 5$	0.272×10^{-02}	0.5518×10^{-02}	0.09496
		$n = 6$	0.143×10^{-02}	0.298×10^{-02}	0.15984
		$n = 7$	0.900×10^{-03}	0.17948×10^{-02}	0.19794
		$n = 8$	0.598×10^{-03}	0.11658×10^{-02}	0.22221
		$n = 9$	0.420×10^{-03}	0.8004×10^{-03}	0.23861
		$n = 10$	0.290×10^{-03}	0.5736×10^{-03}	0.25021
		$n = 11$	0.181×10^{-03}	0.4252×10^{-03}	0.26514

The DR cross sections for each channel were obtained from Eq.4. Figure 3 illustrates the plot of DR cross section (in Megabarn, 1 Mb = 10^{-18} cm^2) versus electron collision energy (in Ryd.) for the DR process: $e^- + 1s^22s \rightarrow 1s^22p7s - 1s^22s7s + h\nu$ in CIV.

In the absence of radiation damping, the cross section for photo-recombination σ_{PR} is related to the cross section for photo-ionization σ_{PI} through detailed balance given by the Milne relation:

$$\sigma_{PR} = \sigma_{PI}\frac{g_i}{g_j}\frac{\alpha^2 E^2_{ph}}{4(E_{ph}-I)} \tag{5}$$

in atomic units. In this equation, g_i and g_j are the statistical weights of the initial and residual ions respectively, E_{ph} and I are the photon energy and ionization potential of the final ionic state, respectively, and α is the fine-structure constant. We have carried out successive non-relativistic and relativistic calculation of the PI cross sections. In the first calculation, the PI calculation from the ^1S state of the Be-like systems:

$$1s^22sns(^1S^0) + h\nu \rightarrow [1s^2nl + e\text{-}(kl')](^1P^0_1) \tag{6}$$

is performed employing the computational package RMATXI [58]. At this level of accuracy we have obtained the total PI cross section from the (^{1}S) to all Li-like states converging to (^{2}P^0) ionization threshold in the vicinity of 2pnl (^{1}P^0) and 2pnd (^{1}P^0) resonances. In order to depict the individual PR cross sections of narrow resonances, we have performed a new set of calculation in vicinity of the only 1s^{2}2pns resonance complexes. The relativistic resonance energies and orbitals were obtained by using the multiconfigurational Dirac-Fock atomic structure code [59].

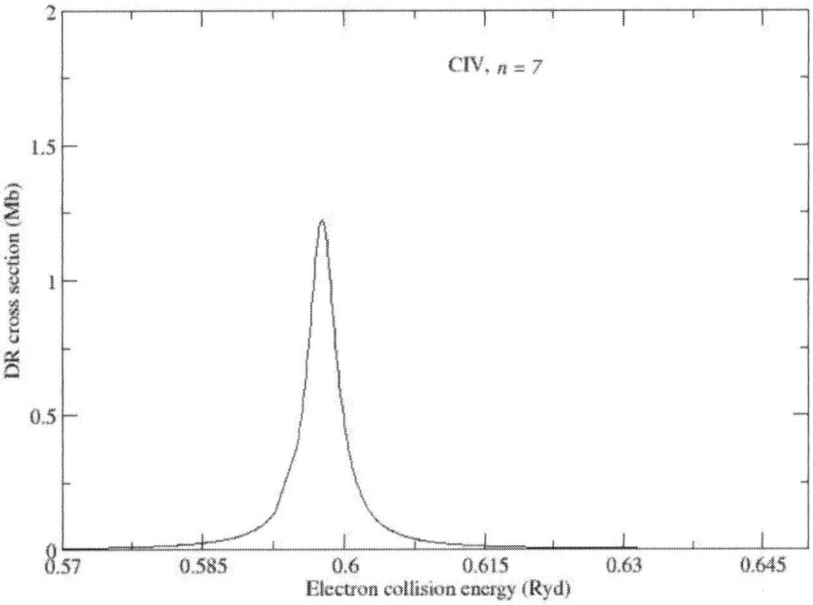

Figure 3. Dielectronic recombination (DR) cross section (Mb) versus collision energy (Ryd)

for $e^- + 1s^22s \rightarrow 1s^22p7s - 1s^22s7s + h\nu$ in the C IV ion.

The PI cross sections have been calculated for all the 1s^{2}2sns(1S) states, with $5 \leq n \leq 12$ belonging to Be-like C ion. For a given *n*, the only ten non-relativistic configuration state functions such that 1s^{2}2s^2, 1s^{2}2p^2, 1s^{2}2sns, 1s^{2}2snd, 1s^{2}2snf, 1s^{2}2sng, 1s^{2}2pns, 1s^{2}2pnp, 1s^{2}2pnf, and 1s^{2}2png have been included in the average level (EAL) calculation, while 1s^2, 2s^2 and 2p^2 shells were kept full. These *jj*-coupled configuration state functions (CSF's) result in a set of 74 levels that are limited to the following symmetry-parity combinations $J\pi = 0^{\pm}, 1^{\pm}, 2^{\pm}, 3^{\pm}, 4^{\pm}, 5^{+}$. To stabilize the order of the levels (odd or even), an initial analysis on each parity has been done. The method of procedure is well described in Ref. [20].

Radiative rates were obtained for the decay of even-parity resonance levels to the configurations: 1s^{2}2s*n*p and 1s^{2}2s*n*f, while the autoionization rates were determined for the decay of the even-parity resonance levels to the single configuration levels involving the even-parity configurations 1s^{2}2sεs, 1s^{2}2sεd and 1s^{2}2sεg, $\varepsilon \leq n$. For the decay of odd-parity resonance levels, the radiative rates were determined to the multiconfiguration levels involving the even-parity configurations 1s^{2}2p^2, 1s^{2}2s*n*s and 1s^{2}2s*n*d, while the autoionizing rates were determined for the decay of the odd-parity resonance levels to the single configuration levels involving the odd-parity configurations: 1s^{2}2snp, 1s^{2}2sns and 1s^{2}2snf levels. To check the accuracy of the resonance positions, we have compared our calculated energies with those reported by the opacity team [60]. Figure 4 presents on logarithmic scale results for the PI and RR cross sections involving the only $1s^22s7s(^1S)$ state in Be-like C ion. In this case, relativistic effects are included into the calculation. The photon energy ranges from 0.295 au to 0.3004 au.

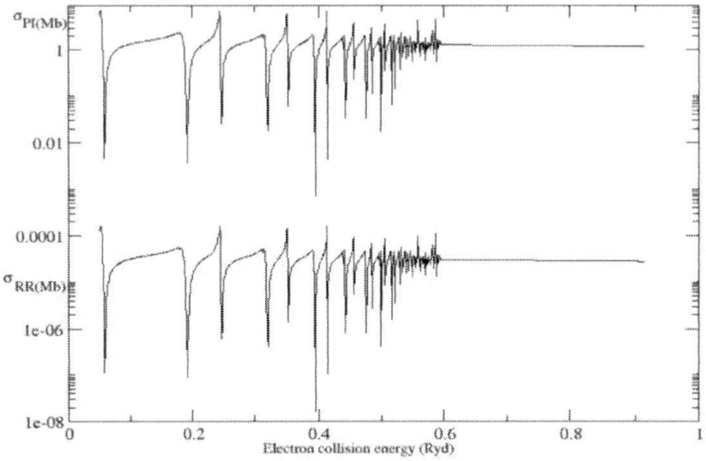

Figure 4. Plot on logarithmic scale of the cross sections (in Mb units) for the photoionization, σ_{PI}, and the radiative recombination, σ_{RR}, for state $1s^22s7s(^1S)$ state to$1s^22p7s(^1P_0)$ state in C^{2+}, versus electron collision energy (in Ryd).

The final term in Eq. (2) has been determined using the Fano parameter evaluated directly from the *2*-state model of Latinne *et al.* [61] Figure 5 shows the energy dependence quantum interference cross section (σ_{int}, Mb) contributing to the PR cross section in $1s^22s7s$-$1s^22p7s$ transition calculated in the vicinity of $1s^22p7s$ autoionizing state, in C^{2+} ion.

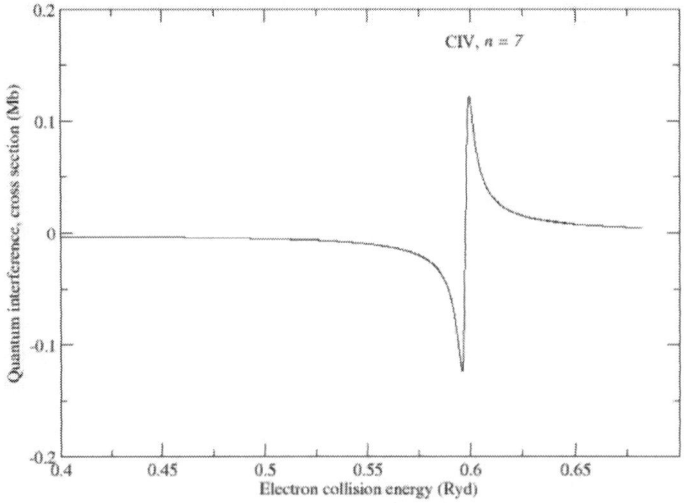

Figure 5. Quantum interference cross section (Mb) versus collision energy (Ryd) for $1s^22p7s$ – $1s^22s7s$ transition in C IV.

Journal of Physics: Conference Series **576** (2015) 012010 doi:10.1088/1742-6596/576/1/012010

The state selective photo-recombination cross section have been evaluated for all $\Delta n = 0$ channels in C^{3+} ion at incident electron energies in the vicinity of the $1s^2 2pns$ autoionizing-configuration complexes. Results are shown in Figure 6.

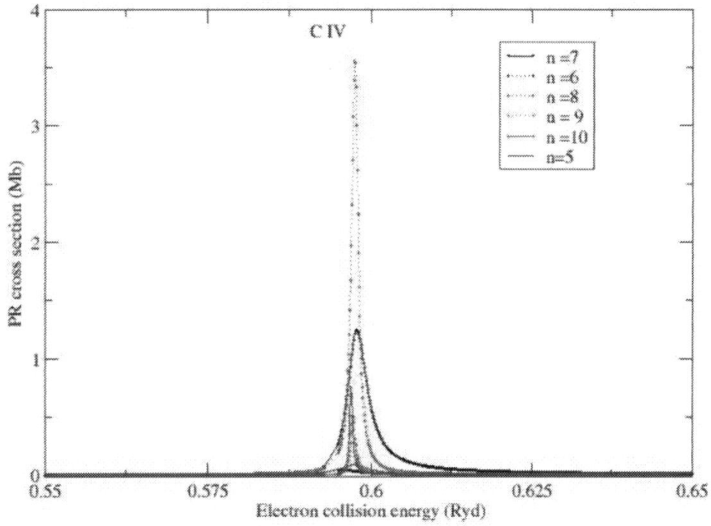

Figure 6. The energy dependence profile of photorecombination(PR) cross section (Mb) in the vicinity of $1s^2 2pns(5 \le n \le 10)$ ($^1P^0$) autoionizing states in CIV ion. Damping radiative effect in dielectronic recombination and quantum interference between dielectronic and radiative recombination processes are included into the calculation

As a result of our analysis[61], we find that for the majority of dielectronic recombination-- resonance transitions considered in the present investigation, only a weak degree of asymmetry was found. This is due to the absence of configuration mixing involving $1s^2 2pnd(^1P^0)$ autoionizing states, which could alter our $\Delta n = 0$ transition rates. In order to obtain more detailed representation for significant and experimentally observable interference effects, the configuration mixing including $1s^2 2pnd(^1P^0)$ autoionizing states should be considered in a further investigation. This investigation could facilitate the scaling analysis for this isoelectronic sequence. Additionally, the plasma environmental conditions are expected to be important and will be considered in our further theoretical work.

4. Concluding remarks

The present theoretical work reports some results from extensive non-relativistic and relativistic calculation of energy levels and oscillator strengths in the neutral carbon atom. In addition, results from detailed five-state close coupling calculation for photo-recombination cross sections in the Li-like C ions have also been reported.

We have attempted to obtain an accurate description of electron correlations both in the initial and final states of electron scattering on carbon atom. The target model considered here has included the configuration interaction allowing for distortion of the initial-state and excited-state orbitals due to electron correlation effects. Furthermore, based on a variety of comparisons among various calculations with GRASP and FAC codes, our fine structure splitting data are in good agreement with the experimental values. Based on these various relativistic calculations more detailed investigations are needed for the radiative rates, oscillator strengths, line strengths and lifetimes.

Photo-recombination cross-sections are important to analyze spectra emitted by fusion plasmas. The present theoretical work extends our previous R-matrix Floquet calculation exploring the role of the quantum interference term on the energy dependence profile of photo-recombination cross section. We have presented results for all the set of photo-recombination transitions involving the $1s^2 2pns$ doubly excited autoionizing states of Be-like C ions. Only a weak degree of asymmetry was found. In order to obtain more detailed representation for significant and experimentally observable interference effects, the configuration mixing including $1s^2 2pnd(^1P^0)$ autoionizing states should be considered in a further investigation.

Acknowledgement

This work, supported by the European Communities under the contract of Association between EURATOM and MEdC, was carried out within the framework of the European Fusion Development Agreement. The views and opinions expressed here do not necessarily reflect those of the European Commission. This work was carried out as a part of an IAEA Coordinated Research Project on "Light Element Atom, Molecule and Radical Behaviour in the Divertor and Edge Plasma Regions".

References

[1] Galavis ME, Mendoza C and Zeippen CJ 1997 Astron. Astrophys. Suppl. Ser. 123 159.
[2] Hibbert A, Biemont E, Godefroid M and Vaeck N 1993 A&A 99 179.
[3] Froese Fischer C and Saha HP 1983 Phys. Rev. A 28 3169.
[4] Froese Fischer C and Saha HP 1985 Phys. Scr. 32 181.
[5] Glass R 1979 J. Phys.B: At.. Mol. Phys. 12 1633.
[6] Berrington KA 1985 J. Phys.B: At. Mol. Phys. 18 L395.
[7] Mitnik DM, Griffin DC, Balance CP and Badnell NR 2003 J. Phys. B:At. Mol. Opt. Phys. 36 713.
[8] Berrington K A, Burke V M, Burke PG and Scilla S 1989 J. Phys. B:At. Mol. Opt. Phys. 22 665.
[9] Muller A 2009 Eur. Phys. J. Special Topics 169 35.
[10] Wilson NJ and Bell KL 2002 Mon. Not. R. Astron.Soc.337 1027.
[11] Wilson NJ, Bell K L and Hudson C E 2005 A & A 432 731.
[12] Wilson NJ, Bell K L and Hudson C E 2007 A&A 461 765.
[13] Nahar SN 2002 Phys. Rev. A 65 052702.
[14] Lawson KD, Coffey IH, Agaarwal KM, Keenan FP and JET-EFDA Contributors 2013 J. Phys. B: At. Mol. Opt. Phys. 46 035701.
[15] Burke PG, Francken P and Joichain CJ 1991 *J. Phys. B.: At. Mol. Opt. Phys*. **24**, 761.
[16] Stancalie V 2005 *Phys. Plasmas* **12** 043301.
[17] Stancalie V 2005 *Phys. Plasmas* **12** 100705.
[18] Robicheaux F, Loch SD, Pindzola MS, Balance CP 2010 *Phys. Rev. Lett.* **105** 233201.
[19] Pindzola MS, Loch SD, Robicheaux F 2011 *Phys. Rev.* A **83** 042705.
[20] Stancalie V 2012 *Nucl. Instr. Meth. Phys. Res. B* **279** 147.
[21] Stancalie V 2013 *Laser and Particle Beams* **31** 1.
[22] Stancalie V 2000 *Phys. Scr.* **61** 459.
[23] Berrington KA 1985 J. Phys. B: At. Mol. Phys.18 L395.
[24] Teng H, Knopp H, Ricz S, Schippers S, Berrington KA, Muller A 2000 Phys. Rev. A 61 060704 (R).
[25] Temkin A 1957 *Phys. Rev.* **107** 1004.
[26] Henry R J W 1968 *Phys. Rev.* **172** 99.
[27] Henry R J W, Burke P G and Sinfailam A-L 1969 Phys. Rev.178 218.

[28] Burke PG and Smith K 1962 Rev. Mod. Phys. 34 458.
[29] Burke PG and Seaton MJ 1971 Meth. Comput. Phys. 10 1.
[30] Saraph HE 1973 J. Phys. B 6 L243.
[31] Rountree SP, Smith ER and Henry RJW, J Phys B 7 L167.
[32] Le Dourneuf M 1976 Thesis unpublished
[33] Le DurneufM, Vo Ky Lan and Burke P G 1977 Comment. At. Mol. Phys.7 1.
[34] Thomas L D, Oberoi RS and Nesbet R K 1975 Phys.Rev. A 10 1605.
[35] Thomas L D, Oberoi RS and Nesbet R K 1975 Phys.Rev. A 12 2378.
[36] Nesbet RK 1973 Comput. Phys. Commun. 6 275.
[37] Dunseath K M, Fon W C, Burke V M, Reid RHG and Noble C J 1997 J. Phys. B: At. Mol. Opt. Phys 30 271.
[38] Zatsarinny O, Bartshat K, Bandurina L and Gedeon V 2005 *Phys. Rev.* A **71** 042702.
[39] Liu J, Wang Y and Zhou Y 2006 *J. Phys. B: At. Mol. Opt. Phys.* **39** 861
[40] Burke VM and Noble CJ 1995 *Comput. Phys. Commun.* **85** 471.
[41] Suno H and Kato T 2006 *At.Data and Nucl. Data Tables* **92** 407.
[42] http://www.nist.gov
[43] Norrington PH and I.P. Grant IP 1987 *J. Phys. B: At. Mol. Phys.* **20** 4869.
[44] http://web.am.qub.ac.uk/DARC/.
[45] http://sprg.ssl.berkeley.edu/~mfgu/fac/
[46] Summers HP 1994 *Adv. At. Mol. Opt. Phys.* **33** 275.
[47] Davies PCW and Seaton MJ 1969 *J. Phys. B: At. Mol Phys* **2** 757 .
[48] Bell RH and Seaton MJ 1985 *J. Phys. B:At. Mol. Phys.* **18** 1589.
[49] Pradhan AK and Seaton MJ 1985 J. *Phys. B: At. Mol. Phys.* **18** 1589.
[50] Zabaydullinh O and J. Dubau 2012 *J. Phys. B: At. Mol. Opt. Phys.* **45**, 115002.
[51] Robicheaux F, Gorczyca TW, Pindzola MS, and Badnell NR 1995 *Phys. Rev. A.* **52**, 1319.
[52] Badnell NR, O'Mullane MG, Summers HP, Altun Z, Bautista MA, Colgan J, Gorczyca TW, Mitnik DM, Pindzola MS, Zatsarinny O, 2003 *A&A* **406**, 1151.
[53] LaGattuta K J 1988 *Phys. Rev.* A**38**,1820.
[54] Haan SL, Jacobs VL 1989 *Phys. Rev.* A. **40**, 80 -94.
[55] Jacobs VL and Behar E 2000 *J. Quant. Spectrosc. Radiat. Transf.* **65**, 317.
[56] Badnell NR, Pindzola MS 1992 *Phys. Rev* A **45**, 2820(1992).
[57] Behar E , Jacobs VL , Oreg J, Bar-Shalom A , Haan SL 2004 *Phys. Rev.* A **69**, 022704.
[58] Berrington KA, Eissner WB, Norrington PH 1995 *Comput. Phys. Commun.* **92** 290 .
[59] Dyall KG, Grant IP, Johnson CT, Parpia FA, Plummer EP 1989 *Comput. Phys. Commun.* **55**, 425.
[60] The Opacity Project Team "The Opacity Project Vol.1" (IOPP, 1995) ISBN 0 7503 02887.
[61] Stancalie V 2013 *Eur. Phys. J. D* **67**, 223.

Journal of Physics: Conference Series **576** (2015) 012011 doi:10.1088/1742-6596/576/1/012011

Kinetic energy distribution for the ionization and dissociation process of C_2H_4 by electron impact

X. Wang[1, 2], Z. Chen[1, 2], B. Wei[1, 2], R. Hutton[1, 2] and Y. Zou[1, 2]

[1]Applied Ion Beam Physics Laboratory, Fudan University, Key Laboratory of the Ministry of Education, Shanghai 200433, China
[2]Institute of Modern Physics, Department of Nuclear Science and Technology, Fudan University, Shanghai 200433, China

E-mail: brwei@fudan.edu.cn

Abstract. The dissociative ionization of ethylene molecules following impact by electrons in the energy range of 20 to 200 eV has been studied with a cold target recoil-ion momentum spectrometer (COLTRIMS). The relative partial cross sections for the ionic fragments $C_2H_n^+$ (n=0~3) with respect to $C_2H_4^+$ were plotted as a function of the incident energy. By measuring the time of flight and position of the fragment ions, the kinetic energy distribution has been deduced. Taking advantage of the supersonic jet expansion, the thermal motion contributing to the kinetic energy distribution of the recoil-ions was reduced. The average kinetic energy of the recoil ions as a function of the incident electron energy has also been plotted. Furthermore, the average kinetic energy release for the dissociation to $C_2H_3^+$ and $C_2H_2^+$ were estimated as 0.76 eV and 0.73 eV respectively.

1. Introduction

Atomic and molecular data plays a crucial role in the development of Magnetic Confinement Fusion (MCF), and various types of experimental data on the interaction of Hydrocarbon molecules with electrons are required for understanding of the behavior of these molecules in the edge plasma region [1-3]. Ethylene (C_2H_4) is the simplest unsaturated hydrocarbon with a C=C bond. The cross sections for electron impact ionization of ethylene have been investigated [4-6]. More recently, with time-of-flight (TOF) mass spectrometry and two dimensional ion-ion coincidence techniques, the study of multiply ionized molecules provides a method to identify the contributions for different dissociation pathways [7, 8]. In fact, electron impact ionization and dissociation of hydrocarbon molecules has been widely studied in the past two decades, see, for example, a recent review by D. Reiter and R. T. Janev [9].

During the interaction between electrons and molecules, the ionic and neutral fragments produced often carry away a certain amount of kinetic energy (KE) which is released from the excitation energy left with the parent particle. Investigations of kinetic energy release (KER) are useful to understand such dissociation processes. The previous studies showed that the same fragment ion can be formed in different dissociation processes, while the dissociation process depends on the excitation energy of the parent ions [10-15]. Therefore, fragment ions from different dissociation processes have distinct kinetic energy distributions. Thus to build a model environment for a plasma, we require not only about fragment ion production efficiencies but also their KE distributions [16]. However, the kinetic energy of product ions is sometimes very small, and can even be below the thermal motion energy at room temperature. This indicates that the thermal motion should be taken into account for precise measurements of the kinetic energy.

In this work, we report the measurement of the kinetic energy distribution of fragmental ions produced through the interaction between an electron beam and a cold hydrocarbon (C_2H_4) target beam. The experiments were carried out by using cold target recoil-ion momentum spectroscopy (COLTRIMS). The incidence beam energies ranged from 20 to 200eV. Taking advantage of supersonic expansion, the thermal contribution of the kinetic energy of the target molecules is much smaller than in previous studies, and, as a result, kinetic energies are recorded more precisely for the dissociative processes. According to the accurate kinetic energy measurement, the mean KER was estimated and used to investigate the different dissociation processes via electron impact.

2. Experiment setup

The present experiment was performed by using COLTRIMS at Fudan University in Shanghai. A schematic diagram of the apparatus used for the present measurements is shown in figure 1. The setup comprises a pulsed electron gun, a three-stage supersonic gas jet, a recoil ion detector with delay-line anode, and a time-of-flight spectrometer. The pulsed electron beam (pulse length ≈ 1 ns, repetition rate 10 kHz), produced by a standard thermal cathode [17], crosses and ionizes a C_2H_4 beam. The cold molecular beam provided by a supersonic gas jet, in which high pressure gas (2 bar) flowed through a 10 μm nozzle into the first stage of the vacuum chamber at room temperature. A 0.1 mm skimmer at a variable distance of about 10 mm downstream of the nozzle was used to extract a geometrically well-defined gas beam. A multistage structure was utilized to keep high vacuum in the collision chamber. The background vacuum in the collision chamber was better than 3×10^{-10} torr during the experiment. A set of Helmholtz coils were used to reduce the earth's magnetic field.

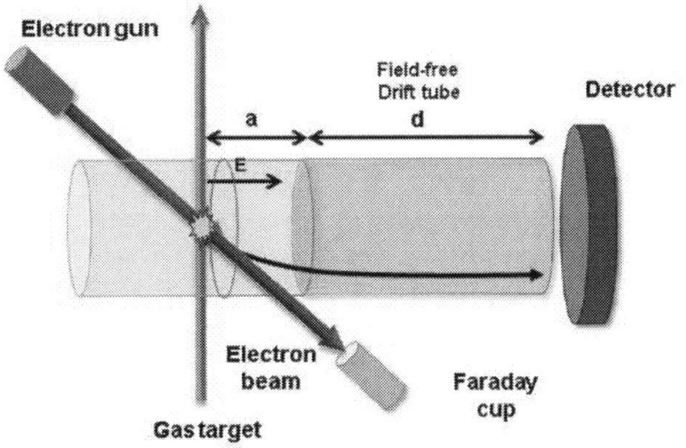

Figure 1. schematic diagram of the apparatus.

The produced recoil ions were accelerated by a pulsed electric field of 25 V/cm and then passed through a field-free drift tube, and finally detected by a 75mm delay-line anode position sensitive detector. The length of the field-free drift tube $d = 2a$ is employed here to reduce the influence of the target size, where a (=10 cm) is the length of the extraction field. A VME64x system was used for data acquisition, which contains a Multi-hit Time to Digital Convertor (TDC). During the experiment, the recoil ion signals produced the start pulse and the signal from the ultra-fast signal generator was used as a common stop. The coincidence measurements were performed in event by event mode, and the information of the position and timing was recorded. The momentum of the recoil ion can be reconstructed according to the measured position and time information in off-line data analysis.

3. Results and discussion

A typical mass spectrum of the fragments of C_2H_4 impacted by 70 eV electrons is shown in figure 2. The observed ions include the parent ion $C_2H_4^+$ and singly charged fragments such as $C_2H_n^+$ (n = 0~3). The peaks of CH_3^+, CH_2^+, CH^+ and C^+ are broader because these ions carry higher kinetic energy. We cannot resolve double charged $C_2H_4^{2+}$ from CH_2^+ since they have the same m/q value. The peak marked as '29^+' in the spectrum originated from of $^{13}CH_2$-CH_2^+, since the natural abundance of ^{13}C is about 1.1%. Both HO^+ and H_2O^+ originated from the background.

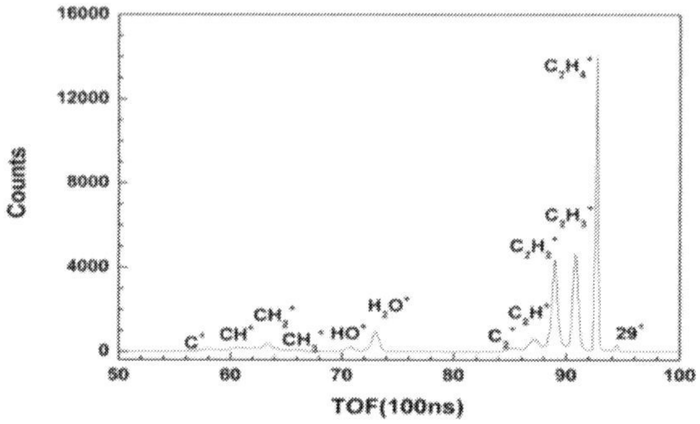

Figure 2. Mass spectrum of C_2H_4 at electron energy of 70 eV.

The cross sections for forming $C_2H_3^+$, $C_2H_2^+$, C_2H^+ and C_2^+ relative to the formation of $C_2H_4^+$ are deduced from the TOF spectra. In figure 3, the relative partial ionization cross sections of $C_2H_n^+$ (n=0-3) to $C_2H_4^+$ are plotted as a function of electron energies in the range of 20 to 200 eV. The uncertainty in the present measurement is less than 10% [18], which is mainly due to the counting statistics and overlaps of the peaks.

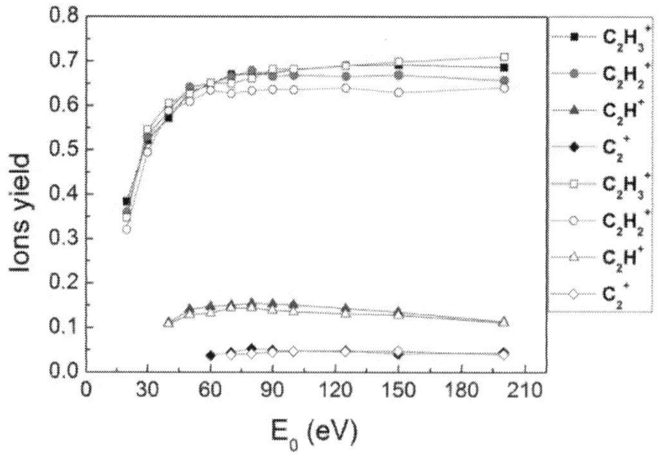

Figure 3. The relative partial ionization cross section for forming $C_2H_3^+$, $C_2H_2^+$, C_2H^+ and C_2^+ following electron ionization of C_2H_4. The data of the present experiment represented by solid symbol and the open symbol given by Ref. [4]

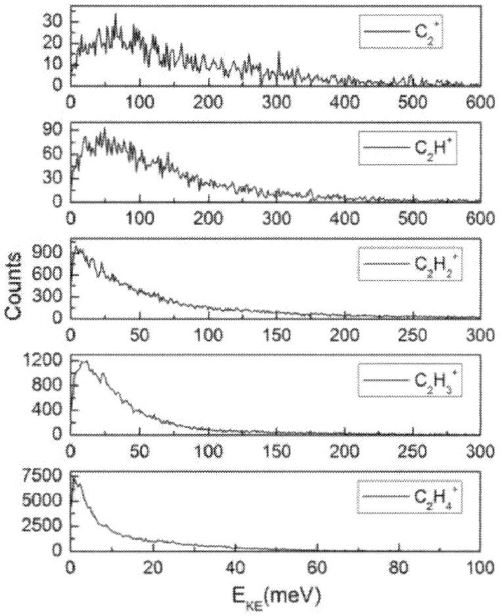

Figure 4. The ion kinetic energy distributions of $C_2H_4^+$, $C_2H_3^+$, $C_2H_2^+$, C_2H^+ and C_2^+ produced in the collision between ethylene molecules and 70 eV electrons. Please note the changed scales for different fragment ion species.

In the present work, we are interested in the kinetic energy release from the dissociation processes related to the ions $C_2H_n^+$ (n = 0~3. Figure 4 shows the kinetic energy distribution of the recoil ions produced in the collision between ethylene molecules and 70 eV electrons. The ground state ethylene molecule has the electronic configuration $(1a_g)^2(1b_{1u})^2(2a_g)^2(2b_{1u})^2(1b_{2u})^2$ $(3a_g)^2(1b_{3g})^2(1b_{3u})^2$ [5].The ionization of an outermost orbital ($1b_{3u}$) electron (binding energy of 10.51 eV) produces a stable ethylene ion in its ground electronic state (X^2B_{3u}). In contrast, the ionization of an electron in a deeper lying orbital can lead to the production of the parent ion $C_2H_4^+$ in an excited state, which is usually unstable and result in a dissociation process. For example, the ejection of a ($1b_{3g}$) electron (binding energy of 13.74 eV) can produce $C_2H_4^+$ ion in its first excited electronic state (A^2B_{3g}) and this has an excitation energy of 3.23 eV. For dissociation of the $C_2H_4^+$ ion via the three channels shown below, the extra energy required above ionization is 2.58 eV 2.72 eV and 7.25 eV for paths (1), (2) and (3) respectively [2].

$$C_2H_4^+ \rightarrow C_2H_3^+ + H + 2e \quad (1)$$

$$C_2H_4^+ \rightarrow C_2H_2^+ + H_2 + 2e \quad (2)$$

$$C_2H_4^+ \rightarrow C_2H_2^+ + 2H + 2e \quad (3)$$

Ethylene ions in the state (A^2B_{3g}) can dissociate following the pathway (1) and (2). However, the dissociation energy for path (2) is 2.72 eV which is higher than 2.58 eV for path (1). Correspondingly, the KER for path (2) is lower than path (1), and this is shown in figure 4, namely, the distribution of $C_2H_3^+$ has a maximum at about 10 meV while the distribution of $C_2H_2^+$ has a maximum at about 5 meV. To be specific, the ethylene ions can be excited to even higher states and undergoes dissociation for producing ions $C_2H_3^+$ and $C_2H_2^+$ with higher kinetic energies. The reaction channel (3) also can produce $C_2H_2^+$ions and this gives a contribution to the broader tail of the kinetic energy distribution for $C_2H_2^+$. For the ions C_2H^+ and C_2^+, the kinetic energy distributions become more broader and Coulomb explosion processes need to be taken into account.

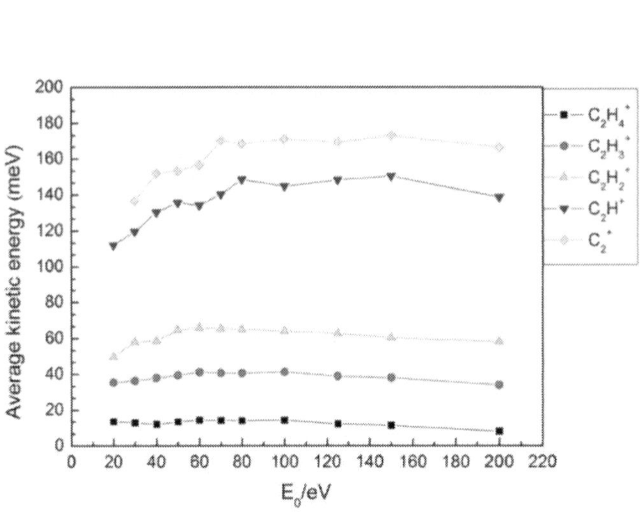

Figure 5. The average kinetic energy of different fragment ions as a function of the incident energy

The average kinetic energy for ions $C_2H_n^+$ (n = 0-4) has been deduced and shown in figure 5 as a function of the incident electron energy. With increasing the incident electron energy, more reaction channels with higher dissociation energies become open. Accordingly, the average kinetic energy for ions $C_2H_2^+$, C_2H^+, C_2^+ increases at first and finally becomes saturated. For $C_2H_3^+$ions the average kinetic energy has no dependence on the incident electron energy at least as seen in the present measurement. However, the value we obtained (~ 40 meV) is much smaller than the previous measurement [6].

The total kinetic energy release can be estimated from the average kinetic energy [6]. If the excitation of a precursor ion above the dissociation limit leads to decay into two fragments. $\overline{E}_0$ and $\overline{E}_1$ are the average kinetic energies of the precursor ion with mass m_0 and the product ion with mass m_1 respectively. Then, the total kinetic energy release $\overline{E_K}$ is given by:

$$\overline{E_K} = \frac{\overline{E}_1 \cdot m_0}{m_0 - m_1} - \frac{\overline{E}_0 \cdot m_1}{m_0 - m_1} \quad (4)$$

For dissociation pathway (1) and (2), the KERs are 0.76 eV and 0.73 eV, respectively. The energy distribution of $C_2H_4^+$ give a direct view of the experimental energy resolution, which is an upper limit as an result of all experimental uncertainty factors including thermal energy. For the uncertainty of KER, as we deduced it from the average kinetic energy of fragment ions, with contribution of statistic uncertainty for averaging, we estimated the final uncertainty is about 15%.These values are in agreement with the previous measurement performed by N. Endstrasser et al [6] within their experimental uncertainty.

4. Conclusion

The electron impact dissociative ionization of ethylene has been studied experimentally. In this paper, we report data for the formation of $C_2H_n^+$ (n = 0 ~ 3) relative to that of $C_2H_4^+$ as a function of incident electron energy from 20 to 200 eV. Good agreements were achieved with previous studies. By measuring the TOF and position on the detector, the kinetic energy of the fragment ions was deduced. Taking advantages of supersonic gas-jet system, the thermal contribution for kinetic energy is dramatically reduced and the resolution of the present measurement is about 12.5 meV. For the excited state of (A^2B_{3g}), the ethylene ions can dissociate to $C_2H_3^+$ or $C_2H_2^+$. However, taking into account the different dissociation energies, the kinetic energy as well as the KER for those two ions are also different. This is indicated through the kinetic energy distributions, although these distributions contain contributions from all exited states. The average KERs for the dissociation pathways leading to the production of $C_2H_3^+$, $C_2H_2^+$ have been estimated to be 0.76 eV and 0.73 eV, respectively.

Acknowledgements
This work was carried out as a part of an IAEA Coordinated Research Project on "Light Element Atom, Molecule and Radical Behaviour in the Divertor and Edge Plasma Regions". This work was also supported by the National Magnetic Confinement Fusion Program under grant No. 2009GB106001, the National Science Foundation of China under contract No. 10904019 and Shanghai Leading Academic Discipline Project (project No. B107).

References

[1] Janev R K *Contemporary Phys.* 2005, *46*, 121.
[2] Janev R K and Reiter D Collision Processes of Hydrocarbon Species in Hydrogen Plasmas: II The Ethane and Propane Families; Report Forschungszentrum Julich 2002, Jul-4005.
[3] Janev R K and Reiter D *Physics Plasma* 2004, *11*, 780.
[4] Tian C and Vidal C R *Chem. Phys. Lett.* 1998, *288*, 499.
[5] Popović S, Williams S and Vušković L. *Phys. Rev. A* 2006, *73*, 022711.
[6] Endstrasser E; Zappa F; Mauracher A; Bacher A; Feil S; Bohme D K; Scheier P; Probst M and Märk T D *Int. J. Mass Spectrom.* 2009, *280*, 65.
[7] Wang P and Vidal C R *J. Chem. Phys.* 2002, *116*, 4023.
[8] Ward M D; King S J and Price S D *J. Chem. Phys.* 2011, *134*, 024308.
[9] Reiter D and Janev R K *Contrib. Plasma Phys.* 2010, *50*, 986.
[10] Popović S, Williams S, and Vušković L *Phys. Rev. A* 2006, 73, 022711.
[11] Osipov T; Stener M; Belkacem A; Schöffler M; Weber Th; Schmidt L; Landers A; Prior M H; Dörner R and Cocke C L, *Phys. Rev. A* 2010, 81, 033429.
[12] Poll H U; Grill V; Matt S; Abramzon N; Becker K; Scheier P and Märk T D, *Int. J. Mass Spectrom.* 1998, 177, 143.
[13] Feil S; Gluch K; Bacher A; Matt-Leubner S; Böhme D K; Scheier P and Märk T D, *J. Chem. Phys.* 2006, 124, 214307.
[14] Myung H K; Brain D L; Lei S and Arther G S, *J. Chem. Phys. A* 2007, 111, 7472.
[15] JiangY H; Rudenko A; Herrwerth O; Foucar L; Kurka M; Kühnel K U; Lezius M; Kling M F; Tilborg J V; Belkacem A; Ueda K; Düsterer S; Treusch R; Schröter C D; Moshammer R and Ullrich J, *Phys. Rev. Lett.* 2010, 105, 263002
[16] Gluch K, Scheier P, Schustereder W, Tempnual T, Feketeova L, Mair C, Matt-Leubner S, Stamatovic A and Märk T D *Int. J. Mass Spectrom.* 2003, 228, 307
[17] Lin S, Chen Z, Wei B, Wang X, Lu D, Hutton R and Zou Y, *Phys. Scr.* 2011 T144, 014095.
[18] Wei B, Chen Z, Wang X, Lu D, Lin S, Hutton R, and Zou Y, *J. Phys. B.* 2013 46, 215205

Charge transfer in collisions of Be^{q+} (q=2-4) and B^{q+} (q=3-5) ions with H

Y Wu[1], L Liu[1], C H Liu[2], Y Z Qu[3], J G Wang[1] and R K Janev[4]
[1] Data Center for High Energy Density Physics, Institute of Applied Physics and Computational Mathematics, P. O. Box 8009, Beijing 100088, P. R. China
[2] Institute of Modern Physics, Chinese Academy of Sciences, Lanzhou 730000, P. R. China
[3]College of Material Sciences and Optoelectronic Technology, University of the Chinese Academy of Sciences, P.O. Box 4588,Beijing 100049, P. R. China
[4] Macedonian Academy of Sciences and Arts, P. O. Box 428, 1000 Skopje, Macedonia

E-mail: wu_yong@iapcm.ac.cn

Abstract. The charge transfer reactions in collisions of Be^{q+}(q=2-3) and B^{q+}(q=3-4) ions and atomic hydrogen are investigated by using the quantal molecular orbital close-coupling (QMOCC) and the two-center atomic-orbital close-coupling (TC-AOCC) methods. Total and sub-shells state-selective cross sections are calculated for low and intermediate energy region and compared with other data available. Sets of recommended cross sections, based on the QMOCC and TC-AOCC calculations, are deduced and tabulated for $Be^{2,3+}$ + H and $B^{3,4+}$ + H collisions, which provide important atomic data needed in the charge-exchange-recombination spectroscopy diagnostics in magnetic fusion plasmas investigation.

1. Introduction

Charge transfer processes in atom – multi-charged ion collisions play an important role in magnetic fusion plasmas investigation and the radiation following the charge transfer processes has been widely used as a plasma diagnostic tool, namely the charge-exchange-recombination spectroscopy (CXRS) diagnostics [1]. The electron capture processes of multiply charged impurity ions in magnetic fusion plasma with the neutral hydrogen species are also important in the studies of impurity transport in the edge and divertor plasma regions [2]. In the past three years, as one of the participants in the IAEA Coordinated Research Program on data for light ions, we have studied the collision processes between some basic edge/divertor plasma constituents (H^+, H, He, H_2) and the light element atomic and molecular plasma impurities including Li, Be, B, C, N, O ions [3-19] by using the quantal molecular orbital close-coupling (QMOCC) and the two-center atomic orbital close-coupling (TC-AOCC) methods over a large energy region.

In the present paper, the collisions of $Be^{2,3,4+}$ and $B^{3,4,5+}$ ions with atomic hydrogen have been investigated by using the QMOCC and TC-AOCC methods. Beryllium is the adopted first wall material for the International Thermonuclear Experimental Reactor (ITER) [4]. The charge transfer cross sections for Be^{4+}+H and B^{5+}+H collisions are calculated by using the TC-AOCC method and sets of recommended charge transfer cross sections are proposed for the $Be^{2,3+}$+H, $B^{3,4+}$+H collisions based on our previous work on Be^{2+}, Be^{3+}, B^{3+}, B^{4+} ions collisions with atomic H [6,7].

The QMOCC and TC-AOCC methods are only outlined here. In QMOCC calculations, a coupled set of second-order differential equations is solved and matched to the plane-wave boundary conditions at a large internuclear distance [20-21]. The molecular structure data (potential curves, radial and rotational couplings matrix elements) required in the QMOCC has been calculated by using the multireference single- and double-excitation conffiguration interaction (MRD-CI) method [22-23]. Note that the translation effects have been considered by using appropriate reaction coordinates [24, 25] and the modified radial and rotational coupling matrix elements [26]. In the TC-AOCC calculations [27], the total scattering wave function is expanded over the electronic states centered on the target and on the projectile which are determined by using the variational method with the even-tempered trial functions [27, 28]. One-active electron approximation and model potential are applied in our TC-AOCC calculation and a sufficiently large expansion basis, including a large number of pseudostates, has been used to ensure the convergence of the cross sections for each of collision systems. For the relative nuclear motion, the straight-line approximation has been applied. Due to the limitation of the model potential and the straight-line approximation adopted, the present TC-AOCC method is more applicable in treating the collision processes where the electron-correlation effects are not important and the collision velocity not too slow (larger than a few hundred eV/u). The charge transfer cross sections with different spin-manifolds (*S*- resolved) become especially important in CXRS simulations when different manifold electronic molecular states evolved in the charge transfer processes. The present QMOCC calculations can provide reliable *n, l, S*- resolved cross sections. For example, the state-resolved cross sections presented in the Fig. 8 of the paper of Liu and Wang [7] are comparable for singlet and triplet manifold for the charge transfer in Be^{3+}+H collisions. Only *n, l* - resolved cross sections are presented in this paper and the QMOCC partial cross sections are obtained by statistically averaging different spin-manifolds contributions.

For the convenience of applications in CXRS diagnostics in magnetic fusion plasmas investigation and other related fields, total and state-selective charge transfer cross sections in $Be^{2,3,4+}$ + H and $B^{3,4,5+}$ + H collisions have been calculated and presented by using the QMOCC and TC-AOCC methods. Sets of total and state-selective cross sections have been recommended over a large energy region based on our calculations, and as parts of the recommended data, the state-resolved cross sections are tabulated for $Be^{2,3+}$ + H and $B^{3,4+}$ + H collisions for energies less than 100eV/u.

2. Results and Discussions

In this section, the total and state-resolved cross section for $Be^{2,3,4+}$+H and $B^{3,4,5+}$+H Collisions are presented and compared with the available data. Recommended cross sections for $Be^{2,3+}$+H and $B^{3,4+}$+H Collisions are deduced based on our previous work on these systems. For Be^{4+}+H and B^{5+}+H collisions, the total and state-resolved cross sections are computed by using the TC-AOCC and excellent agreements have been obtain in comparison with the available data.

2.1. Total charge transfer cross sections for $Be^{2,3+}$+H and $B^{3,4+}$+H Collision Systems

In Figure 1-4, the total QMOCC and TC-AOCC cross sections for $Be^{2,3+}$+H and $B^{3,4+}$+H collisions are presented and compared with the theoretical and experimental data available. It can be found that there are good agreements between our QMOCC and TC-AOCC calculations and the data available in a large energy range, and good agreements have been achieved in the overlapping energy region between the QMOCC and TC-AOCC calculations. Combining the QMOCC results with the TC-AOCC data, reliable cross sections can be obtained over a large energy range from 0.1 eV/u to 100 keV/u. Note that the QMOCC and TC-AOCC results in Fig.1-4 have been reported in the papers of Liu et al. [6, 7] and the detail of the methods applied can be found there. The good consistence between the QMOCC and TC-AOCC calculations provides a good foundation to deduce the recommended data over a larger collision energy region.

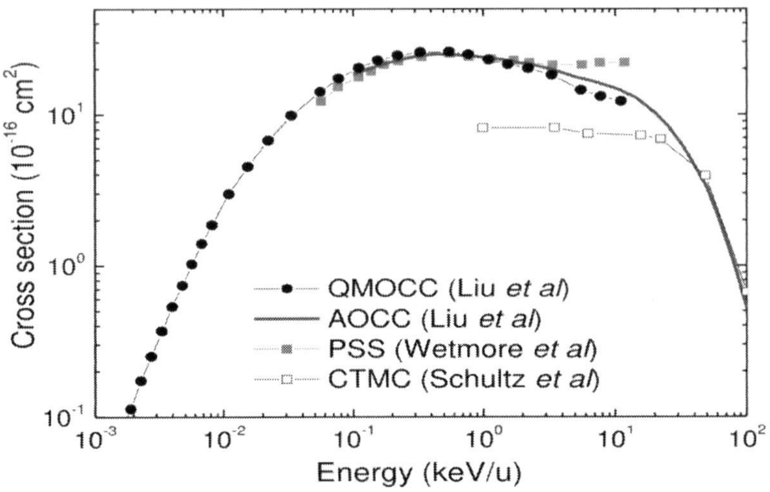

Figure 1. (Color online) The total charge-transfer cross sections for the $Be^{2+}(1s2)$ + H(1s) collision. The QMOCC calculation of Liu et al. [3] (solid line with filled circles); The TC-AOCC results of Liu et al. [6] (solid line); The perturbed stationary state (PSS) results of Wetmore et al. [29] (solid line with filled squares); The classical trajectory Monte Carlo (CTMC) results of Schultz et al. [30] (solid line with open squares).

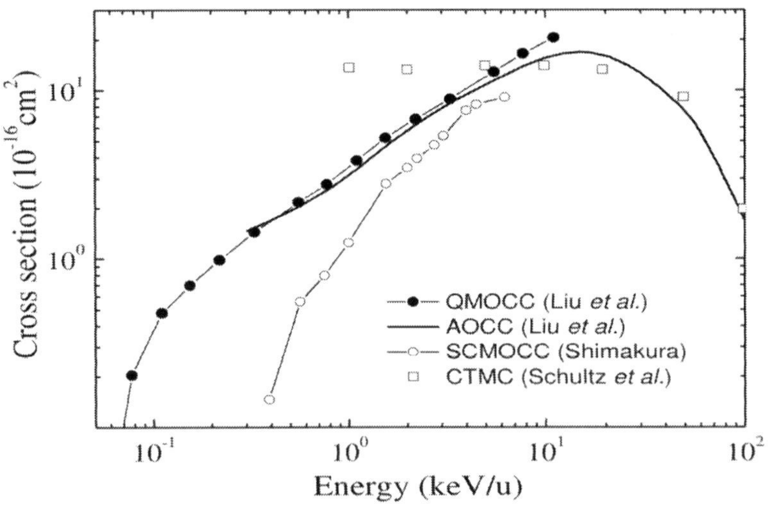

Figure 2. Total charge-transfer cross sections for the Be^{3+} +H collision. The QMOCC calculation of Liu et al. (solid line with filled circles) [7]; The TC-AOCC calculation of Liu et al. (solid line) [6]; SCMOCC results of Shimakura [31] (solid line with open circles); and CTMC results of Schultz et al. [30] (open squares).

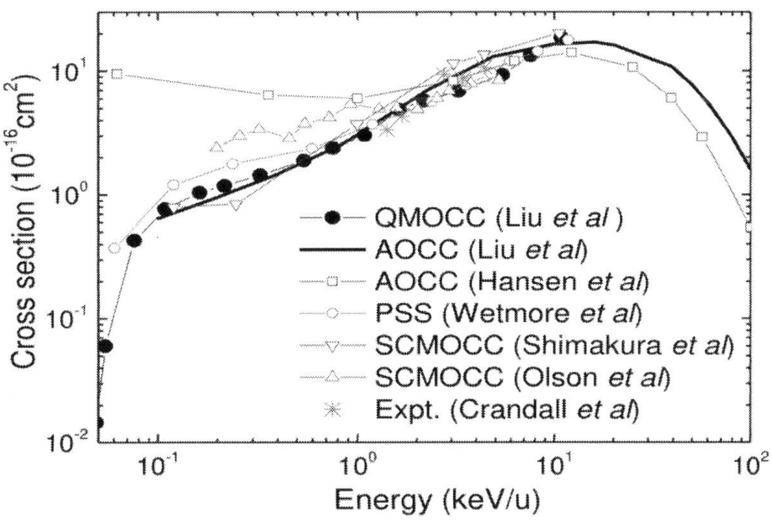

Figure 3. (Color online) Total charge-transfer cross sections for $B^{3+}(1s^2)$ + H(1s) collisions. The QMOCC results of Liu et al. (solid line with filled circles)[3]; The TC-AOCC results of Liu et al.(solid line) [6];AOCC results of Hansen et al. [32] (solid line with open squares); PSS results of Wetmore et al. [29] (solid line with open circles); SC-MOCC results of Shimakura et al. [33] (solid line with downward triangle); SC-MOCC results of Olson et al. [34] (solid line with upward triangle); and experiment results of Crandall et al. [35] (stars).

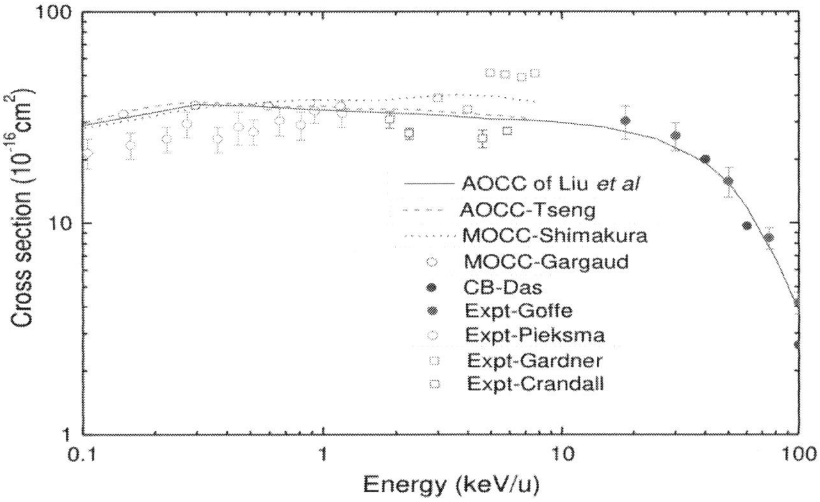

Figure 4. (Color online) The total charge transfer cross sections for B^{4+}-H(1s) collisions as functions of collision energy. Theory: (—) The TC-AOCC results of Liu et al [6]; (----) AOCC results of Tseng and Lin, [36], (○) The quantal molecular orbital close-coupling (QMOCC) results of Gargaud and McCarroll, [37], (.....)The Semi-classical molecular

Journal of Physics: Conference Series **576** (2015) 012012 doi:10.1088/1742-6596/576/1/012012

orbital close-coupling (SCMOCC) results of Shimakura et al., [38] (•) The boundary corrected continuum intermediate state approximation results of Das et al., [39]. Experiment: (•) Goffe et al.,[40] (□) Gardner et al., [41] (□) Crandall et al.,[35], (○) Pieksma and Havener, [42].

2.2. Recommended cross sections for $Be^{2,3+}$+H and $B^{3,4+}$+H Collisions

In Figure 5-8, the state selective cross sections of QMOCC and TC-AOCC calculations are presented for $Be^{2,3+}$+H and $B^{3,4+}$+H collisions. Only *n, l* - resolved cross sections are presented in the present paper and the QMOCC partial cross sections are obtained by statistically averaging different spin-manifolds contributions. It can be found that there are good agreements between the QMOCC and TC-AOCC calculations even for the sensitive *n,l*-resolved cross sections, which provides a good foundation for deducing the *n,l*-resolved cross sections.

For the convenience of application, the state-selective cross sections for $Be^{2,3+}$+H and $B^{3,4+}$+H collisions are evaluated and recommended data are proposed based on the two close-coupling calculations and the data available. In our evaluation, the QMOCC results are preferred for the energy less than about 1keV/u and the TC-AOCC results are chosen for energy larger than a few keV/u. While discrepancies appear in the overlapping energy region, usually the smooth connection between two sets of calculations is preferred in our deduction. If experimental data is available, it serves as the benchmark data in our evaluation. Based on such a criterion, recommended charge transfer cross sections are deduced for $Be^{2,3+}$+H and $B^{3,4+}$+H collisions. As shown in the figures 5-8, the recommended data are represented by the read solid lines. For the dominant reaction channels, there are excellent agreements between two close coupling calculations and the deduction is straightforward, for example the 2s cross sections in $Be^{2,3+}$+H collisions and 3s, 3p cross section in $B^{3,4+}$+H collisions. However, for the 3s capture cross sections presented in the first panel of figure 7, the TC-AOCC results merge to the QMOCC data around the energy of 300eV/u and the TC-AOCC results are adopted to about 300eV/u in the recommendation due to the good agreement between our TC-AOCC results and the data of Hansen and Dubois [32]. In Figure 7 and 8, only the *n, l*-resolved cross sections are proposed for the dominant n=3 channels and *n*-resolved cross sections are presented for the less important channels of *n*=2 and *n*=4. For the clarity of Figures 5-8, the recommended cross sections are only presented for energy larger than 10eV/u. For lower energy region, the QMOCC calculations are more reliable and chosen as recommended data. In table 1-4, the QMOCC cross section for $Be^{2,3+}$+H and $B^{3,4+}$+H collisions are presented. Note that the recommended data in the whole energy range and the *S*-resolved cross sections can be obtained when required.

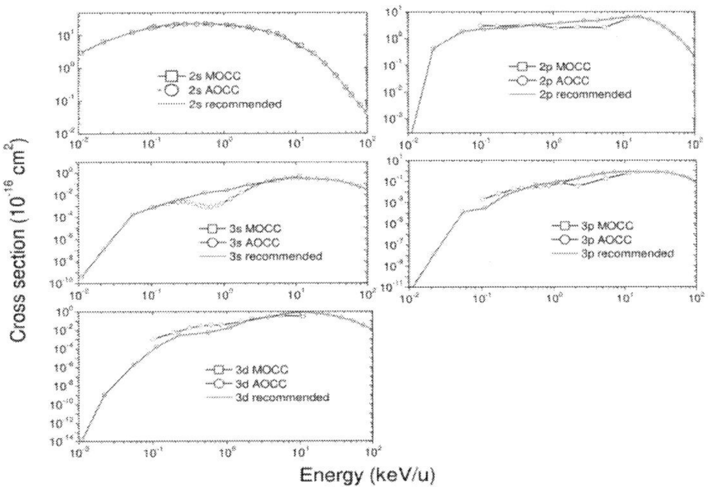

Figure 5. (Color online) The *n, l*-resolved charge-transfer cross sections for the $Be^{2+}(1s^2)$ + H(1s) collision. The QMOCC results: black solid lines with unfilled squares; The TC-AOCC results: black solid lines with unfilled circles. Recommended data: red solid lines.

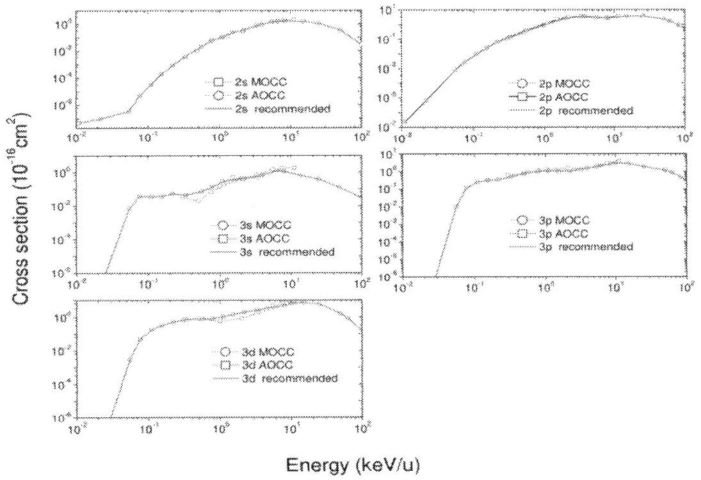

Figure 6. (Color online) The *n, l*-resolved charge-transfer cross sections for the $Be^{3+}(1s)$ + H(1s) collision. The QMOCC results: black solid lines with unfilled squares; The TC-AOCC results: black solid lines with unfilled circles. Recommended data: red solid lines.

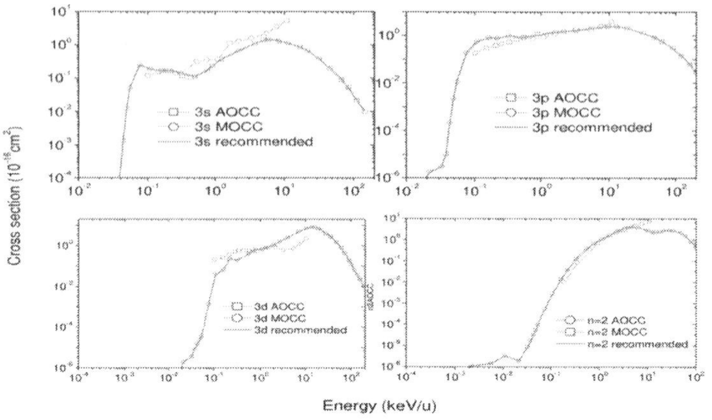

Figure 7. (Color online) The *n, l*-resolved charge-transfer cross sections for the $B^{3+}(1s^2)$ + H(1s) collision. The QMOCC results: black solid lines with unfilled squares; The TC-AOCC results: black solid lines with unfilled circles. Recommended data: red solid lines.

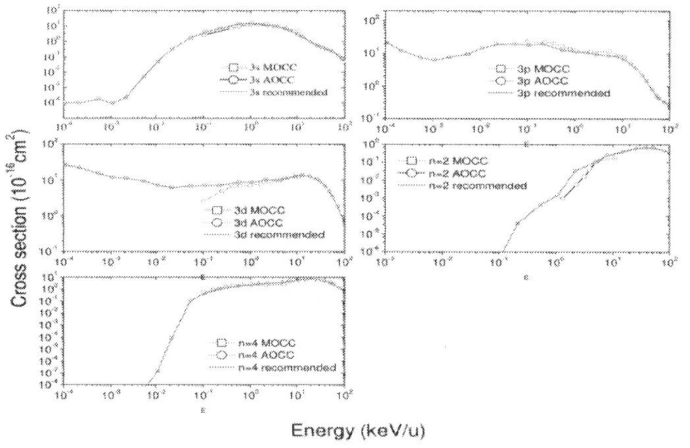

Figure 8. (Color online) The *n, l*-resolved charge-transfer cross sections for the $B^{4+}(1s)$ + H(1s) collision. The QMOCC results: black solid lines with unfilled squares; The TC-AOCC results: black solid lines with unfilled circles. Recommended data: red solid lines.

2.3. *Total and state-resolved cross sections for Be^{4+}+H and B^{5+}+H Collisions*

The TC-AOCC method has been applied to investigate Be^{4+}+H and B^{5+}+H collisions and the obtained total and state-resolved cross sections are compared with the late work of Das *et al* [39] and Errea *et al* [43, 44], as shown in figure 9 and figure 10. For Be^{4+}+H collisions, there are excellent agreement between the present TC-AOCC calculations and the semi-classical MOCC results of Errea *et al* [43] for total cross sections and for the *n,l*-resolved cross sections available for the dominant n=3,4,5 channels, as shown in figure 9. The present TC-AOCC results converge to the boundary corrected continuum intermediate state approximation (CB) calculations of Das *et al.* [39] at higher energy where the CB approximation is expected to be reasonable. For B^{5+}+H collisions, excellent agreements are also achieved over a large energy region between the present TC-AOCC calculations and the recommended data of Errea *et al* [44]. In Errea's evaluation, the Semi-classical MOCC results are preferred for low energy side and the classical CTMC results are used for high energy side. The present TC-AOCC results and the recommended data of Errea *et al* [44] both converge to the CB results of Das *et al.* [39] at energy larger than 100keV/u. For the energies less than a few hundred eV/u, the QMOCC method becomes more applicable. The QMOCC calculations will be performed in the near future and recommended data can be proposed over a larger energy region.

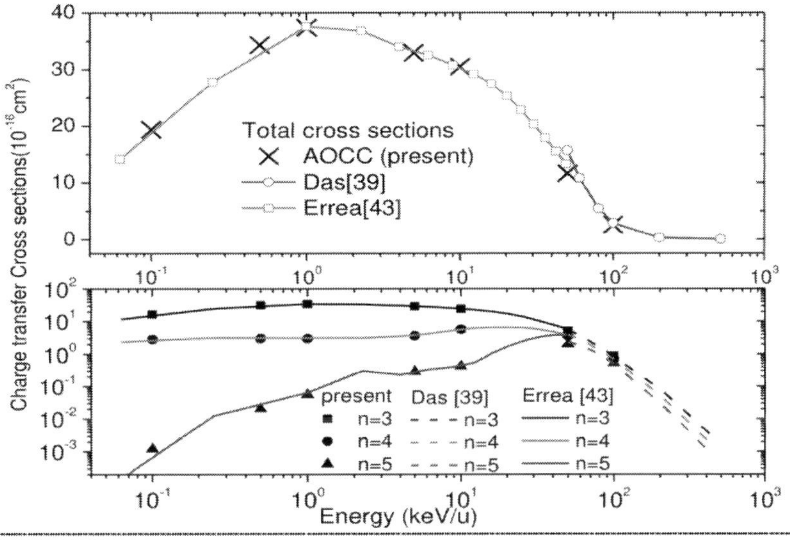

Figure 9. (Color online) The total and *n*-resolved charge transfer cross sections for Be^{4+}-H(1s) collisions as functions of collision energy.

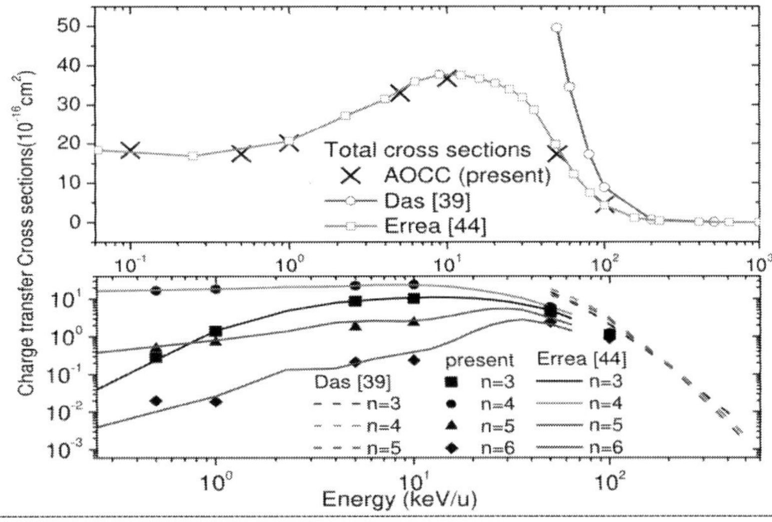

Figure 10. (Color online) The total and *n*-resolved charge-transfer cross sections for the B^{5+}+ H(1s) collision.

3. Conclusions

In the present work, the Be^{q+}-H (1s) (q=2, 3, 4) and B^{q+}-H (1s) (q=3, 4, 5) collision have been investigated by using QMOCC and TC-AOCC methods. Good agreements have been obtained between the present QMOCC and TC-AOCC calculations and the experiment and theoretical data available. For $Be^{2,3+}$-H and $B^{3,4+}$-H collisions, recommended cross sections are deduced over a large energy region, as shown in figure 5-8 and Table 1-4. For Be^{4+}-H and B^{5+}-H collisions, TC-AOCC calculations are performed and excellent agreements have been achieved in comparison with the data available.

Table 1. State-selective charge transfer cross sections (in units of $10^{-16}cm^2$) for Be^{2+}-H calculated by using QMOCC method at selected collision energies. (a (b) stands for $a\times10^b$).

Be^+ state	0.1eV/u	0.5eV/u	1.0eV/u	5.0eV/u	10eV/u	50eV/u	100eV/u
$1s^22s$	4.09(-4)	6.50(-3)	2.80(-2)	8.15(-1)	2.611	12.28	17.23
$1s^22p$					8.80(-5)	1.623	2.23
$1s^23s$						1.33(-4)	3.76(-4)
$1s^23p$						1.12(-4)	4.86(-4)
$1s^23d$						0	1.62(-4)

Table 2. State-selective charge transfer cross sections (in units of $10^{-16}cm^2$) for Be^{3+}-H calculated by using QMOCC method at selected collision energies. (a (b) stands for $a\times10^b$).

Be^{2+} state	0.1eV/u	0.5eV/u	1.0eV/u	5.0eV/u	10eV/u	50eV/u	100eV/u
$1s^2$	4.69(-3)	1.95(-3)	1.52(-3)	1.16(-3)	1.04(-3)	7.25(-4)	5.43(-4)
1s2s	5.74(-4)	2.33(-4)	1.69(-4)	1.07(-4)	9.67(-5)	1.06(-4)	1.09(-4)
1s2p	4.05(-4)	1.64(-4)	1.11(-4)	5.38(-5)	3.86(-5)	5.57(-4)	7.21(-3)
1s3s						2.00(-3)	3.70(-2)
1s3p						2.52(-3)	0.227
1s3d						7.69(-4)	0.135

Table 3. State-selective charge transfer cross sections (in units of $10^{-16}cm^2$) for B^{3+}-H calculated by using QMOCC method at selected collision energies. (a (b) stands for $a\times10^b$).

B^{2+} state	0.1eV/u	0.5eV/u	1.0eV/u	5.0eV/u	10eV/u	50eV/u	100eV/u
$1s^22s$	3.08(-4)	1.38(-4)	9.98(-5)	5.58(-5)	5.79(-5)	7.25(-5)	6.41(-5)
$1s^22p$	6.74(-4)	2.85(-4)	1.91(-4)	9.77(-5)	7.78(-5)	1.15(-4)	2.13(-3)
$1s^23s$	8.38(-5)	4.79(-5)	3.95(-5)	3.81(-5)	4.22(-5)	0.021	0.218
$1s^23p$	0	0	0	0	0	3.77(-3)	0.462
$1s^23d$	0	0	0	0	0	2.48(-5)	0.031

Table 4. State-selective charge transfer cross sections (in units of $10^{-16}cm^2$) for B^{4+}-H calculated by using QMOCC method at selected collision energies. (a (b) stands for a×10^b).

B^{3+} state	0.1eV/u	0.5eV/u	1.0eV/u	5.0eV/u	10eV/u	50eV/u	100eV/u
1s2s	0	0	0	0	0	0	0
1s2p	0	0	0	0	0	0	0
1s3s	1.00(-4)	1.67(-4)	9.21(-5)	5.70(-3)	0.052	1.778	4.078
1s3p	21.22	7.522	6.226	9.721	14.82	19.70	18.26
1s3d	25.95	14.97	11.77	9.206	7.168	6.731	7.044
1s4p	0	0	0	0	0	8.34(-3)	0.020
1s4p	0	0	0	0	0	0.045	0.082
1s4d	0	0	0	0	0	4.78(-3)	0.038
1s4f	0	0	0	0	0	0.052	0.306

References

[1] Boileau A, Hellermann M G von, Horton L D, Spence J and Summers H P 1989 *J. Phys. B* **31** 779
[2] Mattioli M, Michelis C De and Pequet A L 1998 *Nucl. Fusion* **38** 1629
[3] Liu C H, Liu L, Qu Y Z, Wang J G and Janev R K 2010 *Phys. Rev. A* **82** 022710
[4] Zhao Y Q, Liu L, Xue P, Wang J G and Janev R K 2010 *J. Phys. B* **43**, 185202
[5] Liu C H, Wang J G and Janev R K 2010 *J. Phys. B* **43**, 144006
[6] Liu L, Jakimovski D, Wang J G and Janev R K 2010 *J. Phys. B* **43**, 144005
[7] Liu C H and Wang J G 2013 *Phys. Rev. A* **87**, 042709
[8] Liu L, Liu C H, Wang J G and Janev R K 2011 *Phys. Rev. A* **84** 032710
[9] Liu C H, Wang J G and Janev R K 2012 *Phys. Rev. A* **85** 042719
[10] Wu Y, Stancil P C, Liebermann H P, Funke P, Rai S N, Buenker R J, Schultz D R, Hui Y, Draganic I N and Havener C C 2011 *Phys. Rev. A* **84** 022711
[11] Yan L L, Qu Y Z, Liu C H, Wang J G and Buenker R 2012 *The Journal of Chemical Physics* **136** 124304
[12] Liu L, Jakimovski D, Wang J G and Janev R K 2012 *J. Phys. B* **45** 225203
[13] Liu L, Wang J G and Janev R K 2012 *J. Phys. B* **45**, 015202
[14] Yan L L, Qu Y Z, Liu C H, Zhang Y, Wang J G and Buenker R 2012 *Chin. Phys. B* **21** 063401
[15] Wu Y, Stancil P C, Schultz D R, Hui Y, Liebermann H.P, Funke P and Buenker R J 2012 *J. Phys. B* **45**, 235201
[16] Yan L L, Qu Y Z, Liu C H, Wang J G and Buenker R 2012 *The Journal of Chemical Physics* **136** 124304
[17] Buenker Robert J, Liebermann Heinz-Peter, Zhang Yu, Wu Yong, Yan Lingling, Liu Chunhua, Qu Yizhi and Wang Jianguo 2013 *Journal of Computational Chemistry* Doi:10.1002/jcc.23215
[18] Zhang C L, Hong H X, Wang F, Wu Y and Wang J G 2013 *Phys. Rev. A* **87** 032711
[19] Yan L L, Wu Y, Qu Y Z, Wang J G and Buenker R J 2013 *Phys. Rev. A* **88** 022706
[20] Kimura M and Lane N F 1990 *Adv. At. Mol. Opt. Phys.* **26** 79
[21] Zygelman B, Cooper D L, Ford M J, Dalgarno A, Gerratt J and Raimondi M 1992 *Phys. Rev. A* **46** 3846
[22] Buenker R J and Phillips R A 1985 *J. Mol. Struct. Theochem.* **123** 291
[23] Krebs S and Buenker R J 1995 *J. Chem. Phys.* **103** 5613
[24] Gargaud M, McCarroll R and Valiron P 1987 *J. Phys. B* **20**, 1555
[25] Bransden B H and McDowell M R C 1992, Charge Exchange and the Theory of Ion-Atom Collisions (Oxford: Clarendon)

[26] Bacchus-Montabonel M C and Ceyzeriat P 1998 *Phys. Rev. A* **58** 1162
[27] Kuang J and Lin C D 1997 *J. Phys. B* **30** 101
[28] C. M. Reeves, J. Chem. Phys. 39, 1 (1963)
[29] Wetmore A E, Cole H R and Olson R E 1986 *J. Phys. B* **19** 1515
[30] Schultz D R, Krstic P S and Reinhold C O 1996 *Phys. Scr. T* **62** 69
[31] Shimakura N 1988 *J. Phys. B* **21** 2485
[32] Hansen J P and Dubois A 1996 *Phys. Scr. T* **62** 55
[33] Shimakura N, Suzuki S, Murakami Y, Gu J P, Hirsch G, Buenker R J, Kimura M and Shimamura I 1996 *Phys. Scr. T* **62** 39
[34] Olson R E, Shipsey E J and Browne J C 1978 *J. Phys. B* **11** 699
[35] Crandall D H, Phaneuf R A and Meyer F W 1979 *Phys. Rev. A* **19** 504
[36] Tseng H C and Lin C D 2000 *Phys. Rev. A* **61** 034701
[37] Gargaud M and. McCarroll R 1996 *Physica Scripta. T* **62** 49
[38] Shimakura N, Suzuki S and Kimura M 1993 *Phys. Rev. A* **47** 3930
[39] Das M, Purkait M and Mandal C R 1998 *Phys. Rev. A* **57** 3573
[40] Goffe T V, Shah M B and Gilbody H B 1979 *J. Phys. B* **12** 3763
[41] Gardner L D, Bayfield J E, Koch P M, Sellin I A, Pegg D J, Peterson R S and Crandall D H 1980 *Phys. Rev. A* **21** 1397
[42] Pieksma M and Havener C C 1998 *Phys. Rev. A* **57** 1892
[43] Errea L F, Gorfmkiel J D, Harel C, JoUin H, Macias A, MCndez L, Pons B and Riera A 1996 *Physica Scripta.* **T62** 27
[44] Errea L F, Guzmán F, Illescas Clara, Mndez L, Pons B, Riera A and Su´arez J 2006 *Plasma Phys. Control. Fusion* **48** 1585

Acknowledgments
This work was supported by the International Atomic Energy Agency (Vienna, Austria) (Research Contracts Nos. 15689/R0 with RKJ and 15700/R0 with JGW), National Natural Science Foundation of China (Grants Nos. 10875017, 10974021, 10979007 and 10734140) and the National Key Laboratory of Computational Physics Foundation (Grant No. 9140C6904030808).

AUTHOR INDEX